Water and Wastewater Management

Global Problems and Measures

Series Editors

Müfit Bahadir, Institut für Ökologische und Nachhaltige Chemie, Technische Universität Braunschweig, Braunschweig, Germany

Andreas Haarstrick, Leichtweiss-Institut für Wasserbau, Exceed, Technische Universität Braunschweig, Braunschweig, Germany

Water and wastewater management are among the greatest challenges of our century and the challenges posed by climate change will become even greater. Unfortunately, however, most efforts, especially in developing countries but also in the so-called developed countries, have been less than optimal or not optimal at all. In particular, there are still too many people who have to live without clean water and decent sanitation. Today, 2.2 billion people lack access to safely managed drinking water and wastewater, and 4.2 billion people lack safely managed sanitation services. The question, why this is the case - especially in developing countries - as well as other urgent water and wastewater management issues, are discussed in this book-series. Contributions therein present in more detail critical reviews, discussions, and analysis of the water and wastewater situation and management aspects in different regions and countries worldwide.

Müfit Bahadir · Andreas Haarstrick ·
Kemal Güneş · Bülent Topkaya ·
Abbas Al-Omari
Editors

Agricultural Water Management

Springer

Editors
Müfit Bahadir
Institut für Ökologische und Nachhaltige Chemie
Technische Universität Braunschweig
Braunschweig, Germany

Andreas Haarstrick
Leichtweiß Institut
Technische Universität Braunschweig
Braunschweig, Germany

Kemal Güneş
Climate and Life Sciences
TÜBİTAK Marmara Research Center
Gebze, Türkiye

Bülent Topkaya
Department of Environmental Engineering
Akdeniz University
Antalya, Türkiye

Abbas Al-Omari
Water, Energy and Environment Center
The University of Jordan
Amman, Jordan

ISSN 2731-3166 ISSN 2731-3174 (electronic)
Water and Wastewater Management
ISBN 978-3-031-96133-5 ISBN 978-3-031-96134-2 (eBook)
https://doi.org/10.1007/978-3-031-96134-2

This Springer imprint is published by the registered company Springer Nature Switzerland AG
The registered company address is: Gewerbestrasse 11, 6330 Cham, Switzerland

Foreword

Agricultural water management is one of the key components that will increasingly impact food security, especially in the present and future decades. With climate change, it is expected that dry areas will become even drier in the future, and erratic rainfall events will become more intense. As a result, agricultural water management will not only complicate the management of food security but also pose challenges in preserving ecosystems and managing energy supply. The awareness of food security has increased, particularly during the COVID-19 period, highlighting the need for integrated management of all aspects related to agricultural water management for the common future of our planet.

The saying "Water is Life" is a valid concept for the entire ecosystem and our future. Therefore, it is crucial to integrate existing technologies into water management, starting with the agricultural sector, to ensure sustainable water use. Considering that water consumption in agriculture is much higher compared to other sectors (the global average is around 70%), reducing this proportion to optimal levels would benefit the planet in many ways. More efficient water consumption in agriculture would not only contribute to food supply but also increase the amount of water stored in reservoirs for energy production, consequently reducing reliance on fossil fuel-based energy sources. In other words, reducing carbon emissions into the atmosphere would help minimize the effects of climate change. Moreover, efficient water consumption in agriculture would save energy required to deliver water to crops, leading to a reduction in energy inputs for agricultural mechanization. This would contribute to reducing the global share of energy used in agriculture to below 20%.

There is a direct relationship between sustainable agricultural water consumption and food security. By utilizing water in accordance with the needs of plants, it would become possible to practice irrigated agriculture in larger areas, produce more crops, and rapidly feed the growing population. Additionally, efficient water consumption could lead to significant increases in the amount of water allocated to ecosystems, contributing to the preservation of biodiversity, erosion prevention, and the improvement of aquatic and terrestrial habitats, thus creating more livable environments.

In agricultural water management, it is crucial to develop integrated solutions and create sustainable conditions, covering various aspects such as dam construction, irrigation systems, plant species, agricultural mechanization, climate change, and the sociological structure of agriculture. In our world, where sensor technologies and remote sensing systems have significantly advanced, it is possible to provide producers and decision-makers with parcel-based or even localized water management systems, rather than relying on regional or general solutions.

Furthermore, one way to consume less freshwater is to make more extensive use of treated domestic wastewater, particularly in the production of industrial crops and green spaces. This approach would reduce the pressure on freshwater resources and create a less polluted ecosystem through the concepts of reuse and recycling.

For a more livable world,

Florence, Italy

Fabio Masi

Preface

Currently, the world population is about eight billion and is expected to reach approximately ten billion in 2050. It is inevitable that the pressure on agricultural water resources will reach a new stage due to the increasing food supply in parallel with the growing world population and the climate change scenarios projected for the 2050s. According to a United Nations Report (WHO, 6 July 2022), the number of people affected by hunger globally increased by 150 million compared to 2019, reaching about 830 million in 2021. The same report indicated that 2.3 billion people (30% of the world's population) face moderate or severe food insecurity. If the increasing trend in the number of people affected by hunger continues linearly, it is likely that approximately 2.8 billion people in the world will face hunger in 2050. For this reason, the food security of people in the world is directly related to climate change, access to agricultural water and efficient use of agricultural water together with the increasing population. Currently, about 70% of the total water consumption in the world belongs to the agricultural sector. Due to the priority given to human consumption when sharing water with the increasing population, it is imperative to quickly reduce the agricultural water usage rate to reasonable levels. Serious planning is required, starting from now, to transform agricultural production from crop-based water management to water-based agricultural production in order to ensure sustainable water management in agricultural fields. To achieve this goal, it is essential to change farmers' habits as well as to create relevant legal regulations and to ensure the success of the strategies to change product patterns in agricultural areas. This situation will also bring about significant socio-economic changes. To change farmers' habits in addition to state-provided subsidies, intensive agricultural education and demonstration studies should be implemented at the same time. Furthermore, rapid action should be taken to incorporate new agricultural technologies at every stage of agricultural production, especially in irrigation.

It is necessary to use alternative water sources more effectively in areas that will be most affected by climate change. Firstly, after treating domestic wastewater, it should be used for irrigating city green areas and watering various industrial crops that consume substantial amounts of water, such as cotton, sugarcane, and sugar beet. Additionally, the concept of separating wastewater from buildings into black, grey, and yellow water should be implemented while planning new settlement areas. This approach would allow efficient treatment of wastewater according to different types of reuse purposes that are energy efficient. Moreover, infrastructure and legislation arrangements aimed at maximizing the benefits from rainwater-harvesting should be rapidly implemented.

Along with other measures, technologies such as vertical farming and satellite-assisted agricultural input control, which are common applications of digital agriculture, are crucial for food security. However, in managing agricultural water resources, sustainability should be the focus, around which all activities revolve. In this context, the integration of Industry 5.0 into the agricultural sector will facilitate the maximization of environmental, social, and economic benefits, which are the core principles of sustainability. Considering that food security will become an even more pressing global issue in the future, various methodologies should be developed through the United Nations and its affiliated organizations to transfer new management technologies for water resource from developed countries to less developed ones.

With hope for a world without hungry children and a happier future.

Gebze, Türkiye — Assoc. Prof. Dr. Kemal Güneş
Antalya, Türkiye — Prof. Dr. -Ing. Bülent Topkaya
Amman, Jordan — Prof. Dr. Abbas Al-Omari

About This Book

This special volume entitled *Agricultural Water Management* offers a comprehensive exploration of the principles, strategies, and technologies involved in efficiently managing water resources within agricultural systems. This book delves into various aspects of irrigation, water conservation techniques, and sustainable practices that are crucial for addressing global water scarcity issues in farming. It provides an in-depth analysis of water use efficiency, crop water requirements, and the impact of climate change on agricultural water supply. Through a combination of theoretical knowledge and practical insights, this book serves as an essential resource for researchers, students, and professionals in the field of agriculture, environmental sciences, and water management, aiming to promote sustainable farming practices that ensure food security and ecological balance.

This book is part of a book series (special volumes) based on the publication of the book *Water and Wastewater Management—Global Problems and Measures*, published by Springer in 2022 (ISBN 978-3-030-9528-7). The part about *Agricultural Water Management* that was first presented there as an own chapter will now be deepened in this special volume in terms of technological topics.

Contents

Fate of Agricultural Water and Agricultural Water Use Efficiency

Water Footprint Methodology and Its Role in Agricultural Water Management 3
Abdullah Muratoglu and Muhammed Sungur Demir

Understanding Environmental Impacts and Trade-Offs of Irrigation in Horticulture Crop Production in Türkiye using a Multi-Indicator Life Cycle Assessment 17
Yeşim Ahi, Kledja Canaj, Hüseyin T. Gültaş, Mladen Todorovic, and Andi Mehmeti

Adaptation to Climate Change: Irriagation Management Under Water Scarcity and Introduction of Drought-Resistant Crops (Quinoa) in Semi-Arid Regions 35
Servet Tekin, Attila Yazar, and Yeşim Bozkurt Çolak

Plant and Agricultural Water Use Technologies

Irrigation Scheduling Techniques for Sustainable Irrigated Agriculture under the Era of Climate Change 49
Attila Yazar and Servet Tekin

Technologies for Applications of Smart Irrigation 73
Mehmet Metin Ozguven

Determination of Kc Curves of Wheat, Barley, Sunflower, Maize (Grain), Sugar Beet, and Onion by using Landsat 8–NDVI Time Series and Relationships between K_C and NDVI 85
Eyüp Selim Köksal, Emre Tunca, and Sakine Cetin Taner

Land Suitability of Various Irrigation Techniques for More Effective Water Usage and Spatial Estimation of Soil Water Content based on Computational Intelligence 103
Orhan Dengiz

Irrigation Dams, Historical and Modern Rain Water Harvesting

Characterization of an Earth-Filled Irrigation Dam through the Use of Ground Penetrating Radar (GPR) 123
José Tavares Araruna Júnior, Paola Machado Barreto Manhães, Carlos Henrique de Almeida Couto Medeiros, Neil Lennart Anderson, and Norbert Maerz

Sustainable Management of the Loumbila Dam Water Resource in Burkina Faso for Market Gardening: Diagnostic and Prospects, a Case Study 135
Yacouba Sanou and Francois Sawadogo

Determination of Rainwater Harvesting Potential in Greenhouses: A Study in Bursa Uludag University 151
Busra Yayli and Ilker Kilic

Historical Change of Water and Agriculture Practices and Regulatory Frameworks in Yemen 161
Musaed Aklan, Abdulwahab Al-Mujahed, and Raidan Bassah

Alternative Water Resources in Agriculture

Reclaimed Domestic Wastewater Reuse in Irrigation: Circular Economy Approach 185
Lina Abu-Ghunmi and Diana Abu-Ghunmi

Determination of the Amount of Treated Wastewater Potential for Irrigation Purposes: A Case Study of Türkiye 201
Irem Karahan, Serhat Gunaydin, Onur Altun, Yusuf Basaran, Yakup Karaaslan, Afire Sever, and Ilker Angin

Agricultural Irrigation with Reclaimed Municipal Wastewater: Environmental Implications 217
Fatma Bedük, Senar Aydin, Mehmet Emin Aydin, and Müfit Bahadir

Marginal Water Use and Management in Agriculture 233
Berkant Odemis and Dursun Buyuktas

Addressing Climate Change Effects on Freshwater Resources: Case Study on Büyükçekmece Lake Water Budget with Zero Liquid Discharge Prospective 251
Kemal Güneş, Mehmet Beşiktaş, Irfan Yolcubal, Azime Tezer, Ömer Vısalı Sarıkaya, Muhammet Burak Bozçelik, Cihangir Aydoner, Selma Ayaz, Mehmet Şimşek, and Danuta Leszczynska

Water Policies and Management—An Outlook

Cost–Benefit Analysis of Subsidization of Water-Efficient On-Farm Irrigation Systems in the Northern Jordan Valley, Jordan 267
Koichi Noguchi, Kevin Wheeler, and Abbas Al-Omari

The Relation between Water Policies and Strategies and Modernization of the Irrigated Sector in Sudan 287
Tarig El Gamri and Adil M. Elkhidir

Sustainable Irrigation and Irrigation Indices: A Theoretical Review 305
Yusuf Ucar and Otilija Miseckaite

Agricultural Water Management—Global Perspective 321
Havva Eylem Polat

About the Editors

Müfit Bahadir studied chemistry at the Free University Berlin and Bonn University in Germany, received his Ph.D. 1975 from Bonn University and his Associate Professor 1988 from Munich Technical University. He was appointed as Full Professor and founding head of the Institute of Environmental and Sustainable Chemistry at the Technical University Braunschweig 1989. Since 1997, he is an honorary doctor and visiting professor at the Selçuk University Konya in Türkiye. His research fields cover environmental chemistry and analyses, environmental pollution through industrial processes and products, pesticide chemistry and metabolism in soil and water, ecotoxicology, sustainable chemistry, renewable feed stocks, and biodiesel and bio-lubricants. During the last 20 years, his research focused on sustainable water and wastewater management in developing countries. He retired from his university position as professor in 2016, but he is still active as elected member of Braunschweig Society of Sciences (Academy) in conducting R&D on environmental sciences with a focus on water in our vulnerable world.

Andreas Haarstrick studied Chemistry at the Technical University Braunschweig (1983–1989) and received his doctorate in 1992 in bioengineering. Since 2006, he is Professor for Bioprocess Engineering at the Technical University Braunschweig. His teaching and research cover modelling biological and chemical processes in heterogeneous systems, development of models predicting pollutant reduction in and emission behavior of landfills, growth kinetics at low substrate concentrations under changing environmental conditions, Advanced Oxidation Processes

(AOP), and groundwater management. Since 2012, he is the managing director of the DAAD exceed-Swindon Project dealing with sustainable water management in developing countries.

Kemal Güneş graduated 1988 from Ankara University, Faculty of Agriculture. He received his doctorate in 2000 at the same university. In 2003, he conducted his postdoctoral research at Florida State University, FL, USA, in the College of Engineering, Department of Civil and Environmental Engineering, focusing on "Methane Oxidation Through Landfill Cover Soils". In the following years, he implemented numerous pilot-scale and full-scale projects on the treatment of domestic and agricultural return waters using constructed wetland technologies in Türkiye. He also served as the project coordinator in most of the 27 projects aimed at preparing plans for the protection of drinking water basins in Türkiye. Additionally, he worked on the recovery of valuable chemicals. Since 1989, he has been conducting research on "The Impacts of Climate Change on Water Resources" at TÜBİTAK Marmara Research Centre, Climate and Life Sciences Unit.

Bülent Topkaya pursued his civil engineering studies at the Karlsruhe University and completed his Ph.D. in 1984 at Berlin Technical University, both in Germany. In 1998, he was appointed as a full professor and became the founding head of the Department of Environmental Engineering at Akdeniz University, Antalya, Türkiye. Throughout his career, his teaching and research spanned a broad range of topics, including water resources management, integrated solid waste management, environmental impact assessment, life cycle assessment, and the reuse of treated wastewater. After retiring in 2019, he dedicated the final decade of his work to advancing sustainable water management in the Antalya region and developing strategies for climate change adaptation in Antalya City.

Abbas Al-Omari received his Ph.D. degree in Civil and Environmental Engineering from Washington State University in 1999. Since graduation, he worked at the Water, Energy and Environment Centre of the University of Jordan in Amman, Jordan. His research interests are focused on drinking water quality modelling and management, water resources management at the basin and national levels, water and wastewater treatment, and water-energy-nexus. Over his career, he led and participated in more than ten internationally funded research projects in addition to numerous locally funded research projects. The findings of his research are published in internationally recognized and referred journals and conferences. He taught water resources related courses in Jordan, Saudi Arabia, and the United States.

Fate of Agricultural Water and Agricultural Water Use Efficiency

Water Footprint Methodology and Its Role in Agricultural Water Management

Abdullah Muratoglu and Muhammed Sungur Demir

Abstract

The global availability of freshwater resources is significantly constrained and unequally distributed, resulting in severe water scarcity issues. The challenges associated with future water-food security are further intensified by climate change and other human-induced factors. Agriculture, as the primary consumer of water worldwide and a critical input for various sectors and industries, is highly susceptible to water-related issues. Therefore, effective agricultural water management plays a pivotal role in conserving the limited freshwater resources of the world and ensuring food security. In recent years, the field of agricultural water management has witnessed the growing influence of the Water Footprint (WF) methodology. The primary objective of this research is to analyze the role of the WF methodology in agricultural water management. The study provides a comprehensive explanation of the WF methodology and emphasizes its significance in the context of global water resources and the agricultural sector. Additionally, the study examines the impact of various agricultural water management strategies, such as deficit irrigation, mulching, optimization of agricultural crop patterns, and different types of irrigation practices, on the WF. Previous studies have demonstrated that the implementation of these strategies can result in substantial water savings and a significant reduction in the overall WF. These findings underscore the potential advantages of adopting sustainable agricultural practices to address water scarcity, to enhance water use efficiency, and to ensure food security in the agricultural industry.

Keywords

Agriculture · Irrigation management · Optimization · Virtual water · Water footprint

A. Muratoglu (✉) · M. S. Demir
Civil Engineering Department, Batman University, Batman, Türkiye
e-mail: abdullah.muratoglu@batman.edu.tr

1 Introduction

1.1 Water and Scarcity

The Earth is commonly recognized as the "*blue planet*" due to its significant water content. Roughly, 70% of the Earth's surface is covered by water, and it is estimated that there is a total of approximately 1.4 billion km^3 of water on our planet [1]. Nevertheless, humans can only make use of a restricted quantity of this water, which is known as "*freshwater*".

Freshwater refers to the liquid state of water found in rivers, lakes, natural, and artificial

M. Bahadir et al. (eds.), *Agricultural Water Management*, Water and Wastewater Management,
https://doi.org/10.1007/978-3-031-96134-2_1

reservoirs, as well as underground systems. This type of water is suitable for human use and meets specific environmental standards [2]. However, the global availability of freshwater resources is significantly limited. Freshwater resources constitute a mere 2.5% of the total water resources and are predominantly distributed in groundwater and accumulated as ice and glaciers. Surface and other sources account for only 1.2% of the total freshwater reserves. These reserves are primarily found in permafrost regions, rivers, lakes, the atmosphere, and soil moisture. Consequently, the surface water resources available for human use (rivers, etc.) are less than 1/1000 of the total water on Earth [3, 4].

The hydrological cycle plays a crucial role in determining the accessibility of freshwater. Despite the constant total volume of water within the cycle, the distribution of water among different systems, such as groundwater, surface water, atmosphere, and soil moisture, can vary. This non-uniform distribution of limited water reserves presents additional challenges for human use and ecosystems. Thus, the availability of freshwater resources experiences spatial and temporal fluctuations, influenced by several factors, such as geography, meteorology, climate, land cover, and soil texture. The presence of diverse freshwater sources supports natural biodiversity and the formation of distinct geological features. Consequently, water consumption patterns are closely intertwined with its natural distribution of availability in specific regions and time periods.

In recent decades, human activities have further strained the quantity and quality of freshwater on Earth. The growing consumption of freshwater due to industrial operations, intensive agriculture, and increased per capita water usage has posed significant threats to the existing resources. Additionally, human-induced global warming and climate change have jeopardized the spatial and temporal distribution of these resources. These challenges have been worsened by the processes of globalization and international trade, resulting in "*water scarcity*" issues. Essentially, water scarcity refers to the inadequate quantity and quality of water to fulfill the demands of agriculture, households, and industries. The United Nations (UN) reports that 40% of the global population does not have access to water that meets acceptable quality standards. Insufficient assessments of water quality in most countries endanger clean water availability for three billion people around the world. UN Sustainable Development Goal 6 (Water and Sanitation) seeks to "*ensure the availability and sustainable management of water and sanitation for all*". [5]

Given the increasing water scarcity and the challenges associated with maintaining water quality and quantity in multiple sectors, scholars have developed several anthropocentric, ecocentric, and hydrocentric indices to assess the availability of freshwater resources. These indices, such as the Falkenmark indicator, Water Stress Index, Surface Water Availability Index, Water Depletion Index, Available Water Remaining Index, and Green Water Scarcity Index [6], serve as valuable tools for measuring freshwater availability in different contexts. The "*water footprint*" (WF) methodology is a contemporary and extensively utilized assessment tool for evaluating the quality, quantity, and scarcity of freshwater.

1.2 Concept and Definition

The WF is a newly developed method that comprehensively assesses the environmental sustainability of water usage in anthropogenic processes. This methodology provides numerous advantages, such as a comprehensive evaluation of water allocation considering the complete life cycle of crop production. Moreover, it serves as a powerful tool for comparing different products in terms of their water exploitation across various spatial and temporal boundaries. Additionally, the WF methodology allows for the development of impactful policy guidelines and management practices to foster greater awareness and accountability. WF methodology provides a valuable framework for water resource management by facilitating optimization and engineering techniques to conserve water across various sectors and processes. Unlike conventional water use statistics that solely focus on direct water consumption, the WF methodology considers the indirect water consumption associated with the life cycle of products, including "*virtual water*" (VW). Consequently, WF methodology shares a strong connection with the VW approach.

VW denotes the overall quantity of water utilized or allocated during the production process and is also known as "*embedded water*" or "*shadow water*". This term is used to emphasize the intricate connections between a specific product or process and its corresponding water usage. The concept of VW was initially introduced by J.A. Allan in 1993 and laid the groundwork for the development of the WF concept by A. Hoekstra. The Water Footprint Assessment Manual published by Hoekstra et al. [7], represents a key study that serves as the basis for numerous WF analyses.

The WF of a product or process is defined as the sum of direct and indirect water allocation of all items within the supply chain of this product or process. The WF of a geographically delineated area is the sum of all activities of water consumption within that region [7]. Therefore, it is essential to provide a precise portrayal of life cycle analysis or a detailed account of the production chain to accurately compute the WF associated with different products, processes, and geographically delineated areas.

WF serves as a multidimensional indicator assessing the anthropogenic impact on different water resources. In this methodology, human impact is divided into two distinct categories: consumptive and pollutive components. The consumptive component encompasses two different types of water exploitation, namely blue and green water footprints. Similarly, the pollutive component is denoted by grey water. Blue water refers to the surface and groundwater utilized in the production process. Green water, on the other hand, represents the amount of soil moisture consumed by crops. Lastly, the grey water footprint is defined as the overall volume of freshwater required to dilute contaminants. Notably, the WF methodology sets itself apart from other approaches by its unique capacity to integrate both consumptive and pollutive components into a singular metric, quantified in water volume (m^3). Additional insights into the methodology are presented in Sect. 2.

1.3 Agricultural Sector and Water Footprint

Municipal activities, including households, schools, and businesses, account for only 10% of the total global water consumption [8]. The remaining water usage is predominantly linked to the agricultural sector, which accounts for around 70% of freshwater utilization by humans. The agricultural industry involves two main types of water consumption: rainfed and irrigated agriculture. Rainfed agriculture, relying on natural precipitation, is the preferred method for crop cultivation due to the high costs associated with water storage and distribution. However, in regions grappling with water scarcity issues, irrigation becomes necessary, particularly in arid and semi-arid areas. Spanning across an extensive area of approximately 275 million hectares, irrigated agriculture plays a pivotal role in fulfilling 40% of the world's food production requirements [9].

The WF indicator has gained substantial attention in the field of agricultural water management in recent years. Using WF analyses, it becomes feasible to evaluate water usage in agriculture and to devise effective strategies. Prior approaches to water management predominantly centered around optimizing irrigation water solely based on crop evapotranspiration requirements. However, the WF methodology introduces a more comprehensive framework that also considers soil moisture or rainfall utilization. Furthermore, the spatial and temporal optimization of crop production to minimize irrigation and to maximize effective rainfall has substantial implications for water consumption in both regional and global agricultural practices. Previous studies have unveiled notable variations in the WF of agricultural products across different spatio-temporal scales, even within relatively small hydrological basins. According to Muratoglu et al. [10], concentrating strategic crops in regions with a relatively low WF

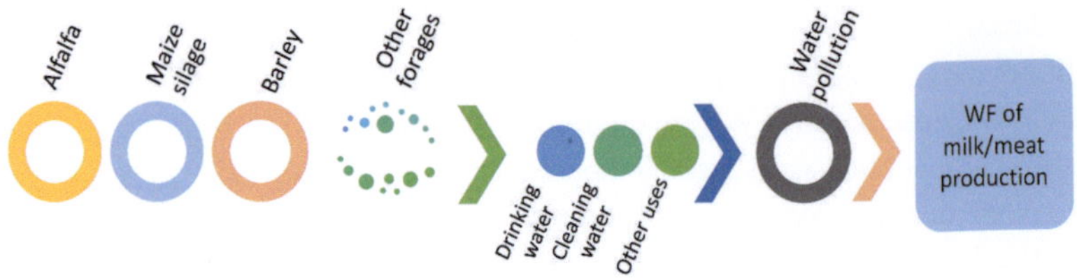

Fig. 1 Sub-products and processes of a typical WF analysis for milk/meat production

per unit product can yield substantial savings in freshwater resources.

1.4 Globalization of Water Through Agriculture

The concept of virtual water (VW), along with the expansion of international trade, has shed light on the global significance of diverse products in different regions across the world. This realization has been particularly pronounced in arid regions, where the recognition that importing water-intensive crops from other countries can contribute to the conservation of their limited freshwater resources has prompted a more strategic utilization of available resources. The process of globalization has underscored the fact that the impacts of consumption, including freshwater use and pollution, persist in the country or region of production, even when a distinct group of individuals consumes or utilizes the product. Consequently, consumers can exert an indirect influence on water resources in other parts of the world. In this context, the WF associated with internationally imported and exported products unravels the hidden links between water availability and demand.

2 Water Footprint Methodology

The methodology used to calculate the WF might vary depending on the specific production process, resulting in different approaches being implemented across different sectors. Agriculture, being the primary source of numerous products, is the sector that undergoes the most extensive WF analyses. The overall WF is connected to all diverse products and processes associated with the final product. For instance, when computing the WF of milk or meat production, the total direct water consumption of animals and the water requirements for processing tasks (cleaning of animals, farms, etc.) are considered. However, direct water exploitation by farm animals represents only a small fraction of the WF (typically less than 1–2%). Conversely, indirect water exploitation, which is linked to the cultivation of fodder crops, is considerably high compared to their direct water consumption. As a result, the WF of final livestock products is determined by considering especially the VW content of various fodder crops utilized throughout the life cycle of farm animals (Fig. 1).

The WF of a particular product is calculated by considering the average WF of all sub-sectors and processes involved in its production. Moving forward, the step-by-step calculation process for specific agricultural products like wheat, maize, and barley will be demonstrated. The WF of agricultural products is primarily shaped by the water consumption required for crop evapotranspiration (ET). Other aspects, such as water losses during transmission (evaporation, leakage, etc.) as well as the WF attributed to energy requirements and transportation can be considered into the analysis to a lesser extent.

Evapotranspiration (ET) is the process by which liquid water on land, sea, and vegetated surfaces is converted into water vapor and released into the atmosphere. Different types of evapotranspiration have been defined in the literature (such as reference ET, crop ET, actual ET, and potential ET). Exploring the specifics of these different ET types falls outside the scope of this study. Evapotranspiration is influenced by various factors, including climate parameters (such as solar radiation, temperature, humidity, and wind speed), crop parameters (such as type, height, roughness, reflection, ground cover, and rooting character), and management and environmental conditions (including salinity, soil zones, management and fertility deficiencies, plant density, and soil water content) [11]. ET usually represents the combined green and blue WF as a consumptive parameter. Finally, incorporating grey WF is also crucial in determining the overall water WF of agricultural products. The total WF is a combination of green, blue, and grey components (Eq. 1). The following sub-sections provide a comprehensive description and quantification of each component.

$$WF_{total} = WF_{blue} + WF_{green} + WF_{grey} \quad (1)$$

2.1 Green WF

Green water plays a significant role in WF analyses, focusing on the soil moisture in the root zone of plants (Fig. 2). Green water term is used to refer to the portion of ET in crops that is met by soil moisture. The calculation of green water usually involves subtracting the volume of water used for irrigation from the total water consumption of the crop cover. There are also several models to predict the green water content of specific soil and crop types at different climate conditions. Accordingly, one can quantify green water using two different approaches. The volume of green water (m^3) can be calculated by converting the total (green) evapotranspiration height (mm) into m^3/ha by a factor of 10. Then, the total WF can be determined by multiplying this component by the total cultivated or harvested area (ha) as shown in Eq. 2. Alternatively, the WF component can be expressed in terms of unit production mass (m^3/ton). In this case, the ET component (m^3/ha) should be divided by the crop yield (ton/ha) as shown in Eq. 3.

$$WF_{green}\left[m^3\right] = 10 \times \sum ET_{green}[mm]xA[ha] \quad (2)$$

$$WF_{green}\left[\frac{m^3}{ton}\right] = \frac{10 \times \sum ET_{green}[mm]}{Y\left[\frac{ton}{ha}\right]} \quad (3)$$

2.2 Blue WF

Blue water represents the most valuable component of the WF. It pertains to the liquid state of freshwater used to meet the water requirements of crops, when the available soil moisture is insufficient. Blue water sources encompass various forms, such as surface water, groundwater, and natural or artificial reservoirs. Notable characteristics of blue water include its high opportunity cost, transportability, and multifaceted utilization across different sectors. Consequently, it is crucial to employ blue water resources wisely, particularly in arid and semi-arid regions. The primary objective of water management analyses should be to minimize the utilization of blue water within the production chain. Estimating blue water entails the application of diverse methodologies. While direct measurement of crop irrigation offers the most accurate quantification of blue water, this approach is often unfeasible, and acquiring irrigation data for large areas can pose challenges.

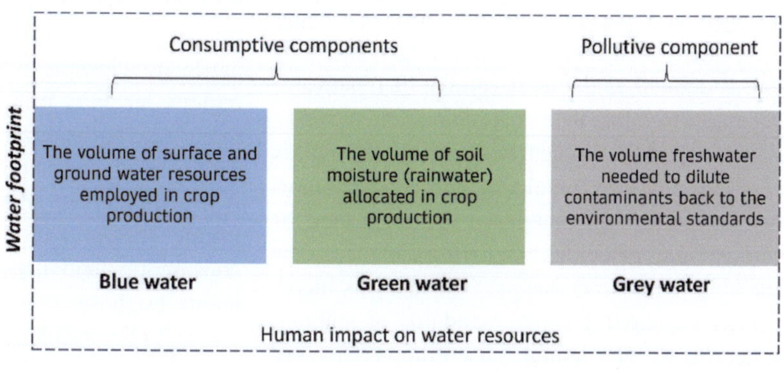

Fig. 2 Three colors of water used in agricultural crop production

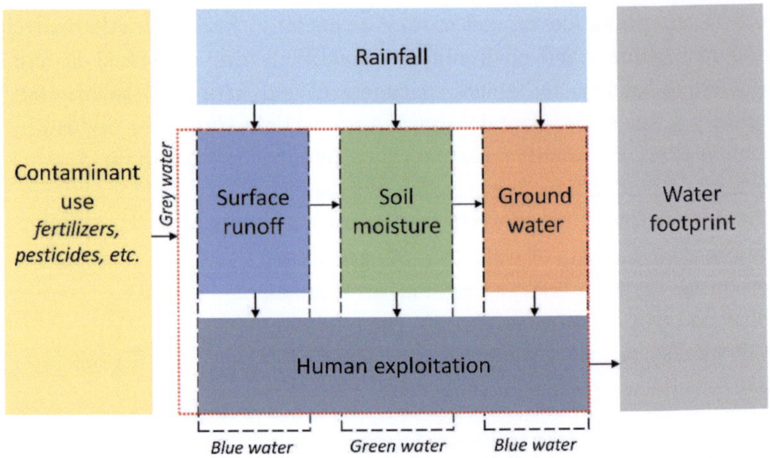

Fig. 3 Schematic representation of various natural and artificial water types

Therefore, approximation methods are frequently employed to estimate the amount of blue water. Commonly utilized methodologies include the soil–water balance method and the effective rainfall approach [12]. A comprehensive exploration of these methodologies exceeds the scope of this study. Upon determining the quantity of blue ET, the assessment of blue WF commonly follows a similar procedure as depicted in Eqs. 2 and 3.

2.3 Grey WF

Grey water, which is the pollutive component of the total WF, differs from the blue and green components and requires a distinct calculation methodology. The widely adopted three-tier methodology proposed by Franke et al. [13] is still considered as the fundamental approach for estimating the grey WF. Grey water is predominantly associated with the use of fertilizers and pesticides in agricultural activities, with nitrogen and phosphate-based fertilizers being the primary contributors. Accordingly, the grey WF is defined as the total volume of freshwater necessary to dilute the contaminant load in freshwater back to the ambient water quality standards [7]. This involves predicting the total contaminant load (L, kg) that leaches into ground or surface water resources employing dimensionless leaching runoff fractions (α). The analysis incorporates the environmental standards of ambient water (Eq. 4). The natural concentration (C_{nat}, kg/m^3) of the contaminant does not contribute to the anthropogenic grey WF, as these nutrients naturally occur in the freshwater. The volume of freshwater required to dilute the contaminants is calculated based on the maximum allowable concentration (C_{max}, kg/m^3). The grey water is expressed in water volumes (m^3) to facilitate its integration with the other WF components (green and blue). Figure 3 presents a visual comparison of three different water colors linked to various water media.

$$WF_{gray}\left[m^3\right] = \frac{L[kg]x\alpha}{(C_{max} - C_{nat})\left[\frac{kg}{m^3}\right]} \tag{4}$$

3 Agricultural Water Management Strategies Based on WF Approach

The increasing pressure on freshwater resources resulting from population growth, industrialization, global warming, and welfare growth [14] has led to widespread water shortages, with some regions reaching critical levels. Therefore, it is imperative to analyze and to optimize the allocation of freshwater resources for human use. A crucial step in addressing this issue is identifying sectors with high anthropogenic water consumption and implementing necessary adjustments in those areas. Notably, the agricultural sector is responsible for the largest share of water consumption globally [15]. Even small improvements in agricultural water use can yield significant benefits for overall water consumption and the preservation of water resources.

Various strategies can be employed to mitigate water stress and to reduce the water footprints associated with agricultural products, with a focus on the blue and grey WF components, which are critical for water stress management. These strategies include optimizing precipitation utilization, adopting high-efficiency irrigation techniques, implementing deficit irrigation strategies, using mulches to minimize water loss, reducing fertilizer wastage, improving crop yields, optimizing crop patterns, and enhancing agricultural practices [15–19]

3.1 Deficit Irrigation

In agricultural practices, efficient and effective water management plays a critical role in sustaining crop productivity and protecting the water resources. Excessive water usage can lead to lower yields and increased water stress, but adopting irrigation practices below the optimal water requirement threshold can result in significant water conservation, with only minor reductions in crop yields. To address water stress at the field level, various strategies for deficit irrigation have been developed. These strategies can be broadly classified into two categories "*continuous deficit irrigation*" and "*adjusted deficit irrigation*". Continuous deficit irrigation involves implementing consistent water restrictions throughout the entire plant growth period, while adjusted deficit irrigation involves applying different levels of water shortages at different phenological stages of the plant development [20, 21].

Analyzing detailed research on crop productivity and modifications in WF is crucial for understanding the functionality of deficit irrigation methods. Several studies have been carried out to measure the impact of deficit irrigation on crop production and WF. The effects of continuous water deficit on garlic cultivation were investigated by Léllis et al. [22]. They implemented varying levels of water restriction (10%, 20%, 30%) and found that a 30% water deficit led to a decrease in crop efficiency by approximately 25%. In this case, the average WF declined by 9.5% (Fig. 4). However, the most notable change was observed in the ratio of blue to green WF, which decreased from 3.8 to 1.8 as a result of employed deficit irrigation practices.

In a separate study conducted by Pardo et al. [23] focusing on barley, it was demonstrated that under the best scenario (20% deficit irrigation), crop yields experienced a decline of approximately 19%, while the WF decreased by roughly 10%. Karam et al. [24] conducted research on corn production in Lebanon and observed that applying a 40% reduction in irrigation led to a decline of

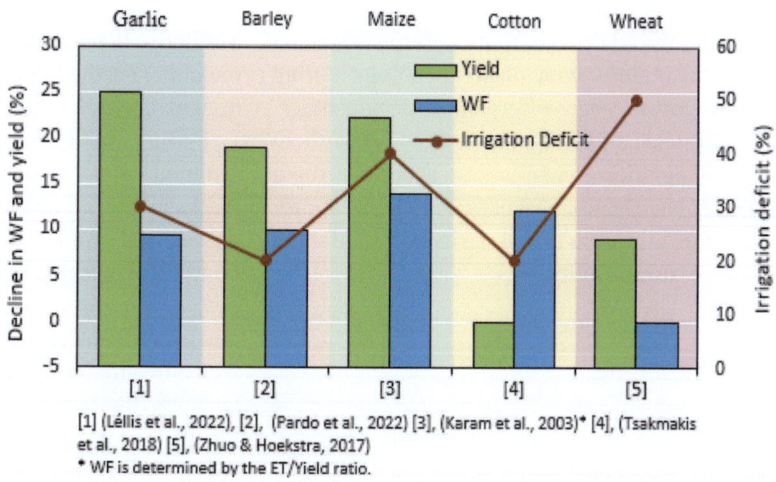

Fig. 4 Impact of deficit irrigation on WF and yield for various crops

roughly 22% in crop yield. Additionally, there was a concurrent decrease in ET of about 33%. A separate investigation by Zhuo and Hoekstra [21] concentrated on wheat production. It was found that implementing a 50% reduction in irrigation resulted in a decrease of around 9% in both ET and crop yield. This resulted in a comparatively smaller fluctuation in the overall WF. Nevertheless, there was a considerable decrease of 38% in the blue WF, leading to a reduced ratio of blue/green WF.

The study conducted by Tsakmakis et al. [25] examined the effects of different irrigation strategies (sprinkler, drip) and water restriction measures (regulated water limitation) on WF of cotton production. Their findings suggest that implementing a 20% deficit irrigation results in an approximate 12% reduction in WF, with an insignificant impact on crop yield. On the other hand, Huang et al. [26] investigated deficit irrigation using WF assessments based on life cycle analysis (LCA). Their study estimates that adopting a 33.3% deficit irrigation could result in a 27.5% decrease in the water scarcity footprint.

The findings presented in Fig. 4 demonstrate that the adoption of continuous deficit irrigation can result in significant reductions in crop yields. Thus, it is crucial to exercise caution when implementing continuous deficit irrigation in regions with limited agricultural land. This is because it can negatively affect agricultural output and overall food supply. However, in areas where agricultural land is plentiful, but water scarcity is a significant issue, it is recommended to prioritize continuous deficit irrigation as it helps minimize water usage. On the other hand, applying adjusted deficit irrigation aligned with plant phenology during various growth stages offers a notable advantage. This approach effectively reduces the WF without causing significant yield losses. Therefore, regulated deficit irrigation practices are the most appropriate choice in terms of both agricultural and water resource sustainability.

3.2 Mulching

Mulching is a strategic technique used in agriculture to reduce water consumption and to alleviate water-related challenges. This practice involves covering either the entire or a portion of the ground surface with any type of material [27]. Mulching is utilized to reduce water loss, to prevent soil erosion, and to improve nutrient management [28]. Employing mulching techniques can effectively lower the rate of evaporation by providing a shield against solar radiation. Moreover, it acts as a natural weed suppressant by depriving them of sunlight, thereby reducing unnecessary transpiration caused by weeds and improving nutrient accessibility for crops. Additionally, the practice of ground covering contributes to maintaining more consistent crop yields, even in the face of extreme climate variations [29].

There are two main types of mulching used in agriculture: organic and synthetic (Fig. 5). Organic mulching utilizes plant residues and does not provide complete ground coverage. Conversely, synthetic mulching covers nearly the entire ground surface and effectively decreases ground evaporation [21]. Typically, WF of crops decreases in the order of no mulching, organic mulching, and synthetic mulching. Correspondingly, the costs increase in the same sequence [28].

Limited research has been conducted on the relationship between mulching and WF. A comprehensive investigation carried out by Nouri et al. [15] assessed the impact of mulching on 10 different crop varieties in the Upper Litani Basin of Lebanon. The study revealed a 3.6% reduction in the blue WF, little effect on the green WF, and an overall decrease of about 3.1% in the total WF. Similarly, Wang et al. [30] observed that plastic mulching resulted in decreased blue and total WF in maize, wheat, and potato cultivation.

Further studies have examined the impact of different types of mulching on WF. Zhuo and Hoekstra [21] conducted a study on the effects of organic and synthetic mulching on wheat production. They revealed that organic mulch covering 80% of the ground surface reduced soil evaporation by 50%. In contrast, synthetic mulch covering the entire ground surface reduced soil evaporation by 80%. Interestingly, neither type of mulching resulted in a decrease in yield. Synthetic mulching led to a 6% and 17% reduction in total and

Fig. 5 Organic (left) and synthetic (right) mulching

blue WFs, respectively, while organic mulching achieved a reduction of 3% and 8%, respectively. Similarly, Chukalla et al. [28] found that organic mulching reduced the WF by 8–9%, whereas synthetic mulching achieved a more substantial reduction of approximately 18%.

Finally, Adetoro et al. [31] investigated the effects of thick and thin mulching on the WF of sugarcane. Their results indicated that thick mulching had a slightly greater impact on reducing WF compared to thin mulching. In contrast, Wang et al. [30] discovered that using plastic mulch in combination with medium and high levels of nitrogen fertilization could enhance the effectiveness of plastic mulch in reducing WF.

3.3 Agricultural Pattern Optimization

WF of crops significantly differs due to variations in agricultural growth phases, diverse climate conditions, and fluctuating water requirements throughout plant development. Furthermore, cultivating crops in unsuitable areas and unfavorable conditions not only decreases crop yields but also results in an increase in the WF. Therefore, it is crucial to establish suitable agricultural patterns that prioritize both food security and the sustainability of water resources. In this context, researchers are actively exploring ways to optimize agricultural practices and alleviate water stress on a global and regional level.

Multiple studies have forecasted potential global water savings under various crop optimization scenarios. According to Davis et al. [32], the optimization of the global distribution of 14 major crops has the potential to increase food production sufficiently to feed an extra 825 million individuals, while simultaneously reducing rainfall and irrigation water consumption by 14% and 12%, respectively. Similarly, Chouchane et al. [33] highlighted that the optimization of global crop patterns has the potential to achieve a 21% reduction in the blue WF.

Several other research have also been conducted at the regional level to explore crop water optimization. Mojtabavi et al. [34] assessed 12 different scenarios for optimizing cropping patterns in Iran. They found that the scenario with the greatest water savings, achieved by adjusting the minimum proportion of agricultural land, resulted in a reduction of approximately 23% in total WF. In another study conducted in the same region, Etedali et al. [19] reported that by optimizing the distribution of maize, barley, and wheat, the national WF could be reduced by 4.5%. This included a 3.5% and 2.9% reduction in irrigated and rainfed wheat, respectively, a 6% and 15% reduction in irrigated and rainfed barley, and an 8% reduction in irrigated maize. As another study, Dai et al. [35] demonstrated that implementing an optimal cropping pattern resulted in a decrease in blue and grey WF in nearly all study regions, with a total WF reduction of approximately 7.4% (1.38 Gm^3). Shi et al. [36] aimed to minimize WF in Northern China by exploring various scenarios involving four practices: reducing fertilization, decreasing leaching fraction, implementing deficit irrigation, and optimizing cropping patterns. Optimizing the cropping patterns in their study resulted in reductions of 1.3% in total WF, 8% in the blue WF, and 1.9% in the grey WF. The optimization also led to a decrease in the cultivated area for wheat and an increase in the cultivated area for maize, which in turn resulted in a 4.2% increase in green WF.

The studies discussed above underscore the significance of adjusting agricultural patterns to safeguard food and water security. However, there are limitations associated with such adjustments. (i) Farmers, driven by economic considerations, often prioritize issues other than water stress and food security. Thus, the successful implementation of agricultural pattern adjustments necessitates the support of government-backed policies. (ii) Consumption patterns vary across nations and regions, and in cases where local production falls short, imports are required. Such scenarios call for consideration of factors like virtual water transfer, economic implications, and carbon emissions during transportation. (iii) Strategic food products play a vital role in national food supply. Consequently, any changes in the agricultural pattern for these products should align with state policies and plans. (iv) The availability of freshwater resources for irrigation purposes should be considered when devising cropping patterns. (v) To ensure the viability of cropping pattern optimization, a comprehensive and interdisciplinary approach involving researchers, local and national authorities, and sector representatives is imperative.

3.4 Irrigation Practices

In the context of agricultural production, effective irrigation management, including the identification of appropriate irrigation timing and quantity, plays a crucial role in both crop yield and water consumption. However, it is important to note that an adequate and suitable irrigation schedule may not always yield optimal results. The choice of irrigation method also has considerable impacts on crop productivity and water consumption. Currently, furrow, drip, and sprinkler irrigation are the most preferred methods in agricultural production [25]. Existing literature suggests that, in certain scenarios, the implementation of drip irrigation can lead to yield increases of up to 30% while reducing water consumption by approximately one-third [37]. Similarly, another research on maize production suggested that the adoption of drip irrigation could potentially result in a shortened growth period, leading to a reduction in ET of up to 10% [38].

The influence of irrigation techniques on WF was evident in several studies comparing different irrigation methods (Table 1). Wang et al. [39] conducted a study assessing the effects of furrow,

Table 1 Comparison of literature studies about the impact of various irrigation methods on WF

Study	Reference irrigation method	Applied irrigation method	Rate of reduction in WF (%)
Wang et al. [39]	Sprinkler	Micro	19.3
	Sprinkler	Furrow	11
Zhuo and Hoekstra [21]	Furrow	SSD	9.2
	Furrow	SSD	5.1
Chukalla et al. [20]	Furrow	Drip	12
	Furrow	SSD	5
Tsakmakis et al. [25]	Sprinkler	Drip	5

sprinkler, and micro irrigation methods. Their results indicated that micro irrigation reduced the WF of wheat by approximately 19.3% compared to sprinkler irrigation, while furrow irrigation reduced it by around 11%. It is also worth noting that the highest crop yields were observed in the order of furrow, micro, and sprinkler irrigation, respectively. Zhuo and Hoekstra [21] reported that subsurface drip (SSD) irrigation resulted in a reduction of about 9.2% in green WF and 5.1% in blue WF compared to furrow irrigation. Moreover, Chukalla et al. [20] found that when maize, potatoes, and tomatoes were irrigated using drip or SSD irrigation instead of furrow irrigation, WF decreased by 12% and 5%, respectively. Finally, Tsakmakis et al. [25] reported that drip irrigation reduced cotton WF by 5% compared to sprinklers.

4 Conclusions

The incorporation of the water footprint (WF) method and its implementation in agricultural water management is crucial for understanding and improving water utilization in agricultural practices. The WF approach considers different sources and media, including green, blue, and grey water, to assess the overall water consumption and the impact of agricultural activities. By analyzing and quantifying these factors, strategies can be developed to mitigate water stress and to reduce the WF associated with agricultural products.

Several agricultural water management strategies can be implemented to optimize water usage and to minimize the WF. These strategies include deficit irrigation, mulching, optimizing agricultural patterns, and employing appropriate irrigation practices. Deficit irrigation techniques, such as continuous and adjusted deficit irrigation, allow for water conservation while minimizing yield losses. Mulching, whether organic or synthetic, reduces evaporation, prevents soil erosion, and improves nutrient management. Optimizing agricultural patterns and crop distribution can result in significant reductions in the WF while ensuring food and water security. Additionally, selecting suitable irrigation practices, such as drip or micro-irrigation, can enhance crop productivity while reducing water consumption.

It is crucial to highlight that the successful implementation of these strategies necessitates a holistic and collaborative effort involving researchers from several disciplines, policy-makers, farmers, and other relevant stakeholders. Moreover, when applying these strategies, it is vital to consider the availability of freshwater resources and their economic implications for both regional and global boundaries.

Overall, the incorporation of the WF method and the execution of agricultural water management strategies present significant opportunities in reducing water stress, improving the sustainability of water resources, and safeguarding food security. Ongoing research and collaborative efforts are imperative to refining and expand the utilization of the WF approach in agricultural practices.

Acknowledgements The authors extend their appreciation to Prof. Dr. Rafet ASLANTAŞ (Eskisehir Osmangazi University) for generously contributing visual aids depicting mulching practices (Fig. 5). The authors also express their thanks to Prof. Dr. İlker ANGIN (Ataturk University), Gonca KARACA BİLGEN (General Directorate of Agricultural Research and Policies, Türkiye), Zehra AVANOZ (Sinop University) and Zehra SAKINÇ (Gaziantep University) for their contributions.

References

1. Cassardo C, Anthony J, Jones A. Managing water in a changing world. Water (Switzerland). 2011;3:618–28. https://doi.org/10.3390/w3020618.
2. Muratoglu A. Water footprint: concept and methodology. In: Reference module in earth systems and environmental sciences. Elsevier; 2021. https://doi.org/10.1016/B978-0-12-821139-7.00034-9.
3. USGS. Water science school. In: United States geological survey. 2023. (November 24).
4. Shiklomanov I. World fresh water resources. In: Gleick PH, editor. Water in crisis: a guide to the world's fresh water resources. 1993.
5. UN-Water. Summary progress update 2021–SDG 6–water and sanitation for all. Geneva, Switzerland;2021
6. Xu H, Wu M. Water availability indices–a literature review. 2017

7. Hoekstra AY, Chapagain AK, Aldaya MM, Mekonnen MM. The water footprint assessment manual. Water footprint network; 2011. 978-1-84971-279-8.
8. Postel S. Last oasis: facing water scarcity. Am J Altern Agric. 1993;8:94–6. https://doi.org/10.1017/S0889189300005063.
9. UNESCO. World water assessment programme (UNESCO WWAP). 2021.
10. Muratoglu A, Iraz E, Ercin E. Water resources management of large hydrological basins in semi-arid regions: spatial and temporal variability of water footprint of the Upper Euphrates River basin. Sci Total Environ.2022; 846:157396. https://doi.org/10.1016/j.scitotenv.2022.157396.
11. Allen RG, Pereira LS, Raes D, Smith M. Crop evapotranspiration: guidelines for computing crop water requirements (FAO Irrigation and Drainage paper No: 56). FAO. 1998. https://doi.org/10.1016/j.eja.2010.12.001.
12. Muratoglu A, Bilgen GK, Angin I, Kodal S. Performance analyses of effective rainfall estimation methods for accurate quantification of agricultural water footprint. Water Res. 2023; 238:120011. https://doi.org/10.1016/j.watres.2023.120011.
13. Franke NA, Boyacioglu H, Hoekstra AY. Grey water footprint accounting: tier 1 supporting guidelines. Delfth, The Netherlands: UNESCO-IHE Institute for Water Education;2013.
14. Liu J, Yang H, Gosling SN, Kummu M, Flörke M, Pfister S, Hanasaki N, et al. Water scarcity assessments in the past, present, and future. Earth's Futur. 2017;5:545–59. https://doi.org/10.1002/2016EF000518.
15. Nouri H, Stokvis B, Galindo A, Blatchford M, Hoekstra AY. Water scarcity alleviation through water footprint reduction in agriculture: the effect of soil mulching and drip irrigation. Sci Total Environ. 2019; 653: 241–52. https://doi.org/10.1016/J.SCITOTENV.2018.10.311. (Elsevier).
16. Chapagain AK, Hoekstra AY. Water footprint of nations. Volume 1 : Main report. Value of water research report series 1. UNESCO-IHE Institute for water education;2004, pp. 1–80.
17. Deihimfard R, Rahimi-Moghaddam S, Collins B, Azizi K. Future climate change could reduce irrigated and rainfed wheat water footprint in arid environments. Sci Total Environ. 2022; 807: 150991. https://doi.org/10.1016/j.scitotenv.2021.150991. (Elsevier B.V.).
18. Ababaei B, Etedali HR. Water footprint assessment of main cereals in Iran. Agric Water Manag. 2017, 179: 401–11. https://doi.org/10.1016/J.AGWAT.2016.07.016. (Elsevier).
19. Ramezani Etedali H, Ahmadaali K, Gorgin F, Ababaei B. Optimization of the cropping pattern of main cereals and improving water productivity: Application of the water footprint concept. Irrig Drain. 2019; 68:765–77. https://doi.org/10.1002/ird.2362.
20. Chukalla AD, Krol MS, Hoekstra AY. Green and blue water footprint reduction in irrigated agriculture: effect of irrigation techniques, irrigation strategies and mulching. Hydrol Earth Syst Sci. 2015;19:4877–91. https://doi.org/10.5194/hess-19-4877-2015.
21. Zhuo L, Hoekstra AY. The effect of different agricultural management practices on irrigation efficiency, water use efficiency and green and blue water footprint. Front Agric Sci Eng. 2017; 185–94. https://doi.org/10.15302/J-FASE-2017149.
22. Léllis BC, Martínez-Romero A, Schwartz RC, Pardo JJ, Tarjuelo JM, Domínguez A. Effect of the optimized regulated deficit irrigation methodology on water use in garlic. Agric Water Manag. 2022. https://doi.org/10.1016/j.agwat.2021.107280.
23. Pardo JJ, Léllis BC, Martínez-Romero A, Schwartz RC, Tarjuelo JM, Domínguez A. Effect of the optimized regulated deficit irrigation methodology on water use in barley under semiarid conditions. Agric Water Manag. 2022; 260. https://doi.org/10.1016/j.agwat.2021.107280.
24. Karam F, Breidy J, Stephan C, Rouphael J. Evapotranspiration, yield and water use efficiency of drip irrigated corn in the Bekaa Valley of Lebanon. Agric Water Manag. 2003; 125–37. https://doi.org/10.1016/S0378-3774(03)00179-3.
25. Tsakmakis ID, Zoidou M, Gikas GD, Sylaios GK. Impact of irrigation technologies and strategies on cotton water footprint using AquaCrop and CROPWAT models. Environ Process. 2018;5:181–99. https://doi.org/10.1007/s40710-018-0289-4.
26. Huang J, Xu C, Ridoutt BG, Chen F. Reducing agricultural water footprints at the farm scale: a case study in the Beijing region. Water (Switzerland). 2015;7:7066–77. https://doi.org/10.3390/w7126674.
27. Rashidi M, Keshavarzpour F. Effect of different irrigation methods on crop yield and yield components of cantaloupe in the arid lands of Iran. World Appl Sci J. 2011;873:876.
28. Chukalla AD, Krol MS, Hoekstra AY. Marginal cost curves for water footprint reduction in irrigated agriculture: Guiding a cost-effective reduction of crop water consumption to a permit or benchmark level. Hydrol Earth Syst Sci. 2017;21:3507–24. https://doi.org/10.5194/hess-21-3507-2017.
29. Mao L, Zhang L, Li W, van der Werf W, Sun J, Spiertz H, Li L. Yield advantage and water saving in maize/pea intercrop. Field Crop Res. 2012;138:11–20. https://doi.org/10.1016/j.fcr.2012.09.019.
30. Wang L, Li L, Xie J, Luo Z, Sumera A, Zechariah E, Fudjoe SK, Palta JA, Chen Y. Does plastic mulching reduce water footprint in field crops in China? A meta-analysis. Agric Water Manag. 2022; 260:107293. https://doi.org/10.1016/j.agwat.2021.107293. (Elsevier B.V.).
31. Adetoro AA, Abraham S, Paraskevopoulos AL, Owusu-Sekyere E, Jordaan H, Orimoloye IR. Alleviating water shortages by decreasing water footprint in sugarcane production: the impacts of different soil mulching and irrigation systems in South Africa. Groundw Sustain Dev. 2020;11. https://doi.org/10.1016/j.gsd.2020.100464.

32. Davis KF, Rulli MC, Seveso A, D'Odorico P. Increased food production and reduced water use through optimized crop distribution. Nat Geosci. 2017; 10:919–24. https://doi.org/10.1038/s41561-017-0004-5. (Springer US)
33. Chouchane H, Krol MS, Hoekstra AY. Changing global cropping patterns to minimize national blue water scarcity. Hydrol Earth Syst Sci. 2020;24:3015–31. https://doi.org/10.5194/hess-24-3015-2020.
34. Mojtabavi SA, Shokoohi A, Etedali HR, Singh V. Using regional virtual water trade and water footprint accounting for optimizing crop patterns to mitigate water crises in dry regions. Irrig Drain. 2018;67:295–305. https://doi.org/10.1002/ird.2170.
35. Dai C, Qin XS, Lu WT. A fuzzy fractional programming model for optimizing water footprint of crop planting and trading in the Hai River Basin, China. J Clean Prod. 2021; 278. https://doi.org/10.1016/j.jclepro.2020.123196.
36. Shi X, Zhao J, Jia H, Zhao J, Lu J, Zhao M, Chu Q. Seeking sustainable pathway of crop production by optimizing planting structures and management practices from the perspective of water footprint. Sci Total Environ. 2022; 843. https://doi.org/10.1016/j.scitotenv.2022.157091.
37. Sammis TW. Comparison of sprinkler, trickle, subsurface, and furrow irrigation methods for row crops. Agron J. 1980;72:701–4. https://doi.org/10.2134/agronj1980.00021962007200050002x.
38. Qin S, Li S, Kang S, Du T, Tong L, Ding R. Can the drip irrigation under film mulch reduce crop evapotranspiration and save water under the sufficient irrigation condition. Agric Water Manag. 2016;177:128–37. https://doi.org/10.1016/j.agwat.2016.06.022.32.
39. Wang W, Zhuo La, Li M, Liu Y, Wu P. The effect of development in water-saving irrigation techniques on spatial-temporal variations in crop water footprint and benchmarking. J Hydrol. 2019; 577. https://doi.org/10.1016/j.jhydrol.2019.123916.

Understanding Environmental Impacts and Trade-Offs of Irrigation in Horticulture Crop Production in Türkiye using a Multi-Indicator Life Cycle Assessment

Yeşim Ahi, Kledja Canaj, Hüseyin T. Gültaş, Mladen Todorovic, and Andi Mehmeti

Abstract

Agriculture is the largest user of water in Türkiye, accounting for 75% of total consumption. The use of freshwater for irrigation has significant environmental consequences that must be considered in the sustainability analysis of crop production. Life-cycle assessment (LCA) is increasingly used for this purpose to comprehensively account for freshwater use related environmental impacts from a life-cycle perspective as well as to evaluate the trade-offs and synergies between water and other agricultural practices in crop production. This study is the first of its kind to examine a multitude of environmental impacts on several irrigated perennial crops cultivated in Türkiye. A streamlined LCA and multi-indicator model were used to assess the magnitude of environmental impacts and to quantify the impact of water consumption and irrigation on crop life cycle performance. Citrus, lemon, peach, apple, cherry, persimmon, kiwi, and pomegranate were analyzed. The results were analyzed at the midpoint, endpoint, and single score levels to provide a holistic picture of the sustainability of each crop. The life cycle water consumption varied greatly, ranging from 4,121 to 13,562 m^3/ha, or from 189 to 705 m^3/ton, with energy for irrigation water supply being more important than crop water use. According to the single score LCA results, the water consumption potential indicator accounts for 41% (range 35–51%), while irrigation as a process account for 37% (range 17–85%) of the total environmental impact. According to the findings, the majority of water consumption and environmental impacts are caused by emissions and resource use (electricity and fertilizer production) prior to crop cultivation, i.e., background subsystem. This study confirms the complex relationship between water and energy in irrigated agricultural production in Türkiye, with both direct and indirect environmental consequences. Water conservation, along with electricity and fertilizer resource efficiency, needs to be prioritized to reduce crop production's energy consumption and environmental impact.

Y. Ahi (✉)
Water Management Institute, Ankara University, Ankara, Türkiye
e-mail: ysmahi@ankara.edu.tr

K. Canaj
Department of Management, Finance, and Technology, LUM Giuseppe Degennaro University, Casamassima, Italy

H. T. Gültaş
Department of Biosystems Engineering, Faculty of Agricultural and Natural Sciences, University of Bilecik Şeyh Edebali, Bilecik, Türkiye

M. Todorovic · A. Mehmeti
International Center for Advanced Mediterranean Agronomic Studies (CIHEAM-Bari), Valenzano, Italy

M. Bahadir et al. (eds.), *Agricultural Water Management*, Water and Wastewater Management,
https://doi.org/10.1007/978-3-031-96134-2_2

Keywords

Agricultural sustainability · Carbon footprint · Irrigation management · Life cycle assessment (LCA) · Water footprint

1 Introduction

Türkiye is the 19th-largest economy [1], and the seventh-largest agricultural producer in the world [2], although the country is situated in an arid and semi-arid environment, making agricultural production heavily reliant on irrigation water supply and availability. Irrigated agriculture consumes the most water in Türkiye, accounting for 75% of total consumption [3]. As a result, one of the top priorities is the sustainability of the country's food system.

Since the 1950s, Türkiye has invested in a variety of programs to transition agriculture from rain-fed to irrigated one as well as from subsistence to commercial agriculture. As a result, farm and agricultural operations rely highly on purchased variable inputs such as irrigation water, diesel, electricity, fertilizer, plant protection products, and machinery. While agricultural output has increased, continued intensification and the resultant use of large amounts of farm inputs have resulted in numerous and significant environmental impacts, including air, soil, and water pollution [4]. Irrigated agriculture greatly affects resource consumption as well as the environment due to excessive groundwater extraction for irrigation. As water-saving technologies require increased pumping and pressurizing, large amounts of energy consumption are associated with freshwater supply, increased energy use, and economic costs [5]. Rain-fed agriculture is associated with large amounts of fertilizer use and machinery operations, which, if not managed properly, can result in significant social, economic, and environmental costs [6]. Water use and freshwater scarcity, land-use change, global warming, acidification, eutrophication, eco-toxicity of pesticides used, and damage to human health and ecosystem quality are impacts of agri-food production systems [7–9]. Land use changes, plant types, fertilizers and pesticide doses, irrigation water amount and quality, irrigation frequency, irrigation method, soil characteristics and quality, mechanical equipment use, and energy source and consumption have an impact on water demand, crop yield, and environmental burden. Additional indirect environmental impacts are associated with the resource use and emissions that take place prior to crop cultivation on the field, such as those associated with the production of agricultural inputs. As a result, life cycle thinking and a system-level perspective are required for a comprehensive evaluation of all upstream and downstream water and energy inputs, resource consumptions, and emissions into the environment [10]. Life cycle assessment (LCA) is the method most widely used to assess the environmental impacts of agricultural products [11]. LCA analyzes sustainability benefits and trade-offs resulting from complex systems and allows the examination of alternative future scenarios for strategic planning. Rain-fed and irrigated olives in Greece [12], paddy rice production in Thailand [13], sunflowers in Portugal [14], wheat in Spain [15] and Italy [16], corn and soybeans in Uruguay [17], maize and wheat in Albania [18], and apple production in China [19] are analyzed and compared using LCA. Some research has focused on cropping systems that use conventional and alternative water resources. Some examples include performance evaluations of corn, soybean, and wheat systems in China [20], cucumber in Algeria [21], and vineyards in Italy [22]. LCA is also used for the environmental impact assessment of irrigation projects [23], irrigation equipment [24], the selection of the most sustainable irrigation system for crop production [25], and the assessment of various development scenarios for irrigated areas [26]. The LCA is applied to investigate the environmental impacts of agricultural and horticultural crop production in Germany [27], Belgium [28], and Iran [29]. The monetary valuation of LCA results is increasingly used by researchers to convert environmental damage into monetary terms [30, 31]. Although research on agricultural LCA began in the mid-1990s [32], an increasing number of publications appeared after 2010. Since ISO (International Organization of Standardization) published the standard for LCA (series ISO 14040–14,044), the credibility of LCA has grown significantly. Nowadays, the LCA is increasingly seen as a key concept for ensuring a transition towards more sustainable production and consumption patterns [33]. LCA is advocated by the European Commission as the best framework for assessing the potential environmental impacts of products currently available.

Several studies have examined the sustainability of production systems in Türkiye, particularly in terms of energy performance [34–38]. However, no studies have been conducted to estimate indicators of environmental pressures (e.g., water consumption, climate change, resource depletion, toxicity, eutrophication, acidification, human health effects) associated with crop production from a life cycle perspective. To address the gap, we used a multi-indicator life cycle assessment (LCA) model to assess the cradle-to-farm gate environmental footprint of irrigated perennial crop production systems in Türkiye. The findings contribute to the establishment of scientifically supported numerical evidence of environmental implications for Turkish perennial crops, the exploration and classification of direct and indirect environmental impacts of crop production systems, and the assessment of the potential environmental relevance of the irrigation process in crop environmental performance.

Table 1 LCA stages explained

Goal and scope definition	The goal definition of the LCA study involves the determination of a goal that sets the scope, including purpose, implementation, and audience, as well as whether a critical review of that goal is required. This phase includes defining the product or service one wants to evaluate, choosing a functional baseline for comparison, and defining the level of detail required [39]
Life Cycle Inventory (LCI)	The final inventory, in other words, provides a list of all inputs and outputs related to the life cycle of the product or service
Life cycle impact assessment (LCIA)	In the impact assessment phase, resource use and emissions produced are classified according to their potential impacts and are quantified for a limited number of impact categories. LCIA consists of four stages: classification, characterization, normalization, and weighting. Classification and characterization are defined as mandatory elements of an impact assessment, while normalization and weighting are defined as optional elements
Life cycle interpretation	The final phase of the LCA involves results interpretation, which discusses and systematically evaluates the results of the inventory analysis and impact assessment [40]

2 Materials and Methods

2.1 Life Cycle Assessment Framework

This LCA is based on four main phases: (1) goal and scope, (2) inventory analysis, (3) impact assessment, and (4) interpretation. Table 1 provides an overview of these stages.

2.1.1 Goal and Scope

The main goal of this analysis is to assess the potential environmental impacts of irrigated perennial crops in Türkiye using an attributional LCA (A-LCA). The reason for carrying out this LCA study is to provide insight into potential trade-offs and related impacts of irrigation and other farming practices (e.g., fertilization, plant protection, and mechanized operations) in irrigated crop production. The results are intended to be presented to key Turkish stakeholders as well as to international researchers working on LCAs of crop production systems. Midpoint, endpoint, and single score information were generated to obtain a comprehensive overview of the environmental sustainability of each crop. Citrus, lemon, peach, apple, cherry, persimmon, kiwi, and pomegranate were investigated (Table 2).

A cradle-to-farm gate LCA system boundary was used for assessment. It covers all the input

Table 2 Crops investigated in this LCA study

Crop	Location	References
Citrus	Adana	[34]
Lemon	Adana	
Pomegranate	Antalya	[35]
Apple	Karaman	[36]
Kiwi	Mersin	[37]
Peach	Çanakkale	[41]
Cherry	Çanakkale	
Persimmon	Adiyaman	[38]

and output flows of materials and energy up to the farm exit gate (Fig. 1). The majority of direct (foreground) and indirect (background) emissions are included. Indirect emissions are those that occur in upstream processes, such as purchased inputs used in farm crop production. The decision-maker cannot directly influence these processes. Direct field and farm emissions are substances emitted from an agricultural area or directly from the farm. As a result, the model includes cradle-to-farm gate burdens for all relevant input materials used in the cultivation process (chemical fertilizers, plant protection chemicals, fuel, machinery, irrigation water, irrigation infrastructure, and electricity). The model includes direct emissions from water applications, fertilizer application, pesticide application, and fuel combustion. The nursery stage and orchard establishment are excluded due to a lack of reliable data regarding this phase. The analysis is based on a multi-year approach, considering that each orchard system is assumed to have been established on land previously.

A multi-year approach is strongly recommended when the functional unit is related to the kg produced [42]. Carbon dioxide (CO_2) sequestration occurring in orchard soils was neglected. To fully comprehend the environmental impacts, two functional units (FU) were established: one ton (mass-based) of the freshly harvested crop at the farm exit gate and one hectare (ha) of cultivated land (area-based). Results expressed per 1 ha reflect a farm's impact intensity, whereas mass-based analysis is used for benchmarking and demonstrating the eco-efficiency of crop products [43]. Co-product credits play an important role in estimating life cycle environmental impacts [44]. However, since the main purpose of this study was to compare the crops on a common basis, no allocation was performed.

2.1.2 Life Cycle Inventory (LCI) Analysis

Life Cycle Inventory Analysis (LCI) is the data collection portion of LCA. It looks at the inputs and outputs of a product or service. The inputs included chemical fertilizers, plant protection

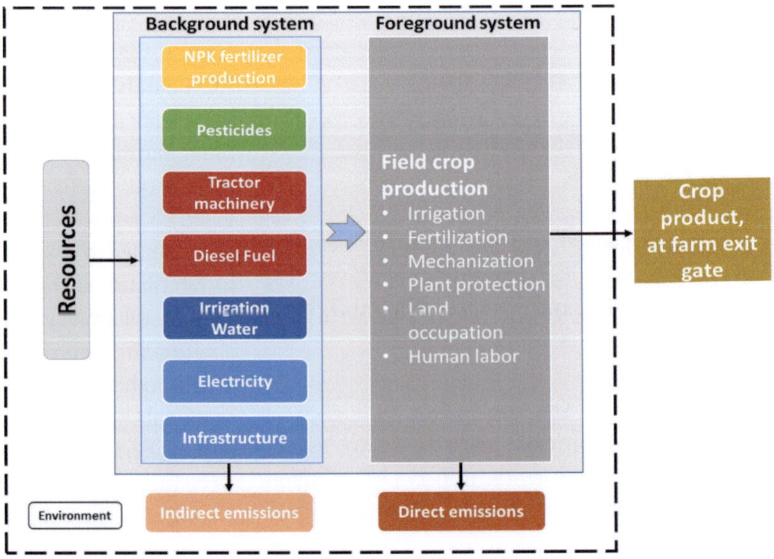

Fig. 1 System boundaries of cradle-to-farm gate LCA assessment (modified from [18])

chemicals, fuel consumption, machinery time, irrigation water, and irrigation energy (Table 3). The farm input–output data were obtained from previous studies conducted in various geographical regions of Türkiye. The data are of high quality because the input for each reference study was obtained directly from producers via surveys and interviews. The temporal scope was from 2010 to 2022. The overall geographic scope of the model was Türkiye and the European Union.

The irrigation infrastructure was modeled using generic infrastructure data, considering a lifetime of 10 years. The different fertilizer emissions addressed in this study were: (i) N-fertilizers generated emissions to air (N_2O, NH_3, NO_x) and nitrate-nitrogen leaching to water (NO_3^-); (ii) P_2O_5-fertilizers emissions to water (P, PO_4^{3-}) and soil. N_2O emissions were calculated as 1% of all nitrogen available. Nitrogen oxide (NO_x) emissions into the air were calculated at 21% of the direct N_2O. The fraction of N synthetic fertilizers that volatilizes as NH_3 and NO_x ($FracN_{GASF}$) was 10% (kg N volatilized/kg of the applied N). The fraction of N lost through leaching and runoff ($FracN_{LEACH}$) was assumed to be 0.3 (kg N/kg N additions). Emissions from human labor were calculated considering a coefficient of 0.7 kg CO_2 per hour of work [45]. Data from the Ecoinvent database was used to calculate direct emissions from fuel combustion and pesticide application. For background processes, the study considers generic data from the database Ecoinvent v 3.1 [46].

2.1.3 Life Cycle Impact Assessment (LCIA)

The LCIA method used in this study is the ReCiPe 2016 midpoint and endpoint hierarchist (H). All 18 midpoints and three indicators in the ReCiPe method [47] were considered in the study (Fig. 2). This LCIA method was chosen because of its breadth of coverage of environmental issues, popularity in addressing problem-oriented (midpoint) and damage-oriented (endpoint) indicators, and ability to include spatially differentiated information [48]. Moreover, it enables a single LCA score to be calculated for the total environmental impact based on the calculated effects. Country-specific factors for Türkiye retrieved from Huijbregts et al. (2017) [49] were included in the LCIA model datasets to spatially differentiate characterization factors for ammonia, nitrogen oxides, phosphorus, phosphates, and water use. They directly influence human health ozone formation potential, fine particulate matter formation potential, ecosystem ozone formation potential, terrestrial acidification potential, freshwater eutrophication potential, and

Table 3 Resource input and crop yield data per 1 hectare for the selected crops in Türkiye [retrieved from 34–38, 41]

Input–output	Crops							
	Citrus	Lemon	Pomegranate	Apple	Kiwi	Peach	Cherry	Persimmon
References	[34]	[34]	[35]	[36]	[37]	[41]	[41]	[38]
Crop Yield (ton)	34.9	35	19.7	42.5	17.5	11	14	39.2
Irrigation (m^3)	4380	4680	56.9	3200	1162	1320	1260	4140
Electricity (kWh)	1280	1390	933.7	770	2096	335	285	938
Machinery (hour)	26.7	27.8	85.3	108	2	118.2	135.1	53.7
Diesel (liter)	76	81	35.7	561.6	4.5	198	161	53.8
Lubricating oil (kg)	2.7	2.8	3	10.8	0.2	11.8	13.4	2.7
Synthetic rubber (kg)	2.4	2.5	2.6	9.5	0.2	10.4	11.9	2.4
Nitrogen (kg N)	284	313.5	255	215	66	147	208	286
Phosphorus (kg P_2O_5)	366	396.5	165	73	22	162	231	234.5
Potassium (kg K_2O)	230	276	131	62	94	47	99	151.5
Pesticides, unspecified (kg)	35	37	72.2	137	0.1	37.1	32.5	35
Human labor	834.3	815.8	912	2928	423.5	726.5	832.1	2653.1

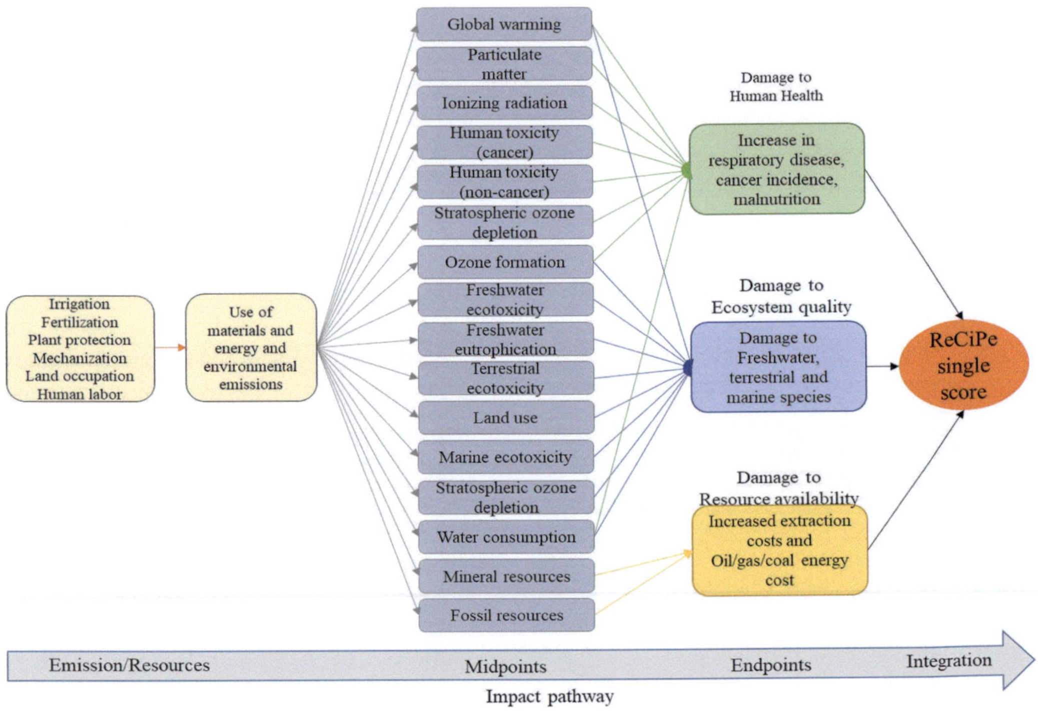

Fig. 2 Recipe 2016 impact pathway from inventory to aggregation to a single score [Elaborated from authors based on the ReCiPe model: https://www.rivm.nl/en/life-cycle-assessment-lca/recipe]

water consumption potential. The software used to analyze life cycle impact was OpenLCA v. 1.10.3 [50]. Ecoinvent v.3.1 was used for the modeling of the secondary impacts of the investigated systems [46]. The environmental impacts of the electricity system were assessed using datasets of Türkiye.

3 Results and Discussion

3.1 Midpoint and Endpoint Impact Categories for 1 ha of Land (Farming Intensity) and 1 ton of Product (Farming Efficiency)

Table 4 shows the findings of the assessment based on 1 hectare of land (impact intensity of a farm). Indicators in the table of impact factors are Ecosystem ozone formation (EOFP), Ecosystem quality (EQ), Freshwater eutrophication (FEP), Freshwater ecotoxicity (FETP), Fossil fuel scarcity (FFP), Global Warming (GWP), Human health (HH), Human health ozone formation (HOFP), Human carcinogenic toxicity (HTP_c), Human non-carcinogenic toxicity (HTP_{nc}), Ionizing radiation (IRP), Land use (LU), Marine eutrophication (MEP), Marine ecotoxicity (METP), Stratospheric ozone depletion (ODP), Particulate matter formation (PMPF), Resource availability (RA), Mineral resource scarcity (SOP), Terrestrial acidification (TAP), Terrestrial ecotoxicity (TETP) and Water consumption (WCP). The LCIA profile was calculated using OpenLCA v. 1.10.3. Lemons, persimmons, and apples has the greatest environmental impact due to the highest farming intensity. These crop products had a global warming potential of more than 10,000 kg CO_2-eq/ha. Nonetheless, there was a significant difference in cropping systems.

The impacts per 1 ha ranged from 3,785.8 to 12,306.8 kg CO_2-eq for global warming potential, 2.2 to 3.3 kg P-eq for freshwater eutrophication, 34.1 to 116.5 kg SO_2-eq for acidification, and 728.7 to 2,270.5 kg oil-eq for fossil resource scarcity. The life cycle water consumption varied greatly,

Table 4 Cradle-to-gate environmental footprint per 1 hectare of land for perennial crops grown in Türkiye (graded color scale: green = lowest impact, orange = midpoint, red = highest impact). Calculated using OpenLCA v. 1.10.3

Indicator	Unit	Citrus	Lemon	Pomegranate	Apple	Kiwi	Peach	Cherry	Persimmon
Midpoint impacts									
PMPF	kg PM2.5 eq	40.0	43.5	27.4	37.6	30.2	18.0	22.3	35.7
FFP	kg oil eq	1617.9	1743.4	1331.3	2270.5	728.7	927.3	1161.6	1568.9
FETP	kg 1,4-DCB eq	254.9	277.5	215.1	223.0	98.9	126.5	163.4	233.1
FEP	kg P eq	3.21	3.34	2.79	3.03	2.19	2.31	2.52	2.98
GWP	kg CO_2 eq	9682.5	10445.5	8070.5	12306.8	3785.8	5422.5	7035.5	10881.3
HTP_c	kg 1,4-DCB eq	213.8	231.2	167.5	238.0	105.7	132.9	164.6	202.8
HTP_{nc}	kg 1,4-DCB eq	6479.2	7063.4	5318.0	5277.0	2513.9	3182.6	4160.8	5792.0
IRP	kBq Co-60 eq	640.7	680.9	540.8	628.2	154.5	324.3	417.0	606.0
LU	m^2a crop eq	571.1	597.6	539.7	534.5	387.4	415.6	465.8	845.0
METP	kg 1,4-DCB eq	346.4	377.4	286.0	286.3	135.6	169.9	222.0	312.3
MEP	kg Neq	54.5	60.1	54.9	41.7	13.1	28.1	39.5	54.7
SOP	kg Cu eq	79.6	86.2	76.8	87.7	11.4	55.8	69.1	76.2
HOFP	kg NOx eq	22.2	24.1	17.5	46.8	7.4	15.4	20.4	20.7
EOFP	kg NOx eq	51.6	55.8	38.3	112.6	17.1	35.6	48.0	46.3
ODP	kg CFC11 eq	0.13	0.15	0.14	0.11	0.03	0.07	0.10	0.14
TAP	kg SO_2 eq	106.4	116.5	98.8	100.6	34.1	55.0	74.6	103.9
TETP	kg 1,4-DCB eq	20493.9	22423.6	19633.8	20709.9	5224.6	11809.0	15509.9	20098.6
WCP	m^3 consumed	12504.0	13562.1	4414.8	8442.4	12345.3	4121.0	4368.8	10555.6
Endpoint impacts									
HH	DALY	0.076	0.083	0.055	0.079	0.056	0.039	0.047	0.073
EQ	species.yr	2.62E-04	2.84E-04	1.76E-04	2.90E-04	1.93E-04	1.40E-04	1.62E-04	2.57E-04
RA	USD2013	670.9	728.2	599.6	934.5	227.4	419.4	540.8	662

ranging from 4,121 to 13,562.1 m^3 consumed. The total toxicity impact ranged from 8,078.8 to 30,379.9 kg 1,4 dichlorobenzene (kg 1,4 DCB) eq, where the major contributor is terrestrial toxicity. Damage to the area of protection range from 0.039 to 0.083 DALY (Disability Adjusted Life Year) for damage to human health, 1.4E-04 to 2.9E-04 species.yr for damage to ecosystems, and 227.4 to 935.5 USD2013 for damage to resource availability.

Table 5 depicts the cradle-to-gate impacts per ton of product, as well as the trade-offs between inputs and productivity, i.e., the influence of productivity (crop yield) across crops. Cherries and peaches have the highest environmental footprint per ton of product as they require higher amounts of energy, pesticides, fertilizers, and water for the given crop productivity. For 16 out of 18 impact categories, Kiwi production had the lowest footprint. The impacts per 1 ton of product range from 216.3 to 502.5 kg CO_2-eq for global warming potential, water consumption ranges 189.1 to 705.4 m^3, freshwater eutrophication ranges from 0.071 to 0.21 kg P-eq, acidification potential ranges from 1.9 to 5.3 kg SO_2-eq, and fossil resource scarcity ranges from 41.6 to 84.3 kg oil-eq. The total toxicity impact ranged from 440.3 to 1444.7 kg 1,4-DCB-eq. Damage to the area of protection ranges from 0.0019 to 0.0036 DALY for damage to human health, 6.83E-04 to 1.27E-05 species.yr for damage to ecosystems, and 13 to 38.6 USD2013 for damage to resource availability.

3.2 Hotspots Analysis

Except for kiwi production, fertilization emerged as the primary cause of global warming, freshwater eutrophication, terrestrial acidification, and toxicity-related effects, as depicted in Fig. 3. In terms of water consumption potential, blue water (irrigation water withdrawal) accounted for the most significant contribution, ranging from 41

Table 5 Cradle-to-gate environmental footprint per 1 ton of product for irrigated perennial crops grown in Türkiye (graded color scale: green = lowest impact, orange = midpoint, red = highest impact). Calculated using OpenLCA v. 1.10.3

Indicator	Unit	Citrus	Lemon	Pomegranate	Apple	Kiwi	Peach	Cherry	Persimmon
Midpoint impacts									
PMPF	kg PM2.5 eq	1.15	1.24	1.17	0.89	1.73	1.63	1.59	0.91
FFP	kg oil eq	46.36	49.81	57.02	53.42	41.64	84.30	82.97	40.02
FETP	kg 1,4-DCB eq	7.30	7.93	9.21	5.25	5.65	11.50	11.67	5.95
FEP	kg P eq	0.09	0.10	0.12	0.07	0.12	0.21	0.18	0.08
GWP	kg CO_2 eq	277.40	298.40	345.60	289.60	216.30	493.00	502.50	277.60
HTPc	kg 1,4-DCB eq	6.13	6.61	7.17	5.60	6.04	12.08	11.76	5.17
HTPnc	kg 1,4-DCB eq	185.65	201.81	227.75	124.17	143.65	289.33	297.20	147.76
IRP	kBq Co-60 eq	18.36	19.45	23.16	14.78	8.83	29.48	29.79	15.46
LU	m^2a crop eq	24.56	25.24	35.36	19.49	39.22	63.78	53.70	21.56
METP	kg 1,4-DCB eq	9.93	10.78	12.25	6.74	7.75	15.44	15.85	7.97
MEP	kg Neq	1.56	1.72	2.35	0.98	0.75	2.55	2.82	1.39
MRS	kg Cu eq	2.28	2.46	3.29	2.06	0.65	5.07	4.94	1.94
HOFP	kg NOx eq	0.64	0.69	0.75	1.10	0.42	1.40	1.46	0.53
EOFP	kg NOx eq	1.48	1.59	1.64	2.65	0.98	3.24	3.43	1.18
ODP	kg CFC11 eq	0.004	0.004	0.006	0.003	0.002	0.006	0.007	0.003
TAP	kg SO_2 eq	3.05	3.33	4.23	2.37	1.95	5.00	5.33	2.65
TETP	kg 1,4-DCB eq	587.22	640.68	840.85	487.29	298.55	1073.54	1107.85	512.72
WCP	m^3 consumed	358.30	387.50	189.10	198.60	705.40	374.60	312.10	269.30
Endpoint impacts									
HH	DALY	0.0022	0.0024	0.0023	0.0019	0.0032	0.0036	0.0034	0.0019
EQ	species.yr	7.51E-06	8.11E-06	7.53E-06	6.83E-06	1.11E-05	1.27E-05	1.15E-05	6.56E-06
RA	USD2013	19.20	20.80	25.70	22.00	13.00	38.10	38.60	16.90

to 95%. The combustion of diesel fuel was the primary source of ozone formation. In terms of identifying the main impacts and hotspots, our findings broadly confirm those of other relevant crop-based LCA studies [12, 18, 51].

Endpoints (or damages) typically refer to three areas of protection: human health, ecosystem quality, and resource availability (Fig. 4).

Fertilization and irrigation were the primary contributors to environmental damage at the endpoint level. Irrigation, as a process, has a greater impact on human health and ecosystem quality (both terrestrial and freshwater quality). Irrigation contributed 17–85% of the damage to human health, 17–86% of the damage to ecosystem quality, and 9–59% of the damage to resource availability. This was mainly due to electricity for water applications. Fertilization was responsible for most of the damage to resource availability, with a contribution ranging from 28 to 67%. Diesel fuel consumption and mechanization had the greatest impact on resource availability due to the high consumption of diesel fuel for cherries, apples, and peaches.

3.3 Single Score LCA Results

Figure 5 depicts the LCA results as a single score. The total environmental impacts per ton of a product ranged from 35.24 to 67.23 points for citrus and peaches, respectively. Among irrigated perennial crops, the average environmental impact was 51.23 points/ton. The impacts per hectare ranged from 740 points for peaches to 1557 points for lemons, with an average impact of 1201 points.

Irrigation had the lowest and the highest effects on total environmental impact for pomegranates

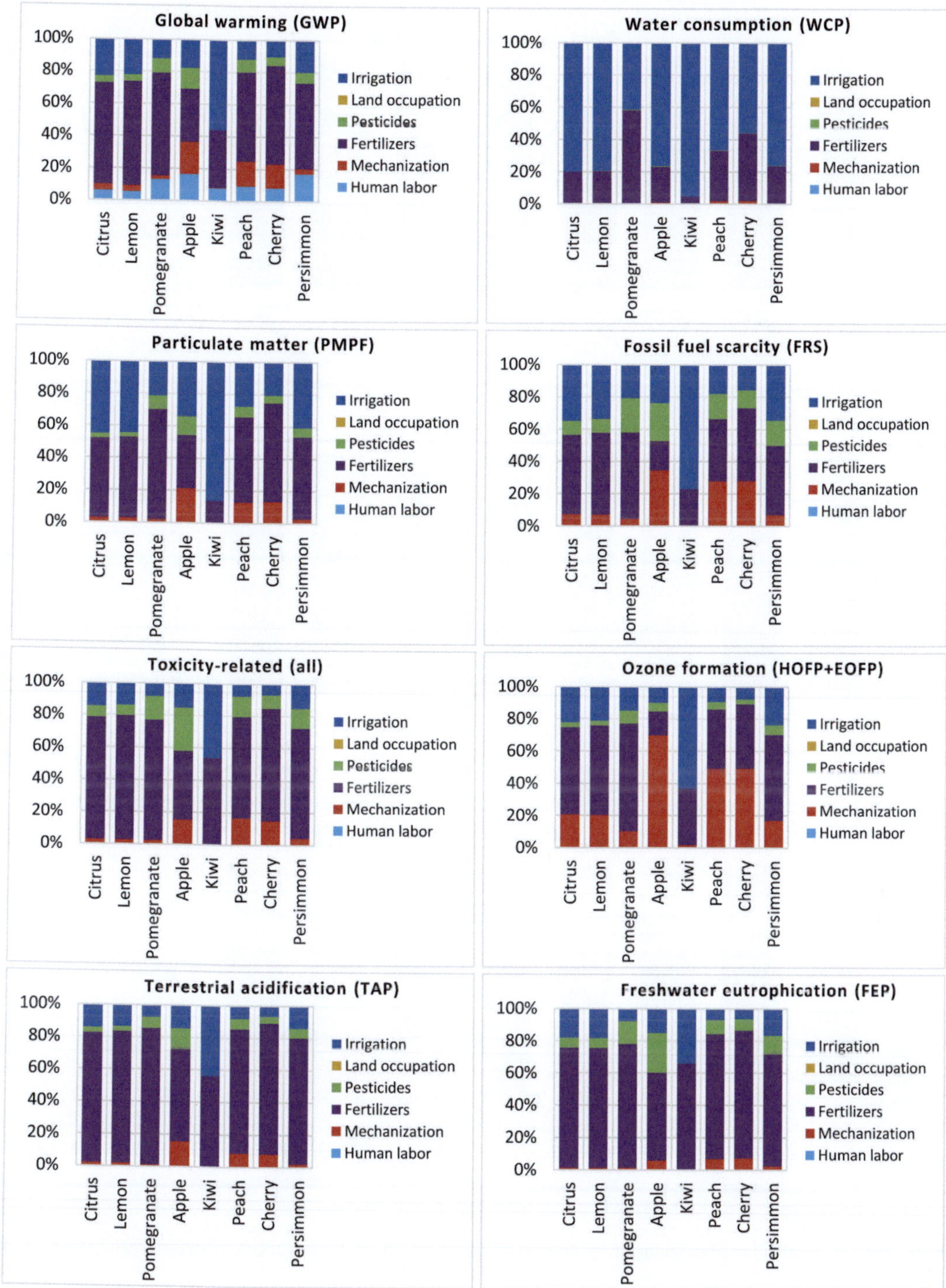

Fig. 3 Process contribution to some midpoint environmental impacts for irrigated perennial crops grown in Türkiye. Calculated using OpenLCA v. 1.10.3

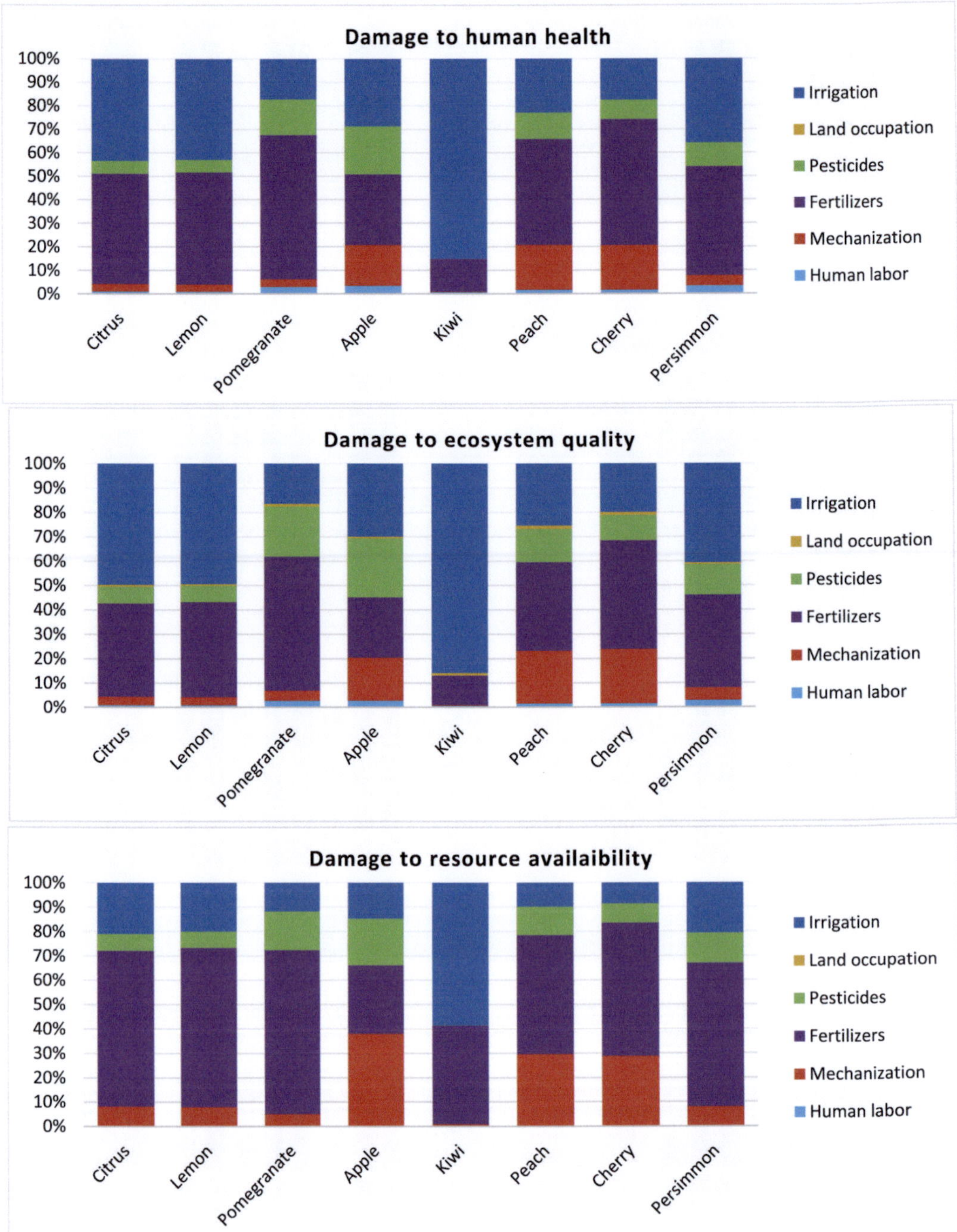

Fig. 4 Process contribution to endpoint environmental impacts for crops grown in Türkiye. Calculated using OpenLCA v. 1.10.3

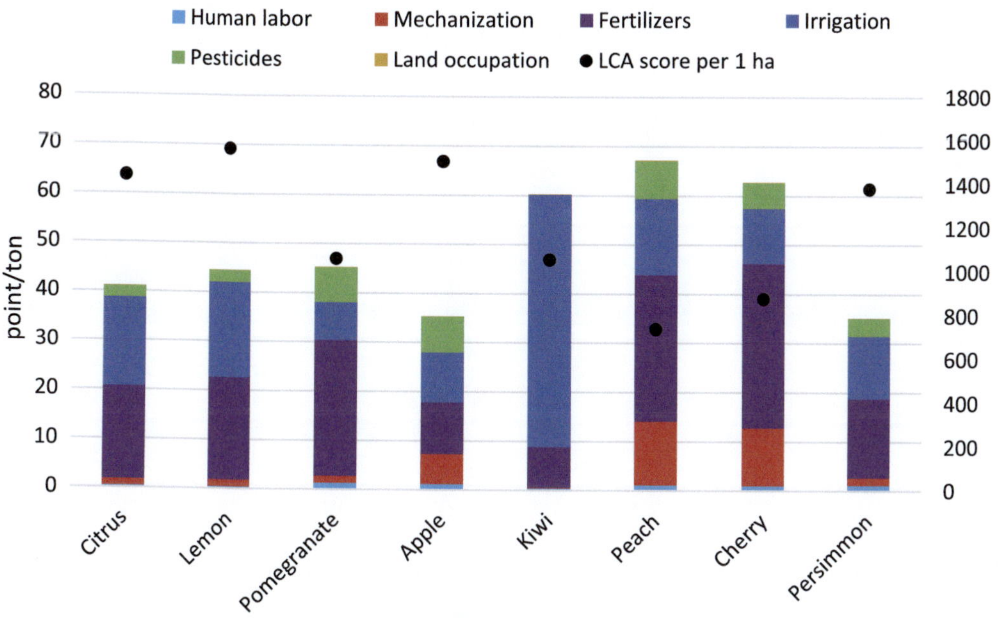

Fig. 5 Cradle-to-gate environmental footprint expressed as a single score of 1 ton of product and 1 ha of land for perennial crops grown in Türkiye. Calculated using OpenLCA v. 1.10.3

and kiwi, at 17.2% and 85.1%, respectively. The average share of irrigation as a process in the total environmental impact of 1 ton of perennial crops was 37%. Fertilizers contributions ranged from 14% for kiwi to 61% for pomegranate. The average contribution of fertilization as a process to the total environmental impact of one ton of perennial crops was 42.5%.

Diesel fuel consumption and machinery for mechanized operations contributed from 3 to 19% of the total environmental impact (average contribution 9%). Pesticide production and application had an effect ranging from 0.02% for kiwi to 21% for apple production.

At the indicator level (Fig. 6a), at the midpoint level, the major impact arises from water consumption (41%), particulate matter formation (average of 40%), and global warming (average of 14%). All three indicators were influenced by electricity for irrigation, fertilization, and mechanization (Figs. 4 and 5). At the endpoint level (Fig. 6b), damage to human health had the highest influence on total environmental impact mainly influenced by water consumption and global warming. Looking at contributions of the sub-systems to the total environmental impact (Fig. 6c), the background sub-system causes more than 75% of impacts for each crop. This means that the greatest impacts are caused by emissions and resource use prior crop cultivation. The majority of background impacts are caused by the production of electricity and nitrogen-based fertilizers.

3.4 Comparison with Results of Other Studies

We conducted a literature review of several international studies to compare our LCA results across studies (Table 6). It is advantageous to highlight the differences between Türkiye and other countries.

The citrus production results were consistent with the findings of Mostashari-Rad et al. [29] for global warming, particulate matter formation, and acidification, but there were significant differences in terms of water consumption, freshwater eutrophication, and endpoint scores. The results for kiwi production were similar for resource damage, but there were significant differences for other impact categories. Nicolò et al. [52] assessed

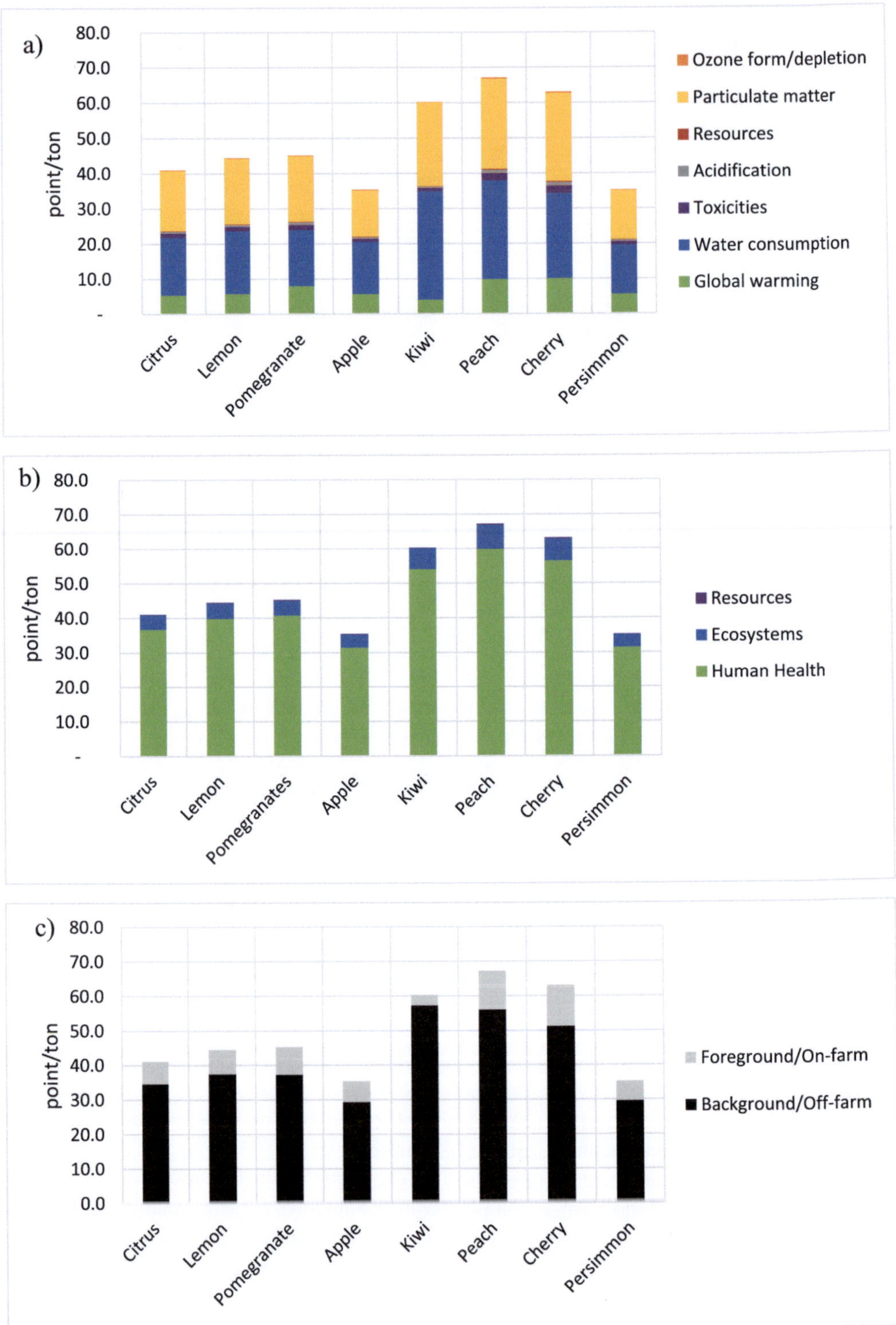

Fig. 6 Cradle-to-gate environmental footprint with the midpoint (**a**), endpoint (**b**), and sub-system (**c**) contribution for crops grown in Türkiye. Calculated using OpenLCA v. 1.10.3

Table 6 Comparison of LCA results per 1 ton of product obtained in this study (indicated in brackets) with international literature

Crop/country	Method used (as reported)	GWP (kg CO_2-eq)	WCP (m^3)	PMPF (kg PM2.5 eq)	AP (kg SO_2-eq)	FEP (kg P eq)	HH (DALY)	EQ (species.yr)	RA (USD2013)	Authors
Citrus, Italy	CML-2001	637 (277.4)	–	–	5.8 (3.05)	1.78 (0.09)	–	–	–	[52]
Citrus, Spain	CML-2001	303 (277.4)	–	–	2.44 (3.05)	0.96 (0.09)	–	–	–	
Citrus, Iran	ReCiPe	206 (277.4)	3.78 (358.3)	0.549 (1.15)	3.1 (3.05)	0.0746 (0.09)	0.001 (0.0022)	1.73E-06 (7.51E-06)	9.3 (19.2)	[29]
Kiwi, Iran	ReCiPe	319 (216.3)	3.9 (705.4)	0.779 (1.73)	4.67 (1.95)	0.091 (0.12)	0.01 (0.0032)	2.36E-06 (1.11E-05)	12.86 (13)	
Kiwi, Iran	–	152.18 (216.3)	–	–	3.53 (1.95)	9.15 (0.12)	–	–	–	[53]
Cherry, Italy	CML-2001	201.51 (502.5)	–	–	3.08 (5.33)	–	–	–	–	[54]
Cherry / Apple, Norway	CML-IA baseline v.3.05	270/170 (502.5 / 289.6)	–	–	7.7 (5.33)	5 (0.18)	–	–	–	[55]
Cherry, Portugal	CML 2001 + ReCiPe 2008	104.6–290 (502.5)	–	–	0.06–0.192 (5.33)	1.7 - 5 (0.18)	–	–	–	[56]
Lemon, Spain	CML-2001	310 (298.4)	200 (387.5)	–	2 (3.33)	–	–	–	–	[51]
Peach, Spain	CML-2001	162 (493)	215 (374.6)	–	1.05 (5)	0.039 (0.21)	–	–	–	
Peach, Spain	ReCiPe	250 (493)	321 (374.5)	–	–	–	–	–	–	[42]
Peach, Iran	ReCiPe 2016	–	–	–	–	–	0.008 (0.0036)	7.40E−06 (1.27E-05)	9 (38.1)	[45]
Apple, Iran	ReCiPe 2016	–	–	–	–	–	0.008 (0.0019)	7.10E − 06 (6.83E-06)	8 (22)	
Apple, Italy	ILCD 2011	588–612 (289.6)	252–259 (198.6)	0.296–0.319 (0.89)	3.82–3.97 (2.37)	0.11–0.13 (0.07)	–	–	–	[57]

LCA citrus growing in Italy and Spain, reporting a 56% and 8% higher global warming potential than the results obtained in this study. Acidification was estimated to be 47% higher in Italy and 25% lower in Spain, while eutrophication was much higher in both countries. The variability of GWP scores for apples is substantial throughout the reviewed literature. It can vary from 40 kg CO2-eq/ton in New Zealand [58] to 3008.53 kg CO_2-eq/ton in Iran [59]. The latter results were greatly influenced by the high amount of diesel fuel consumed during irrigation. For peach production, the GWP of this study was 29% and 38% higher than the results of Martin-Gorriz et al. [51] and Vinyes et al. [42], respectively. However, the WCP scores were 70% and 156% lower, respectively. In comparison to Martin-Gorriz et al. [51], the GWP and WCP results for lemon production were 29% and 150% lower, respectively. Vázquez-Rowe et al. [60] demonstrated that the carbon footprint of pomegranate production in Peru in different years ranged from 5.67 to 20.61 kg CO_2-eq for a 3.9 kg box. This equates to 928 to 5284 kg CO_2-eq per ton of pomegranates, which is 3 to 17 times the GWP potential estimated in this study. Despite significant variability in numerical scores across the reviewed literature, the main hotspots among studies, similar to what has been observed here, remain fertilizer production, irrigation, and diesel fuel for farming operations.

Martin-Gorriz et al. [51] estimated that 38%, 28%, and 25% of global warming of fruit and vegetable production in the region of Murcia was due to fuel combustion to produce electricity for irrigation, diesel combustion in field operations, and fertilizer production, respectively. For acidification impacts, these shares were 47%, 22%, and 22%, respectively. Fertilizers' production was the operation with the most impact in almost all indicators of citrus growing in Spain and Italy [52] with contributions ranging from 40 to 100%. Mostashari-Rad et al. [29] demonstrated that hotspots of environmental burdens belong to on-orchard emission and nitrogen. Tassielli et al. [54] demonstrated that 46% of global warming, 64% of eutrophication and acidification potential, and 52% of ozone formation of cherry production were dominated by the agricultural step due to the use of fuels and fertilizers. Also, Naderi et al. [59] estimated that diesel fuel contributed 83%, 99%, 95%, and 95% of acidification potential, eutrophication potential, freshwater aquatic ecotoxicity, and terrestrial ecotoxicity.

4 Conclusions

In recent years, the general agricultural growth strategy of Türkiye has been to shift from extensive to intensive and irrigated agriculture. This has resulted in a slew of positive, but also numerous negative environmental consequences. In Türkiye, the necessity for a sustainable food system is already recognized, and sustainable food production will not occur unless irrigation water-energy and other farm inputs are properly managed. As a result, there is a growing interest in metrics for understanding resource demands and the environmental impacts of agri-food systems. In this context, life cycle assessment (LCA) has become a critical component of effective environmental management. To date, the environmental performance of crop production in Türkiye has not been assessed using lifecycle-based metrics. We took a step forward by assessing the lifecycle-based environmental performance of some irrigated perennial crops using a multi-indicator model and a cradle-to-farm gate system boundary. Our model screened the impact of water consumption in the LCA context and assessed its relevance for endpoint damages as well as the total environmental load associated with each crop. The evaluation was carried out on a mass basis (1 ton) to assess production efficiency and on an area basis (1 ha) to assess farming intensity. Per 1 hectare, the results show that apple production has a generally higher impact, while kiwi has the lowest. Results expressed on a mass basis showed that peaches and cherries stood out as the most impactful crops due to lower productivity. According to the final model results, irrigation accounts for most of the water consumption potential for all crops studied. Weighting a wide range of indicators for various environmental impact categories revealed that the water consumption potential indicator contributes 35–51% (average 41%) of the total environmental impact, whereas irrigation as a process contributes 17–85% (average 37%). For all crops considered, the greatest impacts are generated from emissions and resource use taking place before crop cultivation (background subsystem). This confirms that the complex relationship between water and energy in irrigated agricultural production in Türkiye, with both direct and indirect environmental consequences. Quantifying indirect impacts in LCA is crucial for acquiring a comprehensive picture of product sustainability and advocating more sustainable practices throughout the whole life cycle. The results of this LCA highlight that environmental impacts of perennial crop in Türkiye can be reduced by optimizing fertilization, using smart and innovative irrigation practices to reduce water-energy use, and fuel efficiency in agricultural operations.

The findings of this study serve as a starting point for better understanding the environmental impacts of crop production in Türkiye. To gain a complete understanding of the overall environmental and sustainability impacts of Turkish commodities, further research should (i) increase the level of detail for crop production inputs (particularly fertilizer and pesticide inputs); (ii) include uncertainty analysis to improve the reliability of the results; (iii) test different LCIA methods, and (iv) supplement LCA with socioeconomic metrics.

Acknowledgements This study was supported by The Scientific and Technological Research Council of Türkiye -TÜBITAK 2221- Fellowships for Visiting Scientists and Scientists on Sabbatical Leave Support Program when Andi Mehmeti was a visiting scholar at Ankara University in March 2022. The authors are grateful to Prof. Sergio Ulgiati of the Università Degli Studi di Napoli "Parthenope" for providing access to the LCA databases.

References

1. World Bank. The World Bank Data in Türkiye; 2022. https://data.worldbank.org/country/TR
2. OECD. Evaluation of agricultural policy reforms in Turkey. OECD; 2011. https://doi.org/10.1787/9789264113220-en.
3. Cakmak EH, Dudu H, Saracoglu S, Diao X, Roe T, Tsur Y. Macro-micro feedback links of irrigation water management in Turkey. Policy Research Working Paper; No. 4781. © Washington, DC: World Bank; 2008. http://hdl.handle.net/10986/6350. (License: CC BY 3.0 IGO).
4. Garcia A. The environmental impacts of agricultural intensification, vol. 9. SPIA Technical Note; 2020. https://iaes.cgiar.org/spia/publications/environmental-impacts-agricultural-intensification.
5. Belaud G, Mateos L, Aliod R, Buisson MC, Faci E, Gendre S, Ghinassi G, Gonzales Perea R, Lejars C, Maruejols F, Zapata N. Irrigation and energy: Issues and challenges. Irrig Drain. 2020;69(S1):177–85. https://doi.org/10.1002/ird.2343.
6. FAO. Water for sustainable food and agriculture water for sustainable food and agriculture. A report produced for the G20 Presidency of Germany; 2017. https://www.fao.org/3/i7959e/i7959e.pdf.
7. Caffrey KR, Veal MW. Conducting an agricultural life cycle assessment: challenges and perspectives. Sci World J. 2013. https://doi.org/10.1155/2013/472431.
8. Brentrup F. Life cycle assessment of crop production. In: Green technologies in food production and processing. Food engineering series. 2012. pp. 61–82. https://doi.org/10.1007/978-1-4614-1587-9.
9. Green A, Nemecek T, Chaudhary A, Mathys A. Assessing nutritional, health, and environmental sustainability dimensions of agri-food production. Glob Food Sec. 2020;26(January): 100406. https://doi.org/10.1016/j.gfs.2020.100406.
10. Pelletier N. Life Cycle Thinking, measurement and management for food system sustainability. Environmental science and technology. 2015. https://doi.org/10.1021/acs.est.5b00441.
11. van der Werf HMG, Knudsen MT, Cederberg C. Towards better representation of organic agriculture in life cycle assessment. Nat Sustain. 2020. https://doi.org/10.1038/s41893-020-0489-6.
12. Fotia K, Mehmeti A, Tsirogiannis I, Nanos G, Mamolos AP, Malamos N, Barouchas P, Todorovic M. LCA-based environmental performance of olive cultivation in Northwestern Greece: from Rain-fed to Irrigated through conventional and smart crop management practices. Water. 2021;13(14):1954. https://doi.org/10.3390/w13141954.
13. Thanawong K, Perret SR, Basset-Mens C. Eco-efficiency of paddy rice production in Northeastern Thailand: a comparison of rainfed and irrigated cropping systems. J Clean Prod. 2014;73:204–17. https://doi.org/10.1016/j.jclepro.2013.12.067.
14. Figueiredo F, Castanheira ÉG, Freire F. Life-cycle assessment of irrigated and rainfed sunflower addressing uncertainty and land use change scenarios. J Clean Prod. 2017;140:436–44. https://doi.org/10.1016/j.jclepro.2016.06.151.
15. Galán-Martín Á, Vaskan P, Antón A, Esteller LJ, Guillén-Gosálbez G. Multi-objective optimization of rainfed and irrigated agricultural areas considering production and environmental criteria: a case study of wheat production in Spain. J Clean

Prod. 2017;140:816–30. https://doi.org/10.1016/j.jclepro.2016.06.099.
16. Todorović M, Mehmeti A, Cantore V. Impact of different water and nitrogen inputs on the eco-efficiency of durum wheat cultivation in Mediterranean environments. J Clean Prod. 2018. https://doi.org/10.1016/j.jclepro.2018.02.200.
17. Darré E, Cadenazzi M, Mazzilli SR, Rosas JF, Picasso VD. Environmental impacts on water resources from summer crops in rainfed and irrigated systems. J Environ Manag. 2019;232:514–22. https://doi.org/10.1016/j.jenvman.2018.11.090.
18. Canaj K, Mehmeti A. Analyzing the water-energy-environment nexus of irrigated wheat and maize production in Albania. Energy Nexus. 2022;7(May): 100100. https://doi.org/10.1016/j.nexus.2022.100100.
19. Zhang Z, Zhao J, Hou L, Xu X, Zhu Y, Zhai B, Liu Z. Comparative assessment of environmental impacts, mitigation potentials, and economic benefits of rainfed and irrigated apple production systems on China's Loess Plateau. Science of the Total Environment, vol. 869. (Oct 2022). p. 161791. https://doi.org/10.1016/j.scitotenv.2023.161791.
20. Romeiko XX. A comparative life cycle assessment of crop systems irrigated with the groundwater and reclaimed water in Northern China. Sustainability. 2019;11(10):2743. https://doi.org/10.3390/su11102743.
21. Azeb L, Hartani T, Aitmouheb N, Pradeleix L, Hajjaji N, Aribi S. Life cycle assessment of cucumber irrigation: unplanned water reuse versus groundwater resources in Tipaza (Algeria). J Water Reuse Desalin. 2020;10(3):227–38. https://doi.org/10.2166/wrd.2020.015.
22. Canaj K, Morrone D, Roma R, Boari F, Cantore V, Todorovic M. Reclaimed water for vineyard irrigation in a mediterranean context: life cycle environmental impacts. Life Cycle Costs Water (Switzerland). 2021;13(16):2242.
23. Chen X, Thorp KR, van Oel PR, Xu Z, Zhou B, Li Y. Environmental impact assessment of water-saving irrigation systems across 60 irrigation construction projects in northern China. J Clean Prod. 2020;245: 118883. https://doi.org/10.1016/j.jclepro.2019.118883.
24. Guiso A, Ghinassi G, Spugnoli P. Carbon footprint of three different irrigation systems. In: Icid. 2015. p 2–5.
25. Khanali M, Ghasemi-Mobtaker H, Varmazyar H, Mohammadkashi N, Chau K, Nabavi-Pelesaraei A. Applying novel eco-exergoenvironmental toxicity index to select the best irrigation system of sunflower production. Energy. 2022;250: 123822. https://doi.org/10.1016/j.energy.2022.123822.
26. Rogy N, Roux P, Salou T, Pradinaud C, Sferratore A, Géhéniau N, Hélias A, Loiseau E. Water supply scenarios of agricultural areas: Environmental performance through Territorial Life Cycle Assessment. J Clean Prod. 2022;366(July): 132862. https://doi.org/10.1016/j.jclepro.2022.132862.
27. Soode E, Lampert P, Weber-Blaschke G, Richter K. Carbon footprints of the horticultural products strawberries, asparagus, roses and orchids in Germany. J Clean Prod. 2015;87(1):168–79. https://doi.org/10.1016/j.jclepro.2014.09.035.
28. Goossens Y, Annaert B, De Tavernier J, Mathijs E, Keulemans W, Geeraerd A. Life cycle assessment (LCA) for apple orchard production systems including low and high productive years in conventional, integrated and organic farms. Agric Syst. 2017;153:81–93. https://doi.org/10.1016/j.agsy.2017.01.007.
29. Mostashari-Rad F, Ghasemi-Mobtaker H, Taki M, Ghahderijani M, Kaab A, Chau KW, Nabavi-Pelesaraei A. Exergoenvironmental damages assessment of horticultural crops using ReCiPe2016 and cumulative exergy demand frameworks. J Clean Prod. 2021;278: 123788. https://doi.org/10.1016/j.jclepro.2020.123788.
30. Canaj K, Mehmeti A, Morrone D, Toma P, Todorović M. Life cycle-based evaluation of environmental impacts and external costs of treated wastewater reuse for irrigation: a case study in southern Italy. J Clean Prod. 2021;293: 126142. https://doi.org/10.1016/j.jclepro.2021.126142.
31. Canaj K, Mehmeti A, Berbel J. The economics of fruit and vegetable production irrigated with reclaimed water incorporating the hidden costs of life cycle environmental impacts. Resources. 2021;10(9):90. https://doi.org/10.3390/resources10090090.
32. Fan J, Liu C, Xie J, Han L, Zhang C, Guo D, McConkey BG. Life cycle assessment on agricultural production: a mini review on methodology, application, and challenges. Int J Environ Res Public Health. 2022;19(16):9817. https://doi.org/10.3390/ijerph19169817.
33. Notarnicola B, Sala S, Anton A, McLaren SJ, Saouter E, Sonesson U. The role of life cycle assessment in supporting sustainable agri-food systems: a review of the challenges. J Clean Prod. 2017; 140:399–409. https://doi.org/10.1016/j.jclepro.2016.06.071.
34. Yilmaz H, Aydin B. Comparative input-output energy analysis of citrus production in Turkey: case of Adana Province. Erwerbs-Obstbau. 2020;62(1):29–36. https://doi.org/10.1007/s10341-019-00447-z.
35. Ozalp A, Yilmaz S, Ertekin C, Yilmaz I. Energy analysis and emissions of greenhouse gases of pomegranate production in Antalya province of Turkey. Erwerbs-Obstbau. 2018;60(4):321–9. https://doi.org/10.1007/S10341-018-0380-Z.
36. Dilay Y, Özkan A, Gülcan B. Energy input-output analysis of apple production in Turkey. a case study for Karaman Province. Erwerbs-Obstbau. 2021;63(4):375–80. https://doi.org/10.1007/s10341-021-00594-2.
37. Gökdoğan O. Energy and economic efficiency of kiwi fruit production in Turkey: a case study from Mersin Province. Erwerbs-Obstbau. 2022;64(1):55–60. https://doi.org/10.1007/s10341-021-00610-5.
38. Baran MF. Determination of energy use efficiency and greenhouse gas (GHG) emissions of persimmon

(Diospyros kaki L.) production in Turkey (A case study in Adıyaman Province). Erwerbs-Obstbau. 2022; 1–7. https://doi.org/10.1007/S10341-022-00639-0.
39. ISO 14045. ISO 14045:2012-Environmental management-Eco-efficiency assessment of product systems–Principles, requirements and guidelines. 2012.
40. Zampori L, Saouter E, Schau E, Cristobal J, Castellani V, Sala S. Guide for interpreting life cycle assessment result. Publications Office of the European Union, 60 2016. https://doi.org/10.2788/171315.
41. Aydın B, Aktürk D. Energy use efficiency and economic analysis of peach and cherry production regarding good agricultural practices in Turkey: a case study in Çanakkale province. Energy. 2018;158:967–74. https://doi.org/10.1016/j.energy.2018.06.087.
42. Vinyes E, Gasol CM, Asin L, Alegre S, Muñoz P. Life cycle assessment of multiyear peach production. J Clean Prod. 2015;104:68–79. https://doi.org/10.1016/j.jclepro.2015.05.041.
43. Cerutti AK, Beccaro GL, Bruun S, Bosco S, Donno D, Notarnicola B, Bounous G. Life cycle assessment application in the fruit sector: state of the art and recommendations for environmental declarations of fruit products. J Clean Prod. 2014. https://doi.org/10.1016/j.jclepro.2013.09.017.
44. Kendall A, Marvinney E, Brodt S, Zhu W. Life cycle-based assessment of energy use and greenhouse gas emissions in almond production, Part I: Analytical framework and baseline results. J Ind Ecol. 2015. https://doi.org/10.1111/jiec.12332.
45. Ordikhani H, Parashkoohi MG, Zamani DM, Ghahderijani M. Energy-environmental life cycle assessment and cumulative exergy demand analysis for horticultural crops (Case study: Qazvin province). Energy Rep. 2021. https://doi.org/10.1016/j.egyr.2021.05.022.
46. Ecoinvent Database 3.1. Ecoinvent Database 3.1. Ecoinvent Centre. 2014. https://doi.org/10.4018/978-1-59140-342-5.ch003.
47. Huijbregts MAJ, Steinmann ZJN, Elshout PMF, Stam G, Verones F, Vieira M, van Zelm R. ReCiPe2016: a harmonised life cycle impact assessment method at midpoint and endpoint level. Int J Life Cycle Assess. 2017;22(2):138–47. https://doi.org/10.1007/s11367-016-1246-y.
48. Canaj K, Mehmeti A, Cantore V, Todorović M. LCA of tomato greenhouse production using spatially differentiated life cycle impact assessment indicators: an Albanian case study. Environ Sci Pollut Res. 2019. https://doi.org/10.1007/s11356-019-07191-7.
49. Huijbregts MAJ, Steinmann ZJ, Elshout PMF, Stam G, Verones F, Vieira MDM, van Zelm R. ReCiPe 2016 v1.1 A harmonized life cycle impact assessment method at midpoint and endpoint level Report I: characterization. RIVM Report 2016-0104a. 2017. https://www.rivm.nl/sites/default/files/2018-11/ReportReCiPe_Update_20171002_0.pdf.
50. GreenDelta. openLCA 1.10.3. 2022. https://www.openlca.org/download/.
51. Martin-Gorriz B, Gallego-Elvira B, Martínez-Alvarez V, Maestre-Valero JF. Life cycle assessment of fruit and vegetable production in the Region of Murcia (south-east Spain) and evaluation of impact mitigation practices. J Clean Prod. 2020. https://doi.org/10.1016/j.jclepro.2020.121656.
52. Nicolò BF, De Salvo MC, Ramirez-Sanz C, Estruch V, Sanjuan N, Falcone G, Strano A. Life cycle assessment applied to different citrus farming systems in Spain and Italy. Agroecol Sustain Food Syst. 2018. https://doi.org/10.1080/21683565.2018.1490942.
53. Nikkhah A, Emadi B, Soltanali H, Firouzi S, Rosentrater KA, Allahyari MS. Integration of life cycle assessment and Cobb-Douglas modeling for the environmental assessment of kiwifruit in Iran. J Clean Prod. 2016. https://doi.org/10.1016/j.jclepro.2016.07.151.
54. Tassielli G, Notarnicola B, Renzulli PA, Arcese G. Environmental life cycle assessment of fresh and processed sweet cherries in southern Italy. J Clean Prod. 2018;171:184–97. https://doi.org/10.1016/j.jclepro.2017.09.227.
55. Svanes E, Johnsen FM. Environmental life cycle assessment of production, processing, distribution and consumption of apples, sweet cherries and plums from conventional agriculture in Norway. J Clean Prod. 2019;238: 117773. https://doi.org/10.1016/j.jclepro.2019.117773.
56. Gaspar PD, Godina R, Barrau R. Influence of orchard cultural practices during the productive process of cherries through life cycle assessment. Processes. 2021; 9(6):1065. https://doi.org/10.3390/PR9061065.
57. Longo S, Mistretta M, Guarino F, Cellura M. Life Cycle Assessment of organic and conventional apple supply chains in the North of Italy. J Clean Prod. 2017. https://doi.org/10.1016/j.jclepro.2016.02.049.
58. i Canals LM, Burnip GM, Cowell SJ. Evaluation of the environmental impacts of apple production using Life Cycle Assessment (LCA): case study in New Zealand. Agric Ecosyst Environ. 2006; 114(2–4): 226–38. https://doi.org/10.1016/j.agee.2005.10.023.
59. Naderi S, Ghasemi Nejad Raini M, Taki M. Measuring the energy and environmental indices for apple (production and storage) by life cycle assessment (case study: Semirom county, Isfahan, Iran). Environ Sustain Indic. 2020; 6:100034. https://doi.org/10.1016/J.indic.2020.100034.
60. Vázquez-Rowe I, Kahhat R, Santillán-Saldívar J, Quispe I, Bentín M. Carbon footprint of pomegranate (Punica granatum) cultivation in a hyper-arid region in coastal Peru. Int J Life Cycle Assess. 2017;22(4):601–17. https://doi.org/10.1007/S11367-016-1046-4.

Adaptation to Climate Change: Irriagation Management Under Water Scarcity and Introduction of Drought-Resistant Crops (Quinoa) in Semi-Arid Regions

Servet Tekin, Attila Yazar, and Yeşim Bozkurt Çolak

Abstract

Extreme climatic phenomena such as drought, salinity, and water scarcity will likely impact the global living conditions. These problems are particularly noticeable in arid and semi-arid regions that cover the majority of the Earth's land surface. Lack of water is one of the most crucial restraints on crop production in the Mediterranean Region. The shortage of freshwater resources, their uneven distribution, irregular rainfall, and climate change have significantly impacted sustainable agriculture in these regions. Therefore, efficient water management and cultivating drought- and salt-tolerant crops such as quinoa and amaranth seem essential for sustainable crop production in water-scarce regions. Cultivating new crops under harsh conditions is one of the approaches often used to preserve agricultural output in such locations for food security. This chapter's purpose is to provide information on how to enhance agricultural crop production in arid and semi-arid areas that are subject to a variety of abiotic (water, salt and heat) conditions. For this, agricultural water management strategies under water scarcity was discussed. Correspondingly, this research evaluated quinoa yield and water productivity responses to regulated deficit irrigation, partial root zone drying, conventional deficit irrigation, and full irrigation utilizing above and below-ground drip systems. This research has shown that deficit irrigation was very beneficial at different trial sites. The results suggest that quinoa responds positively to full saline and freshwater irrigation compared to deficit irrigation. Ultimately, considering its stress tolerance, adaptation to a variety of agricultural circumstances, multiple cultivars that can be cultivated from sea level to highlands, nutritional component, and economic importance, quinoa can be recommended as an alternative crop for marginal agricultural areas. This chapter also examines the challenges of introducing quinoa in this region, the recent condition and prospects of crop dissemination in Türkiye, applications and markets, and dissemination concerns and problems.

S. Tekin (✉)
Faculty of Agriculture, Department of Biosystems Engineering, Kahramanmaraş Sütçü İmam University, Kahramanmaraş, Türkiye
e-mail: servettekin@yahoo.com; servettekin@ksu.edu.tr

A. Yazar
Faculty of Agriculture, Department of Irrigation and Agricultural Structures, Çukurova University, Adana, Türkiye

Y. B. Çolak
Faculty of Agriculture, Department of Biosytems Engineering, Malatya Turgut Özal University, Malatya, Türkiye

M. Bahadir et al. (eds.), *Agricultural Water Management*, Water and Wastewater Management,
https://doi.org/10.1007/978-3-031-96134-2_3

Keywords

Deficit irrigation · Food security · Mediterranean region · Partial root zone drying · Semi arid areas · Quinoa · *Chenopodium quinoa* willd

1 Introduction

Sustainable management and effective use of land and water resources are strategically important worldwide, as they are essential to food security and agriculture. Water is a crucial factor in agricultural production and essential to food security. To meet the food needs of the rising population in the world, agricultural operations are predicted to provide 60% of food globally by 2050, with developing countries producing twice as much. Irrigated agriculture stands for 20% of total agricultural territory and 40% of total food production worldwide. Irrigated agriculture is at least twice as productive per unit of land as rainfed agriculture, allowing for greater production intensification and crop diversification [1]. However, irrigated agriculture is still practiced in many areas of the world, disregarding basic principles of resource conservation and sustainability. Water scarcity is among the most significant impediments to agricultural production in the Mediterranean Region, and notably in the world's semi-arid regions. Freshwater scarcity, its uneven distribution, erratic rainfall, and climate change have a profound impact on sustainable agriculture in these locations [2]. Therefore, in a time of water scarcity, water management must be as efficient as possible to save water and to maximize agricultural productivity.

Water withdrawal projections on a global scale foresee a substantial demand rise in the future to satisfy the needs of the urban, industrial, and environmental sectors. These demands and needs are insufficient because more than a billion people still require access to fresh water or sanitation, and natural ecosystems' water needs have received little consideration. Agriculture uses more water than any other source and wastes much of it through inefficiency. Climate change alters weather and water patterns worldwide, causing scarcity and drought in some areas and flooding in others. Climate change, depletion of water resources, soil degradation, and land abandonment are some of the obstacles that need to be overcome in the new paradigm for competitive and sustainable agriculture. Among the possible strategies, introducing drought-tolerant crops with high profitability and implementing water-saving strategies such as deficit irrigation should be considered [3].

Given that water shortage is the most prevalent water problem worldwide [4], there is much uncertainty regarding how much water will be accessible for future generations. Irrigated agriculture is the primary user of diverted water worldwide, reaching over 70–80% of total water in arid and semi-arid zones. Therefore, it is not surprising that irrigated agriculture is considered the primary water consumer in these areas, especially in drought emergencies [1, 5].

Environmental conditions are shifting, and the worldwide population is projected to surpass 10 billion by 2050. Increasing population and climate change will cause serious challenges in the supply of food worldwide, environmental protection, and biodiversity conservation. Agriculture must adapt to meet these obstacles; one essential factor is the production of "future-proof" crops. These crops must be high-yielding, immune to future climate change, and efficient in their use of precious resources like water, phosphorus, and essential minerals [6]. To maintain productivity in the face of climate change's negative effects, such as more frequent extreme temperatures, drought, or salt, the future climate-proof crops must be more resilient [7]. Future-proof crops also help to offset the effects of climate change by increasing below ground carbon sequestration and contributing to high soil health, erosion resistance, and fertility [8].

There are countless conversations in the scientific literature about the observed and anticipated implications of global climate change on species and ecosystems. In response to the projected effects of climate change, conservation organizations and government agencies are developing "adaptation strategies" to aid human existence and biological systems adjust to altering climate regimes. There is, therefore, an urgent need to implement and to adopt measures/policies to mitigate the impacts caused by climate variability. This need can be achieved by adopting climate-smart agriculture, aligning with the FAO's sustainable food and agriculture vision. This vision includes developing resilient crop varieties that withstand sudden temperature and precipitation stresses and adopting sustainable agricultural practices [9]. In this review, some climate change adaptation measures are discussed, such as irrigation management during water scarcity and the introduction of drought-resistant crops (quinoa) under Mediterranean climatic conditions.

2 Adaptation Measures to Climate Change

Weather patterns change and human activities are considered as the most severe risks to crop productivity enhancing water stress in agricultural systems [10]. Climate change has become an increasing threat to human and ecosystem sustainability and is expected to worsen under future climate change conditions [11]. A World Food Program (WFP) report shows that the increase in crop yields per hectare is much slower than population growth [9]. While climate change and agriculture are intricately intertwined, abrupt changes in climatic circumstances at such a rate jeopardize food security on a worldwide scale. Water is the foundation for sustainable development. As a result of the rising global shortage of water resources and the high cost of irrigation, precise water-saving irrigation methods that limit water usage in crop production have indeed been developed [12]. There is broad agreement that climate change adaptation is best addressed in the environmental sustainability framework. Water management cannot be tailored to climatic volatility. There are options to improve water management that contribute to adaptation to climate change and other changes while boosting growth, depending on local circumstances, requirements, and priorities. These prospects often combine and utilize the most viable ideas, tools, and technology to assist fragile rural communities in instilled and long-term development [13].

The Mediterranean climate is marked by hot, dry weather and chilly, rainy winters [14]. The Mediterranean climate is widespread between 30 and 45° on the west coasts of all continents due to global atmospheric circulation. The Mediterranean region is the most significant part of the Mediterranean climate, accounting for 60% of the total area [15]. Climate change will lead to drier and hotter summers and increase soil salinity problems. Winter precipitation in central and eastern Spain and northern Italy will increase by 2050, whereas it will fall by 10–15% in the southern Mediterranean countries. In winter, the average temperature in the Mediterranean area will rise by 1.25–2.5 °C [16]. Most countries with water scarcity have promoted better agricultural and water management practices, such as soil mulching to reduce unfavorable soil evaporation and pressurized irrigation to reduce water demand [17, 18]. As a result, preventive measures should be put in place to limit the adverse impacts climate change will have on crop yield in the Mediterranean region.

3 Irrigation Management Under Water Scarcity: Deficit Irrigation, Partial Root-Zone Drying, Regulated Deficit Irrigation, Supplemental Irrigation

3.1 Deficit Irrigation

In water-scarce areas, technologies to save water in irrigated areas and hence to enhance water use efficiency are crucial. Because of the rising global scarcity of water resources and the high expense of irrigation, precise water-saving irrigation methods that limit water use in crop production have now been developed [19]. Deficit irrigation (DI), i.e., irrigation below the total evapotranspiration of the crop (ET), can improve efficiency and maximize profit by reducing capital and operating costs. DI is characterized as "deliberate and systematic under-irrigation of crops" [20, 21], and an analytical framework is developed to estimate profit-maximizing levels of water consumption. The author implicitly included economic components of DI in the concept, assuming that under specific conditions, the maximum achievable income for an irrigated land might be achieved. Reference [20] noted that the analysis is dealing with water scarcity, a situation in which DI may be the most profitable approach. Lecler [22] expanded the English definition in his study: DI is an optimization algorithm that maximizes net yields by limiting the amount of irrigation water given to a crop to a level that results in some crop failure due to water strain. Although DI dates to the 1970s, scientists and practitioners have not adopted this technique as a practical alternative to complete irrigation [23].

At DI, amount of irrigation is reduced concerning that needed to meet the maximum ET [20]. Therefore, water demand for irrigation can be reduced, and the water saved can be used for other purposes. Even though DI is merely a technique aimed at optimizing economic performance when water is scarce, reducing the amount of irrigation in an area leads to many adjustments in the farming system. Thus, multi-faceted DI practices lead to technical, socio-economic, and institutional changes. Yet, the objective of this paper is to offer biologists interested in plant-water relationships a better knowledge of the DI idea, which will lead to broader challenges affecting the optimization of a scarce resource in crop production [5].

Water-conserving irrigation approaches include deficit irrigation (DI), partial root-zone drying (PRD), and regulated deficit irrigation (RDI), which were established for restricted irrigation management [12]. Deficit irrigation (DI) is a common strategy for optimizing water use and thus saving money by allowing plants to endure mild water stress with little or no yield decrease or quality [20]. Partial root-zone drying (PRD) is an extension of DI. PRD is a watering method that involves soaking and drying opposite portions of the plant root system alternately [24]. Regulated Deficit Irrigation (RDI) has lately been recognized as one of the most effective water-saving technologies in agriculture. Another type of DI is RDI. RDI is commonly defined as an irrigation practice, in which a crop is irrigated with less than the amount of water required for optimal plant growth during non-critical growth phases without generating substantial losses in yield [25].

3.2 Supplemental Irrigation

In the wet and sub-humid zones, irrigation has been used sometimes as a strategy during drought to stabilize production and supplement rainfall. This practice has been termed supplemental irrigation [26, 27]. Although it uses limited amounts of water due to the relatively high levels of rainfall, the goal is to maximize yields and to eliminate yield variability caused by water shortages. Supplemental irrigation in dry areas is a phrase that has recently been used to describe the practice of applying tiny amounts of irrigation water to winter crops that are generally produced under rain-fed environments [28]. Because maximum returns are not sought in this situation, it is a type of DI. As a result, the phrases deficit irrigation and supplemental irrigation are not interchangeable, and each circumstance should be characterized in terms of water availability relative to maximum yield. The implementations include a thorough picture of DI's agronomic influence. Total fresh matter and total production are generally reduced below DI. The impacts on dry matter and product quality, on the other hand, are positive, particularly in crops where high soil water availability can drastically affect fruit size, color, or composition (grapes, tomatoes, mango, etc.). Consequently, the economic efficacy of DI changes based on the crop [12, 23].

3.3 Partially Drying the Root Zone (PRD) and Regulated Deficit Watering (RDI)

The placement of irrigation water and the moisture deficit within the root zone are modified in PRD and RDI techniques to increase the water use efficiency of the plants. The primary distinctions between PRD and RDI are related to local soil moisture and plant water status variables [29]. In response to alterations in the abiotic environment, both hydraulic and biochemical signals modulate stomatal and plant production rates [30, 31]. In PRD, sub-regions of the plant root zone are alternately dried and moistened to enhance biochemical signals while maintaining the plant's water status [32, 33]. PRD strategies thus maintain the water status of the plant and produce a favorable physiological response through increased biochemical signals. Regulated deficit irrigation (RDI or CDI—Controlled Deficit Irrigation), as proposed by [20], is an irrigation technique that focuses exclusively on reducing irrigation volumes during stages of the crop cycle [30]. In RDI, moisture availability is reduced throughout the plant's root zone, reducing plant water status. Plant water ese efficiency (WUE) is improved by RDI by keeping the plant's water level within prescribed deficit limits relative to maximum water potential [34].

PRD approaches seek to maintain plant water availability and status steady while boosting biochemical signals (raising abscisic acid (ABA) and alkalinizing Superabsorbent Polymer (SAP) pH). It has been figured out that steadily rising ABA is associated with a partial reduction in stomatal conductance and a differential effect on both reproduction and vegetative production [35], which in turn result in a boost in water use efficiency in fruiting plants [32, 33].

4 Introduction of Drought-Resistant Crops: Quinoa

According to FAO, droughts are today's most significant threat to food security. Drought is defined as precipitation in an area is less than expected. Less rainfall means decreased water supply and less crop yields due to lack of water. At least 70% of potential crop yields are lost due to droughts and water scarcity. Therefore, drought-resistant crops have been developed to save water, to increase crop yields and to support food security [36].

Quinoa is a pseudocereal crop that is well adapted to a wide range of climatic conditions. It has the ability to greatly boost production as a fresh crop in the Mediterranean and other regions around the globe, like northern Europe, North America, Asia, and Africa. Quinoa is highly appreciated for its superior nutritive values. Climate change alters crop growing conditions, notably through salinization and drought. Drought and salt are two typical negative environmental circumstances that influence plant growth, determine the global geographical distribution of vegetation, and limit agricultural output in farming [37–39]. Food sources in the southern Mediterranean Region are impeded by insufficient water resources, dryness, and saltiness. Cereals are often farmed in low-yielding monocultures or in combination with fallow land under semi-arid and arid conditions in Mediterranean countries, which are affected by several abiotic stressors (water stress, salinity stress, drought stress etc.) aggravated by climate change. Food crops in arid and semi-arid locations (notably the semi-arid Mediterranean region of Turkey) can be improved by diversifying production and introducing new climate-resilient crops and varieties with improved stress tolerance, such as quinoa. Abiotic factors in the Mediterranean region are becoming more apparent because of global warming, which is expected to result in drier conditions, rising temperatures, and more unpredictability, resulting in desertification [40]. Quinoa (*Chenopodium quinoa* Willd.) is a halophyte that has the potential to become a key crop in drylands and saline environments while serving an expanding worldwide market [41].

Quinoa is a genetically varied Andean plant that has gained wide interest for its nutritional and health benefits, as well as its tolerance to contrasting habitats such as nutrient-poor and saline soils as well as drought-stressed marginal agricultural systems [42–44]. This traditional Andean seed crop has been grown for over 5000 years in the Peruvian and Bolivian Andes [45]. It spreads in poor soil and climatic settings [46] and is gaining popularity worldwide because of its hardiness and excellent nutritional content [47]. The seeds are strong in protein, have a balanced mix of key amino acids (for example, lysine), and are packed with minerals and vitamins [48]. It is robust with strong resistance to frost [49], dryness [40, 44, 50], and soil salinity [41, 47, 51, 52]. Quinoa responds to severe weather. It can grow in humidity levels ranging from 40 to 88% and temperatures ranging from –4 to 38 °C [53]. It can withstand a lack of moisture in the soil. Quinoa provides reasonable results even under climate settings with total annual rainfall of 100–200 mm [53, 54]. Quinoa develops unique water stress resistance mechanisms, helping it to adapt to harsh desert and semi-arid conditions [54, 55]. The plant's deep ramified root structure, specific vesicular glands, small and thick-walled cells, reduced leaf area, and dynamic stomata activities help it resist dryness [55]. Quinoa is a salt-tolerant plant that can sustain 100% seawater treatment [56]. The plant can store salt ions in its tissues and change leaf water potential to regulate cell turgor and transpiration [46, 47, 57]. Quinoa yield is higher under intermediate salinity levels (10–20 dS/m) than under lower salinity in the root zone, with substantial varietal differences. As a result, instead of being a glycophyte, this plant is categorized as a facultative halophyte [51, 54].

In Türkiye, quinoa is a young crop. Quinoa was introduced as part of the European Union-supported project "Sustainable water usage securing food production in dry parts of the Mediterranean Region" from 2008 to 2013. The SWUP-MED project has contributed to the creation of climate-resistant food crops that make much better use of agricultural land in Mediterranean climate-change-affected countries. The outcomes of quinoa research in Türkiye's Mediterranean region are presented in the next paragraphs.

5 Case Studies on Climate-Resistant Crops (Quinoa)

Several field experiments were conducted in Türkiye's Mediterranean region, testing climate-resistant crops (quinoa) in various locations with varying challenges.

Case study-1

From 2010 to 2012, field experiments were conducted at the Cukurova University, Irrigation and Agricultural Structures Department in Adana, Türkiye, to assess the crop yield of quinoa (*Chenopodium quinoa* Willd. cv. Titicaca) to treatment with saline and freshwater in a Mediterranean climate. Irrigation treatments in 2010 and 2011 included full irrigation with freshwater, full irrigation with saline water of varying salt concentrations (40, 30, 20, 10 dS/m), deficit irrigations with freshwater (50%, 75% of total irrigation), partial root-zone drying, and deficit irrigation with saline water of 40 dS/m. In addition to the full irrigation treatments, two deficit irrigation levels of 67 and 33% of total irrigation with fresh or salty (30, 20, 10 dS/m) water were considered in 2012. The results showed that irrigation water salinity ≤30 dS/m affected grain yields marginally compared to freshwater irrigation. Both salinity and drought stress reduced crop grain and biomass develop better. Yet, salinity stress had no effect on grain or biomass yield; hence, quinoa can be classed as a salt-tolerant crop. Above-ground biomass, seed yield, and harvest index indicated that quinoa cv. Titicaca responded well to Mediterranean habitats [58].

Case study-2

During the 2012 quinoa growing season, field experiments were undertaken in Tarsus, Türkiye, to investigate the influence of watering with drainage water and planting dates on quinoa yield and yield elements. The climate in the trial region is typical for the Mediterranean. The experimental field is located at 37 01′ N and 35 01′ E, ten meters above mean sea level. Arikli silty clay loam, which has a relatively good water-holding capacity, is the soil at the experimental site. Three weeks before planting dates, quinoa seeds were given to seedling production greenhouses. On April 11 and April 30, 2012, seedlings were transplanted at 20 cm in a row and 50 cm in row spacing for normal and late planting, respectively. At both planting dates, 70 kg/ha of 20-20-20% N-P-K composite fertilizer was applied and introduced into the soil. On May 15, 2012, urea at a rate of 50 kg/ha (46% N) was applied. The project was intended to test a line-source irrigation system. There were four irrigation levels and a rain-fed treatment evaluated (I1: full irrigation; I2: 20%; I3: 50% and I4: 80% deficit irrigation levels). Quinoa Titicaca was planted on April 11 for regular planting and April 30 for late planting. The quality of the drainage water ranged from 1.27 to 1.68 dS/m. Different water levels and planting periods resulted in considerably variable yield qualities, according to the data. Late planting time has beaten standard planting time. Standard planting plots yielded more quinoa than late planting plots. The top soil layer (30 cm) salinity rose from 1.15 dS/m at planting to 1.46 dS/m at harvest. The salinity of the soil declined at depth. As a result of the research, drainage water can be used to water quinoa during times of water scarcity [59].

Case study-3

This study was carried out again in the experimental field (latitude: 3 6o 59′ N, longitude: 35o 18′ E, mean sea level: 50 m) of the Cukurova University. A 2-years field experiment was undertaken in 2016 and 2017 to investigate the physiological and yield response of quinoa cv Titicaca to various deficit irrigation strategies utilized with surface drip (SD) and subsurface drip systems (SSD) under Mediterranean climate patterns. The treatments included regulated deficit irrigation (RDI), partial root-zone drying (PRD50), standard deficit irrigations (DI50, DI75), and full irrigation (FI) under SD and SSD. Rainfed treatment was also available. In the experimental design, split plots with four replications were used. DI75 and DI50 earned 75 and 50% of the FI, respectively. PRD50 plots received 50% of FI in each application but from different laterals. RDI received 50% of the FI requirement until blooming, then 100% of the water needs. RDI resulted in 23 and 21% water savings for SD and SSD systems, accordingly, in both trial years, and RDI provided a statistically equal yield to FI treatment. DI75 treatment saved 16% of the water used by both drip methods in the first year and 10 and 25% of the water used by SD and SSD systems in the second year, accordingly. Hence, for long-term quinoa cultivation under Mediterranean conditions, RDI and DI75 treatments appear to be feasible alternatives to FI. The leaf water potential (LWP) values in FI plots with both drip systems were higher, and the crop water stress index (CWSI) values were lower than in deficit irrigation treatment plots. Significant second-order polynomial connections between CWSI and LWP were identified as drip systems. The leaf area index (LAI), LWP, and CWSI declined as the drought progressed. CWSI was shown to be significantly (P 0.01) and inversely linked with grain yield, dry matter yield, LAI, and average soil water content, showing that as CWSI increased, quinoa grain yield declined. All these relationships are best described by significant second-order polynomial equations. Quinoa should indeed be irrigated for high yields at LWP values in the range between –1.35 and –1.60 MPa, with an overall CWSI value of roughly 0.35, according to the data [12].

Case study-4

Field studies were done in 2009, 2010, and 2011 to investigate the yield response of quinoa to salty and freshwater under Mediterranean climate patterns. This research was conducted again at Cukurova University. The coordinates for the station are 36° 59' N, 35° 18' E, and 50 m above sea level. The soil at the experimental site is of the Mutlu soil series (Palexerollic Chromoxeret) and has a clay texture across the soil profile. The 2009 study examined four irrigation treatments: full freshwater irrigation (FIF), full saline water irrigation (FIS), deficit watering (DI), and partial root-zone drying (PRD). The DI and PRD treatments were irrigated with freshwater. PRD dries out half of the root zone while wetting the other half. The irrigation treatments in the 2010 experiment were as follows: full irrigation with freshwater (FIF), full irrigation with salty water at various salinity levels (FIS-40 dS/m; FIS-30 dS/m; FIS-20 dS/m; FIS-10 dS/m); deficit irrigation (DIF-50, DIF-75, DIS-40); and dryness treatment. There were eight irrigation treatments in the 2011 experiment: full irrigation with freshwater (FIF); full irrigation with salty water at varying salinity levels (FIS) 30 dS/m; FIS-20 dS/m; FIS-10 dS/m); deficit irrigation 471 (DIF-50; DIF-75); PRD; and dry treatment. Saline water was produced by diluting seawater with fresh water. The seeds of 'Titicaca,' a day-length-neutral cultivar selected at the University of Copenhagen from material originating in southern Chile, were employed (Christiansen et al. 2010). Quinoa seeds were planted by hand on April 10, 2009, March 26, 2010, and March 28, 2011 (3–4 cm apart with 50 cm row spacing). At planting, a composite NPK fertilizer (15-15-15) was applied and introduced into the soil at a rate of 75 kg/ha N, P_2O_5, and K_2O. Quinoa plants in the row were trimmed to be around 15 cm apart. Irrigation treatments commenced after drip irrigation systems were established in the plots. Salinization was triggered at the beginning of floral bud development. In 2009, Quinoa received 302 mm under FIF and 151 mm under DI and PRD. In 2010, full irrigation treatments received 320 mm of fresh and saline irrigation water; DIF-50 and DIS-50 received 160 mm, while DIF-75 received 240 mm. In 2011, only single irrigation was applied to treatments due to above average rainfall during the growing season. When quinoa was irrigated with salty water under Mediterranean climatic conditions, it was discovered that irrigation water salt of 40 dS/m had no adverse impact on plant growth when compared to freshwater irrigation. Water stress, on the other hand, drastically reduced plant biomass yield under DI treatments. Soil salinity was highest in the top 10 cm soil layer of FIS-40 treatment plots, followed by DIS-50 and FIS-30 treatments. As soil depth grew, salt accumulation reduced. The salinity of the soil increased with the salt of the irrigation water. The top layer accumulated the most salt when irrigated with high saline water. Moreover, the high surface evaporation and the high salt content of the applied irrigation water caused the plant to not receive enough water. Yield parameters (for example, above-ground biomass, seed yield, and harvest index (HI) showed that 'Titicaca' responded well to the Mediterranean environment. Quinoa may be classed as a drought- and salinity-tolerant plant, since salt and water stress had no effect on crop production [40].

6 Conclusions

In regions, where non-salty water is readily accessible during the early growing season but too little to meet the water needs of crops throughout the irrigation season, the reuse of saline drainage water may be likely viable. The quality of underground water in dry and semi-arid regions is low, and the availability of fresh water is not enough to satisfy the irrigation needs of the entire region. Reusing drainage water for crop cultivation may be advantageous under such circumstances. Production systems based on salt-tolerant plant species and utilizing drainage waters have the potential to transform such waters from an environmental problem to an economic asset. A policy like this would promote drainage waters to be disposed of inside the irrigated districts, where they are produced, rather than exporting them to other regions by discharge into main irrigation canals, local streams, or rivers. These environmentally and economically sustainable techniques could be the key to future agricultural and economic prosperity as well as to social wealth, in places with salt-affected soils and saline-sodic drainage waters. Quinoa may be recommended as a novel alternative crop for semi-arid and arid Mediterranean settings influenced by several abiotic stress factors because of its stress-tolerant features, adaptability to varying agroecological conditions, and nutritional and economic values.

From these studies it can be concluded that full irrigation is not an option for increasing quinoa yield in water-stressed areas; regulated deficit irrigation, partial root-zone drying, and deficit irrigation are processes that reduce the amount of water used during the growing season without reducing yield, and these may be useful alternatives. As a result, RDI, in which irrigation water is applied by reducing crop water requirements by 50% at the vegetative development stage, is advised alongside standard deficit irrigation in the semi-arid Mediterranean region. Furthermore, the results reveal that quinoa planting timings, irrigation water volume administered with a sprinkler system, and water use are critical in achieving higher quinoa yields under the climate patterns of Türkiye's Tarsus plain. Planting periods affect quinoa yield; yet deficit watering reduces crop yields by reducing grain weight. Quinoa deficiency irrigation is, therefore, not suggested for the region. The sprinkler system enabled accurate irrigation application control. Sprinkler irrigation with careful management can eliminate some application losses that are unavoidable with surface irrigation. This study provides an insight into farmers' understanding and attitudes of quinoa as well as for which reason this is crucial for the crop's sustainability in Turkey. The primary conclusions and suggestions derived from this study are as follows: (i) Economic considerations dominate farmers' worries, with decreased crop prices and high input costs being the most prominent ones. Administrative inefficiencies, political meddling, and a lack of pricing transparency are social issues, whereas pests, diseases, and drought are environmental issues. (ii) Financial advantages are critical in determining whether farmers will use quinoa. Market availability, financial help, cultivation instruction, high yield, acceptable prices, and growth parameters are the important elements influencing the decision-making processes for innovation. These economic reasons also influenced prior post-adoption behavior of farmers. (iii) there is no difference between quinoa trial and non-trial farmers' evaluations of new technologies. They are all enthusiastic about new ideas and are ready to try quinoa. Yet, an examination of their post-adoption behavior suggests that quinoa cultivation will be abandoned if it fails to suit their demands. (iv) Before quinoa is implemented on a big scale in Türkiye, the findings on knowledge and perception should be included into an educational program for farmers. Since the domains of sustainable development are interconnected rather than distinct, the sustainable construction analyzed by this investigation can only work if the financial and biological aspects complement it. (v) Crop trials are critical, as they allow farmers, extension agents (farmers training services), and interested parties to supervise project progress and to carry out successful educational initiatives.

Further trials should be established in partnership with the Ministry of Agriculture to build and to strengthen farmers' awareness. Training programs should be developed to assess farmers' local knowledge and their ability to rely on prior information. Excellent knowledge will help farmers learn the precise image of the crop, which is crucial in influencing the choice of those producers who have yet to embrace the crop. (vi) Urgent actions should be taken to build a market for quinoa to move forward with large-scale trials. The existing small-scale trials cannot disclose all the potential problems in large-scale production. (vii) Future studies should concentrate on the economic viability of quinoa in Türkiye, in addition to how the project can indeed be executed to satisfy all stakeholders.

References

1. World Bank. Water in agriculture. 2023. https://www.worldbank.org/en/topic/water-in-agriculture. Accessed 16 Mar 2023.

2. Yazar A, İnce Kaya Ç. A new crop for salt affected and dry agricultural areas of Turkey: quinoa (Chenopodium Quinoa Willd.). Turkish J Agric Nat Sci. 2014; Special Issue(2):1440–1446.
3. Gutiérrez-Gordillo S, Lipan L, Durán-Zuazo VH, Carbonell-Barrachina AA, Rodríguez BC, Rubio-Casal AE, Carbonell-Bojollo R, Ordoñez-Fernández R, García-Tejero IF. Linking sustainability and competitiveness of almond plantations under water scarcity and changing climate. In: Kumar S, Meena RS, Jhariya MK, editors. Resources use efficiency in agriculture. Springer, Singapore; 2020.
4. Jury W, Vaux H. The emerging global water crisis: managing scarcity and conflict between water users. Adv Agron. 2007;95:1–76.
5. Fereres E, Soriano A. Deficit irrigation for reducing agricultural water use. J Exp Bot. 2007;58(2):147–59.
6. Harbinson J, Parry MAJ, Davies J, Rolland N, Loreto F, Wilhelm R, Metzlaff K, Lankhorst RK. Designing the crops for the future. CropBooster Program Biol. 2021;10:690.
7. Ray D, Gerber J, MacDonald G, West PC. Climate variation explains a third of global crop yield variability. Nat Commun. 2015;6:5989.
8. Amelung W, Bossio D, de Vries W, Kögel-Knabner I, Lehmann J, Amundson R, Bol R, Collins C, Lal R, Leifeld J, Minasny B, Pan G, Paustian K, Rumpel C, Sanderman J, van Groenigen JW, Mooney S, van Wesemael B, Wander M, Chabbi A. Towards a global-scale soil climate mitigation strategy. Nat Commun. 2020;11:5427.
9. Arora NK. Impact of climate change on agriculture production and its sustainable solutions. Environ Sustain. 2019;2:95–6. https://doi.org/10.1007/s42398-019-00078-w.
10. Wang Q, Huang K, Liu H, Yu Y (2023) Factors affecting crop production water footprint: a review and meta-analysis. Sustain Prod Consumption, p. 207–216.
11. Gerten D, Heck V, Jägermeyr J, Bodirsky BL, Fetzer I, Jalava M, Kummu M, Lucht W, Rockström J, Schaphoff S, Schellnhuber HJ. Feeding ten billion people is possible within four terrestrial planetary boundaries. Nat Sustain. 2020;3:200–8.
12. Bozkurt Çolak Y, Yazar A, Alghory A, Tekin S. Yield and water productivity response of quinoa to various deficit irrigation regimes applied with surface and subsurface drip systems. J Agric Sci. 2021;159(1–2):116–27.
13. McCornick P, Smakhtin V, Bharati L, Johnston R, McCartney M, Sugden F, Clement F, McIntyre B. Tackling change: future-proofing water, agriculture, and food security in an era of climate uncertainty. Colombo, Sri Lanka: International Water Management Institute (IWMI); 2013. p. 36.
14. Ceccarelli S, Grando S, Baum M. Participatory plant breeding in water-limited environments. Exp Agric. 2007;43(4):411–35.
15. Joffre R, Rambal S. Mediterranean ecosystems. Encycl Life Sci. 2002; 1–7.
16. Ragab R, Prudhomme C. Climate change and water resources management in arid and semi-arid regions: prospective and challenges for the 21st century. Biosys Eng. 2002;81(1):3–34.
17. Ali T, Huang J, Wang J, Xie W. Global footprints of water and land resources through China's food trade. Glob Food Sec. 2017;12:139–45.
18. Nakawuka P, Langan S, Schmitter P, Barron J. A review of trends, constraints and opportunities of smallholder irrigation in East Africa. Glob Food Sec. 2018;17:196–212.
19. Jones HG. Irrigation scheduling: advantages and pitfalls of plant-based methods. J Exp Bot. 2004;55:2427–36.
20. English MJ. Deficit irrigation: an analytical framework. J Irrig Drainage Eng ASCE. 1990;116(3):399–412.
21. English MJ, Raja SN. Perspective on deficit irrigation. Agric Water Manag. 1996;32:1–14.
22. Lecler NL. Integrated methods and models for deficit irrigation planning. In: Pert RM, Curry RB, editors. Agricultural systems modelling and simulation. Marcel Dekker, Inc.; 1998.
23. Capra A, Consoli S, Scicolone B. Deficit irrigation: theory and practice. In: Alonso D, Iglesias HJ, editors. Agricultural irrigation research progress, ISBN 978-1-60456-579-9. Chapter 4; 2008. p. 53–82.
24. Marsal J, Mata M, Arbones A, Del Campo J, Girona J, Lopez G. Factors involved in alleviating water stress by partial crop removal in pear trees. Tree Physiol. 2008;28:1375–82.
25. Chai Q, Gan Y, Zhao C, Xu HL, Waskom RM, Niu Y, Siddique KHM. Regulated deficit irrigation for crop production under drought stress—a review. Agron Sustain Dev. 2016;36:3.
26. Cabelguenne M, Jones CA, Williams JR. Strategies for limited irrigations of maize in southwestern France: a modelling approach. Trans ASAE. 1995;38:507–11.
27. Debaeke P, Aboudrare A. Adaptation of crop management to water-limited environments. Eur J Agron. 2004;21:433–46.
28. Oweis T, Pala M, Ryan J. Stabilizing rainfed wheat yields with supplemental irrigation and nitrogen in a Mediterranean-type climate. Agron J. 1998;90:672–81.
29. Ruiz-Sanchez MC, Torrecillas A, Perez-Pastor A, Domingo R. Regulated deficit irrigation in apricot trees. Acta Hort. 2000;537:759–66.
30. Chalmers DJ, Mitchell PD, van Heek L. Control of peach tree growth and productivity by regulated water supply, tree density and summer pruning. J Am Soc Hortic Sci. 1981;106:307–12.
31. Davies WJ, Zhang J. Root signals and the regulation of growth and development of plants in drying soil. Annu Rev Plant Physiol Plant Mol Biol. 1991;42:55–76.
32. Loveys B, Stoll M, Davies WJ. Physiological approaches to enhance water use efficiency in agriculture: exploiting plant signalling in novel irrigation practice. In: Bacon M, editor. Water use efficiency in

plant biology. Oxford: Blackwell Publishing; 2004. p. 113–41.
33. Stoll M, Loveys B, Dry P. Hormonal changes induced by partial rootzone drying of irrigated grapevine. J Exp Bot. 2000;51:1627–34.
34. Kriedemann PE, Goodwin I. Deficit irrigation and partial rootzone drying. An overview of principles and applications. Irrigation insights no. 4. Land and Water Australia; 2003.
35. Davies WJ, Wilkinson S, Loveys B. Stomatal control by chemical signalling and the exploitation of this mechanism to increase water use efficiency in agriculture. New Phytol. 2002;153:449–60.
36. Dinar A, Tieu A, Huynh H. Water scarcity impacts on global food production. Glob Food Sec. 2019;23:212–26.
37. Gregory PJ. Food production under poor, adverse climatic conditions. In: Proceedings IX ESA congress. Warsaw, Poland, 2006; p.19.
38. Lin KH, Chao PY, Yang CM, Cheng WC, Lo HF, Chang TR. The effects of flooding and drought stresses on the antioxidant constituents in sweet potato leaves. Bot Stud. 2006;47:417–26.
39. Schulze ED, Beck E, Müller-Hohenstein K. Plant ecology. Heidelberg: Springer; 2005.
40. Yazar A, İncekaya Ç, Sezen S.M, Tekin S. Quinoa experimentation and production in Turkey. In: Bazile D, Bertero D, Nieto C, editors. FAO &CIRAD. State of the Art Report of Quinoa in the World in 2013, by Rome. Chapter: 6.1.4 p. 466–477.
41. Hariadi Y, Marandon K, Tian Y, Jacobsen SE, Shabala S. Ionic and osmotic relations in quinoa (*Chenopodium quinoa* Willd.) plants grown at various salinity levels. J Exp Botany. 2011; 62(1):185–193.
42. Jacobsen SE, Jensen CR, Lui F. Improving crop production in the Mediterranean climate. Field Crop Res. 2012;128:34–47.
43. Riccardi M, Pulvento C, Lavini A, d'Andria R, Jacobsen SE. Growth and ionic content of quinoa under saline irrigation. J Agron Crop Sci. 2014;200(4):246–60.
44. Hinojosa L, González JA, Barrios-Masias FH, Fuentes F, Murphy KM. Quinoa abiotic stress responses: a review. Plants. 2018;7(4):106.
45. Pearsall DM. The origins of plant cultivation in South America. In: Cowan CW, Watson PJ, editors. The origins of agriculture. Smithsonian Institute Press: Washington D.C; 1992. p. 173–205.
46. Garcia, M. Agroclimatic study and drought resistance analysis of quinoa for an irrigation strategy in the Bolivian Altiplano. Dissertationes de Agricultura 556. Belgium: Faculty of Applied Biological Sciences, K.U. Leuven; 2003.
47. Jacobsen SE. The worldwide potential for Quinoa (*Chenopodium quinoa* Willd). Food Rev Intl. 2003;19:167–77.
48. Comai S, Bertazzo A, Bailoni L, Zancato M, Costa CVL, Allegri G. The content of proteic and nonproteic (free and protein-bound) tryptophan in quinoa and cereal flours. Food Chem. 2007;100(4):1350–5.
49. Jacobsen S-E, Monteros C, Christiansen JL, Bravo LA, Cor-cuera LJ, Mujica A. Plant responses of quinoa (*Chenopodium quinoa* Willd.) to frost at various phenological stages. Eur J Agron. 2005;22:131–9.
50. Geerts S, Raes D, Garcia M, Vacher J, Mamani R, Mendoza J, Huanca R, Morales B, Miranda R, Cusicanqui J, Taboada C. Introducing deficit irrigation to stablize yields of quinoa (*Chenopodium quinoa* Willd.). Eur J Agron. 2008;28:427–36.
51. Sanchez B, Lemeur H, Van Damme P, Jacobsen SE. Ecophysiological analysis of drought and salinity stress of quinoa (*Chenopodium quinoa* Willd.). Food Rev Intl. 2003;19:111–9.
52. Ruiz KB, Biondi S, Oses R, Acuna-Rodriguez IS, Antognoni F, Martinez-Mosqueira EA, Coulibaly A, Canahua-Murillo A, Pinto M, Zurita-Silva A, Bazile D, Jacobsen SE, Molina-Montenegro MA. Quinoa biodiversity and sustainability for food security under cilimate change: a review. Agron Sustain Dev. 2014; 34(2):349–359.
53. FAO. Quinoa: An ancient crop to contribute to world food security. FAO Regional Office for Latin America and the Caribbean 2011. http://www.fao.org/alc/file/media/pubs/2011/cultivo_quinua_en.pdf. Accessed 10 Nov 2022
54. Jacobsen SE, Mujica A, Jensen CR. The resistance of quinoa (*Chenopodium quinoa* Willd.) to adverse abiotic factors. Food Rev Int. 2003,19.99–109. https://doi.org/10.1081/FRI-120018872.
55. Jensen CR, Jacobsen SE, Andersen MN, Núñez NS, Andersen D, Rasmussen L, Mogensen VO. Leaf gas exchange and water relation characteristics of field quinoa (*Chenopodium quinoa* Willd.) during soil drying. Eur J Agron. 2000;13:11–25.
56. Koyro H-W, Eisa SS. Effect of salinity on composition, viability and germination of seeds of *Chenopodium quinoa* willd. Plant Soil. 2008;302:79 90.
57. Gómez-Pando LR, Alvarez-Castro R, Eguiluz-de la Barra A. Effect of salt stress on Peruvian germplasm of *Chenopodium quinoa* Willd: a promising crop. J Agron Crop Sci. 2010;196:391–6.
58. Yazar A, Ince Kaya Ç, Sezen SM, Jacobsen SE. Saline water irrigation of quinoa (*Chenopodium quinoa*) under Mediterranean conditions. Crop Pasture Sci. 2015;66:993–1002.
59. Yazar A, Sezen SM, Bozkurt Çolak Y, İnce Kaya Ç, Tekin S. Effect of planting times and saline irrigation of quinoa using drainage water on yield and yield components under the mediterranean environmental conditions. Int J Res Agric Forestry. 2017;4(8):8–16.

Plant and Agricultural Water Use Technologies

Irrigation Scheduling Techniques for Sustainable Irrigated Agriculture under the Era of Climate Change

Attila Yazar and Servet Tekin

Abstract

Increasing global water shortages and rising irrigation costs due to climate change have prompted the development of irrigation scheduling methods that minimize water use and maximize efficiency. Irrigation scheduling is crucial for conserving water, improving crop performance, and ensuring sustainability of irrigated agriculture. It becomes especially sensitive under limited water resources, where water shortages necessitate a refined timing of water application to minimize yield reductions. This chapter reviews various irrigation scheduling methods, contrasting traditional methods, such as water balance and soil moisture-based approaches, with those based on plant responses to water deficits. Plant-based methods, such as dendrometry, fruit gauges, tissue water content sensors, growth, sap flow, and stomatal conductance measurements, are outlined and evaluated. Recent advances, particularly in infrared thermometry and thermography, have also been highlighted. This chapter discusses the suitability of different approaches for specific crops and climatic conditions, their strengths and weaknesses, and their suitability at different spatial and temporal scales. This section also examines the potential of soil- and crop-based planning methods.

Keywords

Irrigation scheduling techniques · Soil crop-based planning · Sustainable irrigated agriculture

1 Introduction

Improving irrigation management is crucial for sustainable agricultural practices as consumers demand safe and environmentally friendly food products. However, concerns about freshwater availability and quality are increasing worldwide, particularly in arid and semi-arid areas, where irrigation water is scarce. This requires wise use of water and agrochemicals, and collaboration with other water-user sectors [1–6].

Water is essential for plant growth and food production and is provided by rainfall, irrigation, or a combination of these sources. As competition for water increases between

A. Yazar (✉)
Faculty of Agriculture, Department of Irrigation and Agricultural Structures, Çukurova University, Adana, Türkiye
e-mail: yazarat@cu.edu.tr

S. Tekin
Faculty of Agriculture, Department of Biosystems Engineering, Kahramanmaraş Sütçüimam University, Kahramanmaraş, Türkiye

M. Bahadir et al. (eds.), *Agricultural Water Management*, Water and Wastewater Management,
https://doi.org/10.1007/978-3-031-96134-2_4

municipal and industrial users and agriculture, an accurate estimation of irrigation water needs is essential for planning and managing water projects. The main objective of irrigation is to maintain crop evapotranspiration (ET) when rainfall is insufficient and soil reservoirs are depleted, thereby reducing crop productivity. To accurately predict the water requirements for irrigation, the soil–plant-atmosphere continuum must be considered as a dynamic system with interactive transport processes. Monitoring and control systems based on real-time feedback measurements of plant and soil components and meteorological variables can maximize the efficiency and productivity of water use. However, irrigation scheduling based on a single variable has led to a significant increase in water use efficiency [1, 3, 7–9]. This chapter deals with the use of sensors and methods to determine the timing and amount of irrigation by measuring the soil, plant, and meteorological variables.

Irrigation scheduling is critical as water supplies become scarce, competition for agricultural water increases, and water prices increase. Poor irrigation management, including the use of pesticides and fertilizers, causes groundwater and soil pollution. To optimize crop productivity and protect the environment, agriculture must use water more efficiently and apply the correct amount of water at the right time. Irrigation scheduling is crucial for crop production in arid and semi-arid regions because it ensures a sufficient supply of moisture to the soil to minimize water stress on plants during critical growth stages. Irrigation scheduling determines the optimal timing and amount of water for crop production to ensure a timely and sufficient water supply, alleviate water shortages in the soil, and enable efficient implementation [13–18].

Irrigation planning and management cover the design of irrigation schedules and water management at all scales. This includes setting water delivery schedules and deciding when, where, and how much water to irrigate. Irrigation scheduling decisions are based on soil–plant-atmosphere systems and can be based on long-term data or real-time information. Information about the crop, soil, climate, irrigation system, water deliveries, and management objectives must be considered to adapt irrigation scheduling to specific situations [15, 19–22].

When developing an irrigation scheduling strategy, farmers should consider their management objectives, such as maximizing net income, minimizing costs, maximizing yield, allocating limited water supply, minimizing groundwater pollution, or optimizing production when the irrigation system capacity is limited. Available water supply and irrigation costs are critical factors in determining an appropriate management objective. In the case of a limited area, the optimum irrigation depth is typically equal to the maximum yield. Therefore, an appropriate irrigation strategy is to avoid water stress on the crops throughout the growing season [14, 16, 20, 23–25]. In situations where water supply is insufficient for maximum yield, the irrigated area or seasonal application depth must be reduced. Farmers must distribute limited water over the entire season, but deficit irrigation planning is uncertain because of varying prices. Traditional irrigation scheduling practices are ineffective for deficit irrigation (DI). The management strategy focuses on optimizing the production per unit of water applied rather than maximizing the yield per unit area. Limited irrigation increases the overall yield of farms and efficiency of water use [14, 26–29].

This chapter discusses different methods of irrigation scheduling, contrasting traditional methods such as water balance and soil moisture-based approaches with those based on crop responses to water deficits. Plant-based methods such as dendrometry, fruit meters, tissue water content sensors and measurements of growth, sap flow and stomatal conductance will be evaluated. Recent advances in infrared thermometry and thermography are highlighted. This chapter discusses the suitability of different approaches for specific crops and climatic conditions, their strengths and weaknesses, and their suitability at different spatial and temporal scales. The chapter also explores the potential of soil- and plant-based scheduling techniques.

2 Irrigation Scheduling Techniques

Water scarcity and competition for high-quality water require sustainable agriculture. Poor management leads to high water losses, low yields, a reduction in irrigated areas and environmental problems. More efficient water use requires a coordinated approach at the water source, in the transportation system and at the farm and field level [30]. Irrigation scheduling is crucial for optimizing production and minimizing environmental impacts. It ensures that the right amount of water is applied at the right time and in the right quantity, and prevents under-watering and over-irrigation. This helps to conserve water, improve irrigation performance and ensure sustainability in irrigated agriculture. Decision making can range from personal experience to computerized methods based on soil water measurements, meteorological data or plant stress monitoring. In plant physiology research, scheduling methods based on leaf turgor pressure, stem diameter and sap flow have been developed to replace traditional methods of measuring soil water content or evapotranspiration rate [2, 16, 30–34]. Newer approaches are based on the reaction of plants to water deficits but have not yet become established. Global water scarcity and high irrigation costs have led to a focus on minimizing water use [2, 16, 35, 36]. Precise irrigation methods such as drip irrigation have reduced the water requirements of crops. However, new methods for accurate irrigation scheduling and control are needed, even for less demanding systems such as flood irrigation.

2.1 Monitoring Methods

Scientific irrigation scheduling is a management technique that determines the timing and amount of water for irrigation to control plant water stress during critical growth periods. Methods include soil–water measurements, soil–water balance calculations, stress-sensing techniques [16], and weather-based methods [37, 38]. Research studies aim to optimize crop water needs and productivity [3, 39].

Irrigation scheduling involves monitoring soil water and crop growth using computer programs. It updates field information, forecasts future irrigation dates, and monitors soil, plants, and microclimates for effective irrigation management [14–16, 20, 40]. The following paragraphs provide an overview of these methods.

Irrigation scheduling is influenced by factors like plant water status, quantity, growth stage sensitivity to water stress, and application frequency. Accurate soil and crop monitoring can provide valuable information, but more quantification is needed [2, 10, 16].

Monitoring methods involve assessing soil water content, plant temperature, and soil water balance to determine irrigation schedules, considering factors like rooting depth and crop evapotranspiration [14, 41–43].

Scheduling practices often rely on monitoring crops or soil, causing irrigation when they reach a critical level. Accurate monitoring is crucial for decision-making but may not be available during irrigation. Timed measurements are essential to avoid stress. Past data must be extrapolated to predict irrigation dates, but climate changes and crop development can introduce errors. Weather forecasts should guide irrigation timing [14, 16, 43, 44]. Monitoring is essential for efficient irrigation management, but it can be time-consuming, costly, and inaccurate if personnel are not trained. It should be synchronized with irrigation and adjusted to ensure the schedule is followed. Periodic monitoring before irrigation can provide better management information. However, monitoring methods can face issues like changing critical values based on location, time, and climate, and not providing information on irrigation depth [2, 14–16].

Table 1 summarizes the most important methods for irrigation scheduling that are based on measuring soil water or calculating the soil water balance. Soil water measurement directly measures the moisture status of the soil to determine irrigation needs, while water balance calculation estimates the moisture status of the soil using a water balance approach. The change in

Table 1 A summary of the main classes of irrigation scheduling approaches, indicating their main advantages and disadvantages [16]

Irrigation Scheduling	Advantages	Disadvantages
I. Soil water measurement (I-a) Soil water potential (tensiometers, psychrometers, etc.)	Water content measures are precise and easy to apply, with many commercial systems available and some sensors, particularly capacitance and time domain sensors, being easily automated	Soil heterogeneity necessitates expensive sensors or extensive monitoring programs, challenging to select root-zone positions, and often does not measure water status at root surface due to evaporative demand
(I-b) Soil water content (gravimetric; capacitance/TDR; neutron probe)		
II. Soil water balance calculations (Require estimate of evaporation and rainfall)	Easy to apply in principle; indicate 'how much' water to apply	Evapotranspiration estimates require accurate local precipitation/runoff estimates, good crop coefficient estimates, and regular recalibration due to cumulative errors
III. Plant "stress" sensing (Includes both water status measurement and plant response measurement)	Measures the plant stress response directly; integrates environmental effects; potentially very sensitive	The system is currently in the research/development stage, not yet widely used for routine agronomy, except for thermal sensing in some situations
(III-a) Tissue water status	Leaf water status is often considered the most accurate measure for physiological processes like photosynthesis, but this argument is generally incorrect as it overlooks root-shoot signaling	All measures are influenced by homeostatic regulation, particularly leaf water status, and are not sensitive to environmental conditions, which can cause short-term fluctuations greater than treatment differences
(i) Visible wilting	Easy to detect	Not precise; yield reduction often occurs before visible symptoms; hard to automate
(ii) Pressure chamber	This widely accepted reference technique is particularly useful for estimating stem water potential (SWP) using bagged leaves or suckers	The process is slow, labor-intensive, and expensive, particularly for predawn measurements, making it unsuitable for automation
(iii) Psychrometer	Valuable, thermodynamically based measure of water status; can be automated	Requires sophisticated equipment and a high level of technical skill, yet still unreliable in the long term
(iv) Tissue water content (RWC, leaf thickness [c- or b-ray thickness sensors], fruit or stem diameter)	Tissue water content changes are easier to measure and automate than water potential measurements and are more closely related to physiological function than total water potential	The complexity and cost of instrumentation make it challenging to replicate accurate measurements, and the sensitivity of leaf thickness is limited by lateral shrinkage
(v) Pressure probe)	The pressure component of water potential is crucial for xylem flow and cell function, such as growth	Only suitable for experimental or laboratory systems
(vi) Xylem cavitation	Can be sensitive to increasing water stress	Cavitation frequency is influenced by stress prehistory, and the cavitation-water status curve demonstrates hysteresis, with most cavitations occurring during drying, indicating unsuccessful rehydration

(continued)

Table 1 (continued)

Irrigation Scheduling	Advantages	Disadvantages
((III b) Physiological responses	Measures of tissue water status, particularly leaf water, may be more sensitive than others	Often requires sophisticated or complex equipment; require calibration to determine 'control thresholds'
(IIIb-i) Stomatal conductance	Generally, a very sensitive response, except in some anisohydric species	Large leaf-to-leaf variation requires much replication for reliable data
– Porometer	Accurate: the benchmark for research studies	The process is labor-intensive and not suitable for commercial use, and it is not easily automated, despite some attempts
– Thermal sensing	Remotely used, capable of scaling up large crop areas, imaging effectively averages leaves, and simple thermometers are cheap, portable, and well-suited for crop monitoring	Calibration of canopy temperature is necessary due to its impact on environmental conditions and stomatal aperture, which can be achieved using wet and dry reference surfaces.)
– Sap-flow sensors	Sensitive	The method indirectly estimates conductance changes due to atmospheric conditions, requires complex instrumentation and technical expertise, requires calibration for each tree, and defines irrigation control thresholds
(IIIb-ii) Growth rate	Probably the most sensitive indicator of water deficit stress	Instrumentation is delicate and generally expensive

soil moisture over time is the difference between inputs and loss. The methods for measuring soil moisture and estimating evapotranspiration have been extensively studied [45, 46], including detailed methods for determining crop water requirements and calculating the water balance for different crops and climatic zones [47]. Although the water balance approach is not very accurate, it is robust under different conditions. However, errors accumulate over time, requiring regular recalibration based on actual soil or plant response measurements. Table 1 also lists the advantages and disadvantages of the different approaches to irrigation scheduling.

Soil water-based approaches face a problem because plant physiology responds directly to the water status in the plant tissue and not to the water content of the soil. Plant response to soil moisture varies as a complex function of evaporative demand, which affects its ability to take up water. The study suggests [35] that "plant stress sensing" could allow for greater precision in irrigation by relying on plant responses rather than measuring soil water status. This approach could be used to schedule irrigation decisions more effectively.

2.1.1 Soil Water Monitoring

Soil moisture measurement is the oldest method for irrigation scheduling, gaining popularity owing to irrigators' trust in soil texture and color. Monitoring crops or atmospheric factors offers additional guidance but does not provide an unambiguous scheduling criterion [13, 20, 42, 48, 49]. Soil moisture content can be measured using various methods such as gravimetric sampling, tensiometers, resistance blocks, or neutron moisture meters. Time domain reflectometry (TDR) may replace neutron meters, but only a few commercial models are available [50]. Irrigations are scheduled by allowing crops to deplete available soil water in the root zone. Water applications can be estimated or calculated based on crop root zone requirements [15, 20, 43, 51].

The mass basis water content of a field soil is crucial for comparing soil volume changes such as tillage; however volumetric water content is essential for irrigation, crop water use, and water use efficiency. The symbols used for these are θm and θv.

If the volume of the soil sample (Vs, m^3) is known, then the volumetric water content (θv, m^3/m^3) can be calculated by converting the mass of water lost on drying, Mw, to a volume, and then dividing by the sample volume.

$$\theta v = (\text{volume of water lost } (Mw/\rho_w))/(\text{total soil volume } (Vs)) \quad (1)$$

where ρw is the density of water (typically assumed to be 1 Mg/m^3).

Plants extract water based on root length per unit soil volume, with the highest root density in the surface soil layer. Soil water content varies with depth; therefore, measurements should be performed at multiple depths within the root zone. The root zone water content (Wrz) can be calculated by adding θv at each depth and soil layer depth.

$$\text{Wrz} = \theta_{v1} x d_1 + \theta_{v2} x d_2 + \theta_{v3} x d_3 \quad (2)$$

where θ_{v1}, θ_{v2} and θ_{v3} are volumetric water contents at three soil depths representing the root zone; d_1, d_2 and d_3 are the thickness of each of the three soil layers sampled; and Wrz has the units of d. The summation of soil water content over a depth range can be performed using various methods, including Eulerian summation.

Direct soil sampling, using oven-drying techniques, is a useful method but is destructive, labor-intensive, slow, and costly. It cannot measure water content twice, adding variability to data. Direct sampling is beneficial in areas where labor costs are not a concern because it avoids accuracy problems. Alternative methods, such as calibration, measure θv using a "surrogate" that changes with soil water changes (Table 2), aiming to estimate the probable value of θv through the relationship between the surrogate measurement and soil water content [43]. This is usually expressed as a graph or a formula. Sometimes, it is a simple linear relationship, such as

$$\theta v = ay + b \quad (3)$$

Table 2 Surrogate measures used by different θv sensors [43]

Method	Surrogate	Measurement explanation
Neutron moisture meter	Count of slow neutrons around the fast neutron source	Radioactive sources emit fast neutrons (5 MeV) that lose energy when they collide with hydrogen, which calibrates the slow neutron concentration. Water is the only rapidly changing source of hydrogen in the soil
Thermal sensors	Heat conductivity or heat capacity of soil	Heat pulses generate and measure temperature changes in soil over time. Soil is a poor heat conductor, whereas water is a good conductor; therefore, the heat transmission rate is closely related to θv
Time-domain reflectometer (TDR)	Travel time of an electromagnetic pulse	A fast rise-time electromagnetic pulse is injected into a soil-buried waveguide, determining its time along metal rods based on the soil's electrical permittivity, with true TDR involving waveform capture and frequency analysis
Campbell FDR	Repetition time for a fast-rise electromagnetic pulse	The method of TDR sensors relies on the reflected pulse reaching a set voltage, which is influenced more by the bulk permittivity (BEC) and temperature
Capacitive sensors	Frequency of the oscillating circuit	A circuit induces an oscillating current, which affects the electrical permittivity and capacitance of the soil. The capacitor electrodes are arranged in a dielectric medium that is influenced by an electromagnetic field. This affects the frequency of oscillation and the electrical permittivity of the soil
Conductivity sensors (e.g., granular matrix sensors and gypsum blocks)	Electrical conductivity of a porous medium in contact with the soil	An alternating current voltage is applied to two electrodes in a porous material in contact with soil, determining the conductivity and water content between them, which is used to estimate the soil water tension (suction), not θv
Tensiometers	Matric and gravitational soil–water potential components	Soil pores create negative pressure through capillary forces, holding water in a porous cup and tube and estimating soil water tension (suction) using a vacuum gauge

where y is the value of the surrogate measurement and the slope (a) and intercept (b) are constants determined by calibration. However, this relationship is often complex.

Non-destructive methods for soil water readings are advantageous because they are recordable and require user knowledge for calibration. However, these methods introduce new sources of error and are affected by factors other than soil water [43]. For example, neutron moisture meters (NMM) are affected by soil hydrogen, chloride, boron, and density. Electromagnetic (EM) methods, such as capacitive, time-domain reflectometry (TDR), and frequency-domain reflectometry (FDR), are affected by salinity, temperature, and metallic soil components. EM systems are sensitive to soil volumes smaller than the REV of the soil, resulting in a small-scale variability of θv. The electronics of these systems are relatively insensitive to temperature changes; however, soil water readings from EM sensors are sensitive to temperature changes. Additional problems include faulty equipment, wear and tear and water and soil entering the electronics, sometimes causing unresolved faults [43]. Some surrogate measures, such as heat dissipation methods, are effective over a specific range of θv but insensitive over another range, such as the heat capacity or heat conductivity of water, which works well between 0 and 0.3 m^3/m^3 water content.

How much water can soil hold?

This chapter explores soil–plant-water relations, focusing on the wilting point and field capacity.

Table 3 Typical field capacity and wilting point values (volumetric, m^3/m^3, %) for different soil textures

Soil texture	Field capacity	Wilting point	Available water
Coarse sand	0.06	0.02	0.04
Fine sand	0.10	0.04	0.06
Loamy sand	0.14	0.06	0.08
Sandy loam	0.20	0.08	0.12
Light sandy clay loam	0.23	0.10	0.13
Loam	0.27	0.12	0.15
Sandy clay loam	0.28	0.13	0.15
Clay loam	0.32	0.14	0.18
Clay	0.40	0.25	0.15
Self-mulching clay	0.25	0.25	0.20

The wilting point occurs when the soil water content is too low for plants, whereas field capacity is the maximum water content that can be maintained without rapid water draining. Excess water usually drains back to the field capacity level. The difference between these is known as available water storage. The water content reaches the wilting point depending on the clay content and soil structure. Table 3 lists the common values for some of the soil types.

The "refill point" is a soil water reference point where plant production decreases due to water stress. Water content depends on the soil type, evaporation conditions, crops, and management practices. Some crops, such as wine grape vines, are of better quality under mild water stress. Plants near the refill point may show signs of wilting, indicating soil drying in the zone adjacent to roots. The refill point is the water content intermediate between the field capacity and the wilting point. Volumetric soil water content sensing ranges from 0.04 m^3/m^3 (%) in loamy sand to 0.09 m^3/m^3 (%) in clay loam. The accuracy of water content estimates should be 0.01–0.02 m^3/m^3 (%). Methods for collecting soil moisture include neutron probe, time domain reflectometry, gravimetric, Aquafer probe, tensiometers, electrical resistance blocks, and the hand feel method. To schedule irrigation, a monitoring site is selected, the soil water content is determined, full and refill points are set, and record-keeping is established.

Soil water measurements should be performed frequently, especially for frequently irrigated high-value crops. For other crops with lower irrigation frequencies, the measurements should be performed less frequently. Measurements should be taken at the depth of maximum root activity [52] and below the root zone to detect excessive water application and leaching [14, 43].

Soil moisture monitoring is often used in scheduling programs to improve soil water balance calculations. To improve the scheduling accuracy, data filtering can be used to combine the results from previous field monitoring. This reduces the uncertainty in point estimates and improves irrigation schedule dependability. Expanding the available data sources could lead to less frequent monitoring, as demonstrated by English et al. [53].

Soil sampling for determining water content is simple, time-consuming, and expensive because of spatial variability [54]. Tensiometers are simple and inexpensive but limited to potentials better than 0.8 bar and require frequent service. They measured the potential only in the immediate vicinity, requiring multiple tensiometers to obtain a reliable spatial average [42]. Tensiometers also face difficulties in swelling clays, as the ceramic cup can break contact with the soil as it dries [55]. A tensiometer is a closed tube with a ceramic tip attached to one end that records soil moisture levels as it dries and draws water from the tube. The tube is filled with water and sealed, and the ceramic tip is in contact with the soil. A pilot hole is augured, and a soil–water slurry is mixed and pushed into the tensiometer tip. The gauge registers the soil tension when the ceramic tip reaches moisture equilibrium with the soil. Tensiometers are responsive to soil water tensions of 5–70 cbars, with a practical operating range of 0–75 cbars [63]. Tensiometers are limited in their use in coarse textured soils or high-frequency irrigation regimes with higher soil moisture levels, such as drip or center pivot irrigation. Calibration is necessary to establish the relationship between the soil water content and tension, and routine maintenance is crucial for successful use.

Gypsum blocks are simple and inexpensive tools used to measure crop water use. However, their calibration is challenging owing to the uncertainty in the relationship between electrical resistance and water potential. This relationship varies from block to block and may also vary from soil to soil, owing to soil moisture characteristics. Despite this, gypsum blocks can be useful as a relative measure of crop water use but require multiple blocks for accurate representation [26].

Neutron probe is a tool used for soil moisture measurement; however, its use for irrigation scheduling is relatively recent. It produces fast neutrons from a radioactive source that are scattered and moderated by soil water. The probe is inserted into the soil through an access tube, and the water content is directly related to the number of neutrons scattered back to a detector. The calibration of the probe is relatively constant from soil to soil, making it easy to use and directly measure soil water [20, 42, 43]. The neutron meter is a useful tool for irrigating soil, allowing the continuous measurement of soil water content at different depths in the field. Its built-in safety features and manufacturer instructions minimize radiation risks. However, their high cost is a major disadvantage.

The introduction of time-domain reflectometry (TDR) in the 1980s [56] enabled the accurate estimation of soil water content using an EM sensor. TDR instruments operate in a frequency range of ~1 GHz, with minimal interference from the bulk electrical conductivity (σa, dS/m). However, some soils are susceptible to the effects of temperature. Evett et al. [57] demonstrated that σa measurements and temperature measurements can correct minor interferences. Lower-cost alternatives based on capacitive measurements operate at frequencies of a decade or smaller than TDR, and interference from σa changes is more significant [58]. Evett et al. [58, 59] found that capacitance type down-hole soil water sensors in access tubes are inaccurate and negatively impact the performance of these EM sensors, as they are not directly in contact with the soil.

Studies show that θv variability increases with sample size, suggesting that spatial variability in EM sensors may not be due to actual θv variability but rather a non-uniform EM field response within the sampling volume. Logsdon's work [60] with the CS616 EM sensor confirmed this preference, a fact established in other fields of study [61]. Mazahrih et al. [62] conducted a field calibration of down-hole sensors in the San Joaquin Valley of central California. They found that the sensor calibrations changed rapidly with depth and were relatively inaccurate even after calibration, likely dependent on σa.

Gypsum blocks and tensiometers are simple tools used in fields to measure soil water potential, whereas neutron gauges are accurate but expensive and require special training because of their radioactive source, necessitating frequent field visits for irrigation purposes [64]. Granular matrix blocks, a new generation of electrical resistance blocks, measure the resistance between electrodes in a sand-like material and a synthetic porous material. A high resistance indicates a lower block water content, causing higher soil water tension. Calibration of a specific soil type is necessary to establish this relationship. These blocks are lightweight, require minimal maintenance, and can withstand soil water tensions of 40–125 cbars, making them ideal for heavier soils [65]. Soil moisture monitoring should be conducted in a field that represents most conditions, considering multiple sites in areas with multiple soil types. Deep-rooted plants should have two monitoring devices per site, one at 30 cm and the other at 60 cm. Soil moisture should be measured at 2/3 of the distance from the trunk to the canopy drip line with tensiometers and watermarks installed at different depths. Irrigation systems should be scheduled using representative information.

One of the simplest irrigation scheduling techniques is the Checkbook method. Checkbook schedules rely heavily on soil moisture monitoring due to evapotranspiration procedures, simplifying crop water use, potentially sacrificing accuracy, and requiring time for updating information [13].

Water budget methods

Water budget methods calculate crop evapotranspiration (ET) and soil reservoir water loss by applying irrigation with a net amount equal to accumulated ET losses, recharging soil, and restarting the water cycle. If a full recharge is not possible, a new balance can be determined using net irrigation or field observations. However, this method may not work in locations where the contributions of the water table or other sources to crop ET cannot be quantified [66].

Water budget management involves managing soil and crop data, including yield threshold depletion and soil moisture. Farmers must irrigate before reaching the depletion level, with management allowable depletion levels (MAD) triggering irrigation. Irrigation scheduling alerts farmers when a field approaches the MAD; however, it is not the sole ruler of irrigation timing. Water budgets represent water movement; however, modeling complex processes requires extensive data and experience.

Smart irrigation scheduling is a technology that aids farmers in determining an optimal watering schedule for their crops. It uses weather stations, soil probes, and moisture sensing devices to measure various factors (air temperature, humidity, wind speed, rainfall, soil moisture depth, temperature, and salinity), enabling more efficient water use, improved yields, and remote control of irrigation systems, which are often paired with software.

Understanding soil water balance is crucial for effective irrigation management, particularly scheduling. It helps predict the irrigation timing and amount if all terms are known or estimated [14, 44, 67]. This method estimates water availability in the crop root zone at any time by integrating soil water availability, plant indicators, and climatic effects for efficient irrigation scheduling and economic production [55, 68]. Hodnett et al. [69] highlights the drawbacks of irrigation scheduling based on soil moisture budgeting, as it can cause soil wetting or drying, potentially impacting yield and water consumption. They proposed a more effective method for maintaining soil moisture potential.

2.1.2 Monitoring the Plant

Plant observation is the oldest method of scheduling irrigation, indicating water stress in plants through visible signs such as color changes and leaf curling. This method requires no data gathering and can lead to growth and yield reduction. Modern methods use physiological states, such as stomatal closure, leaf water potential, pressure bombs, and canopy temperature, which change only after stress has reduced yield. These methods are not routinely used for irrigation scheduling [2, 16, 70]. To determine crop irrigation timing, it is crucial to examine the plant itself, as the soil water status and atmospheric demand do not accurately represent the plant's water status. Measuring plant parameters is essential for evaluating general health [71]. Quantitative estimates of plant water stress are crucial for irrigation scheduling. Ideally, warnings should come from the crops themselves, eliminating the need for precise information on soil water, root distribution, and atmospheric evaporative demand. However, these methods are labor-intensive, tedious, and subject to experimental and sampling errors [72].

Various methods have been proposed for monitoring plant water status, including estimating transpiration, determining leaf tissue hydration, observing stomatal apertures, monitoring stem diameter, and measuring leaf water potential [40]. However, not all methods are suitable for irrigation scheduling [2, 16, 70].

When transitioning from soil water-to plant-based irrigation scheduling, it is crucial to consider the most suitable methods, such as direct measurements of plant water status and sensitive plant processes. Direct measurements are the most reliable indicators of irrigation requirements, but the question remains about where to measure it, as plants have some autonomy over their water status [73, 74]. This control is achieved through changes in the leaf area, root extension, leaf angle, stomatal conductance, and hydraulic properties of the transport system. In extreme cases, plants with good endogenous control systems maintain stable leaf water status under various evaporative demands or soil water supplies [75, 76].

Plant-based measures are chosen based on their sensitivity to water deficits. Hsiao et al. [77] identified cell growth, cell wall, and protein synthesis as the most sensitive to tissue water deficits. Stomatal closure is rarely affected by tissue water potential drops, with 1.0 MPa or more being required for closure. Photosynthesis is moderately sensitive because of its dependence on stomatal aperture, but some processes, such as electron transport, are insensitive. Hsiao's classification is now believed to be misleading and underestimates the true sensitivity of stomata, as it is based on the observed responses to leaf water potential alone [77].

Plant-based irrigation scheduling requires defining the reference or threshold values for plants growing under a non-limiting soil water supply [79]. Understanding the behavior of these values as the environmental conditions change is crucial for developing and validating these methods. Plant-based methods do not provide information on 'how much' irrigation to apply, but only 'if' irrigation is needed.

Plant water status

The first approach to using plants as indicators of irrigation was based on visible wilting, but this method may have lost potential yield [80]. Therefore, more sensitive measures of plant water status are required. Relative water content (RWC) [81] is widely used, but water potential, particularly from the leaves, is often argued to be a more rigorous measure [73, 80]. RWC is more closely related to cell turgor than the total water potential [16, 73].

Leaf water status, a crucial indicator of irrigation needs, is often controlled by stomatal closure or other regulatory mechanisms, making its use unsuitable for strong isohydric species. Jones [73] also noted that rapid temporal fluctuations due to environmental conditions, such as passing clouds, can make the interpretation of leaf water potential as an indicator of irrigation unsatisfactory. Despite these concerns, leaf water potential can provide a sensitive index for irrigation control when corrected for diurnal and environmental variation, making it a valuable tool for monitoring and managing soil moisture levels [16, 82].

Workers have proposed using xylem water potential or stem water potential (SWP) as a more reliable indicator of leaf water status. SWP is measured using a pressure chamber on leaves enclosed in darkened plastic bags, allowing it to equilibrate with the xylem water potential [83]. Some studies have suggested measuring pre-equilibrated leaves from root suckers for a more stable measure [35, 84]. These methods are considered preferable as they approach soil water status more closely than leaf water potential but may miss plant-based advantages.

The predawn leaf water potential, as ψ_{leaf} should largely equilibrate with ψ_{soil} by dawn), is a potential estimator of soil water potential, but it is often insensitive to soil moisture content [85] and is not convenient for irrigation scheduling. Jones [86] proposed an indirect estimation of soil water potential at the root surface of transpiring plants using leaf water potential and stomatal conductance during the day, which has been successfully tested by Lorenzo-Minguez et al. [87]. This approach offers significant advantages over predawn measurements and has been successfully tested.

Plant-based methods for irrigation scheduling and control are not suitable for automation due to the challenges in measuring variables. Automated stem or leaf psychrometer [88] are unreliable, and the preferred method is to use plant water status as an indicator of soil water status. Direct measurement of soil water potential can avoid the need for plant-based measurements, but it doesn't necessarily provide a good measure of effective water potential at the root surface during active transpiration [86]. Several indirect methods have been developed as alternatives to direct measurements, but they suffer from the same disadvantages as direct measurements of leaf water status [16, 89, 90]. These methods have been developed into commercial systems in certain circumstances, but they generally suffer from the same disadvantages as direct measurements.

Leaf water potential—the pressure chamber

The chamber principle involves applying pressure to an enclosed leafy shoot to bring the water potential to zero. The chamber measures the gravitational and frictional potential, or the "xylem pressure potential," which is often close to the total potential because the osmotic potential of the xylem sap is close to zero [72, 92]. The leafy shoot is enclosed in a steel pressure chamber, and pressure is increased until xylem sap appears on the cut end. After the balancing pressure was read, the system was vented through the outlet valve.

In order to obtain a good accuracy in the results, it is necessary to take same cautions concerning the way in which the system is operated and the samples are handled; (i) ensure the accuracy of the pressure gauge, (ii) cut the stern or twig with a sharp knife, ensuring the stem outside the chamber isn't too long and maintaining the same length for all measurements, (iii) To prevent transpiration loss from leafy shoots before and during measurement, shorten the time between detachment and measurement, wrap leaves in plastic bags, or cover the inner wall of the cylinder with wet filter paper. During measurement, water losses can be prevented by the same techniques; (iv) the recommended pressure range is 50 J/kg/s to 10 J/kg/s, with a reduced rate of J/kg/s recommended to prevent premature water bubbling while maintaining the same pressure increase rates for all measurements. Leaf water potential fluctuates during the day owing to excess transpiration over absorption, with its minimum at midday and maximum near sunrise. As the soil moisture limits, both the maximum and minimum values change. The maximum value decreased after irrigation, whereas the minimum value is reached sooner owing to the increased stomatal resistance.

Stomatal resistance or conductance

Under dry soil conditions, increased plant water potential leads to decreased relative transpiration due to stomatal closure. The value of the leaf water varies between plants. Stomatal conductance becomes a more sensitive water stress criterion for conservers [72], making it more sensitive than the water potential during midday measurements. Stomatal conductance measurements were performed using a porometer, with a dynamic diffusion porometer being the most popular until 1980. This instrument consists of a chamber with a humidity sensor and fan and calculates stomatal conductance by increasing humidity by a fixed amount. The steady-state diffusion porometer maintains air humidity by balancing the transpired water with water vapor losses through a gauge. Calibration against a known resistance was performed using perforated plates. Measurements and sampling precautions were taken, with the main precaution being shading the chamber to avoid significant temperature differences between the leaf and sensor [16, 93]. To avoid confusion between the light effect, age effect, and water stress in irrigation scheduling, uppermost young sunny leaves were selected for stomatal conductance measurements [94]. The abaxial and adaxial surfaces of amphistomatous leaves were measured. The mean is taken on conductance, as the stomata act in parallel. The main objective of irrigation is to observe sudden changes in diurnal patterns or the day-to-day evolution. The conductance of a single leaf can be the sum of the abaxial and adaxial surfaces [95] or the mean [97]. 2.1.2.4 Stem and fruit diameter Stem and fruit diameters fluctuate diurnally owing to changes in water content, but they can be used to determine irrigation needs. The magnitude of daily shrinkage indicates the water status and growth rate. However, these measures are not useful in controlling high-frequency irrigation systems. However, the use of maximum daily shrinkage (MDS) has shown promising results for low-frequency irrigation scheduling, providing a sensitive measure of plant water stress. For example, Fereres and Goldhamer [79] found MDS to be a more effective approach for automated irrigation scheduling in almond trees, whereas differences in maximum trunk diameter were useful in olives [98. Dendrometry and micro morphometric techniques have been developed for commercial irrigation scheduling systems, with a focus on stem diameter changes. Variations in trunk diameter or stem water potential are more sensitive indicators of irrigation needs than fruit diameter [90]. There is growing interest in the evaluation of these techniques for irrigation scheduling. A dendrometer is a dial gauge used to measure changes in tree diameter as the water status of the plant changes. It determines the need for and amount of irrigation based on the change in diameter over a specific period, similar to leaf thickness sensors. Although it does not detect water stress quickly, research has aimed to relate diameter changes to irrigation requirements.

Sap flow

The use of heat pulse and energy balance thermal sensors for sap-flow measurements in plant stems offers an alternative irrigation scheduling method. This method is sensitive to water deficits and stomatal closure and has been tested in various crops such as grapevines, fruits, olive trees, and greenhouses. However, determining the correct control points for any crop remains challenging [2]. Sap flow indicates changes in transpiration rate, largely determined by stomatal aperture but also by environmental conditions such as humidity. Changes in sap flow can occur without changes in stomatal opening. Although tree size and exposure may vary, the general patterns of change in response to environmental conditions and water status remain consistent [99]. Regular calibration measurements can be used to determine the appropriate sap-flow rates as control thresholds for larger trees. An irrigation scheduling algorithm can be derived based on the diurnal patterns of sap flow, with midday reductions indicating water deficits. However, sap flow may lag changes in transpiration rate owing to the hydraulic capacitance of the stem and other plant tissues, making it a potential issue for precision control [2, 100]. Modern farmers adopt irrigation to ensure profitable yields and to protect commercial enterprises from drought risks. However, there is a need for more accurate scheduling and precise irrigation management in orchards [101]. The scientific community has invested in new irrigation technologies and efficient scheduling methods. Plant-based methods have the greatest potential for irrigation control; however, there are limitations in defining reference values and plant variability [102]. Advances in electronic sensors and automated equipment are helping overcome these limitations [79, 103, 104]. Thermal remote sensing methods can be combined with plant-based methods for precise irrigation of heterogeneous commercial orchards using a manageable number of instrumented plants [105]. Research has shown that sap flow measurements are widely used for the in-situ determination of plant water consumption. These systems are automated and reliable, making them suitable for long-term field applications. Comparisons have been made with other water-stress indicators such as apples [106], grapes [107], lemons [108], and plums [109]. Sap flow records can also be used to study plant transpiration dynamics. Nadezhdina and Cermak [110] utilized the Sap Flow Index (see [106] for details) to regulate irrigation in fruit tree orchards, suggesting a subtle change in the sap velocity profile due to water stress as a suitable indicator. Fernández et al. [111] and Nadezhdina et al. [112] suggested using sap flow ratios in the inner and outer xylem regions to trigger irrigation. Another method is to measure the sap flow in irrigated plants to avoid soil water limitation. Irrigation amounts can be controlled by comparing measurements with representative plants in other parts of the orchard [113]. The reference rate of plant transpiration based on the leaf area and local microclimate can also be calculated [114]. User-friendly models have been developed to control irrigation in commercial orchards [115].

Infrared Thermometry and Crop Water Stress Index (CWSI)

Plant-based methods that have been used to evaluate water stress in crops and to schedule irrigations include thermally based techniques such as elevated crop canopy temperature, the empirical CWSI developed by Idso et al. [116], the theoretical CWSI formulated by Jackson et al. [117] and the TTT (Time Temperature Threshold) method studied in detail by Evett et al. [59, 118] and Peters and Evett [119, 120] and patented by Upchurch [121, 121]. The TTT method is suitable for automatic irrigation scheduling in subsurface drip irrigation [123] and center pivot systems [120]. The most important non-contact method for assessing plant stress is the measurement of leaf temperature with an infrared thermometer (IRT). Early research by Tanner [124] demonstrated the feasibility of this method and modern, portable and lightweight devices are now available.

Plant temperature is used to assess stress as water evaporates through the stomata of the leaves and cools them below air temperature. Stress can occur due to insufficient soil water, high atmospheric demand, insects, disease, saline conditions or low fertility. This decrease in transpiration and increase in leaf temperature is the basis for the Crop Water Stress Index (CWSI). Two approaches have been proposed to define this index: theoretical [117] and empirical [116].

The empirical approach requires non-stressed initial situations for different plant communities in order to determine the difference between leaf and air temperature (VPD) for healthy, well-watered plants. The upper limit depends on net radiation [125] and wind speed, but wind information is not used in this method. Early versions of the CWSI [116] did not require net radiation data.

Jackson et al. [117] developed an analytical expression to calculate aerodynamic resistance (ra) using standard equations for surface energy balance and heat and water vapor transport. Measurements of VPD, net radiation and wind speed are required. However, the determination of ra is difficult due to diurnal and seasonal variations in the upper and lower bounds. At the upper limits, the foliage is warmer and conditions are unstable; at the lower limits, the foliage is cooler and conditions are stable. The value of ra is assumed to be the same for both limits, but its calculation using standard aerodynamic expressions varies considerably between the two conditions.

Plant stress is caused by insufficient soil water, plant diseases, insect infestation, salinity, nutrients and sudden advective conditions. These factors interrupt the transpiration flow and increase the temperature of the leaves. The observer must identify the cause of the change in plant temperature.

An early consideration of short-term variation in leaf temperature due to radiation and wind speed in fields was to relate leaf temperatures to air temperature and integrate these differences (e.g., the stress degree day measure of Jackson et al. [126]), with a significant increase in canopy surface temperature indicating stomatal closure and water deficit stress. The method was transformed into a more practical approach following the introduction of the crop water stress index (CWSI) by Idso [116, 117], where CWSI was obtained from the canopy temperature (T_{canopy}) according to

$$\mathrm{CWSI} = \left(T_{canopy} - T_{nws}\right) / \left(T_{dry} - T_{nws}\right) \quad (4)$$

where T_{nws} is a so-called non-water-stressed baseline temperature for the crop in question at the same atmospheric vapor pressure deficit, and T_{dry} is an independently derived temperature of a non-transpiring reference crop.

The approach uses temperature differences from air temperature to establish standard relationships for Tdry and Tnws. It was developed for arid climate but is less effective and therefore less useful in humid or cloudy climates with lower signal-to-noise ratios. Infrared thermometry is widely used in irrigation scheduling, especially in arid areas [14, 127, 128], and despite its limitations, trapezoidal methods have been developed that combine temperature data with a vegetation index in the visible/near-infrared range [129].

Monitoring foliage temperature using infrared thermometers is a widely tested method, especially in arid/semiarid regions. The temperature difference between leaves and air increases with increasing plant water stress, resulting in a water stress index for the plants. This tool is used for irrigation scheduling [116, 117, 127]. Plant-based methods only indicate when to irrigate, and soil moisture measurements or estimation methods are used to determine water application [16, 20, 51, 119, 128].

The use of inexpensive infrared thermometers (IRT) to measure canopy temperature (Tc) for irrigation scheduling has become practical, although the commercial use of these sensors is relatively recent [128, 131–133], according to reports by Tanner [124] and Monteith and Szeicz [130].

The technique for irrigation scheduling has several shortcomings. These include difficulties in measuring crop temperature (Tc) at early stages of development [117, 128, 134], determining timing but not quantity [135], and sensitivity to the viewing angle of the sensor and the zenith angle of the sun [136, 137]. Standardization and consistency are crucial for the method. The empirical Crop-Water Stress Index (CWSI) is not suitable for scheduling irrigation under variable weather conditions because it cannot account for temperature changes due to radiation and wind speed [138]. This limits the usefulness of the technique, especially under low vapor pressure deficit conditions. Reliable irrigation needs identification may only be possible when stress has developed and CWSI has increased rapidly, potentially causing problems with crops susceptible to water stress and low water holding capacity soils. The theoretical approach [117] for determining CWSI may improve its reliability [128], but its practical application is limited by the need for additional weather information and uncertainty in estimating canopy resistance and rc, as noted by Jackson et al. [138]. This may pose issues for crops with water stress and low water holding capacity soils.

Infrared thermometers (IRTs) are used in agricultural research to measure surface temperatures, develop vegetation stress indexes [117, 139], monitor crop canopy temperatures, schedule irrigations [16, 119] and predict crop yields [140–142]. Historically, stationary industrial IRTs were used, but they can be expensive and impractical for commercial use. Wireless IRTs are a practical alternative, but their availability is limited. Economical radiation thermometric detectors are available, but few studies exist on their use in agricultural settings.

Wireless IRT modules have been evaluated for agricultural applications, with low-cost models like those by Mahan & Yeater [143] and Mahan et al. [144] showing similar accuracy to industrial-quality sensors over a 13–35 °C range. These IRTs are now available and have been compared to commercial wired IRT measurements in a cotton field in Australia. Wang et al. [145] summarize sensor technologies and standards for wireless communications in agriculture.

2.1.3 Monitoring the Weather

Evapotranspiration (ET) is a method used by researchers and irrigators to estimate the water consumption of plants based on weather data. The meteorological demand is tracked, and the irrigation quantity is determined accordingly. The reference evapotranspiration (ETo) is determined based on weather data such as maximum and minimum temperatures, wet and dry temperatures, solar radiation and wind direction.

The weather observation is used for irrigation scheduling, which incorporates evapotranspiration to update the soil water balance and predict future water use. Crop ET is calculated using reference crop ET (ETo or ETc) and a crop coefficient to relate actual water use to the reference crop. Combination-based equations are most accurate for short intervals, but require climate data for air temperature, vapor pressure, wind speed and solar radiation. Simplified methods for estimating ETr have been used to predict crop water use because these data are not available for all locations [14, 44, 47, 146].

The accuracy and variability of calculated ET must be considered when updating the soil water balance, as all ET models require local calibration for accurate estimation [147]. Therefore, frequent monitoring of soil moisture is recommended.

Pan evaporation (Ep) is a method of measuring evaporative demand at weather stations. It differs from crop utilization due to factors such as water transport, reflection of solar radiation and site conditions. A full cover of manure can evaporate 75–85% of the water, while crop water use is less than 75 if the manure does not cover the soil surface. The evaporation pan has become widely used due to its ease of use, simplicity of data and low legal costs. Thom et al. [148] conducted a comprehensive analysis between the Penman model and pan evaporation and concluded that pan evaporation can be adequately described by a combination equation with adjustments in the psychrometric constant and the wind function, which is a meteorological quantification of the adjustments in the Doorenbos and Pruitt [147] tables.

Evaporation pans are widely used worldwide as they can model the potential evaporation (Ep) from free water surfaces in a visible way and relate it to a reference evapotranspiration (grass or alfalfa) using an empirical coefficient. The Class A pan is the official network tool in the US and many other countries. Covering a Class A evaporation pan can reduce evaporation by 10% in semi-arid areas [149]. When installed in standard grass weather stations with proper maintenance, evaporation pans can reliably estimate ET, especially when evaporation is averaged over 7 days or automated hourly measurements are taken. The correlation between ETo or ETr and Ep can be improved by hourly measurements or by extending the time used to correlate the measurements [147].

Technologies have been introduced to automate the reading of evaporation pans for data acquisition systems. Techniques to automate a

Class A evaporation pan have been developed by Phene and Campbell [150] and Phene et al. [151]. However, the cumulative pan evaporation method for irrigation scheduling has the disadvantage that it does not account for soil moisture added by rainfall or shallow water tables, making it suitable for seasons without irrigation [148, 151].

Jensen [152], Jensen et al. [153] and Jensen et al. [154] developed an irrigation scheduling program for the US Department of Agriculture that uses meteorological data to calculate water use and maintain a soil water budget. There are variations of this approach, including a simplified checkbook method [155]. Computerized scheduling is useful for frequent irrigation or large fields but is not widely used in either developed or developing countries. Accurate field data, including rainfall, irrigation water application, meteorological data, soil water status, and crop water conditions [14], are critical for successful irrigation scheduling.

The USDA Soil Conservation Service and Michigan State University have developed a computerized irrigation scheduling program that uses real-time weather station data to calculate the soil water balance for crop root zones [156]. The program is easy to use and uses the latest technology to optimize crop yields and protect groundwater quality. It allows for complex calculations that would not be possible with manual calculations, such as predicting ET, and considers multiple locations within the field to create specific schedules for individual fields that meet different objectives. The program is designed to be user-friendly and effective in achieving agricultural goals.

The advancement of microcomputers has led to various computerized methods for irrigation scheduling. For example, CROPWAT, developed by the FAO [157]; IRSIS, developed by the Katholleke Universiteit Leuven (Belgium); WATERPERFECT, developed by Williams and Pogue [158]; SCS-SCHEDULER, developed by SCS in the USA. The models access databases for soil properties, crop information, meteorological data, water supply, irrigation systems and economic data. Automated weather station networks have facilitated computerized scheduling methods that allow access to and processing regional stations and overcome the limitations of data entry and analysis.

Computerized scheduling has improved the estimation of water use and irrigation, sometimes using crop growth models for irrigation scheduling. However, these models require more crop and soil data that growers have difficulty obtaining, potentially limiting the use of crop simulation models.

Irrigation scheduling is critical for farmers growing irrigated crops but requires frequent field visits and can be time consuming. Computer simulation models can help with irrigation scheduling by incorporating the soil–plant-atmosphere complex. Soler and Hoogenboom [159] conducted a study to determine the effects of different irrigation schedules on the growth and development of cotton and peanuts. They used the CSM-CROPGRO-cotton and CSM-CROPGRO-peanut models to define irrigation treatments and estimate the timing of irrigation and the amount of water to be applied. The irrigation event was triggered when the actual soil water content in the effective root zone fell below a specific threshold of available water content (AWC) threshold. Treatments corresponded to 30, 40, 60, and 90% of the irrigation threshold (IT).

The models required daily weather data, including temperature, solar radiation and precipitation. Soil water content was monitored using time-domain reflectometry (TDR) and a soil profile probe (PR2/6). Growth analysis was performed for both crops every 18 days, and yield and yield components were determined at the last harvest.

In cotton, the 30 and 40% IT treatments had significantly less bolls/m^2, while no significant differences were observed between the 60 and 90% IT treatments. In cotton and peanut, low values for aboveground biomass at harvest and yield were observed in the 30 and 40% IT treatments. The study concluded that dynamic crop growth models can be used for irrigation scheduling, but that variable irrigation depth and characterization of soil properties are required.

The Florida Automated Weather Network (FAWN) has developed the Citrus Irrigation Scheduler, a computerized method of irrigation scheduling for citrus growers. The system uses inter-tree and in-row spacing to estimate canopy volume and root distribution, as well as irrigation system characteristics such as spray diameter, wetting pattern and flow rate to determine application rate and application time. The system also provides a schedule of days between irrigation and hours of irrigation duration and suggests a delay in application for up to one inch of rainfall [160].

Irrigation scheduling and advisory programs are not new to U.S. agriculture. Studies have shown that they are profitable and beneficial when applied with precision. Costs affect the acreage to which irrigation scheduling is applied [161]. Surveys in the late 1990s showed that irrigation scheduling was used for large acreages and high-water costs [162]. The California Irrigation Management Information System (CIMIS) saves water and has other benefits that outweigh the operating costs. However, irrigation scheduling has only become prevalent when water is inexpensive, as has been the case in the western United States for decades [163].

Various data management methods for irrigation decision making have been developed [164–171], but widespread adoption is difficult due to factors such as lack of involvement, lack of consideration of user needs, excessive complexity of inputs, time management concerns, and time required to learn or adopt the program. Some tools have been programmed without considering the limited computer skills of many intended users.

Crop growth models are critical for studying the effects of irrigation on crop growth and yield, determining best practices, and tailoring strategies for specific sites, crops, and economic conditions. However, they need to be validated for the intended applications [172].

CropSyst, a crop growth model, has been modified to evaluate crop response and water management under saline conditions. The model considers the osmotic effect of salt and the effects of salt on root permeability. A sensitivity analysis shows that the model can simulate crop salt tolerance, environmental conditions and the effects of soil water availability on crop response to salinity. Comparison with field data shows that the model predicts salinity and water stress effects on crop yields [173].

Real-time irrigation (RTI) is a method in which water is applied at the plant root level based on the actual soil moisture value. If the moisture is below the optimum value, irrigation is applied; if it is above, irrigation is stopped. The frequency and duration of irrigation varies with the extent and needs of the plants. RTI requires daily processing of climatological data and requires skilled farmers to monitor soil moisture and adjust irrigation at irregular intervals [19]. Local precipitation measurements are crucial for planning a field due to the high spatial variability. To better quantify rainfall variability, several rain gauges should be placed around the farm.

The components of the water balance in the soil such as moisture, precipitation and irrigation depth are not always known with absolute certainty due to spatial variability. This uncertainty can be reduced by selecting representative sites and regularly monitoring the same site. The confidence interval for irrigation depth depends on the uniformity of the irrigation system and the accuracy of depth control. Insufficient uniformity can lead to wide confidence intervals and uncertain irrigation dates. Inaccuracies in the control system can also increase uncertainty. If the confidence interval is narrow, irrigation can be scheduled for longer periods, whereas if the interval is wide, soil water content should be measured [14, 174].

The most effective approach to irrigation scheduling involves a combination of two or more methods, including field information, water budget and evaporative demand, to assess the soil–plant-water balance and prevent unintended leaching of plant nutrients [13, 14, 16, 55, 115].

3 Applicability and Limitations of Irrigation Scheduling Methods and Techniques

Despite various irrigation scheduling methods, adoption by farmers is limited. The identification of constraints and requirements is crucial for the selection of appropriate irrigation scheduling methods. Some of the limitations and difficulties in applying irrigation scheduling tools at the farm level are listed here; (i) Irrigation scheduling becomes sensitive when water resources are limited and requires sophisticated timing to minimize yield losses, as well as knowledge of salinity tolerance under saline conditions, (ii) Irrigation calendar planning is often challenging due to rainfall variability and requires careful consideration of options, requirements and constraints, (iii) Deficit irrigation requires reliable water stress indicators, while accurate knowledge of yield-salinity relationships is critical for saline water management, (v) Farmers benefit from water-efficient irrigation scheduling that includes crop liberalization, water pricing and irrigation profitability, although technology costs may be higher than end users can afford, (vi) Knowledge sharing on the proper use and management of tools is not widely available to farmers, (vii) Farmers often lack an proper understanding of hydrology and water budgets, leading them to prefer to irrigate as usual and lack the necessary technical skills for proper scheduling.

Crop water requirements and irrigation applications have been determined using a variety of methods and tools, including soil and crop monitoring methods, soil–water balance-based planning, and simulation models. Soil monitoring techniques include gravimetric determinations, tensiometers, gypsum blocks, neutron probes, and devices using electromagnetic correlations and dielectric constants. Plant monitoring methods determine stress characteristics such as changes in stem or branch diameter, leaf water potential and sap flow. Canopy temperature measurements using radiometers and infrared devices are used but are mostly limited to industrial operations or irrigation consultants and researchers.

Real-time models for farmers have proven difficult, despite efforts to facilitate data transfer. Simple models using average data are more easily adopted with limited technologies. Most farmers have a good idea of when to irrigate, and scientific irrigation scheduling methods often do not provide enough benefits to offset the cost and inputs. Accurate irrigation scheduling is becoming increasingly important as water supplies become scarce and competition increases. Research is needed to reduce water and energy consumption and increase profitability through better scheduling. Future techniques must consider crop responses to water stress, evapotranspiration rates, management objectives, system constraints and public policy.

To minimize groundwater pollution, agriculture must optimize the use of irrigation water in terms of crop productivity and environmental protection. Maintaining water quality is a challenge and scheduling must consider factors beyond crop evapotranspiration. Current scheduling methods often overlook irregularities in the system and in the fields. Future management decisions will require data from various sensors, including monitoring soil water, measuring canopy temperature and calculating soil water balance. Relationships between climatic variables, soil water and plants should be developed for irrigation in arid and semi-arid areas. The evaluation of the effects of water deficits on the economic yield components of different crops is of crucial importance.

Scientists face the challenge of developing new knowledge to distinguish crop condition and stress types, including water deficiency, salinity, elemental toxicity, disease, insects, and nutrition, from their overall health [1, 2, 16, 40].

The challenge lies in developing scheduling methods that can be easily adopted by irrigators. Due to a lack of economic and social incentives, real-time scheduling methods have progressed slowly. Despite technological advances, growers want simple, cost-effective and accurate irrigation schedules. Management systems for growers need to be developed. Research priorities should be set for each country to address the most important water problems. Educational and technical assistance programs can help reduce the gap between research and actual irrigation practices.

4 Conclusion

Growing populations and rising living standards are increasing the demand for agricultural products, leading to environmental concerns and challenges in decision making on water use in irrigated agriculture. To ensure efficient water application, farmers need immediate information on weather, soil and crop conditions throughout the growing season. Sustainable land and water management combined with innovative agricultural technologies can help mitigate climate change and help farmers adapt to its effects. Irrigation scheduling is critical to efficient crop production, and informed decisions about when to start and stop irrigation are essential. Factors such as soil temperature, moisture status, weather forecast, water restrictions, and evapotranspiration rates should be considered. Monitoring soil moisture helps to make informed decisions and understanding the amount of water applied, its movement through the soil and the different irrigation needs depending on soil type, crop type, climate and irrigation system performance is essential. The integration of innovative technologies such as IoT and machine learning could further improve scheduling methods.

References

1. Phene CJ, Reginato RJ, Itier B, Tanner BR. Sensing irrigation needs. In: Hoffman GJ et al., editors. Management of farm irrigation systems. ASAE monograph. St. Joseph, MI; 1990. p. 207–251.
2. Fernández JE, Green SR, Caspari HW, Diaz-Espejo A, Cuevas MV. The use of sap flow measurements for scheduling irrigation in olive, apple and Asian pear trees and in grapevines. Plant Soil. 2008;305:91–104.
3. O'Shaughnessy SA, Evett SR. Canopy temperature based system effectively schedules and controls center pivot irrigation of cotton. Agric Water Manag. 2010;97(2010):1310–6.
4. Ding GKC. Managing water scarcity—the role of sustainable water management. Elsevier; 2023. https://doi.org/10.1016/B978-0-323-90386-8.00063-2.
5. Ahmed Z, Gui D, Murtaza G, Yunfei L, Ali S. An overview of smart irrigation management for improving water productivity under climate change in drylands. Agronomy. 2023;13(8):2113.
6. Ingrao C, Strippoli R, Lagioia G, Huisingh D. Water scarcity in agriculture: an overview of causes, impacts and approaches for reducing the risks. Heliyon. 2023;9(8): e18507.
7. Postolache O, Pereira JM, Girão PS, Monteiro AA. Smart sensing technology for agriculture and environmental monitoring. New York: Springer; 2012.
8. Culibrk D, Vukobratovic D, Minic V, Fernandez MA, Osuna JA, Crnojevic V. Sensing technologies for precision irrigation. New York: Springer; 2014.
9. Azar AT, Ammar HH, de Brito Silva G, Razali MSAB. Optimal Proportional Integral Derivative (PID) controller design for smart irrigation mobile robot with soil moisture sensor. In: Hassanien A, Azar A, Gaber T, Bhatnagar RF, Tolba M, editors. The international conference on advanced machine learning technologies and applications (AMLTA2019); 2020.
10. Wanjura DF, Upchurch DR, Mahan JR. Evaluating decision criteria for irrigation scheduling of cotton. Trans ASAE. 1990;33(2):512–8.
11. Chai Q, Gan Y, Zhao C, Xu HL, Waskom RM, Niu Y, Siddique KH. Regulated deficit irrigation for crop production under drought stress: a review. Agron Sustain Dev. 2016; 36:3.
12. Parkash V, Singh S. A review on potential plant-based water stress indicators for vegetable crops. Sustainability. 2020;12(10):3945.
13. Cassel DK. Irrigation scheduling. II. Simple low-cost techniques. Crops Soils. 1984; 36:615–618.
14. Martin DL, Stegman EC, Fereres E. Irrigation scheduling principles. In: Hoffman GJ et al., editors. Management of farm irrigation systems. ASAE monograph. St. Joseph, MI; 1990. p. 155–199.
15. Heermann DF, Martin DL, Jackson RD, Stegrnan EC. Irrigation scheduling controls and techniques. In: Stewart BA, Nielsen DR, editors. Irrigation of agricultural crops. Agronomy monographs No:30, ASA. CSSA, SSSA, Madison, WI; 1990. p. 510–533.
16. Jones HG. Irrigation scheduling: advantages and pitfalls of plant-based methods. J Exp Bot. 2004;55(407):2427–36.
17. Zhao A. Discussion on the problems and related measures of water-saving irrigation in irrigation and water conservancy project. Constr Eng. 2018;8:125–6.

18. Liang Z, Liu X, Xiong J, Xiao J. Water allocation and integrative management of precision irrigation: a systematic review. Water. 2020;12(11):3135.
19. Abdulmumin S, Bastiansen J, Smith M, Rijks D, Gbeckor-Kore NA. Application of climatic data for irrigation planning and management. Rome: Training Manual. FAO WMO; 1990.
20. Hillel D. Role of irrigation in agricultural systems. In: Stewart BA, Nielsen DR, editors. Irrigation of agricultural crops. ASA-CSSA-ASA, Madison. WI; 1990. p. 5–29.
21. Garrido-Rubio J, González-Piqueras J, Campos I, Osann A, González-Gómez L, Calera A. Remote sensing–based soil water balance for irrigation water accounting at plot and water user association management scale. Agric Water Manag. 2020;238: 106236.
22. Ferreira A, Rolim J, Paredes P, Cameira MDR. Methodologies for water accounting at the collective irrigation system scale aiming at optimizing water productivity. Agronomy. 2023; 13(7):1938.
23. Koech R, Langat P. Improving irrigation water use efficiency: a review of advances, challenges and opportunities in the Australian context. Water. 2018;10(12):1771.
24. Gu Z, Qi Z, Burghate R, Yuan S, Jiao X, Xu J. Irrigation scheduling approaches and applications: a review. J Irrig Drain Eng. 2020;146(6):04020007.
25. El-Nashar W, Elyamany A. Adapting irrigation strategies to mitigate climate change impacts: a value engineering approach. Water Resour Manag. 2023;37:2369–86.
26. Hillel D. Advances in irrigation, vol. 4. New York, Elsevier: Academic; 1987. p. 1–405.
27. English MJ, Musick JT, Murty VVN. Deficit irrigation. In: Hoffman GJ et al., editors. Management of farm Irrigation systems, ASAE Monograph, St. Joseph, MI; 1990. p. 631–655.
28. Geerts S, Raes D. Deficit irrigation as an on farm strategy to maximize crop water productivity in dry areas. Agric Water Manag. 2009;96:1275–84.
29. Galindo A, Collado-González J, Griñán I, Corell M, Centeno A, Martín-Palomo MJ, Girón IF, Rodríguez P, Cruz ZN, Memmi H, Carbonell-Barrachina AA, Hernández F, Torrecillas A, Moriana A, Pérez-López D. Deficit irrigation and emerging fruit crops as a strategy to save water in Mediterranean semiarid agro systems. Agric Water Manag. 2018;202:311–24.
30. Yazar A, Kanber R, Ozekici B. Irrigation scheduling in agronomic practice. In: Pereira LS et al. editors., Sustainability of Irrigated agriculture. NATO ASI Series E: Applied Sciences, vol. 312; 1995. p. 251–266.
31. Ortuno MF, Conejero W, Moreno F, Moriana A, Intrigliolo DS, Biel C, Mellisho CD, Pérez-Pastor A, Domingo R, Ruiz-Sánchez MC, Casadesus J, Bonany J, Torrecillas A. Could trunk diameter sensors be used in woody crops for irrigation scheduling? a review of current knowledge and future perspectives. Agric Water Manag. 2010;97:1–11.
32. Fernández JE, Moreno F, Martín-Palomo MJ, Cuevas MV, Torres-Ruiz JM, Moriana A. Combining sap flow and trunk diameter measurements to assess water needs in mature olive orchards. Environ Exp Bot. 2011;72(2):330–8.
33. Fernández JE, Cuevas MV, Rodriguez-Dominguez CM, Perez-Martin A, Torres-Ruiz JM, Elsayed-Farag S, Diaz-Espejo A, Martín-Palomo MJ. Sap flow response to olive water stress: a comparative study with trunk diameter variations and leaf turgor pressure. Acta Hort. 2012;951:101–10.
34. Rodriguez-Dominguez CM, Ehrenberger W, Sann C, Rüger S, Sukhorukov V, Martín-Palomo MJ, Diaz-Espejo A, Cuevas MV, Torres-Ruiz JM, Perez-Martin A, Zimmermann U, Fernández JE. Concomitant measurements of stem sap flow and leaf turgor pressure in olive trees using the leaf patch clamp pressure probe. Agric Water Manag. 2012;114:50–8.
35. Jones HG. Plant water relations and implications for irrigation scheduling. Acta Hort. 1990;278:67–76.
36. Schattman RE, Jean H, Faulkner JW, Maden R, McKeag L, Nelson KC, Grubinger V, Burnett S, Erich MS, Ohno T. Effects of irrigation scheduling approaches on soil moisture and vegetable production in the Northeastern U.S.A. Agric Water Manag. 2023; 287:108428.
37. Ko J, Piccinni G. Corn yield responses under crop evapotranspiration-based irrigation management. Agric Water Manag. 2009;96(5):799–808.
38. Gonzalez IDlA. Assessment of the correction of the reference evapotranspiration at nonirrigated weather stations affected by aridity and delimitation of the meteorological conditions that limit its implementation. Environ Process. 2023; 10(40):1–26.
39. Khorsandi A, Hemmat A, Mireei SA, Amirfattahi R, Ehsanzadeh P. Plant temperature-based indices using infrared thermography for detecting water status in sesame under greenhouse conditions. Agric Water Manag. 2018;204:222–33.
40. Hsiao TC. Measurement of plant water stress. In: Stewart BA, Nielsen DR, editors. Irrigation of agricultural crops. ASA-CSSA-ASA, Madison, WL; 1990. p. 244–281.
41. Stegman EC, Musick JT, Stewart JI. Irrigation water management. In: Jensen ME, editors. Design and operation of farm irrigation systems ASAE monograph No: 3 St. Joseph, MI; 1980. p. 763–801.
42. Campbell GS, Campbell MD. Irrigation scheduling using soil moisture measurements: theory and practice. In: Hillel D, editor. Advances in irrigation, vol. 1. New York: Academic Press; 1982. p. 25–41.
43. Evett SR. Neutron moisture meters. In: Evett SR, Heng LK, Moutonnet P, editors. Field estimation of soil water content: a practical guide to methods, instrumentation, and sensor technology. international atomic energy agency. Vienna, Austria, Chapters 2–3; 2008.
44. Jensen ME, Rangeley WR, Dielman PJ. Irrigation trends in world agriculture. In: Stewart BA, Nielsen DR, editors. Irrigation of agricultural crops.

Agronomy Monograph No: 30. ASA-CSSA-SSSA. Madison, WI; 1990. p. 32–63.
45. Smith KA, Mullins CE, editors. Soil and environmental analysis: physical methods, 2nd. New York: Marcel Dekker; 2000.
46. Dane JH, Topp GC, editors. Methods of soil analysis. Part 4: Physical methods. Soil science society of America books series: 5. Madison, WI: Soil Science Society of America; 2002.
47. Allen RG, Pereira LS, Raes D, Smith M. Crop evapotranspiration. FAO Irrigation and Drainage Paper 56. Rome: FAO; 1998.
48. Evett SR. Soil water and monitoring technology. Irrigation of agricultural crops, 2nd ed., Agronomy Monograph no. 30; 2007.
49. Hedley CB, Yule IJ. Soil water status mapping and two variable-rate irrigation scenarios. Precision Agric. 2009;10:342–55.
50. Topp GC, Davis JL. Time-domain reflectrometry and its applications to irrigation scheduling. In: Hillel D, editors. Advances in irrigation, vol. 3. Orlando, FL: Academic; 1985, p. 107–129.
51. Bucks D, Sammis TW, Dickey GL. Irrigation for arid areas. In: Hoffman GJ et al., editors. Management of farm irrigation systems, ASAE Monograph, St. Joseph, MI; 1990. p. 499–542.
52. Hagan RM, Haise HR. Irrigation of agricultural lands. Madison, Wisconsin: American Society of Agronomy, No. 11; 1967. p. 1180.
53. English ME, Gleen M, Van Sickle J. Scheduling for optimum water use. In: Proceeding of the irrigation scheduling conference. ASAE, St. Joseph, MI, 1981. p. 61–72.
54. Warrick AW, Nielsen DR. Spatial variability of soil physical properties in the field. In: Hillel D, editor. Application of soil physics. New York, NY: Academic Press; 1980. p. 219–44.
55. Perrier ER, Salkini AB, editors. Supplemental Irrigation in the Near East and North Africa. London: Kluwer Academic Publishers; 1987.
56. Topp GC, Davis JL, Annan AP. Electromagnetic determination of soil water content: measurements in coaxial transmission lines. Water Resour Manag. 1980;16:574–82.
57. Evett SR, Tolk JA, Howell TA. Time domain reflectometry laboratory calibration in travel time, bulk electrical conductivity, and effective frequency. Vadose Zone J. 2005;4:1020–9.
58. Evett SR, Schwartz RC, Tolk JA, Howell TA. Soil profile water content determination: spatiotemporal variability of electromagnetic and neutron probe sensors in access tubes. Vadose Zone J. 2009;8:926–41.
59. Evett SR, Tolk JA, Howell TA. Soil profile water content determination: sensor accuracy, axial response, calibration, temperature dependence and precision. Vadose Zone J. 2006;5:894–907.
60. Logsdon SD. CS616 calibration: field versus laboratory. Soil Sci Soc Am J. 2009;73:1–6.
61. Panteny S, Stevens R, Bowen CR. The frequency dependent permittivity and AC conductivity of random electric Networks. Ferroelectrics. 2005;1319:199–208.
62. Mazahrih NT, Kabeth-Bader N, Evet SR, Ayars JE, Trout TJ. Field calibration accuracy and utility of four down-hole water content sensors. Vadose Zone J. 2008;7:992–1000.
63. Taber HG, Lawson V, Smith B, Shogren D. Scheduling microirrigation with tensiometers or Watermarks. Int Water Irrig. 2002;22(1):22–6.
64. Blume HR, Kuder LJ, Jantz DR, Shaw AD. Methods in determining crop water usage. ASAE Paper No. 88-2046. St Joseph, MI; 1988.
65. Shock CC. Soil water potential measurement by granular matrix sensors. In: Stewart BA, Howell TA, editors. Encyclopedia of water science. New York, NY: Marcel Dekker, Inc.; 2003. p. 899–903.
66. Harrison K. Irrigation scheduling methods. Cooperative Extension Service, Bulletin: The University of Georgia; 2009. p. 974.
67. Villalobos FJ, Mateos L, Fereres E. Irrigation scheduling using the water balance. In: Villalobos FJ, Fereres E, editors. Principles of agronomy for sustainable agriculture. Switzerland: Springer International Publishing; 2016. p. 269–79.
68. Shock CC, Kimberling K, Tschida A, Nelson K, Jensen L, Shock CA. Soil moisture based irrigation scheduling to improve crops and the environment. Oregon state university agricultural experiment station. In: Special report, vol. 1048; 2003. p. 227–234
69. Hodnett MG, Bell JP, Koon PD, Soopranmen GC, Batchelor CH. The control of drip irrigation of sugarcare using "Index" tensiometers: some comparisons with control by the water budget method. Agric Water Manag. 1990;17:189–207.
70. Turner NC. Plant water relations and irrigation management. Agric Water Manag. 1990;17(1990):59–73.
71. Reginato RJ, Howe J. Irrigation scheduling using crop indicators. J Irrig Drain Eng. 1985;111(2):125–33.
72. Ritchie GA, Hinckley TM. The pressure chamber as an instrument for ecological research. In: MacFadyn A, editor. Advances in ecological research, vol. 9; 1975. p. 165–174.
73. Jones HG. Physiological aspects of the control of water status in horticultural crops. HortScience. 1990;25:19–26.
74. Bates LM, Hall AE. Stomatal closure with soil water depletion not associated with changes in bulk leaf water status. Oecologia. 1981;50:62–5.
75. Tardieu F, Simonneau T. Variability among species of stomatal control under fluctuating soil water status and evaporative demand: modelling isohydric and anisohydric behaviors. J Exp Bot. 1998;49:419–32.
76. Schultz HR. Differences in hydraulic architecture account for near-isohydric and anisohydric behavior

of two field-grown Vitis vinifera L. cultivars during drought. Plant, Cell Environ. 2003;26:1393–405.
77. Hsiao TC. Plant response to water stress. Annu Rev Plant Physiol. 1973;24:519–70.
78. Massacci A, Jones HG. Use of simultaneous analysis of gas-exchange and chlorophyll fluorescence quenching for analyzing the effects of water stress on photosynthesis in apple leaves. Trees. 1990;4:1–8.
79. Fereres E, Goldhamer DA, Parsons LR. Irrigation water management of horticultural crops. HortScience. 2003;38(5):1036–43.
80. Slatyer RO. Plant water relationships. London and New York: Academic; 1967.
81. Barrs HD. Determination of water deficits in plant tissues. In: Kozlowski TT, editor. Water deficits and plant growth, vol. I. London and New York: Academic Press; 1968. p. 235–368.
82. Peretz J, Evans RG, Proebsting EL. Leaf water potentials for management of high frequency irrigation on apples. Trans Am Soc Agric Eng. 1984;27:437–42.
83. McCutchan H, Shackel KA. Stem water potential as a sensitive indicator of water stress in prune trees (*Prunus domestica* L. cv. French). J Am Soc Hortic Res. 1992;117:606–11.
84. Simonneau T, Habib R. The use of tree suckers to estimate root water potential. Plant, Cell Environ. 1991;14:585–91.
85. Garnier E, Berger A. The influence of drought on stomatal conductance and water potential of peach trees growing in the field. Sci Hortic. 1987;32:249–63.
86. Jones HG. Estimation of an effective soil water potential at the root surface of transpiring plants. Plant, Cell Environ. 1983;6:671–4.
87. Lorenzo-Minguez P, Ceulemans RJM, Gabriels R, Impens I, Verdonck O. Effect of short-term water stress on leaf gas exchange processes and plant water potential in Schefflera. HortScience. 1985;20:932–4.
88. Dixon MA, Tyree MT. A new stem hygrometer, corrected for temperature-gradients and calibrated against the pressure bomb. Plant, Cell Environ. 1984;7:693–7.
89. McBurney TA. The relationship between leaf thickness and plant water potential. J Exp Bot. 1992;43:327–35.
90. Selle's G, Berger A. Physiological indicators of plant water status as criteria for irrigation scheduling. Acta Horticulturae. 1990; 286:87–100.
91. Scholander PF, Hammel GT, Bradstreet ED, Hemmingsen EA. Sap pressure in vascular plants. Science. 1965;148:339–46.
92. Boyer JS. Leaf water potentials measured with a pressure chamber. Plant Physiol. 1967;42(1):133–7.
93. Morrow PA, Slatyer RO. Leaf resistance measurements with diffusion porometers: precautions in calibration and use. Agric Meteorol. 1971;8:223–33.
94. Jarvis PG. The interpretation of the variations in leaf water potential and stomatal conductance found in canopies in the field. Philos Trans R Soc Lond Ser B, Biol Sci. 1976; 273(927):593–610.
95. Oosterhuis DM, Walker S. Stomatal resistance measurement as an indicator of water deficit stress in wheat and soybean. South African J Plant Soil. 1987;4(3):113–20.
96. Monteith JL, Unsworth MH. Principles of environmental physics. London: Eward Arnold; 2014.
97. Idso SB, Kimball BA, Mauney JR. Effects of atmospheric CO_2 enrichment on root: shoot ratios of carrot, radish, cotton and soybean. Agr Ecosyst Environ. 1988;21:293–9.
98. Moriana A, Fereres E. Plant indicators for scheduling irrigation of young olive trees. Irrig Sci. 2002;21:83–90.
99. Eastham J, Gray SA. A preliminary evaluation of the suitability of sap flow sensors for use in scheduling vineyard irrigation. Am J Enol Vitic. 1998;49:171–6.
100. Wronski EB, Homes JW, Turner NC. Phase and amplitude relations between transpiration, water potential and stem shrinkage. Plant, Cell Environ. 1985;8:613–22.
101. Fereres E, Evans RG. Irrigation of fruit trees and vines: an introduction. Irrig Sci. 2006;24:55–7.
102. Naor A, Cohen S. The sensitivity and variability of maximum trunk shrinkage, midday stem water potential, and transpiration rate in response to withholding of irrigation in field-grown apple trees. HortScience. 2003;38:547–51.
103. Naor A, Gal Y, Peres M. The inherent variability of water stress indicators in apple, nectarine and pear orchards, and the validity of a leaf-selection procedure for water potential measurements. Irrig Sci. 2006;24:129–35.
104. Naor A. Naor, A. (2006) Irrigation scheduling and evaluation of tree water status in deciduous orchards. Hortic Rev. 2006;32:111–65.
105. Sepulcre-Cantó G, Zarco-Tejada PJ, Jiménez-Muñoz JC, Sobrino JA, de Miguel E, Villalobos FJ. Detection of water stress in an olive orchard with thermal remote sensing imagery. Agric For Meteorol. 2006;136:31–44.
106. Nadezhdina N, Cermak J. Responses of sap flow rate along tree stem and coarse root radii to changes of water supply. In: Stokes A, editor. The supporting roots of trees and woody plants: form, function and physiology. Developments in plant and soil sciences, vol. 87. Springer, Dordrecht; 2000.
107. Escalona JM, Flexas J, Medrano H. Drought effects on water flow, photosynthesis and growth of potted grapevines. Vitis. 2002;41:57–62.
108. Ortuño A, Báidez A, Gómez P, Arcas MC, Porras I, García-Lidón A, Del Río JA. Citrus paradisi and Citrus sinensis flavonoids: their influence in the defence mechanism against *Penicillium digitatum*. Food Chem. 2006;98(2):351–8.
109. Intrigliolo DS, Castel JR. Performance of various water stress indicators for prediction of fruit size response to deficit irrigation in plum. Agric Water Manag. 2006;83:173–80.
110. Nadezhdina N, Cermak J. Automatic control unit for irrigation systems based on sensing the plant water

status. Annais Do Instituto Superior De Agronomia. 1997;46:149–57.
111. Fernández JE, Palomo MJ, Díaz-Espejo A, Clothier BE, Green SR, Girón IF, Moreno F. Heat-pulse measurements of sap flow in olives for automating irrigation: tests, root flow and diagnostics of water stress. Agric Water Manag. 2001;51:99–123.
112. Nadezhdina N, Nadezhdin V, Ferreira MI, Pitacco A. Variability with xylem depth in sap flow in trunks and branches of mature olive trees. Tree Physiol. 2007;27:105–13.
113. Goldhamer DA, Fereres E. Irrigation scheduling protocols using continuously recorded trunk diameter measurements. Irrig Sci. 2001;20(3):115–25.
114. Green SR, Kirkham MB, Clothier BE. Root uptake and transpiration: from measurements and models to sustainable irrigation. Agric Water Manag. 2006;86:165–76.
115. Pereira AR, Green SR, Villa Nova NA. Penman-Monteith reference evapotranspiration adapted to estimate irrigated tree transpiration. Agric Water Manag. 2006;83:153–61.
116. Idso SB, Jackson RD, Pinter PJ, Harfield JL. Normalizing the stress-degree-day parameter for environmental variability. Agric Meteorol. 1981;24:45–55.
117. Jackson RD, Idso SB, Reginato RJ, Pinter PJ. Canopy temperature as a crop water stress indicator. Water Resour Res. 1981;17:1133–8.
118. Evett SR, Howell TA, Schneider AD, Wanjura DF, Upchurch DR. Water use efficiency regulated by automatic drip irrigation control. In: 2001 Proceedings. International Irrigation Show. San Antonio, TX, VA: The Irrigation Association, Falls Church; 2001. p. 49–56.
119. Peters RT, Evett SR. Spatial and temporal analysis of crop stress using multiple canopy temperature maps created with an array of center-pivot-mounted infrared thermometers. Trans ASABE. 2007;50(3):919–27.
120. Peters RT, Evett SR. Automation of a center pivot using the temperature—time-threshold method of irrigation scheduling. J Irrig Drain Eng. 2008;134(3):286–90.
121. Upchurch BL, Throop JA, Aneshansley DJ. Detecting internal breakdown in apples using interactance measurements. Postharvest Biol Technol. 1997;10(1):15–9.
122. Wanjura DE, Upchurch DR, Mahan JR. Establishing differential irrigation levels using temperature—time thresholds. Appl Eng Agric. 2004;20(2):201–6.
123. Evett SR, Howell TA, Schneider AD, Wanjura DF, Upchurch DR. Automatic drip irrigation control regulates water use efficiency. Int Water Irrig. 2002;22(2):32–7.
124. Tanner CB. Plant temperatures. Agron J. 1963;55:210–1.
125. Idso SB, Klawsen KL, Anderson MG. Foliage temperature: effects of environmental factors with implications for plant water stress assessments in the CO_2/climate connection. Water Resour Res. 1986;22:1702–16.
126. Jackson RD, Reginato RJ, Idso SB. Wheat canopy temperature—Practical tool for evaluating water requirements. Water Resour Res. 1977;13(3):651–6.
127. Jackson RD. Canopy temperature and crop water stress. In: Hillel D, editor. Advances in irrigation, vol. 1. New York: Academic; 1982. p. 43–85.
128. Stockle CO, Dugas WA. Evaluating canopy temperature based indices for irrigation scheduling. Irrig Sci. 1992;13:31–7.
129. Moran MS, Clarke TR, Inoue TR, Vidal A. Estimating crop water deficit using the relation between surface-air temperature and spectral vegetation index. Remote Sens Environ. 1994;49:246–63.
130. Monteith JL, Szeicz G. Radiative temperature in the heat balance of natural surfaces. Q J R Meteorol Soc. 1962;88:496–507.
131. Hatfield JL. Measuring plant stress with an infrared thermometer. HortScience. 1990;25(12):1535–8.
132. Gardner BR, Shock CC. Interpreting the Crop Water Stress Index, ASAE Paper No. 89-2642 St. Joseph. MI; 1989.
133. Yazar A, Howell TA, Dusek DA, Copeland KS. Evaluation of crop water stress index for LEPA irrigated corn. Irrig Sci. 1999;18:171–80.
134. Howell TA, Hatfield JL, Yamada H, Davis KR. Evaluation of cotton canopy temperature to detect crop water stress. Trans ASAE. 1984;27:84–8.
135. Nielsen DC. Scheduling irrigations for soybeans with the crop water stress index (CWSI). Appl Agric Res. 1990,23.103–16.
136. Huband NDS, Monteith JL. Radiative surface temperature and energy balance of a wheat canopy. Bound-Layer Meteorol. 1986;36:107–16.
137. Fuchs M. Infrared measurements of canopy temperature and detection of plant water stress. Theoret Appl Climatol. 1990;42:253–61.
138. Jackson RD, Kustas WP, Chaudhury BJ. A reexamination of the crop water stress index. Irrig Sci. 1988; 9: 309–317
139. Colaizzi PD, Barnes EM, Clarke TR, Choi CY, Waller PM, Haberland J, Kostrzewski M. Water stress detection under high frequency sprinkler irrigation with water deficit index. J Irrig Drain Eng. 2003;129:36–43.
140. Pinter PJ, Fry KE, Guinn G, Mauney JR. Infrared thermometry: a remote sensing technique for predicting yield in water-stressed cotton. Agric Water Manag. 1983;6:385–95.
141. Irmak S, Haman DZ, Bastug R. Determination of crop water stress index for irrigation timing and yield estimation of corn. Agron J. 2000;92:1221–7.
142. Ajayi AE, Olufayo AA. Evaluation of two temperature stress indices to estimate grain sorghum yield and evapotranspiration. Agron J. 2004;96:1282–7.
143. Mahan JR, Yeater KM. Agricultural applications of a low-cost infrared thermometer. Comput Electron Agric. 2008;64:262–7.

144. Mahan JR, Conaty W, Neilsen J, Payton P, Cox SB. Field performance in agricultural settings of a wireless temperature monitoring system based on a low-cost infrared sensor. Comput Electron Agric. 2010;71(2):176–81.
145. Wang N, Zhang N, Wang M. Wireless sensors in agriculture and food industry—recent development and future perspective. Comput Electron Agric. 2006;50(1):1–14.
146. Heermann DF. Evapotranspiration research priorities for the next decade-Irrigation. Trans ASAE. 1988;31(2):497–502.
147. Doorenbos J, Pruitt WO. Guidelines for predicting crop water requirements. FAO, Irrigation Drainage Paper, No: 24. FAO, Rome; 1977.
148. Thom AS, Thony JL, Vauclin M. On the proper employment of evaporation pans and atmometers in estimating potential transpiration. Q J R Meteorol Soc. 1981;107:711–36.
149. Howell TA, Miller RJ, Phene CJ, Meck DU. Evaporation from screened Class A-pans in a semi-arid climate. Agric Meteorol. 1983;29:114–24.
150. Phene CJ, Campbell RB. Automating pan evaporation measurements for irrigation control. Agric Meteorol. 1975;15:181–91.
151. Phene CJ, DeTar WR, Clark DA. Real time irrigation scheduling with automated evaporation pans. Appl Eng Agric. 1991;8(6):787–93.
152. Jensen ME. Scheduling irrigations with computers. J Soil Water Conserv. 1969;24(8):193–5.
153. Jensen ME, Robb DCN, Franzoy CE. Scheduling irrigation using climate-crop-soil data. J Irrig Drain Div. 1970;96(1):25–37.
154. Jensen ME, Wright JL, Pruitt BJ. Estimating soil moisture depletion from climate. Crop and soil data. Trans ASAE. 1971; 14(5):954–959.
155. Lundstrom DR, Stegman EC, Werner HL. Irrigation scheduling by the check-book method. In: Proceedings of the irrigation scheduling conference. Chicago: ASAE, St. Joseph, MI; 1981. p. 187–193.
156. Shayya WH, Bralts VF, Olmsted TR. General irrigation scheduling package for microcomputers. Comput Electron Agric. 1990;5(3):197–212.
157. FAO. Irrigation scheduling: from theory to practice. The Internation Commission on Irrigation and Drainage (ICID) and the Food and Agriculture Organization of the United Nations (FAO46th Meeting of the ICID Executive Council in September 1995; 1995.
158. Williams MA, Pogue WR. "WATERPERFECT" Irrigation management program. ASAE Paper No. 92-2551. St. Joseph, MI; 1992.
159. Soler CMT, Hoogenboom G. Determining irrigation scheduling for cotton and peanut using cropping system models. In: Proceedings of the 2007 world environmental and water resources congress. Tampa, Florida Sponsored Environmental and Water Resources Institute (EWRI) of ASCE; 2007.
160. Morgan KT, Hanlon EA, Obreza TA. A web-based irrigation scheduling model to improve water use efficiency and reduce nutrient leaching for florida citrus. SL286, one of a series of the Soil and Water Science Department, Florida Cooperative Extension Service, Institute of Food and Agricultural Sciences, University of Florida; 2009.
161. Dudek DJ, Horner GL, English MJ. The derived demand for irrigation scheduling services. Western J Agric Econo, Western Agric Econ Assoc. 1981;6(2):1–12.
162. Leib BG, Hattendorf M, Elliott T, Matthews G. Adoption and adaptation of scientific irrigation scheduling: trends from Washington, USA as of 1998. Agric Water Manag. 2002;55(2):105–20.
163. Parker D, Cohen-Vogel DR, Osgood DE, Zilberman D. Publicly funded weather database benefits users statewide. Calif Agric. 2000;54:21–5.
164. AzSched. Arizona Irrigation Scheduling System; 2023. Available: https://ag.arizona.edu/crop/irrigation/azsched/azsched.html.
165. Vories ED, Tacker PL. Effect of ET calculation method on irrigation scheduling in Midsouth. ASABE Paper No. 062205. ASABE St. Joseph, MI; 2006.
166. CIMIS. California Irrigation Management Information System; 2010. Available at: https://cimis.water.ca.gov/cimis/welcome.jsp.
167. CoAgMET. Arizona Irrigation Scheduling System; 2023. Available at: https://coagmet.colostate.edu/
168. KanSched. Kansas State Irrigation Scheduling Program; 2023. Available at: https://kansched3.engg.ksu.edu
169. NDAWN. North Dakota Agricultural Weather Network; 2023. Available at: https://ndawn.ndsu.nodak.edu/
170. TXHPET. The Texas High Plains Evapotranspiration Network; 2023. Available at: http://txhighplainset.tamu.edu/
171. TWRI. Texas Water Resources Institute; 2023. Available at: https://twri.tamu.edu/
172. Jones JW, Ritehie JT. Crop growth models. In: Hoffman GJ et al. Editors. Management of farm irrigation systems, St. Joseph, MI: ASAE Monograph; 1990. p. 63–85.
173. Stockle CO, Nelson R, Campbell GS. CropSyst, cropping systems simulation model. User's Manual. Dept. of Biological Systems Engineering: Washington State University, Pullman, Washington, USA; 1993.
174. Jensen ME, Wright JL. The role of evapotranspiration models in irrigation scheduling. Trans ASAE. 1978;21(1):2–87.

Technologies for Applications of Smart Irrigation

Mehmet Metin Ozguven

Abstract

In a world, where the population is constantly increasing, agricultural productivity must be high to feed people. One of the most important factors affecting yield is irrigation. However, today various degrees of water scarcity are experienced in different parts of the world. Considering that water scarcity affects a large population, one of the measures to be taken in this regard is to prevent inefficient irrigation. The technical possibilities offered by technology have enabled the development of smart irrigation applications. In smart irrigation systems, first, data are obtained such as soil moisture status, which are necessary for determining the water requirement of the plant using methods such as sensors, cameras and remote sensing techniques. Then, the obtained data are transmitted wirelessly by methods such as internet of things or radio frequency, and the transmitted data are evaluated by various decision support systems and decisions can be made for the best irrigation management. In this way, irrigation efficiency is increased, contributing to plant productivity, protecting water resources by preventing inefficient water use, and preventing negative effects such as soil erosion, salinity, and loss of soil fertility. In this study, the benefits of smart irrigation applications and the technologies used in smart irrigation applications are explained with case studies and technical details.

Keywords

Drought · Irrigation · Precision irrigation · Smart irrigation · Water scarcity

1 Introduction

The Intergovernmental Panel on Climate Change (IPCC) report predicts that various regional changes in climate such as heat waves, extreme temperatures, heavy rains, drought, and tropical cyclones will occur with global warming. IPCC scientists also point out that extreme temperature thresholds for agriculture and health would be exceeded more frequently [1]. Water is at the center of the numerous impacts of climate change and plays a key role in adaptation to climate change. Climate change has negative consequences on many existing challenges that require good water governance at both local, regional, and global levels [2]. Increasing water scarcity is now one of the leading challenges for sustainable development. In addition, the increase in world population and living standards, and the effects of climate change will lead to more water

M. M. Ozguven (✉)
Faculty of Agriculture, Department of Agricultural Machinery and Technologies Engineering, Ankara University, Ankara, Türkiye
e-mail: mmozguven@ankara.edu.tr

M. Bahadir et al. (eds.), *Agricultural Water Management*, Water and Wastewater Management,
https://doi.org/10.1007/978-3-031-96134-2_5

scarcity [3]. About 10% of the world's population lives in countries struggling with high or critical water stress. Although there is variability at regional, national, and sub-national levels, 18.4% of the world's total available freshwater resources are withdrawn because of different economic activities [4]. The amount of flow from rivers and aquifers to inland renewable water resources is 44,000 km^3 per year, while the amount of water withdrawn exceeds 4,000 km^3 per year. The local effects of water scarcity and freshwater pollution are accelerating and appear in large areas. Decreasing groundwater levels due to increased water withdrawal is an important indicator of water scarcity [5]. Except for some basins with densely populated big cities, the most challenging sector in the extremely high water-stressed basins is agriculture, as can be seen in Fig. 1 [4].

Plants take water from the soil through their roots to maintain their normal development. Nutrients in the root environment are transported to the extreme points with this water taken. Since the continuity of the cycle is provided by the removal of water from the leaves of the plant by transpiration, sufficient moisture level must be constantly provided in the root zone [7]. Irrigation is done when the water needed by the plants cannot be provided by natural precipitation. Insufficient irrigation causes the yield and quality of the plant to decrease. The water resources, from which irrigation water is supplied have been adversely affected by the change in precipitation characteristics due to climate change and the decrease in snowfall caused by the increasing temperature.

About 80% of the global freshwater withdrawals are committed to agriculture. The proportion of irrigated land to total cultivated areas varies greatly regionally. This ratio is at least 4% in Africa and at most 42% in South Asia, 30% in India and 52% in China. Much of the increase in agricultural production necessary to feed the growing world population and to stabilize food production becomes possible with irrigation and drainage. Therefore, irrigation is critical for agricultural productivity, especially in arid, semi-arid, and sub-humid regions. However, misuse of various customary in agricultural production has negative effects such as soil erosion, soil salinization, loss of soil fertility, environmental costs, and health impacts [8]. Therefore, there is a need for agricultural practices that increase agricultural productivity and incomes, reduce soil degradation, and prevent environmental pollution. Such an application is possible with smart irrigation (SI) applications that optimally meet the specific needs of each plant. SI practices have the potential to increase efficiency and quality, to optimize water use, to minimize negative environmental impacts, and to increase economic efficiency.

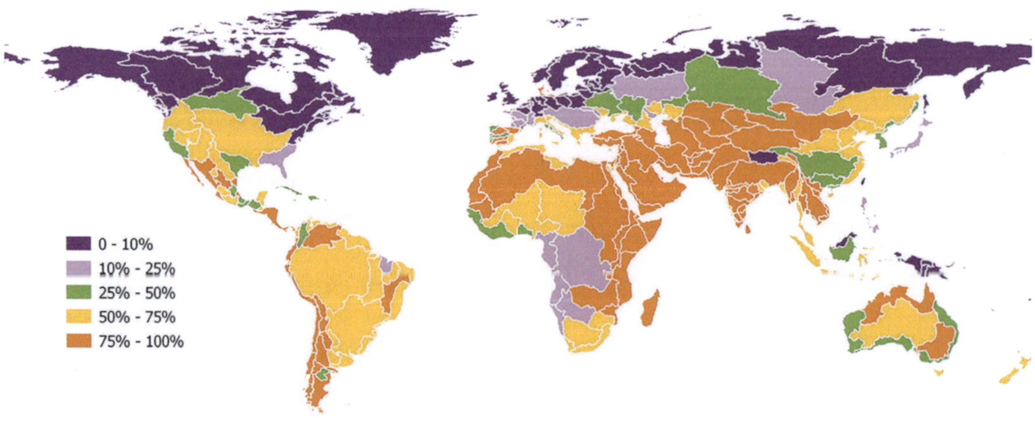

Fig. 1 Global map showing the ratio of agricultural sector freshwater withdrawal to total freshwater withdrawal in 2018 [6]

2 Smart Irrigation Technologies

The main technologies used in SI are explained below by giving sample studies (Fig. 2). These technologies offer the possibility of monitoring various parameters, automation, real-time control, and variable rate irrigation (VRI) in irrigation management.

2.1 Precision Irrigation

In precision irrigation (PI) the amount of water is applied according to soil type, crop type, topography, or field barriers. With this system, real-time data obtained from the field continuously and effectively are analyzed by using advanced hardware and software, and managers are appointed with the opportunity to make more accurate decisions about irrigation management [10]. VRI involves customizing irrigation based upon soil and yield maps, topography and other defined information required for irrigation. Thus, unnecessary irrigation can be prevented by giving the right amount of water to the right place with VRI [11]. Adopting PI techniques helps to improve water use and to maximize plant productivity [12]. Figure 3 shows an example PI system allowing VRI.

In PI applications, the crop is expected to meet its needs in a timely manner, as efficiently and spatially as possible, in a uniform manner. Results from VRI studies in the literature showed water savings ranging from nil to 50% in individual years, and an average of 8 to 20% over several years [14]. Dukes [15] reported that smart irrigation controllers have been developed to reduce excessive irrigation, some experimental studies using these devices resulted in water savings of between 40 and 70%, but in larger pilot-scale studies, less than 10% water savings were achieved. Yule et al. [16] reported that VRI can save 10–15% of the water used in traditional irrigation. Hedley and Yule [17] reported that approximately 25% of water savings can be achieved through improvements in application efficiency achieved with spatial irrigation applications. King et al. [18] conducted a study to measure potato yield and quality under VRI and conventional irrigation practices. As a result of the study, it was determined that the total tuber yield per unit water applied from irrigation and precipitation was 4% higher in 2001 and 6% higher in 2002 under VRI. In addition, the researchers reported that based

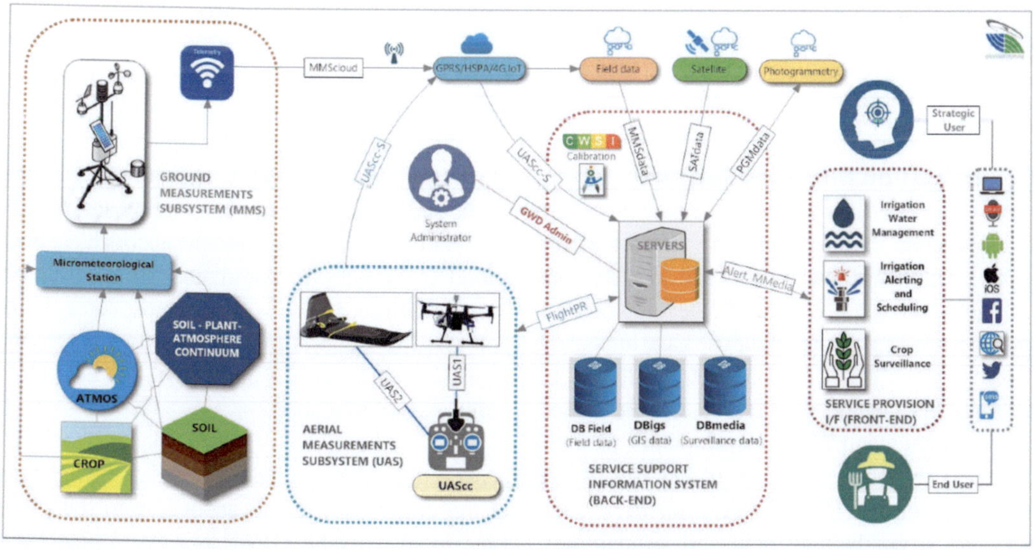

Fig. 2 Main technologies used in SI [9]

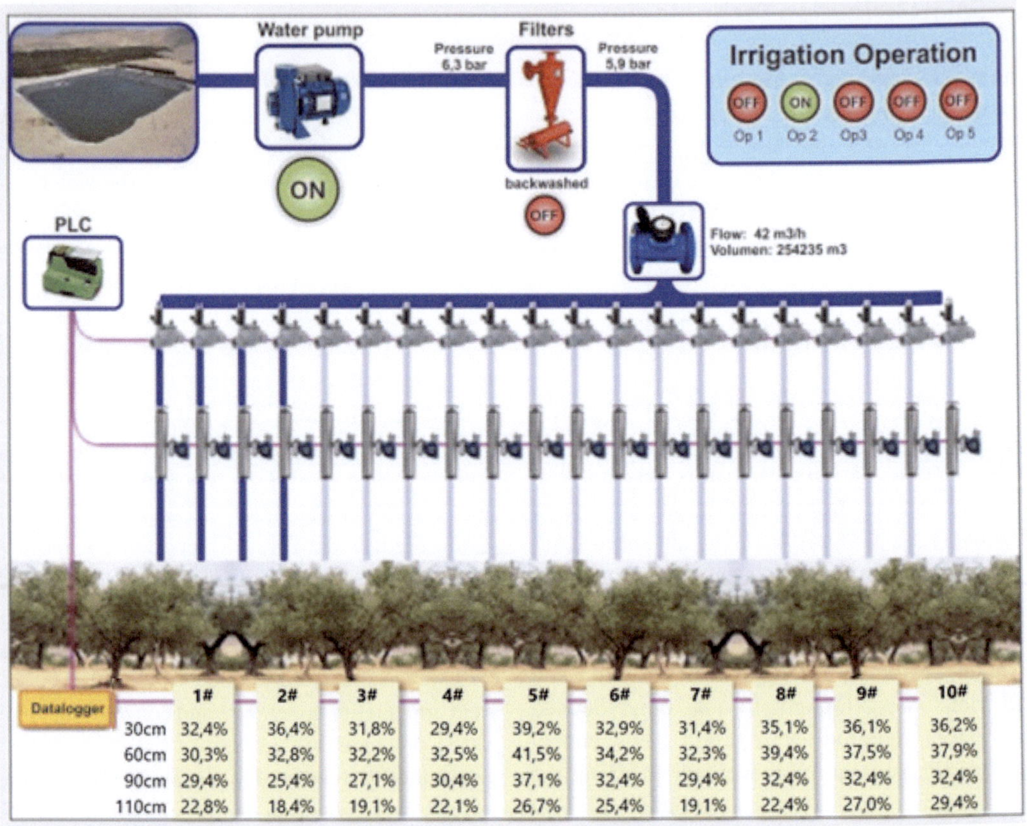

Fig. 3 Example of a web-based system for remote monitoring and management of PI [13]

on the price structure of the potato processing contract adjusted for local tuber quality, the trend in gross income across the field was $159/ha greater with VRI.

2.2 Irrigation Moisture Sensors

Neutronmeter, Time Domain Reflectometry (TDR) or moisture sensors are used to determine the plant water requirement. The development of new moisture sensors and their use have increased in recent years due to the need for calibration, measurement method and high cost of the device in neutronmeter and TDR measurements. In addition, these sensors may consist of a sensor fusion that determines different soil properties such as electrical conductivity and soil temperature, apart from measuring moisture [7]. Shock et al. [19] tested various soil moisture sensors such as neutron thermalization, frequency domain sensor, tensiometer, granular matrix sensors and air permeability of porous ceramics (Irrigas) devices for their sensitivity and usefulness for irrigation programming for PI. It was reported that the tested soil moisture measurement devices generally correlate closely with each other. Neutron probe and Irrigas have not been found suitable for automation. Frequency domain sensors have been found to be easy to automate, but not reliable enough without engineering improvements or soil-specific calibrations. The tensiometer and granular matrix sensors are reported to be easy to calibrate and to read.

Measurements are made by placing moisture sensors at predetermined locations in the field and at the desired soil depth according to the plant. Placement is usually done as a fixed placement in the soil, so that the moisture status of the soil can be monitored regularly throughout the entire

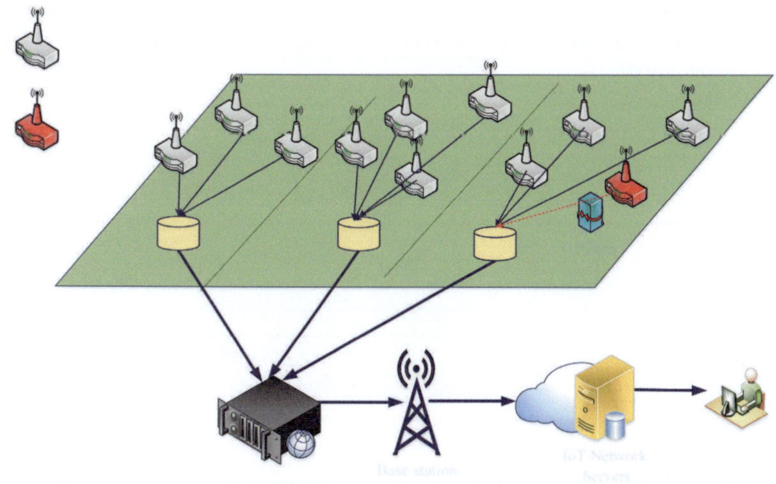

Fig. 4 Transmission and evaluation of data from sensors in the SI system [20]

production season. The measured values are transmitted to a central unit by wireless communication methods such as Internet of Things (IoT), radio frequency (RF), ZigBee, Wi-Fi, etc. (Fig. 4). Then, the data showing the moisture changes in the field are digitized and irrigation maps are prepared. The geostatistics method is widely used to transform the data obtained with moisture sensors into maps. In the geostatistics method, the spatial structure and spatial dependence of the measured humidity values are digitized and the value of the moisture values at the unsampled point is estimated using the obtained relationships. While creating the map, different ranges and colors are selected depending on the user's choice and the data range used. Generally, the red-colored areas on the map indicate the regions with the lowest humidity, and the green-colored areas indicate the regions where the humidity is higher [7].

2.3 Remote Sensing and Geographic Information Systems

In precision agriculture applications, information such as soil properties, soil nutrient level, yield values, topographic data, climate data, disease and weed status are obtained by using various sensors and remote sensing (RS) technologies and application maps are created by associating this information with their actual location in the production area. Then, variable-level applications are made using appropriate hardware and software [21]. RS and geographic information system (GIS) plays a vital role in watershed management, as it helps to analyze land use, soil moisture, precipitation, evapotranspiration, surface runoff, groundwater flow, snow water equivalent, irrigation management, drought management, reservoir sedimentation, flood [22]. New technologies for estimating irrigation crop water requirements (CWR) include RS and dynamic crop models (DCM) for irrigation planning (Fig. 5). DCMs synthesize causal relationships between plant-water-soil relationships, water stress, and its effective components on plant growth and development. Crop models are classified as water-based and radiation-based, depending on the main variable affecting crop growth. Water stress can be calculated from soil water content or lack of transpiration. With RS, information on crop water stress, evapotranspiration and soil moisture can be obtained for CWR assessment. All this information can be used to develop simpler methods integrated into irrigation management DCM, to provide continuous simulations of time-dependent crop-water dynamics, and to predict future crop-water interactions. DCMs provide a better understanding of water stress-related processes but are often better suited for research rather than real applications by farmers. However,

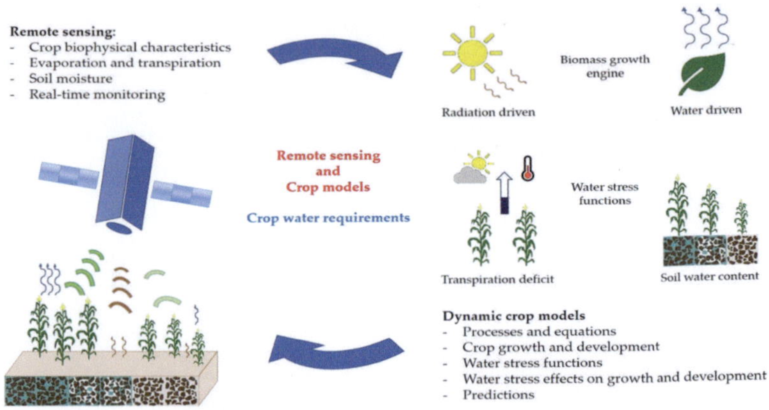

Fig. 5 The use of dynamic crop models (DCM) and remote sensing (RS) in irrigation management [23]

freely available DCMs such as AQUACROP, APSIM, DSSAT, EPIC, CROPSYST, STICS, RZWQM2, and DAISY are an important source of information on plant water stress and the effects of stress on crop growth and development [23].

Quemada et al. [24] reviewed the study of different RS techniques to monitor plant water status found in the literature, to help farmers controlling irrigation management and avoiding unnecessary periods of water scarcity and unnecessary waste of water. The scope of this review article consists of 77 references published between 1981 and 2021. The different approaches collected were categorized as soil (12.2%), canopy (33.8%), leaf (35.1%) or stem (18.9%) according to the part of the plant being measured. The main monitoring technologies were examined, it was reported that the most common detector was the radiometer (20.3%), followed by continuous wave spectroscopy (12.2%), camera (10.8%) and THz time-domain spectroscopy (10.8%). Excluding the two studies reviewed, the minimum coefficient of determination (R^2) obtained in the references was 0.64. This value indicates that the correlation between estimated and measured data for different technologies and monitoring methods is high. In addition, the five most common water indicators in these studies were the normalized difference vegetation index (12.2%), the backscattering coefficients (10.8%), the spectral reflectance (8.1%), the reflection coefficient (8.1%) and the dielectric constant (8.1%).

2.4 Unmanned Aerial Vehicles

The cameras and sensors on the drone are used in a wide variety of activities in agriculture. Some of these activities are creation of orthomosaic maps, derivation of NDVI, determination of plant characteristics, detection of plant diseases and pests, estimation of moisture, monitoring of crop water stress and plant growth, 3D modelling, classification of lands [25]. There are already a wide variety of cameras and sensors in different brands and models that can be used with drones. Drone designs are made with different features according to the needs and suitable sensor(s) and/camera(s) are mounted on it [26]. The water status in a field is often determined by making multi-point measurements. However, this method often fails to accurately define the true field heterogeneity. By using sensors mounted on the unmanned aerial vehicle (UAV) in remote sensing, characterization of all areas is provided with images with sufficient resolution to analyze plants one by one [27].

Soil moisture condition, seasonal fluctuations, precipitation-use efficiency, availability of stored water and the adopted irrigation methodology are important parameters that affect plant growing systems and crop productivity. Although different irrigation methods are used in plant production, irrigation should be observed 24 h in all methods. For this purpose, images obtained using predetermined flight paths with periodic flights with drones with high resolution sensors can help farmers. With their ability to take accurate images using visible color, near-infrared (NIR)

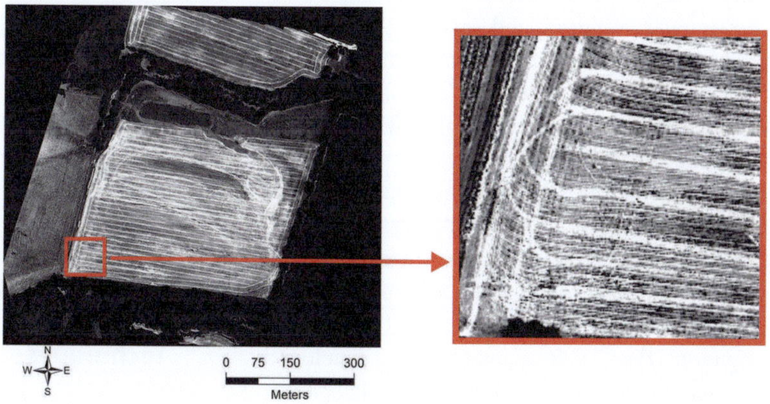

Fig. 6 Orthomosaic from TIR images [29]

and infrared (IR) sensors, drones tend to effectively assess and to map water bodies, to mark irrigation channels, to determine crop water condition, and to analyze drought or flood effects on a large farm. Such data are acquired by drones in a few minutes' flight over the fields. Based on this, management blocks for irrigation can be created. Adopting such water-blocking and PI techniques helps to improve water use and to maximize plant productivity [12]. Despite the diversity of plants and conditions in UAV-based studies in literature, crop water stress index vegetation indices (Normalized Difference Vegetation Index (NDVI), Green Normalized Difference Vegetation Index (GNDVI), etc.), photochemical reflection index, and difference between canopy and air temperatures are determined. The values measured by UAV and ground measurements for the determination of plant water stress are generally related to each other. For this reason, the values obtained by the UAV can be used to estimate the ground values with simple equations. For this purpose, leaf water potential, stem water potential, stomatal conductivity, and water content are measured in ground measurements [28].

Visible color (VIS-C), multispectral (MS) and thermal infrared (TIR) cameras that can be attached to UAVs can determine the locations of agricultural underground drainage systems. Allred et al. [29] conducted aerial surveys of 29 farmland sites to evaluate the potential of this technology for mapping buried drainage pipes using a UAV with VIS-C, MS and TIR cameras. The results of the study show that VIS-C images detected at least some drainage lines in 48% (14 out of 29) areas, MS images detected drainage lines in 59% (17 out of 29) areas, and TIR images detected drainage lines at 69% (20 out of 29) areas. At the end of the study, the researchers concluded that although TIR performed best overall (Fig. 6), it was found that VIS-C or MS were more effective than TIR in mapping subsurface drainage systems, therefore VIS-C, MS and TIR images were combined use increased success.

2.5 The Internet of Things (IoT) Technology

Use of the Internet of Things (IoT) in agriculture has become widespread and begun to replace some traditional methods. The availability of a wide variety of affordable IoT-enabled sensors, the advancement of wireless technology, and the economic and easy accessibility of cloud computing have made it possible to make many agricultural applications. The data obtained with the sensors are transferred to the cloud by wireless communication and this information can be accessed from anywhere in the world at any time. In this way, farmers, producers, and agricultural business managers can follow all the activities carried out and take the necessary decisions instantly. In addition, the use of IoT in agriculture is not limited to producers and businesses. In addition to manufacturing and businesses, it serves a wide audience working in the agriculture and food sectors such as suppliers, government institutions, cooperatives, vendors, and service. Agricultural productivity can be increased by evaluating data from a wide variety of sensors (climatic data, soil temperature, soil moisture, salinity, pH, etc.) from agricultural fields [30].

Ahmed et al. [31] developed agronomic models that program irrigation and fertilization according to plant needs. The models are integrated with a smart system and monitor the soil moisture and maintain the desired soil moisture with automatic irrigation (Fig. 7). Researchers have developed a fertigation support system that includes photovoltaic solar energy and IoT suitable for the PI system. The system is integrated with an agricultural decision support system (DSS), where data on plants, soil and weather conditions are obtained with sensors. Soil moisture radiation, humidity, temperature, and other physical parameters are monitored in the system. This developed system was used to evaluate the estimation of sustainable Reference Evapotranspiration (ETo) and irrigation planning for a three-years' study in a Mediterranean environment. There were seven ETo models in the study, including three temperatures, three radiation, and a combination-based model. It was found that there was a small increase in the accuracy of the ETo estimates, and solar fertigation and the Hargreaves–Samani (H–S) equation were closely correlated with the standard FAO P–M. It was also concluded that hybrid agricultural DSS is suitable for intelligent fertigation planning.

2.6 Artificial Intelligence and Machine Learning Techniques

Artificial intelligence (AI) is the acquisition of perception, learning, experience and thinking ability of human intelligence into computers, machines or systems and it is the ability to make decisions and to take the necessary action in the face of predictable or unpredictable new situations. In this decision-making process, these variables are taught to AI and formulas that can interpret like human mental functions are created. Thus, a thinking and decision-making model is created with computer software, like the human mindset [32]. The goal in machine learning (ML) is to make accurate predictions or decisions for an input that has never been

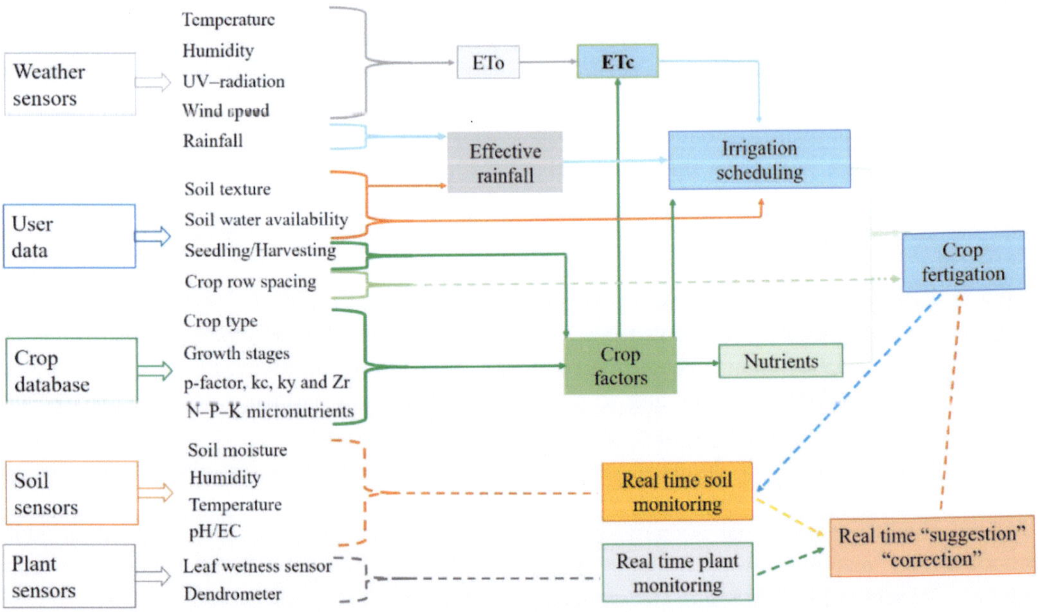

Fig. 7 A hybrid system developed for real-time irrigation planning [31]

seen before, and to develop efficient algorithms that automate this prediction and decision-making processes. During the development of the algorithm, the rules of an expert's decision-making criteria should be considered. In the application of ML methods, there may be complexities such as computational complexity, training complexity and trained algorithm implementation complexity. The performance of an algorithm is evaluated according to the test error. In addition, while an algorithm is running, there may be many test points, and quick decisions are required at these points. Therefore, the test process should have low computational load. In ML, data are randomly divided into training sample, validation sample and test sample [30].

Goap et al. [33] developed a smart system to predict a field's irrigation needs using sensor data from soil, weather, and environmental conditions. In the study, predictions were made with a smart algorithm based on machine learning using soil moisture, soil temperature, air temperature, precipitation, UV radiation, relative humidity and weather forecast data (Fig. 8). Sensor data were collected wirelessly over the cloud using web services and web-based information visualization. The system can perform closed-loop control of the water supply to realize a real-time fully autonomous irrigation scheme.

2.7 Irrigation Systems Automation

Automation consists of two parts hardware and software. The hardware consists of all devices such as sensors, relays, controllers, and valves required for the operation of the automation system. On the other hand, software is the program that enables the hardware to work and control in the desired way. The necessary information for the conditions to be controlled is collected by sensors. Sensor data is transmitted from the field to a central unit via IoT or wireless communication methods such as RF, ZigBee, LoRa or Global System for Mobile Communications (GSM) lines. The sensor data processed in the software are compared with the set values previously introduced to it, and the control decision to be made is transferred to the controller. The controller performs actions in line with these decisions. Today, smart sensors have been developed. With these sensors, data are transmitted at desired intervals. Irrigation automation can be done by various methods such as remote controlled, time based,

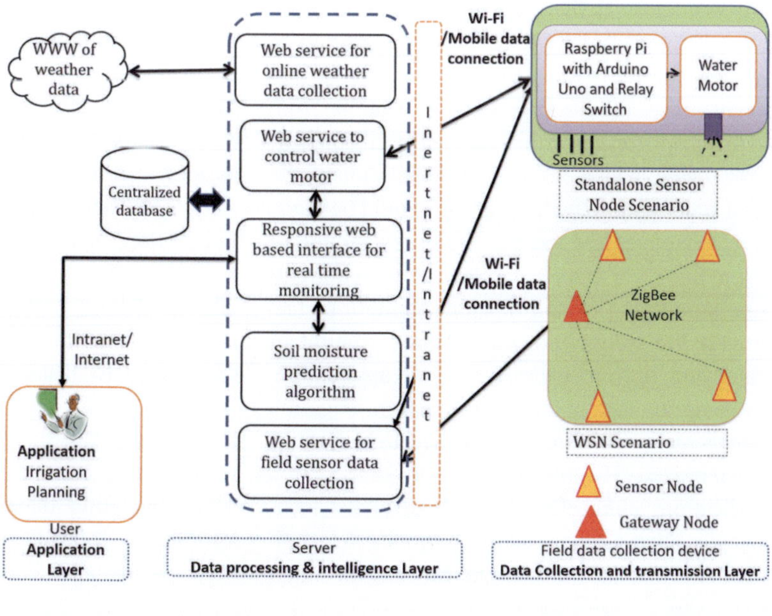

Fig. 8 Architecture of the proposed SI system [33]

volume based, semi- automatic, fully automatic, and ML based. Soil monitoring sensors used in irrigation are soil temperature, soil moisture, soil salinity, pH, EC, and air temperature sensors.

The sensors were previously used only in automation applications in agricultural applications. In these automation applications, often repetitive tasks were performed, in which the data from the sensors were evaluated and the desired output was obtained. The development of data mining and ML techniques has led to new very important applications in this field. In these applications, it has been possible to develop a wide variety of expert systems by using various computational techniques that replace human ability for real-time evaluation of data obtained from sensors and for decision making. The advantages of different techniques in data mining and ML techniques are used for problem solving or increasing system success in various agricultural applications, especially for clustering and classification [30].

3 Discussion

Global warming, climate change and wrong agricultural irrigation practices have led to an increase in the withdrawal of the world's fresh-water reserves. The reports of major organizations indicate that more challenging situations like problems of the global food supply are expected in the near future. In such a case, it is thought that it will cause intense migrations, especially from Africa to Europe. For this reason, in the agricultural sector, which is the sector that causes the most water withdrawal, water use should be made more effective, and various measures should be taken. It is clear that SI practices along with real-time monitoring of soil moisture conditions can provide the plants with the required amount of water and prevent excessive water application. By using the data obtained from RS in irrigation management, the efficiency of irrigation systems is maximized. SI applications can make irrigation applications more effective with advanced technologies such as smart sensors, drones, RS, IoT, VRI, AI, ML, and automation. Most of the smart irrigation studies in literature consist of academic studies. However, to achieve the desired benefit from SI, it must be adopted and widely used. SI applications require different levels of technique and technology using a variety of hardware and software. Being able to use one or more of these applications is directly related to technical and economic possibilities as well as farmers' further education. The possibility of using all these technologies together can be difficult under today's conditions. However, by continuing the development of these systems and introducing alternative products to the market, it will be possible for farmers to purchase and to use them widely. International organizations such as the United Nations, FAO and local governments should be among the priority issues for these practices to become widespread, especially in regions with water shortages. Uncertainties arising in the face of situations such as decreasing water resources, irregular rainfall and increasing water constraints require the use of SI applications as much as possible and within the possibilities. For example, each farmer-enterprise can try to use some SI applications within his own means. The main determination criterion in deciding whether to use SI applications should be the evaluation of whether economic and environmental benefits are achieved because of the implementation.

4 Conclusions

Many difficulties and uncertainties such as diseases, pests, economic crises, decrease in agricultural lands, decrease in natural resources, etc., are struggled during agricultural activities. In addition, natural events such as drought, hail, flood, tornado, etc., caused by global warming and climate change, make agricultural activities even more difficult. Smart farming applications provide the ability to instantly collect and to manage information and data from agricultural activities in real time and then to analyze that data back in the office using farm management software. Thus, by using these technologies together in agriculture, it has become possible to take more effective management decisions for better quality production and to implement more competitive, efficient, and sustainable agriculture.

With the use of smart agriculture technologies in irrigation, it helps both to reduce the effect of drought and to increase food production. For example, adopting PI techniques helps to improve water use and to maximize plant productivity. With RS, information about land use of large agricultural areas, vegetation indices, plant species, plant yield, physical and chemical properties of soils, soil moisture, water resources, etc., can be obtained easily. It is possible to monitor irrigation in a very short time, and to determine the water status and drought effects of the plant with a drone. With the use of IoT, data from a wide variety of sensors can be obtained in real time and sent to the microprocessor for evaluation, and irrigation can be done automatically when desired with the irrigation automation system. The autonomous working opportunity offered by AI and especially ML techniques can be used in the evaluation of large datasets from a wide variety of sources and types and in operating the irrigation system.

References

1. IPCC. Climate change widespread, rapid, and intensifying. The intergovernmental panel on climate change report; 2021. August 9, 2021. https://www.ipcc.ch/2021/08/09/ar6-wg1-20210809-pr/.
2. Rieu-Clarke A, Moynihan R, Magsig B-O. Transboundary water governance and climate change adaptation: international law; policy guidelines and best practice application. Paris, France: WWAP, UNESCO; 2015
3. FAO. Water scarcity–one of the greatest challenges of our time. Rome; 2019. https://www.fao.org/fao-stories/article/en/c/1185405/.
4. FAO and UN-Water. Progress on level of water stress: global status and acceleration needs for SDG Indicator 6.4.2. Rome; 2021. p. 96. http://www.fao.org/3/cb6241en/cb6241en.pdf.
5. FAO. The state of the world's land and water resources for food and agriculture–systems at breaking point. Synthesis report 2021. Rome; 2021. https://doi.org/10.4060/cb7654en.
6. Biancalani R, Marinelli M. Assessing SDG indicator 6.4.2 'level of water stress' at major basins level. UCL Open Environ. 2021;(3):05. https://doi.org/10.14324/111.444/ucloe.000026.
7. Ozguven MM. Contribution of precision agriculture to drought and food security. In: Baig MB, editor. Climate change, food security and climate-smart agriculture. Springer Nature Singapore Pte Ltd; 2023. ISBN (in print).
8. Anonymous. Agricultural irrigated land refers to agricultural areas purposely provided with water, including land irrigated by controlled flooding; 2022. https://databank.worldbank.org/metadataglossary/world-development-indicators/series/AG.LND.IRIG.AG.ZS.
9. Alexandris S, Psomiadis E, Proutsos N, Philippopoulos P, Charalampopoulos I, Kakaletris G, Papoutsi E-M, Vassilakis S, Paraskevopoulos A. Integrating drone technology into an innovative agrometeorological methodology for the precise and real-time estimation of crop water requirements. Hydrology. 2021;8:131. https://doi.org/10.3390/hydrology8030131.
10. Özgüven MM, ve Karaman S. Hassas Sulama Teknolojileri. 27. Tarımsal Mekanizasyon Ulusal Kongresi, 5–7 Eylül Samsun, 2012;S:288–97. (Turkish).
11. Valley (2022). Variable Rate Irrigation (VRI). https://www.valleyirrigation.com/vri.
12. Krishna KR. Agricultural drones: a peaceful pursuit. Apple Academic Press; 2018. ISBN 978–1–315–19552–0.
13. Capraro F, Tosetti S, Rossomando F, Mut V, Vita Serman F. Web-based system for the remote monitoring and management of precision irrigation: a case study in an arid region of Argentina. Sensors. 2018;18(11):3847. https://doi.org/10.3390/s18113847.
14. Zachariah AB. Precision agriculture: the future of farming. Delve Publishing; 2019. ISBN 978–1–77361–806–7.
15. Dukes MD. Water conservation potential of landscape irrigation smart controllers. Trans Am Soc Agric Biol Eng. 2012;55(2):563–9.
16. Yule IJ, Hedley CB, Bradbury S. Variable-rate irrigation. In: 12th Annual symposium on precision agriculture research & application in Australasia. Sydney; 2008.
17. Hedley CB, Yule IJ. Soil water status mapping and two variable-rate irrigation scenarios. Precision Agric. 2009;10:342–55.
18. King BA, Stark JC, Wall RW. Comparison of site-specific and conventional uniform irrigation management for potatoes. Appl Eng Agric. 2006;22(5):677–88.
19. Shock CC, Pereira AB, Feibert EBG, Shock CA, Akin AI, Unlenen LA. Field comparison of soil moisture sensing using neutron thermalization, frequency domain, tensiometer, and granular matrix sensor devices: relevance to precision irrigation. J Water Resour Prot. 2016;8(2). https://doi.org/10.4236/jwarp.2016.82013.

20. Khan AI, Alsolami F, Alqurashi F, Abushark YB, Sarker IH. Novel energy management scheme in IoT enabled smart irrigation system using optimized intelligence methods. Eng Appl Artif Intell. 2022;114: 104996. https://doi.org/10.1016/j.engappai.2022.104996.
21. Özgüven MM. Hassas Tarım. Akfon Yayınları, Ankara; 2018. ISBN: 978–605–68762–4–0. (Turkish).
22. Masud MJ, Bastiaanssen WGM. Remote sensing and GIS applications in water resources management. In: Bakhsh A, Choudhry MR, editors. Applied irrigation engineering. University of Agriculture Faisalabad Pakistan; 2017. ISBN 978–969–8237–97–4.
23. Tolomio M, Casa R. Dynamic crop models and remote sensing irrigation decision support systems: a review of water stress concepts for improved estimation of water requirements. Remote Sens. 2020;12:3945. https://doi.org/10.3390/rs12233945.
24. Quemada C, Pérez-Escudero JM, Gonzalo R, Ederra I, Santesteban LG, Torres N, Iriarte JC. Remote sensing for plant water content monitoring: a review. Remote Sens. 2021;13:2088. https://doi.org/10.3390/rs13112088.
25. Ozguven MM, Altaş Z, Güven D, Çam A. Tarımda Drone Kullanımı ve Geleceği. Ordu Univ J Sci Technol. 2022;12(1):52–70. https://doi.org/10.54370/ordubtd.1097519.
26. Ozguven MM, Yanar Y. The technology uses in the determination of sugar beet diseases. In: Misra V, Santeshwari S, Mall AK, editors. Sugar beet cultivation, management and processing. Springer Nature Singapore Pte Ltd; 2022. ISBN 978–981–19–2729–4.
27. Gonzalez-Dugo V, Zarco-Tejada P, Fereres E. Applicability and limitations of using the crop water stress index as an indicator of water deficits in citrus orchards. Agric For Meteorol. 2014;198–199:94–104.
28. Barbedo JGA. A review on the use of unmanned aerial vehicles and imaging sensors for monitoring and assessing plant stresses. Drones. 2019;3:40. https://doi.org/10.3390/drones3020040.
29. Allred B, Martinez L, Fessehazion MK, Rouse G, Williamson TN, Wishart D, Koganti T, Freeland R, Eash N, Batschelet A, Featheringill R. Overall results and key findings on the use of UAV visible-color, multispectral, and thermal infrared imagery to map agricultural drainage pipes. Agric Water Manag. 2020;232: 106036. https://doi.org/10.1016/j.agwat.2020.106036.
30. Ozguven MM. The digital age in agriculture (1st ed.). Boca Raton: CRC Press Taylor & Francis Group LLC; 2023. ISBN 9781032385808.
31. Ahmad U, Alvino A, Marino S. Solar fertigation: a sustainable and smart IoT-based irrigation and fertilization system for efficient water and nutrient management. Agronomy. 2022;12:1012. https://doi.org/10.3390/agronomy12051012.
32. Özgüven MM. Technological concepts and their differences. In: International Erciyes agriculture, animal & food sciences conference 24–27 April 2019. Kayseri, Turkiye: Erciyes University; 2019 (Turkish).
33. Goap A, Sharma D, Shukla AK, Krishna CR. An IoT based smart irrigation management system using machine learning and open source technologies. Comput Electron Agric. 2018;155:41–9. https://doi.org/10.1016/j.compag.2018.09.040.

Determination of Kc Curves of Wheat, Barley, Sunflower, Maize (Grain), Sugar Beet, and Onion by using Landsat 8–NDVI Time Series and Relationships between Kc and NDVI

Eyüp Selim Köksal, Emre Tunca, and Sakine Cetin Taner

Abstract

Evapotranspiration (ET) constitutes a significant parameter for various operational practices, including water resources management, hydrological calculations, design of large water structures, and design and operation of irrigation systems. Several methods have been developed for estimating potential crop evapotranspiration (ETc) values of field crops. FAO-Penman Monteith's approach, depending on reference evapotranspiration (ETo) and plant coefficients (Kc), is the most common one among these methods. There are several limiting factors in front of the experimental identification of Kc coefficients of field crops in different regions of the world. Previous research revealed significant relationships between spectral vegetation indices (SVI) calculated using remote sensing data and Kc values. However, most of these studies were about the relationships between SVI and Kc rather than the generation of Kc curves, indicating the temporal variations in Kc values to be used in ETc estimation with the aid of SVIs. The primary target of the present study was to generate Kc curves for wheat, barley, grain maize, sunflower, sugar beet, and onion using Normalized Difference Vegetation Index (NDVI) values calculated from Landsat 8 satellite images. Field experiments were conducted in a semi-arid region of Türkiye in the 2015 and 2016 growing seasons. NDVI values were obtained from different fields cultivated under farmers' conditions. NDVI data were used as time series, and the temporal variations in the highest NDVI values of each image ($NDVI_{max}$) were analyzed. The lengths of each Kc phenological stage (initial, crop development, mid, and late seasons) were determined for each crop. Kc coefficients were adapted from FAO 56 handbook using meteorological parameters. In this way, Kc curves of relevant field crops were generated, and statistical relationships between $NDVI_{max}$ and Kc values of each crop were determined. Resultant Kc curves complied well with the previous literature. The linear relationships between Kc and $NDVI_{max}$ yielded an R^2 of 0.98 for sunflowers and an R^2 of 0.97 for the other crops.

Keywords

Crop coefficient · Evapotranspiration · FAO 56 · NDVI · Normalized Difference Vegetation Index · Remote sensing

E. S. Köksal (✉) · S. C. Taner
Faculty of Agriculture, Department of Agricultural Structures and Irrigation, Ondokuz Mayıs University, Atakum, Samsun, Türkiye
e-mail: eselim@omu.edu.tr

E. Tunca
Faculty of Agriculture, Department of Biosystems Engineering, Düzce University, Düzce, Türkiye

M. Bahadir et al. (eds.), *Agricultural Water Management*, Water and Wastewater Management,
https://doi.org/10.1007/978-3-031-96134-2_6

1 Introduction

Potential crop evapotranspiration (ETc) provides significant data for water resources management, water accounting, irrigation system design and operation, and various other water-related issues. In the FAO Irrigation and Drainage Paper No 56 (FAO 56), ETc is defined as the quantity of evapotranspiration (ET) realized under standard conditions [1]. In previous studies, reference evapotranspiration (ETo) and single crop coefficient (Kc) representing standard crop conditions are recommended in the estimation of ETc. The ETc value of a plant for a certain period is obtained by multiplying ETo of that period with Kc values [1–4]. Kc values are mainly indicated with a Kc curve composed of four linear segments representing the initial, crop development, mid, and late seasons. Relevant linear segments represent the average trend in each period [3, 5]. Several experimental studies have been conducted worldwide under different climate and soil conditions to determine ETc and Kc values for different growth stages [1, 5–11]. There are two primary issues pointed out in Kc studies. The first one is the length of the initial, crop development, mid, and late seasons. The second issue is identifying Kc values for each period under standard conditions [1, 12]. The lengths of Kc periods are determined through phenological observations. Kc values were defined as the ratio of crop evapotranspiration (ETc) and reference evapotranspiration (ETo). This way, the length of growth stages and Kc values are used to generate Kc curves [1].

Pereira et al. (2015) [3] indicate that in ETc calculations, Kc represents crop growth stage, amount of vegetation, cultivar type, planting date, crop season length, and canopy architecture. Plant breeders have been developing new cultivars yearly, which mostly exhibit phenological differences from the old ones. Kc-related parameters may show significant differences in climate, soil, and agricultural practice conditions even in the same cultivar. Then it is possible to indicate that Kc, the primary data for the estimation of ETc, may present differences based on period lengths, crop species, climate, soil, regional conditions, and agricultural practices. The Kc data should also be updated or renewed in time. The Kc growth stage lengths presented for several plants in the FAO Irrigation and Drainage Paper No 24 (FAO 24) [2] and FAO 56 may provide a general idea to the users. Such information should also be determined each year based on present cultivars, agricultural practices, and local observations [1, 3, 13], since the accuracy of period lengths is as effective as the accuracy of Kc values on ETc values. Although Kc values are significant for water and agricultural water management, such data are unavailable for several regions, crops, and cultivars worldwide. The reasons for such unavailability include a lack of human resources with expertise on ETc, the costly and time-consuming nature of experimental studies, and out of date or unreliable nature of the reported Kc values [5, 7]. Recent evidence indicated that the prediction of Kc values with the aid of remote sensing technology might eliminate such insufficiencies.

There are two approaches commonly used in the estimation of ET with the aid of remotely sensed data [3]: (i) the models dependent on energy balance algorithms [14–18], and (ii) the models dependent on spectral vegetation index (SVI) values [19–23]. In SVI-dependent ET mapping, Kc values are estimated, and resultant Kc values are multiplied by ETo to get ET. Several studies about using SVIs were obtained through remote sensing techniques to determine Kc values. Most of these studies depend on mathematical correlations between the SVI sensitive to actual vegetation conditions and Kc. Such equations may represent induvial crops or mixed plant covers [24]. The studies on this subject were initiated in the 1980s [19, 22, 24–26].

In previous studies conducted to determine Kc with SVIs, relevant SVIs were initially calculated from hand-held spectroradiometers' data. Kc values were determined using ETc values, calculated with a soil water budget under controlled field experiments (e.g., by using lysimeter systems). Then this approach was employed in research with SVIs obtained from satellite images [20, 23, 27–30]. In the literature, mostly linear regression equations were reported for the correlations between SVI and Kc [29, 31–34]. In some of these studies, calculated cover ratios (fc) were estimated with SVI, and the estimation of Kc values with the use of fc values was recommended [35–37]. In many of these studies, the dual crop coefficient approach was used, and correlations between basal crop coefficient (Kcb) represent the share of transpiration in Kc and SVI. Normalized Difference Vegetation Index (NDVI) and Soil Adjusted Vegetation Index (SAVI) are the most common SVIs used in those studies [38, 39].

Under farm conditions, actual evapotranspiration (ETa) is primarily different from ETc. Such differences are mainly attributed to differences in cultivars, plant density, sowing-planting time, irrigation, pests and disease control practices, climate, soil, salinity, and other conditions. Since under non-standard conditions, the ratio of ETa to ETo is lower than Kc, some Kc-SVIs modeling studies, carried out under lower ETa conditions, report equations with high-level error. According to FAO 56, Kc values should be corrected with a stress factor (Ks), and adjusted Kc values should be calculated under such water stress conditions [1, 3]. Furthermore, there are several studies about the determination of Kc values using remote sensing techniques. However, the number of studies about the determination of the initial, crop development, mid, and late-season lengths are quite limited.

The objectives of the present study were to identify Kc crop growth stages for wheat, barley, sunflower, maize (grain), sugar beet, and onion plants in a semi-arid climate region with the use of NDVI maps obtained from Landsat 8 satellite images as time-series; to generate Kc curves with the use of Kc values calculated based on FAO 56 approaches; and to determine mathematical relationships between Kc values and NDVI values with the aid of linear regression. The goal is to put forth the potential use of satellite images in the generation of Kc curves for standard conditions.

2 Materials and Methods

The materials and methods of this study are presented in a flow chart in Fig. 1. In this study, satellite images, meteorological data and spatial data were used. Kc curves, the primary data for estimation of ETc, were generated for wheat, barley, sunflower, maize (grain), sugar beet, and onion plants using remotely sensed data. Relationships between Kc values and NDVI values of each plant were determined.

2.1 Study Area

The study was conducted over the irrigated lands of Merzifon (40° 47′ 21" N–35° 26′ 25" E; 569 m) town of Amasya province, Türkiye, in 2015 and

Fig. 1 Flow chart of the study

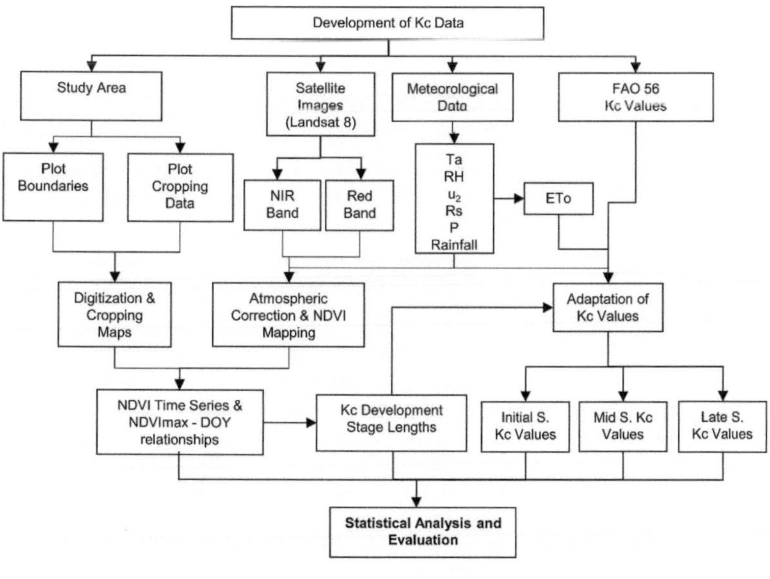

2016. The villages considered in this study have a similar agricultural production pattern. Generally, field crops are cultivated in this area, and wheat, barley, sunflower, maize, sugar beet, and onion are the common crops. The study area has a smooth terrain and is composed of almost flat land. Salinity and alkalinity problems were not encountered in the fields.

Plot borders of agricultural fields were digitized and recorded in vector format. Crops cultivated in each plot in each year were coded on maps in the geographical information system environment, and these data were used as the primary layer. The cropping patterns of 2015 and 2016 are presented in Fig. 2.

2.2 Climate Conditions and Meteorological Data

The study area has a semi-arid climate. According to meteorological observations of the years 2015 and 2016, the average T_{max} was realized in August (2015: 31.5 °C and 2016: 29.6 °C), and the lowest average relative humidity (RH_{min}) occurred in September (2015: 28.1% and 2016: 31.3%). The highest daily solar radiation (R_s) was observed in July 2015 (23.3 MJ/m^2.day) and June 2016 (24.8 $MJ/m^2.day^1$). The study area has a wind speed of around 2 m/s and sometimes over 2 m/s. Although the annual total precipitation of both years (2015: 406.6 mm and 2016: 491.4 mm) was close to long-term averages, the study area received higher precipitation in May 2016 (135.8 mm). The highest ETo values were observed in July and August. Total precipitation was lower, and ETo was higher in 2015 than in 2016 [40].

2.3 Satellite Imagery and NDVI Mapping

Landsat 8 satellite images were used in this study. 19 Landsat 8 images were used for 2015, and 18 images were used for 2016. The distribution of images in 2015 and 2016 based on imagery dates are presented in Fig. 3.

QGIS software was used for image processing. Normalized Difference Vegetation Index (NDVI) values of Landsat 8 images were calculated based on at-surface spectral reflectance ratios (Eq. 1) [14, 39, 41].

$$NDVI = \frac{(NIR - RED)}{(NIR + RED)} \tag{1}$$

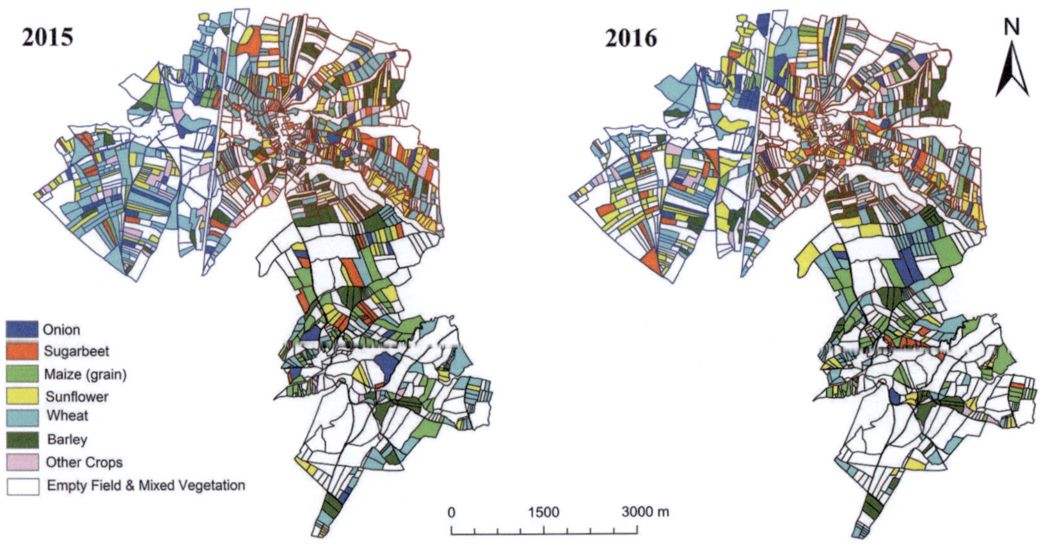

Fig. 2 Plot-based distribution of wheat, barley, sunflower, maize (grain) sugar beet and onion in 2015 and 2016

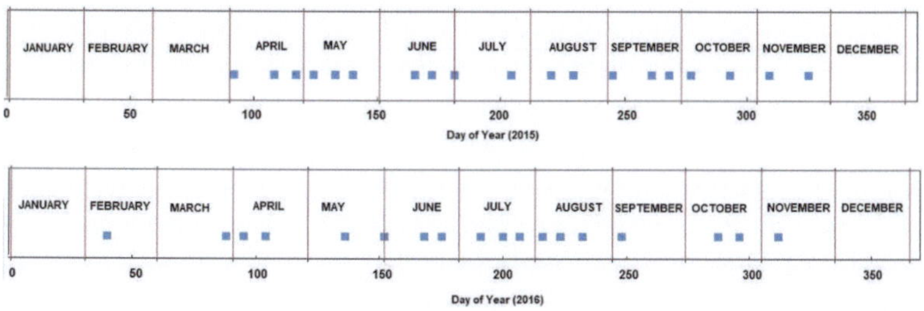

Fig. 3 Distribution of Landsat 8 satellite images in 2015 and 2016

where; NIR and RED represent at surface spectral reflectance values calculated for these bands [42, 43].

2.4 Determination of Kc Development Stage Lengths and Kc Values

In the Kc approach, plant growth stages are divided into four stages: the initial, crop development, mid, and late season [1]. Under standard ETc conditions, the length of these stages varies based on plants, cultivars, sowing-planting dates, agricultural applications, and climate. Generally, the initial stage of field crops covers the period from sowing to 10% coverage, and then the crop development stage starts. The crop development stage is continuous until full coverage. For several crops, effective full cover coincides with the initiation of flowering. For row crops, initiation of effective full cover is considered as the period in which plant leaves of adjacent rows are intermixed. With the end of the crop development stage, mid-season starts. Mid-season covers the period from the effective full coverage until maturity. Maturity can be identified with the symptoms of aging, yellowing, or senescence of leaves. The late season starts with maturity and ends with the harvest or the last harvest [1].

This study used time series NDVI values to identify Kc growth stages. For this purpose, at least five plots with a size greater than 2 ha were selected for each crop separately. For relevant production sites of each crop, NDVI values of the entire area and pixels coinciding into the plot were taken under non-cloudy conditions. Plants exhibited different growth in different fields and different periods. However, conditions without any factors negatively influencing plant growth and development were considered in the standard Kc approach. Therefore, in identifying the growth stages of each crop, concerning the imagery dates, the maximum NDVI (NDVImax) values in all pixels for the corresponding crop were considered.

On the contrary, the requirement for daily satellite imagery arises to observe the daily fluctuations in NDVI (Normalized Difference Vegetation Index) values. However, the temporal resolution offered by Landsat 8 is insufficient to fulfill this demand adequately. To address this limitation and calculate daily NDVI values, a solution was devised by establishing linear regression models between NDVImax values and the corresponding Day of Year (DOY) for each crop's growth stage. In the early and mid-season stages, where the variance in NDVImax vs. DOY values is relatively minor, the approach involved averaging NDVImax values. On the other hand, during the late season, linear regression equations capturing the decline in NDVImax vs. DOY were employed. This method allowed for a more comprehensive understanding of the development of vegetation and changes throughout the growth cycle.

In this way, theoretical increase and decrease lines were obtained for daily $NDVI_{max}$ values of crop development and the late season stages. For all crops considered in this study, the sowing date was regarded as the first day of the initial stage. The DOY coinciding with the average $NDVI_{max}$ value of the initial stage over the line generated for the crop development stage was considered as the end of the initial stage and the start of the crop development stage. The DOY coinciding with the mid-season $NDVI_{max}$ values was regarded as the end of the crop development stage and the start of the mid-season. Similarly, the DOY coinciding average $NDVI_{max}$ value of the mid-season over the line generated for the late season was considered the end of the mid-season and the start of the late season. The harvest date was taken as the end of the late season. Based on this approach, the lengths of each Kc stage were separately identified for wheat, barley, sunflower, maize, sugar beet, and onion.

The initial, mid, and late season Kc values of the present crops were calculated through adaptation of Kc values given in of FAO 56 (Table 12, page 110) with the aid of approaches provided in the same Ref. [1].

3 Results and Discussions

3.1 NDVI Time Series and Crop Coefficient (Kc)

NDVI maps obtained from LANDSAT 8 satellite images taken between April and September of 2015 and 2016 are presented in Figs. 4 and 5. The white lines in the figures show plot borders. In general, it was determined that the seasonal changes of NDVI values in the plots were compatible with the phenological characteristics of the plants grown. Besides, NDVI levels and temporal variations were not similar in all areas cultivated with the same crop. Such differences were attributed to differences in sowing dates, soil conditions, irrigation, fertilization, pests, and disease control practices. According to land observations made in the days of Landsat 8 satellite imagery, there were no significant differences in sowing, harvest dates, and phenological stages of the years. Therefore, temporal variations in NDVI values of 2015 and 2016 were used together in the identification of the Kc growth stages of each crop.

Figure 5 presents variations in NDVI values of different areas in the same date and temporal variations in NDVI values for wheat, barley, sunflower, grain maize, sugar beet, and onion. Considering the variations in $NDVI_{max}$ vs DOY values, it was observed that there were no significant relationships between these two parameters in the initial and mid-season. Still, there were significant linear relationships between NDVI and DOY in crop development and the late seasons. Accordingly, Fig. 6 also includes average $NDVI_{max}$ values for the initial and mid-season, linear regression equations, and determination coefficients (R^2) for crop development and late seasons.

Adapted Kc values using FAO 56 Kc values and relevant adaptation approaches, Kc stage lengths calculated with the aid of the approaches of this study are provided in Table 1 for each crop and each year. Also, Kc curves developed for each crop and each year are presented in Fig. 7. Resultant NDVI and Kc-related findings were assessed below separately for each crop.

(a) Wheat

Wheat had a longer initial stage with low NDVI values than the other crops. In this growth stage, NDVI values of the plots varied between 0.1 and 0.23. Wheat was sown at the beginning of November, and the initial stage lasted for 95 days. The average $NDVI_{max}$ value of this stage was 0.18. Wheat exhibited vegetative growth after about DOY 34, and with such development, $NDVI_{max}$ level reached 0.85, and plants then got into a stable period. Thus, a rapid increase was observed in the crop development stage, which lasted for 61 days (DOY 34–95). In the crop development stage, there was a linear relationship ($R^2 = 0.94$) between $NDVI_{max}$ and DOY (Fig. 6). $NDVI_{max}$ level was 0.85 in the mid-season, and this stage lasted for 48 days. After DOY 143, $NDVI_{max}$ values decreased against DOY, and the two parameters had a significant linear relationship ($R^2 = 0.94$). In the late season, the length was determined as 47 days. As it was indicated in Kc graphs generated for 2015 and 2016 (Fig. 7), the Kc values of both years were calculated as 0.4 for the initial stage of wheat. The Mid-season Kc value was calculated as 1.12 for 2015 and 1.18 for 2016. The late season Kc values were calculated as 0.39 and 0.40 for 2015 and 2016, respectively. Differences in Kc values over the years can be explained by changes, especially in RH and u_2 conditions. Variations in the length of Kc growth stages of wheat comply with the values provided in FAO 56.

Previous studies about wheat Kc values indicated that Kc values varied between 0.2 and 0.8

Fig. 4 NDVI time series maps of the study area generated using Landsat 8 satellite imagery for 2015

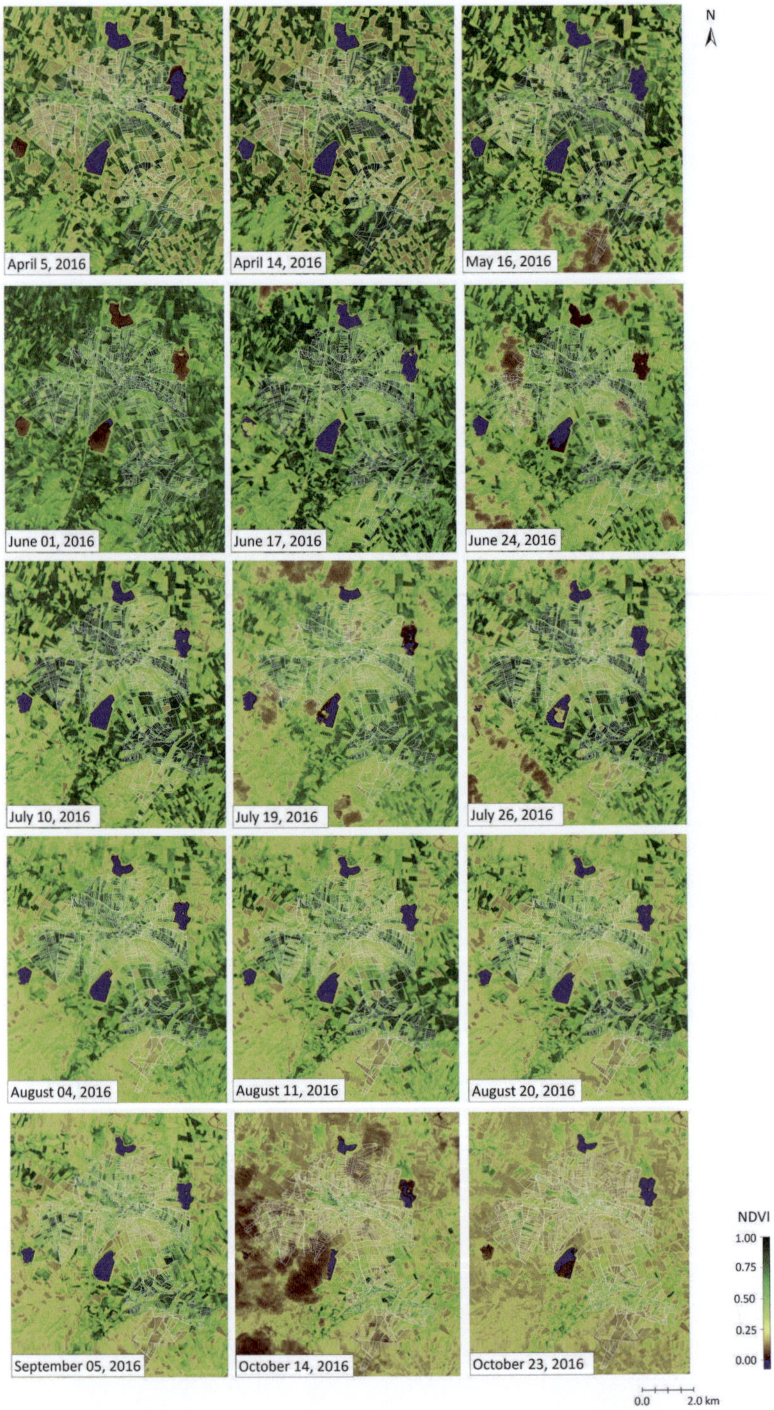

Fig. 5 NDVI time series maps of the study area generated using Landsat 8 satellite imagery for 2016

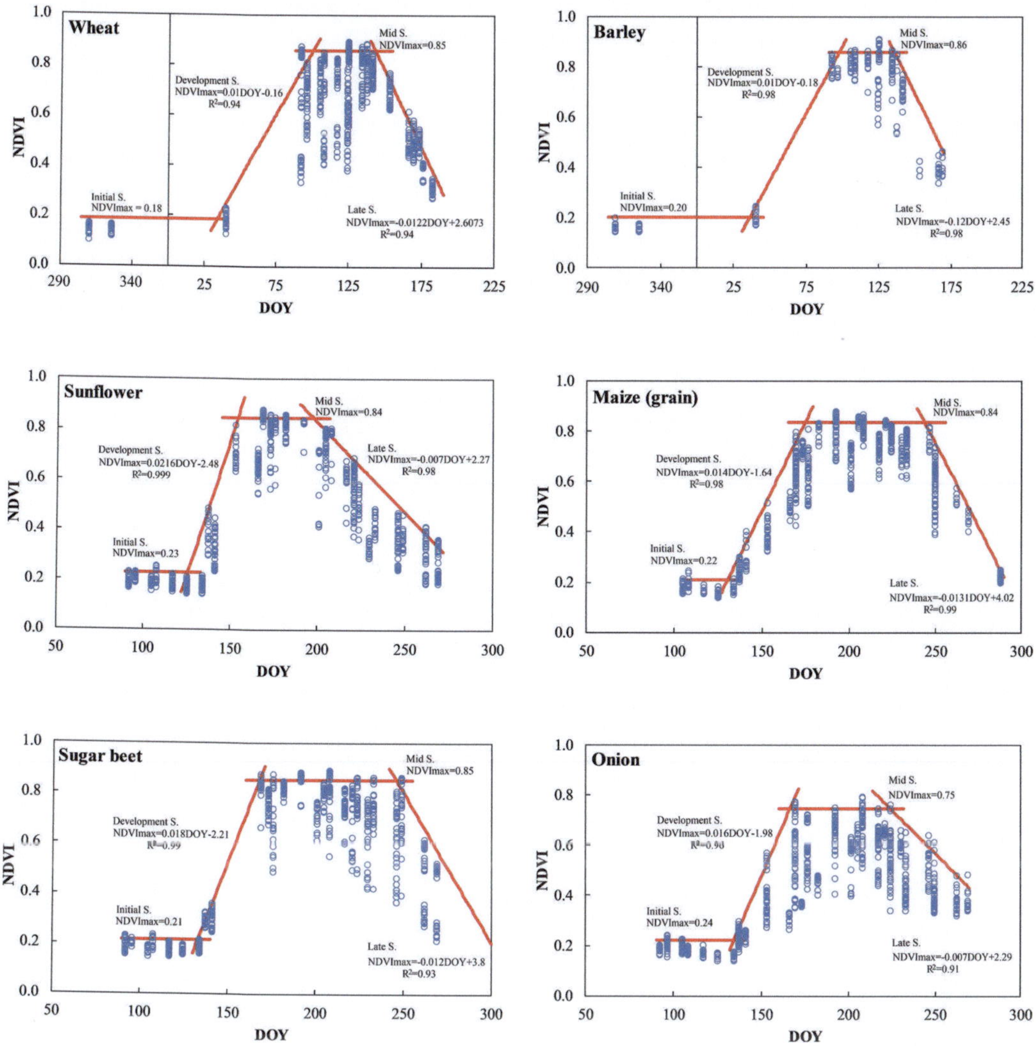

Fig. 6 Variation of pixel NDVI values among day of year (DOY) and linear relationships between NDVImax and DOY for four Kc phenological stages of wheat, barley, sunflower, maize (grain), sugar beet, and onion

for the initial stage, 1.1 and 1.35 for mid-season, and 0.2 and 0.8 for the late season [9, 12, 44–47]. Differences in Kc values were primarily observed in the initial stage and late season. According to Allen et al. [1], the initial Kc values may reach maximums with infiltration's effect. A similar case is also valid for the late-season Kc values. Thus, present Kc values for wheat's initial, mid, and late seasons complied with previous studies findings.

(b) Barley

Barley generally exhibits similar characteristics to wheat. Barley is sown on the same date as wheat in the study area, but harvest is 15 days earlier than wheat. The average NDVImax values of barley-cultivated plots were calculated as 0.2 for the initial stage (97 days). The initial stage ended on DOY 36, and the crop development stage continued until DOY 97. A significant linear relationship ($R^2 = 0.98$) between $NDVI_{max}$

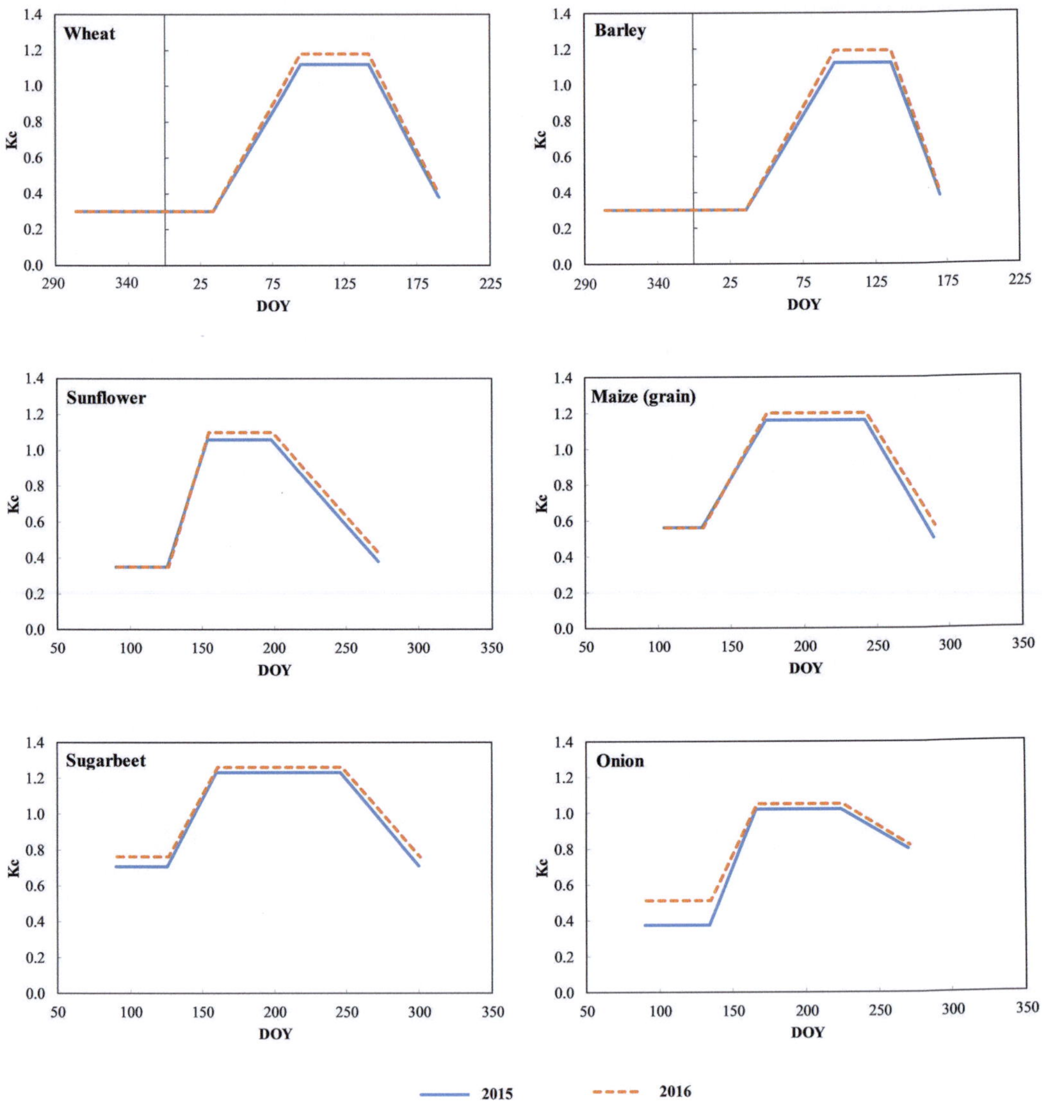

Fig. 7 Kc curves derived from Landsat 8-NDVI time series and FAO 56 for wheat, barley, sunflower, maize (grain), sugar beet, and onion

and DOY values was determined for the crop development stage (61 days). The barley's mid, and late season lengths were identified as 39 and 34 days, respectively. Similar to wheat, the average of $NDVI_{max}$ values of mid-season was calculated as 0.86. In the late season, decreases in $NDVI_{max}$ values against DOY were expressed in a linear regression equation ($R^2 = 0.98$). Kc values of barley were calculated as 0.4 for the initial stages of 2015 and 2016, 1.12 for the mid-season of 2015, and 1.18 for the mid-season of 2016. Kc values of the late season were calculated as 0.38 in 2015 and 0.40 in 2016 (Table 1, Fig. 5). FAO 56 is the most significant source in the literature about Kc values and Kc stages of barley. The values obtained for wheat in previous studies also highlight the values obtained for barley. Present growth stage lengths and Kc values of barley comply with the findings of those studies.

Table 1 Sowing dates, initial, crop development, mid and late-season stages, and Kc values of wheat, barley, sunflower, maize, sugar beet, and onion

Crop	Sowing date	Initial stage			Crop Dev. Stage	Mid-season stage			Late-season stage			Total growing season (day)
		Length (day)	Kc		Length (day)	Length (day)	Kc		Length (day)	Kc		
			2015	2016			2015	2016		2015	2016	
Wheat	Nov. 01	95	0.40	0.40	61	48	1.12	1.18	47	0.39	0.40	251
Barley	Nov. 01	97	0.40	0.40	61	39	1.12	1.18	34	0.38	0.40	231
Sunflower	April 01	35	0.49	0.49	29	44	1.06	1.1	74	0.38	0.42	182
Maize	April 15	26	0.44	0.49	44	68	1.15	1.21	47	0.53	0.56	185
Sugar beet	April 10	33	0.44	0.53	35	78	1.23	1.25	54	0.71	0.77	210
Onion	April 01	44	0.37	0.51	32	58	1.00	1.06	46	0.80	0.82	180

(c) Sunflower

The average sowing date of sunflower was taken on April 1 in the study area. According to $NDVI_{max}$ values, the length of the initial stage was identified as 35 days. The average of $NDVI_{max}$ values of the initial stage was calculated as 0.23. The length of the crop development stage was determined as 29 days, and there was a highly significant correlation ($R^2 = 0.999$) between $NDVI_{max}$ and DOY values of this stage. The average of $NDVI_{max}$ values of the mid-season was calculated as 0.84, and the length of the mid-season was identified as 44 days. The length of the late season was determined as 74 days. There was a significant linear relationship ($R^2 = 0.98$) between $NDVI_{max}$ and DOY values of the late season. In the present study area, sunflowers are grown both irrigated and rainfed. The length of Kc stages was calculated based on $NDVI_{max}$ values of irrigated lands. NDVI values under rainfed conditions differed from the $NDVI_{max}$ values throughout the growing season. Such differences can be seen from the NDVI values below the daily $NDVI_{max}$ values in Fig. 6. However, local growers attributed the extended late season (74 days) to the traditional practice of harvesting sunflowers after the seeds had ripened.

In Allen et al. [1], Kc stage lengths of the initial, crop development, mid, and late seasons were reported as 25, 35, 45, and 25 days, respectively. Tyagi et al. [48] reported Kc stage lengths as 21 days for the initial stage, 35 days for crop development, 14 days for mid-season, and 14 days for the late season. Howell et al. [11] reported lengths of sunflower's initial, crop development, mid, and late seasons as 20, 17, 45, and 18 days, respectively. Present Kc stage lengths comply with the ones reported in FAO 56 handbook but do not comply with the results of other studies. In Tyagi et al. [48], sowing was performed in March, and harvest was completed in June. In Howell et al. [11], sowing was performed at the end of June. For the present study area, March is early for sowing, and sowing after June is quite late. Therefore, differences in the lengths of Kc stages were mainly attributed to differences in cultivars and climate conditions.

The Kc values were calculated as 0.49 for the initial stage of 2015 and 2016, Kc values of mid-season were calculated as 1.06 for 2015 and 1.1 for 2016, the late season Kc values were calculated as 0.38 for 2015 and 0.42 for 2016 (Table 1). Kc value of the initial stage of sunflower was reported as 0.52 by Tyagi et al. [48] and 0.55 by Howell et al. [11]. Considering the differences in climate parameters and sowing dates, the present Kc value for the initial stage of sunflower (0.49) calculated based on FAO 56 complied with the earlier reports. Mid-season Kc values calculated for a semi-arid region were slightly lower than those of Tyagi et al. [48] and Howell et al. [11]. Still, differences in climate parameters and cultivars were sufficient to explain such differences. The late-season Kc values of the present study also exhibited similar compliance with the previous studies as in the initial and mid-season Kc values.

(d) Maize (Grain)

In the study area, maize was sown in mid-April. The average of $NDVI_{max}$ values was calculated as 0.22 for the initial stage and 0.84 for the mid-season. There was a linear relationship ($R^2 = 0.98$) between $NDVI_{max}$ and DOY of the crop development stage (Fig. 6). With the aid of this relationship, daily variations in $NDVI_{max}$ values were estimated and daily $NDVI_{max}$ values allowed to identify the end of the initial stage and start of the crop development stage. Present findings revealed that the initial stage lasted 26 days, and the crop development stage lasted 44 days. Vegetation and, thus, NDVI values decreased in maize with ripening initiation. Such a case indicated the end of mid-season. A significant linear relationship ($R^2 = 0.99$) existed between NDVImax and DOY values in the late season. With the aid of this relationship, the start of the late season was identified, and the mid-season lasted for 68 days. In the present study area, sufficient drying is waiting for the harvest of grain maize. Based on approximate harvest dates and NDVI images, it was determined that the late season of maize lasted for 47 days. According to NDVI distribution graphs given in Fig. 6, similar to the other crops, there were variations in maize vegetation of each pixel belonging to different locations. However, such differences are less in maize than in the other crops of the study. Also, the differences in maize vegetation of various enterprises and different sections of a plot were relatively less in mid-season. Such differences were mainly attributed to differences in irrigation uniformity and the effects of pests and diseases. In previous studies, the length of the initial stage was reported as between 20 and 30 days, the length of the crop development stage was reported as between 30 and 40 days, the length of mid-season was reported as between 30 and 50 days and the length of the late season was reported as between 10 and 50 days [1, 6, 9, 45, 46, 49–51]. At the present Kc stage, the lengths generally comply with those earlier ones, and differences were mainly attributed to differences in cultivars, climate conditions, and agricultural practices. Kc values of maize's initial, mid-season, and late-season were respectively identified as 0.44, 1.15, and 0.53 in 2015 and 0.49, 1.21, and 0.56 in 2016 (Table 1). Previous studies reported Kc values of maize were between 0.28–0.45 for the initial stage, 1.12–1.18 for mid-season, and 0.25–0.40 for the late season [6, 9, 45, 46, 50, 51]. Present Kc values generally comply with those earlier ones.

(e) Sugar Beet

Sugar beet was generally sown within the initial ten days of April. NDVI images of sugar beet-cultivated plots indicated that the initial stage lasted 33 days. The average NDVImax value was 0.21 for the initial stage, reaching 0.85 in the mid-season. There was a significant linear relationship ($R^2 = 0.99$) between $NDVI_{max}$ and DOY values of the crop development stage (Fig. 6). This linear relationship revealed that the crop development stage lasted for 35 days. $NDVI_{max}$ values started to decrease after DOY 246; thus, mid-season lasted for 78 days. Harvest was practiced during the last week of October. In this stage, $NDVI_{max}$ values decreased until 0.38, and the late season lasted 54 days. There was a linear relationship ($R^2 = 0.93$) between $NDVI_{max}$ and DOY values of the late season.

The NDVI values of different sugar beet plots and variations of daily $NDVI_{max}$ values of Kc growth stages are presented in Fig. 6. While there were no significant differences in NDVI values of different locations in the initial and crop development stages, there were significant differences in mid-season and late-season NDVI values of different plots and different sections of a plot. NDVI maps (Figs. 4, 5) and field observations revealed significant differences in the soil textures of sugar beet fields and the irrigation practices of growers. The lengths of Kc growth stages of sugar beet were reported in FAO 56 handbook respectively as 30, 45, 90, and 15 days for the initial, crop development, mid, and late seasons. Thao [52] reported similar growth stages with FAO 56. In previous studies, the length of the initial stage of sugar beet was reported as between

24 and 25 days, the crop development stage was reported as between 26 and 32 days, mid-season was reported as between 65 and 79 days, and the late season was reported as between 20 and 29 days [8, 53]. Present findings for the initial crop development and mid-season complied with the literature, but the late season was longer than the previous literature since growers plan the harvest based on their order of purchase by the sugar factory.

In this way, growers do not store sugar beet and directly transport it to the factory. Kc values of FAO 56 were adapted based on present relative humidity, wind speed, and plant height parameters. Kc values of the initial stages were calculated as 0.44 in 2015 and 0.53 in 2016. Mid-season Kc values were calculated as 1.23 in 2015 and 1.25 in 2016. Kc values of the late seasons were calculated as 0.71 in 2015 and 0.77 in 2016. Thao [52] reported Kc values of drip-irrigated sugar beet as 0.52 for the initial stage, 0.93 for the mid-season, and 0.34 for the late season. In previous studies, Kc values of the initial stage were reported as between 0.42 and 0.70, Kc values of mid-season as between 1.04 and 1.13, and Kc values of the late season as between 0.65 and 1.04 [8, 53, 54]. Present Kc values generally comply with the experimental Kc values of the other studies. Kc values of drip-irrigated plants were lower, especially in the mid and late seasons, than the Kc values of the different stages [52], since less evaporation from the soil is observed in drip irrigation. In the present study region, farmers irrigated sugar beet either by surface or sprinkler irrigation.

(f) Onion

Onion was sown before the end of March. Thus, the first day of April was selected as the start of the growing season. $NDVI_{max}$ values stayed at 0.22 until the middle of May. Then, vegetation levels increased fastly, and $NDVI_{max}$ values reached 0.75 level. Considering the periods with rising and stable vegetation, it was observed that the initial stage may be 44 days, and the crop development stage may be 32 days. There was a linear relationship ($R^2 = 0.96$) between $NDVI_{max}$ and crop development stage DOY values (Fig. 8). The recess in onion vegetation was observed about the middle of August. Onion is harvested mainly in the last ten days of September. There was a significant linear relationship ($R^2 = 0.91$) between $NDVI_{max}$ and DOY values of the late season. Present analysis revealed that the mid-season lasted 58 days and the late season lasted 46 days. $NDVI_{max}$ values decreased to 0.29 in the late season. Daily variations in NDVI values of onion throughout the growing season are presented in Fig. 6. The existence of different NDVI values at the same DOY was attributed to different NDVI values of different pixels of satellite images. Accordingly, except for the initial stage, there were significant differences in the vegetation of onion-cultivated lands. For instance, while $NDVI_{max}$ value was 0.77 at DOY 169 of mid-season, the lowest NDVI was 0.35. Such differences were attributed to recess in plant growth due to soil conditions, irrigation, and other agricultural practices. López-Urrea et al. [55], in a semi-arid region study, reported the initial, crop development, mid, and late season lengths as 52, 62, 30, and 24 days, respectively. In FAO 56 handbook, the length of the same stages was reported as 20, 35, 110, and 45 days. This study's initial stage length findings complied with the results of López-Urrea et al. [55] and crop development, and the late season lengths complied with the values of the FAO 56 handbook. Differences mainly resulted from differences in cultivars, climate parameters, and agricultural practices. The initial, mid, and late season Kc values were 0.37, 1.0, and 0.80 for 2015 and 0.51, 1.06, and 0.82 for 2016, respectively. In previous studies for onion, the initial stage Kc values were reported as between 0.3 and 0.65, mid-season Kc values as between 0.85 and 1.14 and the late season as between 0.56 and 0.75 [10, 50, 55–57].

3.2 Statistical Analysis and Evaluation

The relationships between Kc values calculated for different growth stages of wheat, barley, sunflower, maize, sugar beet, and onion and NDVImax values are presented in Fig. 8. The R2 value of the linear regression between Kc

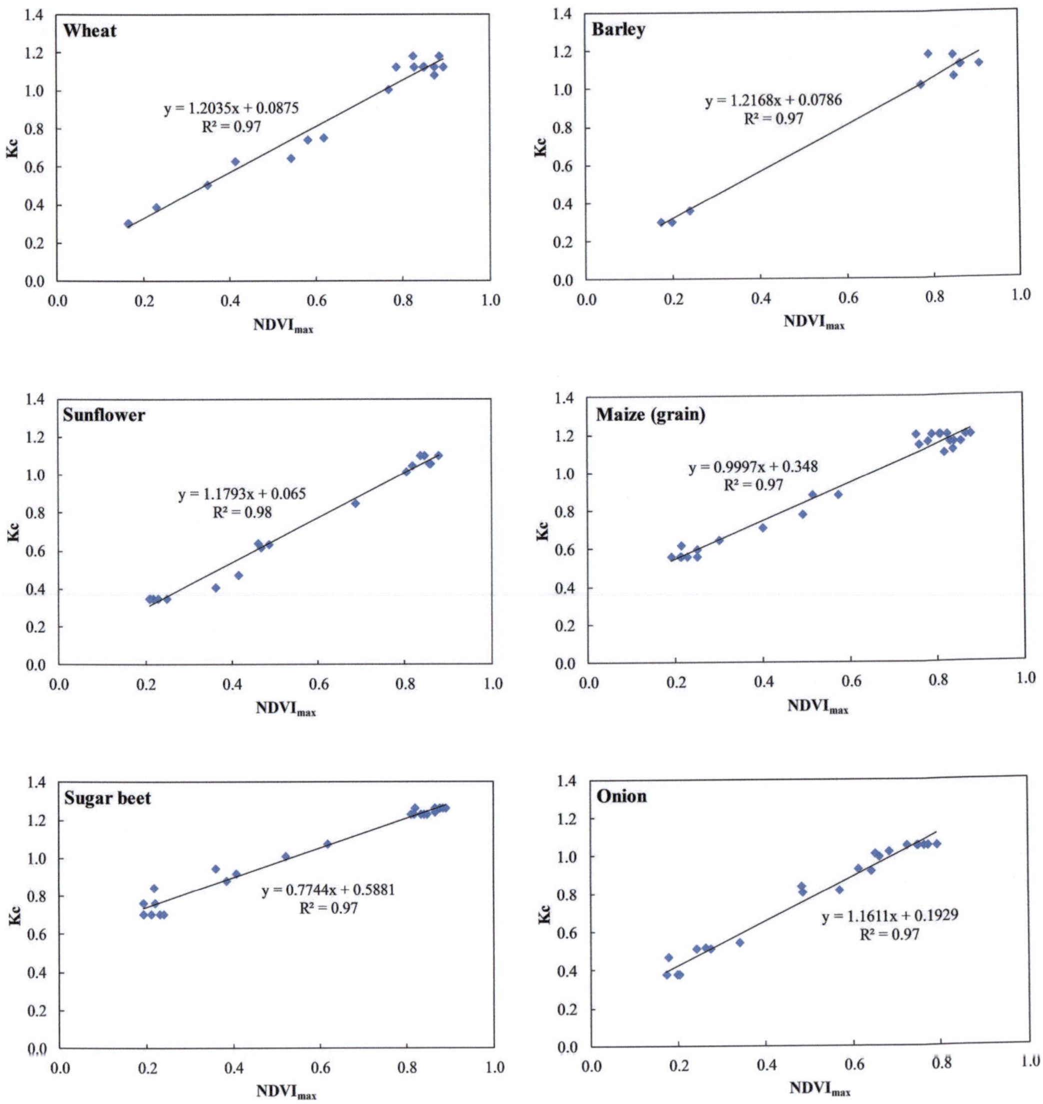

Fig. 8 Relationships between Kc (derived from FAO 56) and NDVImax representing four Kc phenological stages of wheat, barley, sunflower, maize (grain), sugar beet and onion

and NDVImax values was calculated as 0.98 for sunflowers and 0.97 for the other crops. The slope and bias values of the resulting regression equations were compared, and it was found that wheat and barley exhibited similar values, which was anticipated due to their comparable vegetation structures. Conversely, the slope and bias values of the other crops showed notable differences.

Vegetation structures and phenological differences in sunflower, maize, sugar beet, and onion plants explain such differences in linear regression equations. Present equations were considered usable in NDVImax-dependent Kc coefficient estimation under standard conditions. Such a finding complies with the results of earlier studies. Kamble et al. [32] were able to explain the relationships between NDVI and Kc with linear regression equations. The R2 values of that study for the relationships between NDVI and Kc were lower than the present R^2 values, since standard conditions were used in the present study for NDVImax and Kc values. Still, Kamble et al. [32] used lower-resolution satellite images in the NDVI&Kc model (Kc = 1.457 NDVI–0.1725) and non-standard and water stress conditions. However, the error rates in estimating Kc values with NDVI values were still within acceptable limits. Pôças et al. [37] developed an SVI-dependent model for standard Kc values and dual Kc approach and reported similar findings to the present study for barley and maize. Chattaraj et al. [58] reported significant equations for Kc estimations of wheat from hyperspectral SVIs. Toureiro et al. [33] obtained an equation of Kc = 0.918 NDVI + 0.303 to estimate Kc values of maize. Such an equation complies well with the present equation obtained for maize. Belmonte et al. [59] recommended the Kc = 1.25 NDVI + 0.2 for NDVI-based estimation of Kc values for the initial and mid-season. In this study, equations recommended for the initial and mid-season of maize, onion, sugar beet, barley and wheat were defined based on Kc values provided in FAO 56 handbook. The Kc values calculated for the initial stage with the aid of this equation complied with the FAO 56 Kc values provided for these crops, except for onion, and Kc values calculated for mid-season complied with FAO 56 Kc values for all crops. Relevant equations generally comply with the present equations, but the greatest difference was observed for sugar beet. Finally, present linear regression equations identified for NDVImax and Kc values well complied with the literature. With the use of these equations presented in Fig. 6, standard Kc values could reliably be estimated for relevant crops cultivated in semi-arid regions.

4 Conclusions

Plant-based estimation of potential evapotranspiration (ETc) values plays a significant role in region-based agricultural water management and irrigation system design, operation, and management. The method provided in FAO 56 handbook for the estimation of ETc is widely accepted worldwide and commonly used by most researchers. By using this method, the reference evapotranspiration (ETo) for different climates and time zones could be calculated based on meteorological data. In this sense, ETo calculations may always be up to date. However, in ETc calculations, Kc values representing the crops may vary based on several factors. Especially the length of the initial, crop development, mid, and late season representing the vegetative growth stages of the plants vary based on climatic parameters, soil conditions, cultivars, and agricultural practices. Most conventional Kc studies used soil water budget parameters and plant physiological characteristics through field experiments. Although such field experiments constitute the bases of Kc studies, conducting such studies separately for each condition exhibiting differences in the region, plant, cultivar, climate, and soil is impossible.

Earlier research has shown significant promise in utilizing remote sensing techniques to estimate Kc values. However, most of these previous studies focused on establishing the correlation between spectral vegetation indices (SVI) and Kc values. On the contrary, the methods employed to determine Kc values often involve comparing actual evapotranspiration (ETa) under water stress conditions to reference evapotranspiration (ETo), rather than adhering strictly to the standard Kc as outlined in the FAO-56 handbook. These approaches are mostly considered within the context of adjusted Kc.

In the present study, a new approach was developed to use satellite images to generate Kc graphs for wheat, barley, grain maize, sunflower, sugar beet, and onion plants under standard conditions. With this approach, the length of Kc growth stages could be identified with an SVI (i.e., NDVI) obtained through processing satellite images. The Kc coefficients of these growth stages could be calculated using the methods specified in FAO 56 handbook. In this way, Kc graphs could be generated faster and in a cost-efficient fashion. It is also relatively easy to update these Kc graphs. Estimated ETc values using linear relationships between resultant Kc values and NDVI values have great potential to be used in agricultural water management over large areas through remote sensing technologies. Further research is recommended to conduct with the technique developed in this study, and high-resolution satellite images to generate Kc graphs for different plant species in different regions.

Acknowledgements This study was supported by The Scientific and Technological Research Council of Türkiye (TÜBİTAK) under the project number 114O534. Also, this study is based upon work from COST Action CA17134 "Optical synergies for spatiotemporal SENsing of Scalable ECOphysiological traits" (SENSECO), supported by COST (European Cooperation in Science and Technology).

References

1. Allen RG, Pereira LS, Raes D, Smith M. Crop evapotranspiration-Guidelines for computing crop water requirements-FAO Irrigation and drainage paper 56. FAO, Rome. 1998;300:D05109.
2. Doorenbos J, Pruitt W. Crop water requirements. FAO irrigation and drainage paper 24. Land and Water Development Division, FAO, Rome; 1977. p. 144.
3. Pereira LS, Allen RG, Smith M, Raes D. Crop evapotranspiration estimation with FAO56: Past and future. Agric Water Manag. 2015;147:4–20.
4. Wright JL. New evapotranspiration crop coefficients. Proc Am Soc Civil Eng J Irrig Drain Div. 1982; 108: 57–74.
5. Allen RG, Pereira LS, Howell TA, Jensen ME. Evapotranspiration information reporting: I. Factors governing measurement accuracy. Agric Water Manag. 2011;98:899–920.
6. Alberto MCR, Quilty JR, Buresh RJ, Wassmann R, Haidar S, Correa TQ Jr, Sandro JM. Actual evapotranspiration and dual crop coefficients for dry-seeded rice and hybrid maize grown with overhead sprinkler irrigation. Agric Water Manag. 2014;136:1–12.
7. Allen RG, Pereira LS, Howell TA, Jensen ME. Evapotranspiration information reporting: II. Recommended documentation. Agric Water Manag. 2011;98:921–9.
8. Ebrahimipak N, Ghalebi S. Determination of evapotranspiration and crop coefficient of sugar beet using lysimeter and its comparison with experimental methods in Shahrekord, Iran. J Sugar Beet. 2014;30:23–32.
9. Gao Y, Duan A, Sun J, Li F, Liu Z, Liu H, Liu Z. Crop coefficient and water-use efficiency of winter wheat/spring maize strip intercropping. Field Crop Res. 2009;111:65–73.
10. Grattan S, Bowers W, Dong A, Snyder R, Carroll J, George W. New crop coefficients estimate water use of vegetables, row crops. Calif Agric. 1998;52:16–21.
11. Howell TA, Evett SR, Tolk JA, Copeland KS, Marek TH. Evapotranspiration, water productivity and crop coefficients for irrigated sunflower in the US Southern High Plains. Agric Water Manag. 2015;162:33–46.
12. Ko J, Piccinni G, Marek T, Howell T. Determination of growth-stage-specific crop coefficients (Kc) of cotton and wheat. Agric Water Manag. 2009;96:1691–7.
13. Allen RG, Wright JL, Pruitt WO, Pereira LS, Jensen ME. Water requirements. In: Design and operation of farm irrigation systems, 2nd ed. American Society of Agricultural and Biological Engineers;2007. pp. 208–88.
14. Allen RG, Tasumi M, Trezza R. Satellite-based energy balance for mapping evapotranspiration with internalized calibration (METRIC)—Model. J Irrig Drain Eng. 2007;133:380–94.
15. Bastiaanssen WG, Menenti M, Feddes R, Holtslag A. A remote sensing surface energy balance algorithm for land (SEBAL). 1. Formulation. J Hydrol. 1998;212:198–212.
16. Roerink G, Su Z, Menenti M. S-SEBI: a simple remote sensing algorithm to estimate the surface energy balance. Phys Chem Earth Part B. 2000;25:147–57.
17. Senay GB, Bohms S, Singh RK, Gowda PH, Velpuri NM, Alemu H, Verdin JP. Operational evapotranspiration mapping using remote sensing and weather datasets: a new parameterization for the SSEB approach. JAWRA J Am Water Res Assoc. 2013;49:577–91.
18. Su Z. The Surface Energy Balance System (SEBS) for estimation of turbulent heat fluxes. Hydrol Earth Syst Sci. 2002;6:85–100.
19. Bausch WC, Neale CM. Crop coefficients derived from reflected canopy radiation: a concept. Trans Asae. 1987;30:703–0709.
20. Campos I, Neale CM, Calera A, Balbontín C, González-Piqueras J. Assessing satellite-based basal crop coefficients for irrigated grapes (Vitis vinifera L.). Agric Water Manag. 2010;98:45–54.
21. Köksal ES. Evaluation of spectral vegetation indices as an indicator of crop coefficient and evapotranspiration under full and deficit irrigation conditions. Int J Remote Sens. 2008;29:7029–43.
22. Neale C. Development of reflectance-based crop coefficients for corn. Ph.D Dissertation, vol. 292. Agricultural Engineering, Colorado State University, Fort Collins, USA; 1987. p. 1891-9
23. Singh RK, Irmak A. Estimation of crop coefficients using satellite remote sensing. J Irrig Drain Eng. 2009;135:597–608.
24. Neale CM. Monitoring corn development using reflected radiation. Colorado State University;1983.
25. Bausch W, Neale C. Spectral inputs improve corn crop coefficients and irrigation scheduling. Trans ASAE. 1990;32:1901–8.
26. Neale CM, Bausch WC, Heermann DF. Development of reflectance-based crop coefficients for corn. Trans ASAE. 1990;32:1891–900.
27. Gontia NK, Tiwari KN. Estimation of crop coefficient and evapotranspiration of wheat (Triticum aestivum) in an irrigation command using remote sensing and GIS. Water Resour Manag. 2010;24:1399–414.

28. González-Dugo M, Mateos L. Spectral vegetation indices for benchmarking water productivity of irrigated cotton and sugarbeet crops. Agric Water Manag. 2008;95:48–58.
29. Reyes-Gonzalez A, Hay C, Kjaersgaard J, Neale C. Use of remote sensing to generate crop coefficient and estimate actual crop evapotranspiration. In: Proceedings of the 2015 ASABE annual international meeting;2015, p. 1.
30. Trout TJ, Johnson LF, Gartung J. Remote sensing of canopy cover in horticultural crops. HortScience. 2008;43:333–7.
31. Farg E, Arafat S, El-Wahed MA, El-Gindy A. Estimation of evapotranspiration ETc and crop coefficient Kc of wheat, in south Nile Delta of Egypt using integrated FAO-56 approach and remote sensing data. Egypt J Remote Sens Space Sci. 2012;15:83–9.
32. Kamble B, Kilic A, Hubbard K. Estimating crop coefficients using remote sensing-based vegetation index. Remote Sens. 2013;5:1588–602.
33. Toureiro C, Serralheiro R, Shahidian S, Sousa A. Irrigation management with remote sensing: evaluating irrigation requirement for maize under Mediterranean climate condition. Agric Water Manag. 2017;184:211–20.
34. Zolfagharnejad H, Kamkar B, Abdi O. Vegetation index-deduced crop coefficient of wheat (Triticum aestivum) using remote sensing: case study on four basins of Golestan Province Iran. World Acad Sci Eng Technol Int J Biol Biomol Agric Food Biotechnol Eng. 2017;11:498–501.
35. Allen RG, Pereira LS. Estimating crop coefficients from fraction of ground cover and height. Irrig Sci. 2009;28:17–34.
36. Er-Raki S, Chehbouni A, Guemouria N, Duchemin B, Ezzahar J, Hadria R. Combining FAO-56 model and ground-based remote sensing to estimate water consumptions of wheat crops in a semi-arid region. Agric Water Manag. 2007;87:41–54.
37. Pôças I, Paço T, Paredes P, Cunha M, Pereira L. Estimation of actual crop coefficients using remotely sensed vegetation indices and soil water balance modelled data. Remote Sens. 2015;7:2373–400.
38. Huete AR. A soil-adjusted vegetation index (SAVI). Remote Sens Environ. 1988;25:295–309.
39. Rouse Jr JW, Haas R, Schell J, Deering D. Monitoring vegetation systems in the Great Plains with ERTS;1988.
40. MGM. Turkish state meteorological service. 2016. https://www.mgm.gov.tr/. Accessed 12 Oct 2016.
41. Allen RG, Tasumi M, Morse A, Trezza R. A Landsat-based energy balance and evapotranspiration model in Western US water rights regulation and planning. Irrig Drain Syst. 2005;19:251–68.
42. Bannari A, Morin D, Bonn F, Huete A. A review of vegetation indices. Remote Sens Rev. 1995;13:95–120.
43. Glenn EP, Neale CM, Hunsaker DJ, Nagler PL. Vegetation index-based crop coefficients to estimate evapotranspiration by remote sensing in agricultural and natural ecosystems. Hydrol Process. 2011;25:4050–62.
44. Kang S, Gu B, Du T, Zhang J. Crop coefficient and ratio of transpiration to evapotranspiration of winter wheat and maize in a semi-humid region. Agric Water Manag. 2003;59:239–54.
45. Li YL, Cui JY, Zhang TH, Zhao HL. Measurement of evapotranspiration of irrigated spring wheat and maize in a semi-arid region of north China. Agric Water Manag. 2003;61:1–12.
46. Liu Y, Luo Y. A consolidated evaluation of the FAO-56 dual crop coefficient approach using the lysimeter data in the North China Plain. Agric Water Manag. 2010;97:31–40.
47. Zhang B, Liu Y, Xu D, Zhao N, Lei B, Rosa RD, Paredes P, Paço TA, Pereira LS. The dual crop coefficient approach to estimate and partitioning evapotranspiration of the winter wheat–summer maize crop sequence in North China Plain. Irrig Sci. 2013;31:1303–16.
48. Tyagi N, Sharma D, Luthra S. Determination of evapotranspiration and crop coefficients of rice and sunflower with lysimeter. Agric Water Manag. 2000;45:41–54.
49. Jiang X, Kang S, Tong L, Li F, Li D, Ding R, Qiu R. Crop coefficient and evapotranspiration of grain maize modified by planting density in an arid region of northwest China. Agric Water Manag. 2014;142:135–43.
50. Piccinni G, Ko J, Marek T, Howell T. Determination of growth-stage-specific crop coefficients (KC) of maize and sorghum. Agric Water Manag. 2009;96:1698–704.
51. Suyker AE, Verma SB. Evapotranspiration of irrigated and rainfed maize–soybean cropping systems. Agric For Meteorol. 2009;149:443–52.
52. Thao T. Developing crop coefficients (kc) for sugar beet (beta vulgaris l.) grown under drip irrigation using weighing lysimeters. 2017.
53. Abyaneh HZ, Farrokhi E, Varkeshi MB, Ahmadi M. Determination of water requirement and effect of its variations on some quantitative and qualitative traits of sugar beet product. J Sugar Beet. 2012;27:21–7.
54. Wahab AA, Salih AA. Water requirements of sugar beet Beta vulgaris under heavy cracking clay soils. J Agric Sci Technol. 2012;2:865.
55. López-Urrea R, de Santa Olalla FM, Montoro A, López-Fuster P. Single and dual crop coefficients and water requirements for onion (Allium cepa L.) under semiarid conditions. Agric Water Manag. 2009;96:1031–6.
56. Al-Jamal M, Sammis T, Ball S, Smeal D. Yield-based, irrigated onion crop coefficients. Appl Eng Agric. 1999;15:659.
57. Snyder RL, Lanini B, Shaw D, Priott W. Using reference evapotranspiration (ETo) and crop coefficients to estimate crop evapotranspiration (ETc) for agronomic crops, grasses, and vegetable crops. Leaflet-University of California, Cooperative Extension Service (USA);1987.

58. Chattaraj S, Chakraborty D, Garg R, Singh G, Gupta V, Singh S, Singh R. Hyperspectral remote sensing for growth-stage-specific water use in wheat. Field Crop Res. 2013;144:179–91.
59. Belmonte AC, Jochum AM, GarcÍa AC, Rodríguez AM, Fuster PL. Irrigation management from space: towards user-friendly products. Irrig Drain Syst. 2005;19:337–53.

Land Suitability of Various Irrigation Techniques for More Effective Water Usage and Spatial Estimation of Soil Water Content based on Computational Intelligence

Orhan Dengiz

Abstract

Land and soil conservation measures and effective soil–water management in arable lands, as well as monitoring the actual situation are critical. The need for improved water management at the field scale through the adoption of more effective irrigation techniques is great due to the escalating climate instability, which multiplies the effects of traditional surface irrigation methods on land degradation on the one hand, and the rising crop yield demands from expanding populations on the other. Sustainable land use necessitates sustainable irrigation; nevertheless, each soil has a variable vulnerability to irrigation-induced degradation processes. Several scientists suggested a parametric evaluation approach for comparing different irrigation strategies on a certain land. It can be used to determine, which fields are particularly sensitive (e.g., to soil erosion) and, as a result, need to have their irrigation systems altered. Alternate furrow irrigation as well as other low-cost approaches like deficit irrigation and partial root drying can significantly enhance water use efficiency in the field. In addition, in the face of agricultural drought, examining soil–plant–water relationship is a critical step toward effective water management. In this context, soil physico-chemical and hydrological features have a greater impact on water availability in the soil and irrigation practices. Admittedly, in large size locations, intensive soil sampling is unaffordable and uneconomic. That's why a wide range of environmental and agricultural hydrological models such as PTFs, Splintex, LLWR, Hydrus, SWIM, etc., have created and applied to quantify and to integrate the most essential physical, chemical, and hydrological processes in the unsaturated soil zone. Also in recent years, machine learning methods such as ANN, SVM, logistic regression, DT, RF, and DL have also been widely used to determine soil water content. Moreover, understanding the spatial, temporal and seasonal variability in soil water retention qualities is essential for precision water inputs into especially arid and semi-arid cultivated soils. That's why researchers have successfully developed geostatistics and interpolation methods integrated with GIS for spatial estimating soil hydraulic properties.

Keywords

Land suitability · Irrigation techniques · Hydrological models · Computational intelligence · Geostatistics · GIS

O. Dengiz (✉)
Faculty of Agriculture, Department of Soil Science and Plant Nutrition, Ondokuz Mayıs University, Samsun, Türkiye
e-mail: odengiz@omu.edu.tr

M. Bahadir et al. (eds.), *Agricultural Water Management*, Water and Wastewater Management,
https://doi.org/10.1007/978-3-031-96134-2_7

1 Introduction

Since land and water resources are limited, conventional farming applications are being replaced with more effective new technologies to serve the needs of sustainable agriculture. Mihalikova and Dengiz [1] reported that in recent decades, agricultural water management, particularly irrigation, has made a considerable contribution to global food security, food production, and rural farmer incomes. For instance, it has been projected that without irrigation, the global production of grain crops would be 20% lower [2]. Therefore, land conservation measures and effective soil water management in cultivated areas, as well as monitoring the current situation are critical. Due to the rising instability of the climate, which multiplies the soil degradation effects created by traditional surface irrigation techniques on the one hand, and the rising demands for crop products from growing populations on the other, improved water management at the field scale through the implementation of more useful irrigation techniques is in high demand.

Land degradation caused by irrigation practices that involve in particular soil salinization, soil erosion, and groundwater pollution is a major problem worldwide. Furthermore, incorrect agrotechnical practices resulting in substantial water footprints stem from a combination of factors including insufficiently priced water, subsidies, utilization of outdated or inefficient water technologies, and farmers' insufficient awareness about fundamental water conservation principles. As a result, agricultural productivity is on the decline. According to United Nations Educational Scientific and Cultural Organization–World Water Assessment Programme [3], a total of 18% of the world's cropland is used for irrigated agriculture, which uses roughly 70% of all freshwaters to produce around 40% of the world's food. That's why, sustainable land use necessitates sustainable irrigation; nevertheless, each soil has a variable vulnerability to irrigation-induced degradation processes. Several scientists suggested a parametric evaluation approach for comparing different irrigation strategies on a certain land [1, 4–6]. It can be used to determine which fields are particularly sensitive (especially to soil erosion) and, as a result, need to have their irrigation systems altered. Alternate furrow irrigation as well as other low-cost approaches like deficit irrigation and partial root drying, can significantly enhance water use efficiency in the field. As an example, by using parametric evaluation, Dengiz [7] offered a method for comparing several irrigation techniques on a specific field. Also, according to Cakmak [8] and de Fraiture [2], land features were among the factors that needed to be assessed for selecting, designing, and putting into practice the best irrigation and drainage methods in order to provide effective water resource utilization and conservation.

In addition, in the face of agricultural drought, examining soil–plant–water relationships is a critical step toward effective water management. In this context, soil physicochemical and hydrological features have a greater impact on water availability in the soil and irrigation practices. Obviously, intensive soil sampling is neither feasible nor economically in great scaled fields. That's why, a wide range of environmental and agricultural hydrological models such as Pedotransfer functions (PTFs), Splintex, Least Limiting Water Range (LLWR), Hydrus, Soil and Water Integrated Model (SWIM), etc., have been developed and applied to quantify and to integrate the most essential physical, chemical, and hydrological processes in the unsaturated soil zone in order to reduce laborious, time-consuming, and expensive methods. Mihalikova [9] indicated that growth in pedotransfer functions (PTFs) during the past few decades has produced more or less accurate estimates of the necessary modeling inputs. Moreover, in comparison with determining soil moisture constants, LLWR determinations are highly challenging and time-consuming. From the past to the present, pedotransfer functions have been used in research on the LLWR predictability [10–13]. Also, in recent years, machine learning approaches such as Artificial neural networks (ANN), Support vector machines (SVM), logistic regression, Decision Trees (DT), Random Forest (RF), and Deep Learning (DL) have also been widely used to determine soil water content. Moreover, understanding the spatial, temporal and seasonal variability in soil water retention qualities is essential for precision water inputs into especially arid and semi-arid cultivated soils. That's why scientists have successfully developed geostatistics and interpolation methods integrated with GIS for spatial estimating soil hydraulic properties.

As a result, soon, existing water resources will not be able to meet a variety of needs, which will undoubtedly prompt the search for alternative lands for irrigation to guarantee sustainable global food security. Therefore, for a sustainable land management of irrigated agricultural land, it is necessary to determine irrigation methods by considering the soil and land characteristics, and to develop new irrigation technologies. The main goals of this article are to explain spatial estimation of soil water content based on computational intelligence and to contribute to the decision and planning process of irrigation management by a land suitability assessment for different irrigation methods by applying various approaches.

In this article, the main goals of this are to explain spatial estimation of soil water content based on computational intelligence and to contribute to the decision and planning process of irrigation management by a land suitability assessment for different irrigation methods by applying various approaches. Thus, it will be an important resource and provide to help decision makers, scientists related to this field or farmers to make more accurate decisions for sustainable land, soil and water management.

2 Land Suitability for Various Irrigation Techniques

The suitability of a piece of land is determined by how well it can naturally support the intended use. In other words, the evaluation and classification of certain land parcels according to their suitability for a given application is the land suitability classification process. According to the FAO methodology [14], land suitability is closely tied to "land attributes," which include intangible characteristics such as erosion resistance, water availability, and flood threats. These properties are influenced by the quantifiable or estimable land parameters, such as slope degree and length, rainfall, and soil texture. It is also helpful to study the appropriateness using these latter values. Thus, the land characteristics parameters were used to workout land suitability for irrigation. For determining a site's suitability for effluent irrigation and a piece of land's suitability for irrigation water, Bond [15] and Dengiz [7] created classification systems. Both methods take topographic factors and soil characteristics into account when determining whether a piece of land is suitable for irrigation in general. A parametric evaluation methodology for irrigation techniques was proposed by Sys et al. [16], and it was mostly based on the chemical and physical characteristics of the soil [4, 7, 17]. The four divisions they suggest for dividing the variables influencing a soil's suitability for irrigation are as follows:

- Physical characteristics of the soil such as soil texture, permeability and the amount of available water;
- Chemical characteristics such as exchangeable and sodium soluble salts, that affect the salinity/alkalinity state as well as CaCO3 content;
- Drainage characteristics; and
- Topographical elements like slope and drainage features.

For effective irrigation farming, selecting the right irrigation technique is crucial for better nutrient and pesticide control in crop production, efficient water usage, and reduction in land and water damage. However, in the case of intensive water use, irrigation practice has the potential to have a significant impact on the quality of the land and water. Particularly, the extensive use of water affects how water is distributed throughout the environment and has an impact on the movement of pollutants (such as pesticides and nutrients), compaction, salinization, waterlogging, and erosion, among others. Compatibility between the land and water is crucial for irrigation.

The applied irrigation water may adversely affect the chemical and physical characteristics of the soil if they are incompatible. Seelig and

Franzen [18] indicated that a detailed assessment of the soil characteristics, topography, and irrigation water quality is necessary to determine whether a piece of land is suitable for irrigation. Irrigators can more effectively manage their crops, soil, irrigation systems, and water sources, if they have a fundamental understanding of the relationships between water, soil, and plants.

Applying excellent irrigation management strategies will enable to maximize yield with the water available. Additionally, timely water application and good water management may stop the degradation of the land. Some methodological approaches for land suitability of various irrigation techniques have been developed, and two of the most used of these methodological approaches are parametric approach and multi-criteria assessment analysis.

2.1 Parametric Evaluation Approach

Mathematical formulas are used in parametric systems, which have a single category and produce a numerical result. According to McRae and Burnham [19], parametric approaches are widely acknowledged to be the most accurate: "*simple, objective, quantitative, reliable, easy to understand and apply, even by the non-specialist, and easy to modify and adapt to new uses.*" There are three basic types of manipulation that may be distinguished: additive, multiplicative, and complex functions such as Storie [20], Square root [16], Productivity index [21], and so on. In a parametric calculation, every aspect that has an impact on the land use potential is given a numerical value that typically ranges between 0 and 100 or between 1.0 (for the greatest potential) and almost 0.0 (for the smallest potential). The components included in the system (both in terms of nature and number of parameters) and the mathematical manipulation may change.

Many investigators from all around the world have researched land suitability evaluation for various forms of irrigation using the parametric evaluation approach. As an example, Bienvenue [22] assessed land suitability for drip and surface irrigation techniques in Thies, Senegal. According to their findings, the main restricting criteria, which led to S2 (Moderately suitable) to N2 (Non suitable) classes for the type of irrigation system, were demonstrated to be soil drainage and texture. Because, in order to ensure good drainage in soil, soil texture should be of medium texture such as clay loam. In soils with very heavy texture (>45% clay) or coarse texture (such as sandy loam or loamy sand texture), water moves either very slowly or very quickly. Due to the soil air condition and water holding capacity, it generates a bad environment for plant growth in both situations. In addition, in the Tunisian Oued Rmel Catchment, Mbodj [23] examined the appropriateness of the land for surface irrigation and drip irrigation. As a result of the topographic, soil, and drainage restrictions associated with surface irrigation, their findings indicated that drip irrigation was more suitable than surface irrigation. Also, Albaji [17] studied several irrigation techniques using a parametric approach in Iran's west-central Ilam Province's Abbas Plain. Sprinkler and drip irrigation were shown to be more effective and efficient than surface irrigation for increasing land production, according to comparisons of various types of irrigation systems. However, soil calcium carbonate content and drainage were the key limiting factors when using sprinkler/drip and surface irrigation in the research region.

Based on identifying land qualities/characteristics, such as soil physical properties (soil texture, gravel percent, and soil depth), soil chemical properties (calcium carbonate and gypsum content, soil salinity and alkalinity), land terrain (slope and drainage), and climate characteristics, such as wind velocity affecting sprinkler irrigation, the land suitability evaluation in this approach is conducted.

By using this parametric evaluation Dengiz [7] offered a technique to compare various irrigation methods within a particular field, enabling the identification of fields that are particularly susceptible and may require modifications to their irrigation systems. In his study, different irrigation methods were compared including surface and drip irrigation in the Central Research Institute, south of Ankara, Türkiye. In his study, the

Table 1 Suitability classes for the irrigation capability indices (Ci) classes. Dengiz [7]

Capability index	Definition	Symbol
>80	Highly suitable	S1
60–80	Moderately suitable	S2
45–59	Marginally suitable	S3
30–44	Currently not suitable	N1
<29	Permanently not suitable	N2

parametric evaluation technique of Sys [16] was applied to assess the suitability of the area for various irrigation methods utilizing the soil properties. According to Eq. 1, these qualities are appraised and used to determine the irrigation capability index (Ci):

$$\mathrm{Ci} = \mathrm{A} * \mathrm{B}/100 * \mathrm{C}/100 * \mathrm{D}/100 * \mathrm{E}/100 * \mathrm{F}/100 \quad (1)$$

where; Ci = Capability index for irrigation; A = soil texture rating, B = soil depth rating, C = $CaCO_3$ status, D = electro-conductivity rating, E = drainage rating, F = slope rating.

According to the score of the capacity indices, suitability classes are described and given in Table 1.

Dengiz [7] created six thematic layers by digitally encoding all land and soil features in a GIS database along with the corresponding attribute data. The diagnostic factors of each thematic layer were assigned score of factor rating described in Tables 2 and 3.

The resultant layers were created by spatially layering these six layers. Figure 1 shows a schematic diagram of the spatial overlay that displays the features of the land and soil.

According to findings, 51.2% of the study area (Central Research Institute) was extremely suited for the drop irrigation method, compared with 13.1% of the study area for surface and gravity irrigation (Fig. 2).

Also, Dengiz concluded that the drip irrigation method increased land suitability by 38% compared to the surface irrigation method. In the research region, soil salinity, drainage, and soil texture were the most significant limiting factors for surface irrigation, whereas drainage and soil

Table 2 Rating values of the soil texture class for the three irrigation systems. Dengiz [7]

Texture classes		CL	SiL	SCL	L	SiL	Si	SiC	C	SC	SL	LS	S
Surface irrigation system													
Fine gravel (%)	<15	100	100	95	90	90	90	85	85	80	75	55	30
	15–40	90	90	85	80	80	80	95	95	90	65	50	25
	40–75	80	80	75	70	70	70	80	80	75	60	45	25
Coarse gravel (%)	15–40	80	80	75	70	70	70	80	80	75	60	45	25
	40–75	50	50	45	45	45	45	40	40	35	35	25	25
Sprinkler irrigation system													
Fine gravel (%)	<10	100	100	95	90	90	90	85	85	80	75	50	25
	10–35	90	90	85	80	80	80	95	95	90	65	45	25
	35–80	80	80	75	70	70	70	80	80	75	60	35	25
Coarse gravel (%)	10–35	80	80	75	70	70	70	80	80	75	60	35	25
	35–80	50	50	45	45	45	45	40	40	35	35	25	25
Drip irrigation system													
Fine gravel (%)	<15	100	100	95	90	90	90	85	85	95	95	85	70
	15–40	90	90	85	80	80	80	95	95	90	85	75	65
	40–75	80	80	75	70	70	70	80	80	85	80	55	50
Coarse gravel (%)	15–40	80	80	75	70	70	70	80	80	80	75	60	35
	40–75	50	50	45	45	45	45	40	40	35	35	35	35

Table 3 Other parameters' class rating Dengiz [7]

Slope classes (%)	Rating for gravity irrigation		Rating for drop irrigation	
	Non terraced	Terraced	Non terraced	Terraced
0–1	100	100	100	100
1–3	95	95	100	100
3–5	90	95	100	100
5–8	80	95	90	100
8–16	70	85	80	90
16–30	50	70	60	70
>30	30	50	40	50
Drainage classes	C, SiC, SC	Other textures	C, SiC, SC	Other textures
Well drained	100	100	100	100
Moderately drained	80	90	100	100
Imperfectly drained	70	80	80	90
Poorly drained	60	65	70	80
Very poorly drained	40	65	50	65
Drainage status not known	70	80	70	80
EC (dS/m)	C, SiC, S, SC	Other textures	C, SiC, S, SC	Other textures
<4	100	100	100	100
4–8	90	95	95	95
8–16	80	50	85	50
16–30	70	35	75	35
>30	60	20	65	20
$CaCO_3$ (%)				
<0.3	90		90	
0.3–10	95		95	
10–25	100		95	
25–50	90		80	
>50	80		70	
Soil depth (cm)				
<20	30		30	
20–50	60		70	
50–80	80		90	
80–100	90		100	
>100	100		100	

salinity were the main limiting factors for drip or targeted irrigation. Thus, Dengiz demonstrated which parts of the study area comply with surface or drip irrigation techniques, by considering the mentioned land and soil limiting constraints, and he offered solutions for ensuring sustainable use of water and land-soil resources.

2.2 Hierarchical/Multicriterial Assessment

Investigations of land assessment are crucial for directing decisions on how to use land in a way that will maximize its potential and preserve natural resources for future generations [24].

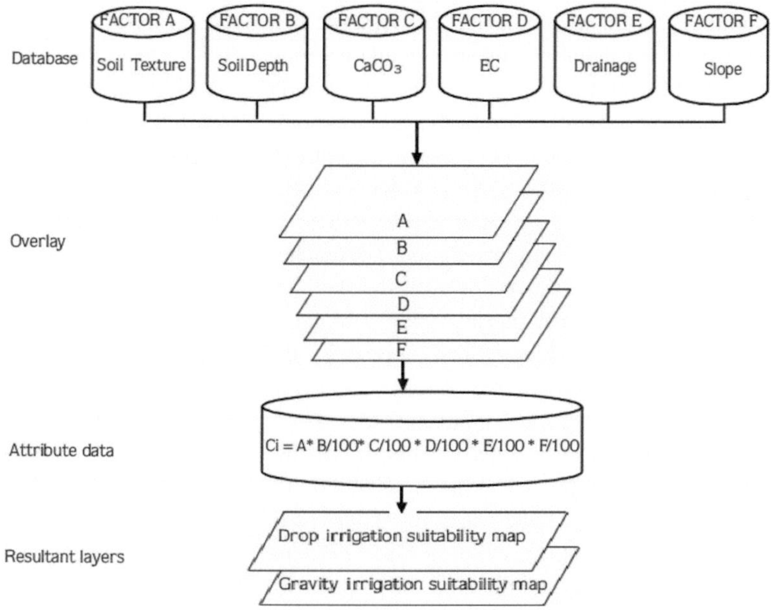

Fig. 1 Schematic diagram of the spatial overlay [7]

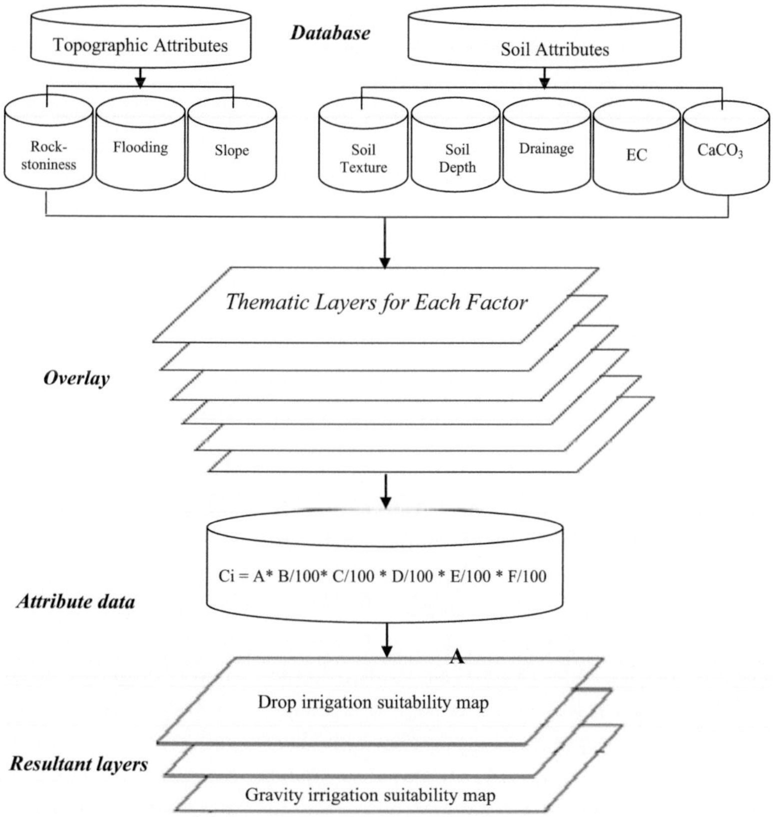

Fig. 2 Drop and gravity irrigation suitability maps of the study area [7]

Additionally, Dengiz and Usul [25] reported that the idea of sustainable land use entails creating high-quality products in a way that is socially acceptable, ecologically friendly, and economically efficient, while also assuring optimal exploitation of the natural soil and water resources available for effective agricultural production. As mentioned before, land evaluation approaches can be separated into two categories, which are the hierarchical/multicriterial and parametric techniques. In the hierarchical approach, the classes are organized into a hierarchy of levels of importance by categorical systems (order, class, subclass, type, etc.). In other words, hierarchic systems, for example the Analytical Hierarchy Process (AHP) [26], Land Capability Class [28], Suitability Class for Agricultural Use [27], and FAO [14] systems, classify land into categories with a different land suitability.

The suitability of some parts of land for different irrigation techniques depends on a variety of factors. The relative relevance levels of these criteria that conflict with one another is, therefore, difficult to ascertain. Multi-Criteria Assessment (MCA) methodologies offer resolving complicated geographic issues by managing and analyzing huge volumes of spatial data for land evaluation of heterogonous natural and system [29]. The AHP method is one of the MCA techniques and is widely employed in literature. AHP is a weighted factor rating approach that can find and consider decision-making discrepancies. Therefore, it has been considered to a variety of decision-making problems including the assessment of alternatives. In a typical AHP approach, specialists must assign a specific number between 1 and 9 to the pairwise comparison to construct the priority vector. As an example, Mihalikova and Dengiz [1] performed a study at an area of 972.2 ha in Çarşamba Delta Plain located in the central Black Sea Region of Türkiye to evaluate considering two approaches, parametric evaluation system (PM) and multi criteria assessment (MCA). These approaches can be employed in the decision and planning process of irrigation management by land suitability evaluation for 3 different irrigation techniques. To recommend an appropriate irrigation system and to compare these approaches, the PM and MCA techniques were both applied for land suitability assessment. Using a geographic information system (GIS), suitability maps for drip, sprinkler and surface irrigation technologies were created after the land and soil characteristics had been examined and assessed (Fig. 3).

Because of the analytical hierarchy approach, MCA strategy was more adaptable and sensitive, while PM approach was quicker and simpler to apply. However, significant differences between the two approaches were not discovered. In both, drip irrigation capability ratings were at their maximum across all land mapping units in soil series maps. Slope, soil depth, soil and texture were the key restrictions for sprinkler irrigation systems as well as for surface irrigation systems. On the other hand, the steep slope was the principal restriction for drip watering.

The comparison of the result maps (Fig. 4) shows that a change in irrigation management tactic would offer the best ecological answer, as the employment of drip and sprinkler irrigation methods would increase the effectiveness of water use and save the soil. Although the surface irrigation method is widely used and the old method in the study area, the other two irrigation systems are advised for sustainable use of water and soil resources, which are essential for the ecohydrological equilibrium of this ecologically significant maritime region. Moreover, when comparing the two assessment techniques, MCA outperformed PM by an average of 36%, 19% and 48% for surface, drip, and sprinkler irrigation systems, respectively. Due to the AHP process' inclusion of an expert knowledge system, it was discovered that the MCA technique was more sensitive and adaptable than the parametric approach, which merely employs rating scores in equations.

Main limiting parameters in using all three irrigation techniques in this field were terrain properties (drainage, slope) and some soil features (depth and texture). Findings demonstrated that arability of 33% or 30% of the study area (according to PM or MCA, respectively) would be improved by applying drip irrigation method instead of surface and sprinkler irrigation methods. The comparison

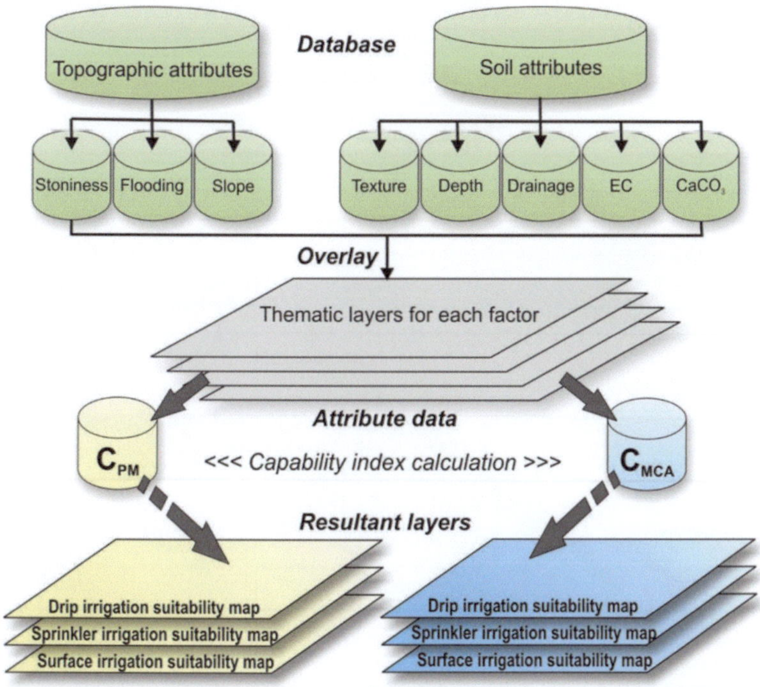

Fig. 3 Follow diagram of GIS application for the three different irrigation techniques suitability maps gained by the two different methods [1]

of the suitability of the three types of irrigation systems revealed that it would be more beneficial to irrigate by drip type, when based on the two methods, as improves all the land suitability by taking precautions based on limitations factors, particularly from currently and permanently non suitable areas for surface and sprinkler irrigation systems to be suitable for drip irrigation. The results demonstrated that by applying drip irrigation methods instead of sprinkler and surface irrigation methods, about the arability of 319 ha in study area would improve (by PM). Moreover, this case also valid for MCA approach that approximately 289 ha of the study area would be more suitable by applying drip irrigation method instead of sprinkler and surface irrigation methods. Thus, soil and water resources can be used efficiently and degradation processes and their effects such as salinization, erosion, etc., can be reduced.

3 Spatial Estimation of Soil Water Content Based on Computational Intelligence

Due to limited land resources, old or traditional farming techniques are being replaced with more effective modern ones to meet the demands of sustainable agriculture. It is crucial to monitor the existing situation and to improve soil conservation measures and practical soil water management in farmed regions. Geographic information systems (GIS) and hydrological and hydro-pedological modeling are being utilized to examine the existing situation and to simulate various scenarios in this subject. But the models need information about the hydraulics of the soil, such as the hydraulic conductivity and soil water retention curve. Mihalikova [11] reported that for the purpose of conservation planning for the sustainable use of soil and water resources, many watershed basins lack appropriate data. Also, they indicated that in a wide study region, extensive soil sampling is neither practical nor cost-effective, so PTFs (pedotransfer functions) and geostatistical or interpolation models play crucial roles in data

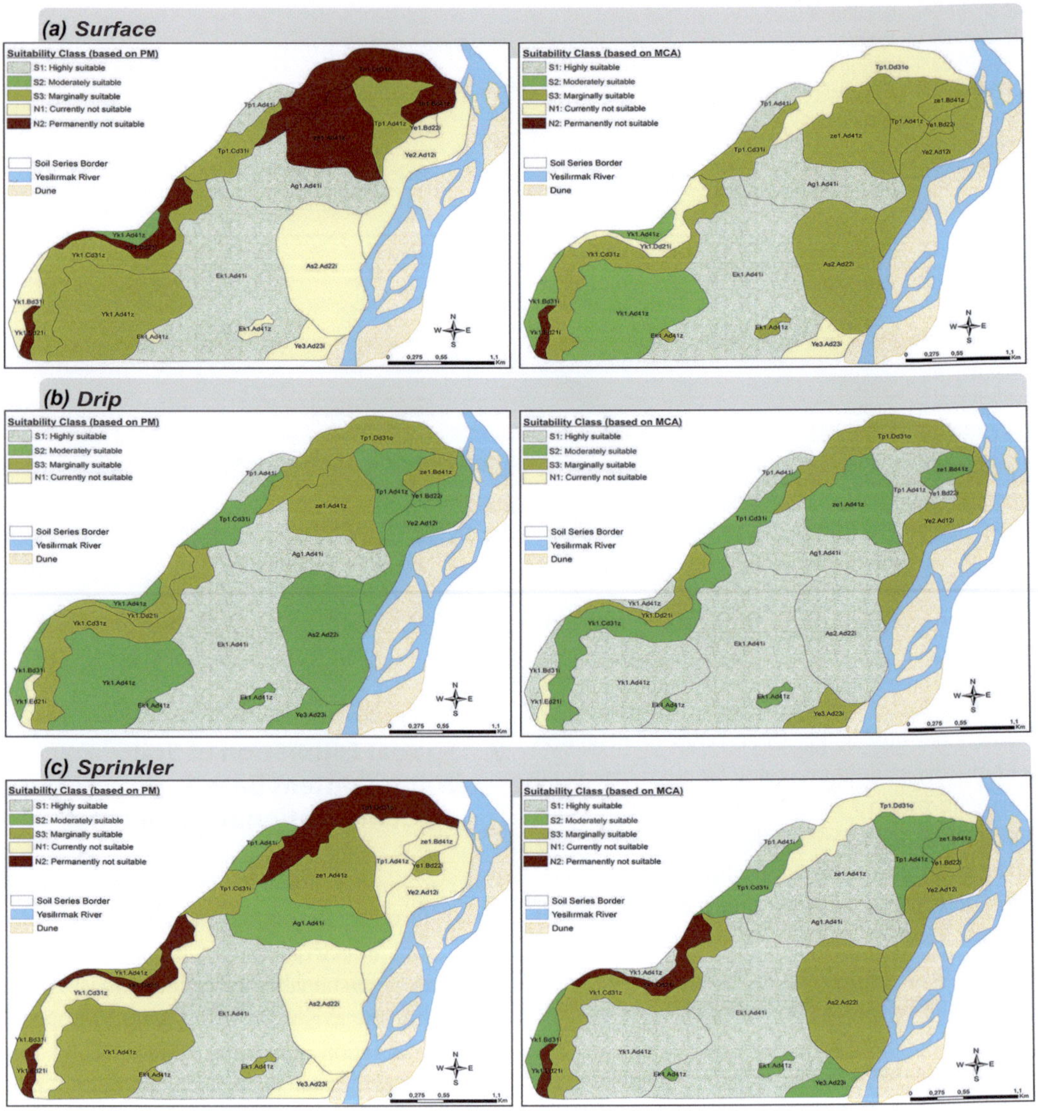

Fig. 4 Land suitability maps for surface, drip and sprinkler irrigation systems [1]

mining. That's why, using several interpolation geostatistical models and PTFs, these researchers investigated to estimate and to evaluate the estimations of fundamental soil hydraulic parameters in Büyükçay Watershed in Türkiye. The results show that ordinary co-kriging produces better results for all soil hydraulic characteristics with lower MAE and MSE values than ordinary kriging and IDW. Although class PTFs shown less sensitivity to spatial variability than geostatistical approaches, soil hydraulic characteristics inside PTFs were estimated with good reliability using a continuous approach.

Mihalikova [9] studied about mapping soil water retention on cultivated areas in Central and Eastern parts of the Black Sea Region in Türkiye. They indicated that due of the soils' susceptibility to available water capacity and water erosion, mapping and risk assessment are becoming increasingly important. For this reason, knowledge of the soil hydraulic properties is necessary, although little of it has been discovered locally. That's why, utilizing 3,400 disturbed soil samples from the surface layer, the cultivated area of 8 provinces (Gümüşhane, Samsun, Ordu, Sinop, Rize, Giresun, Artvin and Trabzon) was studied. For the purposes of estimating field capacity (FC), permanent wilting point (WP) and accessible water capacity (AWC), ten models of PTFs have been tested that are commonly available in the literature. Based on information from five test locations, the accuracy of estimations was examined and evaluated (382 samples) and data from five testing locations were used to evaluate the estimation uncertainty. In addition, to interpolate the predicted values throughout the study area, they used suitable geostatistic methods and created maps of their spatial distribution using GIS. According to their results, it was challenging to predict FC in places with significant clay concentration; most models underestimated it. However, several PTFs approximated WP with acceptable reliability. They concluded that AWC estimation was not appropriate. The software k-Nearest based on the k-Nearest neighbor algorithm was the best-performing PTF in the study. When the original reference data set was replaced with local data (from testing localities), and only sand, silt, and clay concentrations were employed as predictors, they achieved the best correlation coefficient (R), root mean square error (RMSE) and mean absolute error (MAE). Moreover, after separating the data into two subsets, one for estimation and the other for evaluation the reliability of estimation was checked, and the full data set was then used to calculate the data on water retention in the research area. The researchers reported that authorities, decision makers as well as farmers in irrigated fields can use the final maps (Fig. 5) and the calculated findings in the decision-making processes for soil conservation and land management methods.

According to McBratney [30], the rise in PTFs during the past three decades has produced more or less accurate estimates of the necessary modeling inputs. Mohammed [31] examined images of soil profiles, soil horizons and soil samples from the U.S. National Cooperative Soil Survey database and used different shape metrics to quantify the kind of soil structure. The quantified data were then used to the estimation of FC and WP. Also, Toth [32] created a new generation of hydraulic PTFs for Europe that are hierarchical, allow point or parameter estimate depending on the quantity of inputs available, and consider both chemical and physical soil features.

Alaboz et al. [13] stated that a crucial strategy for efficient water management in the face of agricultural drought is to examine the relationships between soil, plants, and water. Also, well recognized is the direct relationship between sufficient water in the plant root zone and plant development and yield. Therefore, the soil's production capacity, which is the environment in which plants flourish, is directly correlated with its physical features, including root penetration resistance, porosity, aeration, and soil water content. Water movement is impacted by variations in pore size distribution brought on by soil surface and root area compression as well as aeration. In this instance, soil bulk density has a greater impact on soil water availability than water content at the wilting point and field capacity.

With the advancement of technology, algorithms, which are used in every sector today, have taken on a central role in our lives. The applications of machine learning algorithms or artificial neural network methods are being researched considering recent advancements in the agricultural sector. These techniques are quick and efficient at resolving complicated situations. When it comes to plant water uptake and root development in agricultural production, the least limiting water range (LLWR) aspect is crucial. An indicator of the suitability of the soil structure for plant growth is the LLWR index. The LLWR is defined as the range in soil water content after rapid drainage has ceased, within which restrictions to plant growth related to water potential, aeration, and mechanical resistance are at their lowest, according to Da Silva and Kay [10]. It is well known that the moisture content at which the aeration porosity is 10% or FC moisture level is the upper limit of the effect on root growth. The mechanical resistance which is considered as the limiting factor for plant growth, known as WP or soil moisture

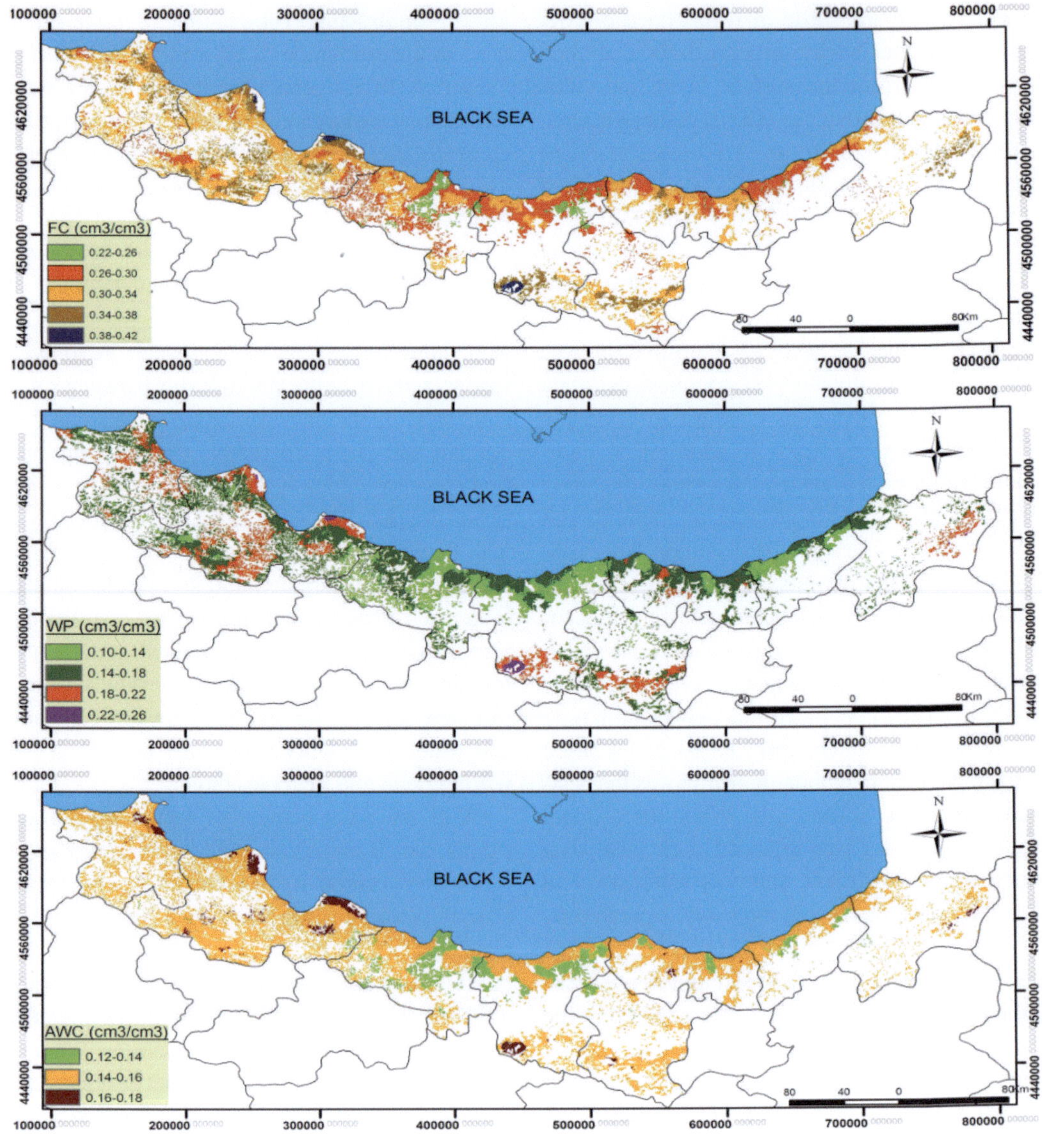

Fig. 5 Spatial distribution maps of the FC, WP, and AWC for in Central and Eastern parts of the Black Sea Region [9]

level, is assessed as the minimum threshold and corresponds to a soil pressure of 2 MPa. Soil compaction ought to have an impact on the LLWR based on the various soil management techniques.

Alaboz examined the predictability of the LLWR properties by investigating deep learning (DL), artificial neural networks, and the k-nearest neighbour (k-NN) algorithm from machine learning methods. The fieldwork for this study was done in the Terme region of the Turkish province of Samsun, in the eastern part of the Çarşamba Plain, where alluvial deposits carried by the Terme River are distributed. In their study, field capacity (θFC), the water content at which the aeration porosity is 10% (θAp), permanent wilting point (θWP), and the soil water content at the wetness when penetration resistance (PR) is equivalent to 2 MPa (θPR) were used to derive four potential ways to calculate LLWR (Fig. 6).

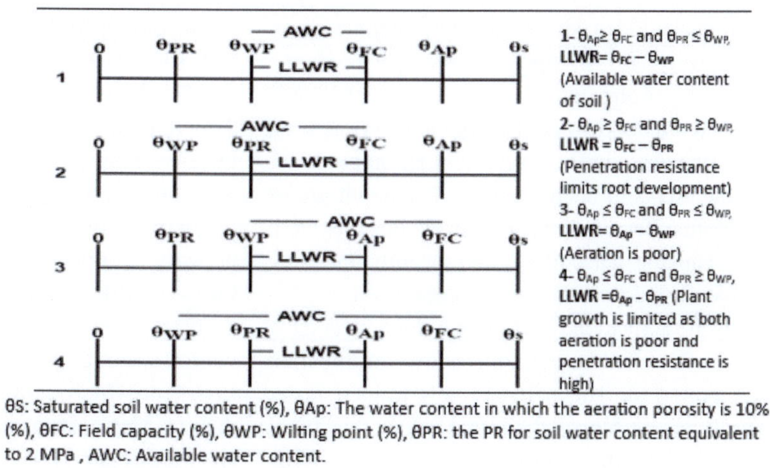

Fig. 6 Possibilities of determining LLWR value [13]

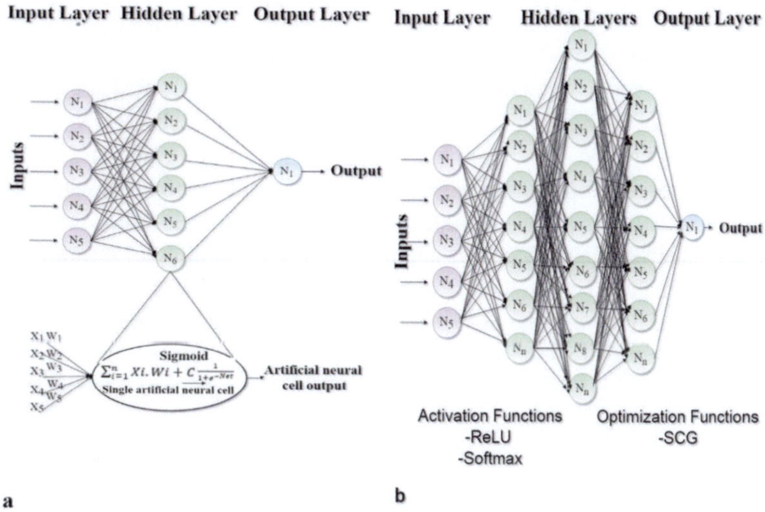

Fig. 7 The architecture of the artificial neural network and deep learning [13]

Alaboz et al. [13] were performed to estimate LLWR using k-Nearest Neighbours algorithm (k-NN), Deep Learning (DL), Artificial Neural Network (ANN) from machine learning methods. Simple processor cells (neurons) make up artificial neural networks (ANNs), which are trained to store data and to operate in parallel and scattered order. An artificial neural network cell has inputs (X1, X2, Xi), weights coded as W1, W2, Wi, a summation function, an activation function, and outputs (Y) (Fig. 7a). The weights and inputs are multiplied to provide the input value (NET). Equation 2 provides the formula that uses "summation" as the NET calculation method with multi-layer feed-forward back propagation networks (Eq. 3) and Levenberg–Marquardt as the training technique, the sigmoid activation function was used in the current study.

$$\mathrm{NET} = \sum_{i=1}^{n} \mathrm{Xi.\ Wi} + \mathrm{D} \tag{2}$$

$$\mathrm{f(NET)} = \frac{1}{1 + e^{-\mathrm{NET}}} \tag{3}$$

where: NET: input value is obtained by multiplying the weights with the inputs, Xi: inputs, Wi: weights, D: bias.

There are three different kinds of layers in DL: the input layer, the output layer, and any additional hidden levels that may include many layers (Fig. 7b). An input layer as well as often a hidden layer and output layer are present in artificial neural networks. The input layer, at least three or more hidden layers, and an output layer are the layers of deep learning. The depth grows as the number of concealed levels does as well. DL is capable of accurately and speedily resolving problems that are extremely complicated.

As the activation function in hidden layers, the ReLU (Rectified Linear Unit) activation function was preferred. The network gets trained faster for all values bigger than zero when this function is used, which is the main advantage. The formula for this function is f(x) = max (x, 0). The outcome is X, if the function's output X is more than zero, and 0 if it is less than zero. In the output layer, the softmax function was utilized (Eq. 4). For issues involving multilayer classification, this function is favored. They used Scaled Conjugate Gradient back propagation (SCG) algorithm as an optimization function. 20 hidden neurons, 40 hidden neurons, and 20 output neurons were all included in the ANN. In their research, they used a batch size of 10 and the SCG optimizer to train the network over 15 epochs.

$$Y = \mathrm{Soft}_{\mathrm{max}} X_{(\mathrm{input})} = \frac{e^{x(\mathrm{input})}}{\sum \mathrm{Xn}\ \ e^{\mathrm{Xn}\ (\mathrm{input})}} \quad (4)$$

In the equation, Y = output of layers or neurons; X = input layer to hidden neurons; Xn = input layer n to hidden neurons.

The K-Nearest Neighbors algorithm determines the nearest neighbor, and the data point class based on the K value. How many closest observations will be utilized to estimate the new observation is indicated by the letter K. Neighborhood distance is most often measured in terms of Euclidean distance (Eq. 5).

$$d(Xi, Yi) = |Xi - Yi| = \sqrt{\sum_{i=1}^{n} (Xi - Yi)^2} \quad (5)$$

where; d: distance, Xi and Yi two input vectors, distances of Xi and Yi points

According to their results, 31.7% of the soils had LLWR contents higher than 15%, while 68.3% of those soils had LLWR contents lower than 15%. In the input layer, the soil parameters FC, clay, bulk density, organic matter, and $CaCO_3$ were employed, since they had a strong connection with LLWR. The DL method's 5: 20:40:20:1 network architecture, along with the best R^2, lowest RMSE, and MAE values, provide the most accurate estimation. It was revealed that ANN estimation accuracy was higher than k-NN. Using the Levenberg–Marquardt backprop learning procedure in ANN, estimated values that were close to DL were obtained. The k-NN estimation approach was shown to have the lowest estimation errors, which were found to be around 2.5%. DL was determined as the best estimation method (MAE: 0.94%; RMSE: 1.45%; R2:0.93) and the worst one was k-NN (MAE: 2.00%; RMSE: 2.55%; R2: 0.77) for LLWR. In addition, LLWR can be estimated with high accuracy by using ReLU and Softmax functions in the DL method even in alluvial terrains with a complicated distribution and significantly varied soil qualities. In addition, a comparison of the spatial distribution map produced using estimated values, and the map produced using observed values revealed a change that was quite similar (Fig. 8). Consequently, researchers reported that deepening the network structure in artificial neural networks resulted in a more accurate estimation. In addition, utilizing the LLWR estimated values acquired with DL, spatial distribution maps in alluvial soils were effectively constructed.

4 Conclusions

Nowadays, it is well known that due to limited land and water supplies, traditional agricultural techniques are being replaced with more efficient contemporary technologies to meet the demands of sustainable agriculture. Therefore, it is essential to take steps to conserve land and to manage soil and water in arable areas as well as to monitor the real situation. The increasing climate instability, which multiplies the effects of conventional surface irrigation methods on land degradation on the one hand, and the rising crop yield demands from expanding populations on the other, make improved water management at the land scale through the adoption of more effective irrigation techniques imperative. In addition, examining the interactions between soil, plants, and water is also a crucial first step toward efficient water management in the face of agricultural drought. In this sense, water availability in the soil and irrigation operations are more influenced by soil physicochemical and hydrological parameters. That's why the most crucial physical, chemical, and hydrological processes in soil have been quantified and integrated using a wide range of environmental and agricultural hydrological models. Artificial neural network techniques have been very popular in recent years for figuring out the water content of soil. Moreover, understanding the spatial, temporal and seasonal changing in soil water retention qualities is essential about precision water inputs into especially arid and semiarid cultivated soils. That's why researchers have successfully developed geostatistics and interpolation methods integrated with GIS for spatial estimating soil hydraulic properties.

In nature, soil–water-plant relationships have a highly complex structure. To ensure the sustainable utilization of these natural resources without compromising their characteristics, a comprehensive understanding of this relationship is necessary. In this article, highlighting the land suitability assessment procedures for various irrigation methods that consider the ideal land and soil conditions for different irrigation methods is the first of the study's primary subjects. Secondly, it is to provide information with studies on recently developed estimation models on issues such as soil–water relationship and soil water content. Thus, a resource will be provided to help decision makers, scientists related this field or farmers make more accurate decisions for sustainable land, soil and water management.

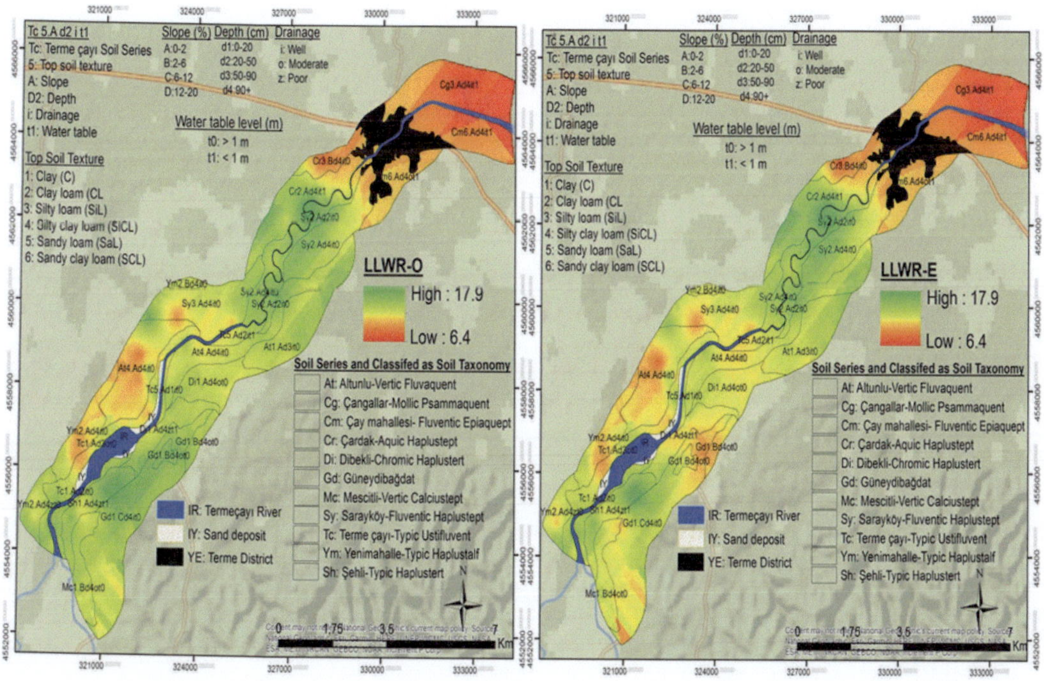

Fig. 8 Spatial distribution maps of the LLWR (observed: LLWR-O and estimated: LLWR-E, [13])

References

1. Mihalikova M, Dengiz O. Towards more effective irrigation water usage by employing land suitability assessment for various irrigation techniques. Irrig Drain. 2019. 0.1002/ird.2349.
2. de Fraiture C, Fayrap A, Unver O, Ragab R. Integrated water management approaches for sustainable food production. Irrig Drain. 2014;63:221–31. https://doi.org/10.1002/ird.1847.
3. UNESCO–WWAP. United Nations Educational Scientific and Cultural Organization–World Water Assessment Programme; 2006.
4. Liu W, Qin Y, Vital L. Land evaluation in Danling county, Sichuan province, China. In: 26th Course professional master. Geometric and natural resources evaluation. 7th Nov 2005–23rd Jun 2006. IAO, Florence, Italy;2007.
5. Albaji M, Golabi M, Nasab SB, Zadeh FN. Investigation of surface, sprinkler and drip irrigation methods based on the parametric evaluation approach in Jaizan Plain. J Saudi Soc Agric Sci. 2013;2015(14):1–10.
6. Efendi A, Dengiz O. Comparison of land suitability for three different irrigation methods determined through a parametric approach. In: AgroEnviron2016 10th international symposium and agricultural and the environment, USA;2016.
7. Dengiz O. Comparison of different irrigation methods based on the parametric evaluation approach. Turk J Agric For. 2006;30(6):21–9.
8. Cakmak B, Unver O, Akuzum T. Agricultural water use in Turkey. Water Int. 2004;29(2):257–64.
9. Mihálikovǎ M, Özyazıcı MA, Dengiz O. Mapping soil water retention on agricultural lands in central and eastern parts of the black sea region in Turkey. J Irrig Drain Eng. 2016;142(12):05016008–11.
10. Da Silva AP, Kay B. Estimating the least limiting water range of soils from properties and management. Soil Sci Soc Am J. 1997;61(3):877–83.
11. Mihalikova M, Başkan O, Dengiz O. Capability of different interpolation models and pedotransfer functions to estimate soil hydraulic properties in Büyükçay Watershed. Environ Earth Sci. 2015;74:2425–37. https://doi.org/10.1007/s12665-015-4246-5.
12. Tavanti RF, Freddi ODS, Tavanti TR, Rigotti A, Magalhaes WDA. Pedofunctions applied to the least limiting water range to estimate soil water content at specific potentials. Engenharia Agrícola. 2019;39(4):444–56.
13. Alaboz P, Baskan O, Dengiz O. Computational intelligence applied to the least limiting water range to estimate soil water content using GIS and geostatistical approaches in alluvial lands. Irrig Drain. 2021. https://doi.org/10.1002/ird.2628.
14. FAO. Food and agriculture organization of The United Nations. Land evaluation. Soils bulletin, vol. 32. Rome;1976.
15. Bond WJ. Assessing site suitability for an effluent plantation. In: McKenzie NJ, Coughlan K, Cresswell H, editors. Soil Physical Measurement and Interpretation for Land Evaluation. CSIRO Publishing;2002. p. 351–9.
16. Sys C, Van Ranst E, Debaveye J. Land evaluation part I Principles in land evaluation and crop production calculations. In: General administration for development cooperation (GADC), Agricultural Publications No. 7, Brussels, Belgium; 1991. p. 40–80.
17. Albaji M. Land suitability evaluation for sprinkler irrigation systems. In: Khuzestan water and power authority (KWPA), Ahwaz, Iran (in Persian);2010.
18. Seelig B, Franzen D. Soil, water and plant characteristics important to irrigation. EB-66, February 1996, Nort Dakota, USDA;1996.
19. McRae SG, Burnham CP. Land evaluation. Oxford, UK: Clarendon Press; 1981. p. 239.
20. Storie RE. An index for rating the agricultural value of soils. Bulletin 556. University of California, College of Agriculture, Agricultural Experiment Station Berkeley, California, USA; 1938. p. 46.
21. Delgado F, Lopez R. Evaluation of soil development impact on the productivity of Venezuelan Soils. Adv GeoEcol. 1998;31:133–42.
22. Bienvenue JS, Ngardeta M, Mamadou K. Land evaluation in the province of Thies, Senegal. In: 23rd Course professional master, geomatics and natural resources evaluation, Nov 8, 2002–June 20, 2003, IAO, Florence;2003. http://www.iao.florence.it/training/geomatics/Thies/senegal23.pdf.
23. Mbodj C, Mahjoub I, Sghaiev N. Land evaluation in the Oued Rmel catchment, Tunisia. In: 24th Course professional master, geomatics and natural resources evaluation, Nov 10, 2003–June 23, 2004, IAO, Florence;2004. http://www.iao.florence.it/training/geomatics/Zaghouan/Tunisia_24lq.pdf.
24. Dengiz O, Bayramin İ, Yüksel M. Geographic information system and remote sensing based land resource assessment, land evaluation of beypazari area soils by ILSEN method. Turk J Agric For. 2003;27:145–53.
25. Dengiz O, Usul M. Multi-criteria approach with linear combination technique and analytical hierarchy process in land evaluation studies. Eurasian J Soil Sci. 2018;7(1):20–9.
26. Saaty T. The analytic hierarchy process: planning, priority setting, resource allocation. McGraw Hill International, New York, USA;1980. p. 287.
27. Klingebiel AA, Montgomery PH. Land capability classification. United States Department of Agriculture. Handbook No. 210, Washington, USA;1966
28. Senol S, Tekes Y. Arazi değerlendirme ve arazi kullanım planlaması amacıyla geliştirilmiş bir bilgisayar modeli. Türkiye Toprak Ilmi Dernegi. İlhan Akalan Toprak ve Çevre Sempozyumu, Cilt 1, Ankara, Turkey;1995. p. 204–10.

29. Dengiz O. Soil quality index for paddy fields based on standard scoring functions and weight allocation method. Arch Agron Soil Sci. 2020;66(3):301–15. https://doi.org/10.1080/03650340.2019.1610880.
30. McBratney AB, Minasny B, Tranter G. Necessary metadata for pedotransfer functions. Geoderma. 2011;160(3–4):627–9.
31. Mohammed A, Hirmas DR, Giménez D, Nemes A. Investigating relationships between soil morphology, classification, and hydraulic properties. In: Proceedings of international annual meeting, ASA-CSSA-SSSA, American Society of Agronomy, Crop Science Society of America, Soil Science Society of America, Madison, WI; 2014.
32. Toth B, Weynants M, Nemes A, Mak´o A, Bilas G, T´oth G. New generation of hydraulic pedotransfer functions for Europe. Eur J Soil Sci. 2015; 66(1):226–38.

Irrigation Dams, Historical and Modern Rain Water Harvesting

Characterization of an Earth-Filled Irrigation Dam through the Use of Ground Penetrating Radar (GPR)

José Tavares Araruna Júnior, Paola Machado Barreto Manhães, Carlos Henrique de Almeida Couto Medeiros, Neil Lennart Anderson, and Norbert Maerz

Abstract

The establishment of highly mechanized, capital-intensive system of agriculture allied to the Cerrado biome climate has led to the development of earth-filled dams to provide irrigation in the MATOPIBA Region in Brazil. In most cases, these dams as-built often differ from as-designed and generally possess a high hazard potential because of their large reservoir capacity. In this study, the usefulness of the use of ground penetrating radar in establishing the as-built condition of the River Balsas dam is investigated. The results showed that is possible to acquire a detailed knowledge of the height above ground, the structural height, the position of the phreatic surface, the thickness of each foundation material and the position of bedrock. The GPR survey was also very useful to delineate the extension of a breach that was affecting the dam's performance and also showed the importance of paleochannels in providing key information about dam leakage.

Keywords

As-built records · Ground penetrating radar · Irrigation dam · Leakage · Aleochannels

1 Introduction

Brazil is one of the world's largest food producers. During the past decades, most of its production comes from the MATOPIBA region, an acronym that stands for the initials of the states of Maranhão, Tocantins, Piauí, and Bahia. This region is situated within the Cerrado biome, a savanna with a semi-humid tropical climate which is limited to two dominant seasons throughout the year, wet and dry, where the dry season lasts for four to six months.

Cattle ranching is particularly prominent in MATOPIBA. It is generally accepted that about half of all the Brazilian cattle are located there [1]. According to Spring [2], MATOPIBA is a major producer of grains (i.e., soya, beans, maize and rice) but also cotton and coffee. The establishment of a highly mechanized, capital-intensive system of agriculture allied to Cerrado climate has

J. T. A. Júnior (✉)
Pontifical Catholic University of Rio de Janeiro, Rio de Janeiro (RJ), Brazil
e-mail: araruna@puc-rio.br

P. M. B. Manhães
GEOCOBA–Projetos de Engenharia Limitada, Rio de Janeiro (RJ), Brazil

C. H. de Almeida Couto Medeiros
CH Medeiros Consultoria Geotécnica, Salvador (BA), Brazil

N. L. Anderson · N. Maerz
Missouri University of Science and Technology, Rolla, MO, USA

M. Bahadir et al. (eds.), *Agricultural Water Management*, Water and Wastewater Management,
https://doi.org/10.1007/978-3-031-96134-2_8

led to the development of dams to provide irrigation. Small dams, such as those used for irrigation purposes, could threaten not only the environment but negative impacts might well arise from their failure, including loss of human life, economic loss, and lifeline disruption. Hence, the integrity of those structures is of fundamental importance to all stakeholders as well as to the engineers who play an important role in the dam conception, design, construction, and operation. So, to achieve adequate safety conditions throughout the life cycle of such dams, stakeholders have to adopt measures of prevention and control proportional to the risks involved.

During the past decade, inspection procedures were developed aiming to reduce dam risks [3]. For example, legislation was approved by the US House Representatives (e.g., US Dam Safety Act of 2006), resolutions were issued by State Agencies and inspection training courses were offered on a regular basis [4]. All these procedures have in common that inspection records should be kept as historical records of the dam's performance. However, in developing countries there is not an engineering culture of safely recording and maintaining data records from conception, design, construction and operation of small earth-filled dams. In addition, the as-built dam is often different than as designed. One also should keep in mind that the as-built records should not only include the dam's drawings but also keep any relevant geotechnical and geological information which might affect its performance. As-built records are usually obtained through a site investigation program that is often carried out by boring and drilling exploratory holes aiming to determine the ground and groundwater conditions at and around earth-filled dams. Therefore, it is important to handle dam investments carefully, establish priorities, consider the technical and economic viability of various options, and implement the best engineering solutions [5]. During the past decade, non-invasive methods, including seismic and electrical resistivity techniques, were also employed to help assess the characteristics and problems arising from dam's performance [6–10]. In addition, geophysical methods, which can gather data to cover the entire study area, with lower cost and more practical measurements, can determine the engineering properties of the foundation materials by minimizing the time-consuming and expensive drilling works in dam sites, considering the site characteristics and project properties [11, 12]. In the case of small earth-filled structures, some geophysiscists prefer the use of ground penetrating radar (GPR) since it enables one to rapidly cover extensive areas with the possibility to inspect the data collected in real time [13–15]. However, as Sentenac et al. [10] pointed out, earth-filled dams are often made of fine-grained soils and with such conductive materials GPR analysis needs expertise and modelling skills.

The main aim of this study was to examine the feasibility and effectiveness of GPR to survey small earth-filled dams, for the detection of piping, since water-filled voids provide an excellent dielectric constant contrast. The possibility of employing GPR to acquire a detailed knowledge of the integrity of these structures could hold a relevant role on dam safety. In this paper the usefulness of the use of ground penetrating radar in establishing as-built of a small irrigation earth-filled dam in MATOPIBA Region (Brazil) is presented.

2 Materials and Methods

2.1 GPR Use in Earth-Filled Dam Characterization

Ground penetrating radar (GPR) is a non-invasive geophysical method based on the propagation of electromagnetic waves in the microwave band of the radio spectrum. According to Annan [16], the GPR principle is like seismic reflection. The difference arises from the fact the energy may be reflected at boundaries where subsurface electrical properties change rather than subsurface mechanical properties as it occurs with acoustic energy. GPR is a time domain system that applies an impulse to the antenna, placed close to the air–ground interface that creates and detects key electromagnetic fields. According to Annan [16], the GPR transmitter antenna translates the excitation voltage into a predictable temporal and spatial distributed field and the receiving antenna detects the temporal variation of a vector component of the electromagnetic field (EM) field and translate it into a recordable signal. The author also points out that a shield could be used to focus or direct EM signal downward. In addition, shielded antennas also minimize the direct transmitter to receiver energy, minimize the energy that escapes into the air and minimize external electromagnetic noise. However, as the author suggests, shielding can create more problems than it solves. Annan points out that signals generated within the shielding box can be large and reverberate for long time intervals increasing the system ring-down. On the other hand, the maximum depth at open sites is obtained with unshielded antennas that are lighter and smaller. GPR surveys can be carried out by varying some methodological aspects regarding variation in the spacing of the antennas (offset) and change in the position of the antennas (polarization). There are two traditional research methodologics in offset function: the prospect for common offset (common-offset), and prospecting for multiple offsets (multi-offset). In the first case, the antennas with fixed geometry (spacing and orientation), perform measurements at intervals of regular spaces. The purpose of the application of this methodology is to map reflective surface. In this case, variations in the amplitude of reflection and delay time indicates change of material. Multiple offsets can be carried out with a common midpoint (CMP) or with wide angle reflection and refraction (WARR). These techniques are primarily used to obtain an estimate of the radar signal speed depth, through the variation of antenna spacing and measuring the change in the measuring time of reading of the signals in both directions. Regarding the polarization issue, the position of the antennas could vary between each other or could vary in relation to the direction of the investigation. The antennas may be parallel to each other and side by side of each other (BD), parallel to each other and a followed by another (EF) and perpendicular to each other (XPOL). In relation to the direction of the investigation, they may be parallel (PL) or perpendicular (PR). The appropriate antenna selection is made in accordance to the anticipated depth of investigation. The higher the antenna frequency the lower the antenna signal penetration through the porous or fractured media.

During a survey, a GPR transmitter emits through an antenna an electromagnetic pulse source into the ground. When this pulse encounters a boundary between materials having different dielectric permittivity, it will be partially reflected or refracted back to the surface where a receiving antenna records the magnitude and arrival time of the turned pulse. In the case of an earth-filled dam, electromagnetic waves will be reflected or refracted back to the surface in the event of a strong electrical impedance caused by: (a) saturation of the porous material, (b) presence of materials with different mineralogy, (c) presence of materials with different degree of compaction and (d) presence of piping.

So, through the readings made and processed by an automatic data acquisition system, it is possible to obtain using calibration/validation procedures, important information about an earth-filled dam including (Fig. 1):

- the position of the phreatic surface;
- the position of the cutoff;
- the position of the toe drain;
- the position of the rock strata;
- the thickness of the foundation layer, and
- the possible existence of piping.

2.2 River Balsas Dam

River Balsas Dam is situated in the municipality of Luís Eduardo Magalhães in the State of Bahia. It was built during the 1990's for irrigation and fish farming purposes [17]. The dam, viewed in Fig. 2, consists of a compacted soil fill with homogeneous section having 850 m of crest length. Its main cross section is seen in Fig. 3 and shows as-designed height of 6.56 m, a crest width of 6.0 m, and upstream slope of 1/3 (V/H) and downstream slope of 1/2.5 (V/H). During construction

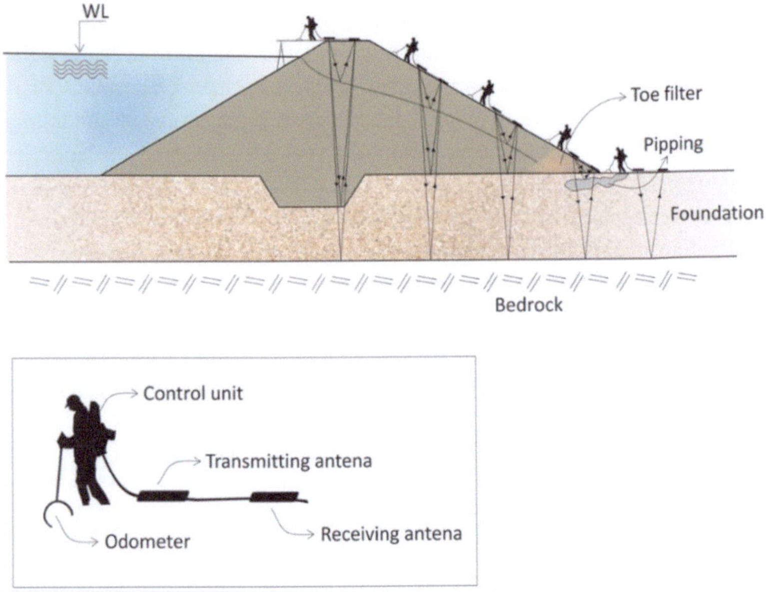

Fig. 1 Schematic diagram on how important information about an earth–filled dam could be obtained

Fig. 2 River Balsas Dam

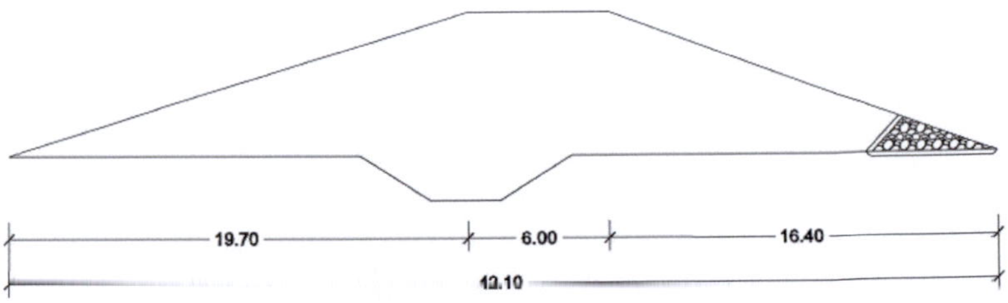

Fig. 3 As-designed main cross section of River Balsas Dam (adapted from [17])

the crest was widened to 10.0 m to accommodate the parallel traffic of two heavy trucks.

According to dam historical records, the earth fill was built using fine sand mixed with silt and clay, possessing a liquid limit of 23% and a plasticity limit of 6% [18]. This soil was classified under ASTM D 2487 [19] as SM-SC. The historical records do not provide any information about the mass-volume-moisture relationship of the soil, nor the degree of compaction adopted on

Fig. 4 Uncontrolled breach at the toe filter

its construction. In 2015, a breach appeared at its toe filter. The uncontrolled release of impounded water was measured by means of a flow meter, as seen in Fig. 4, and reached 200 m^3/h. The water coming from the breach presented no turbidity, but it was a clear sign of abnormality outside the design assumptions and was adversely affecting the dam's primary function of impounding water for irrigation.

This lesser degree of failure led the Brazilian Water Agency (ANA) classify the dam as possessing a medium risk. The Agency considered that despite the breach could progressively lead to or heighten the risk of a catastrophic failure it was generally amenable to corrective action. In addition, the agency categorized the dam as possessing high hazard potential since it has a reservoir capacity of 7.958 × 106 m3 and an unintentional release of stored water from its reservoir would cause adverse consequences such as loss of life, property damage, or other adverse impacts. Ever since the breach was detected, the stakeholders have investigated its cause. A series of measures were carried out, including the installation of standpipes piezometers in the dam body and its foundation and a GPR survey to characterize the dam.

2.3 GPR Survey

The survey was carried out employed a methodology adapted from the TRIAD approach developed by the United States Environmental Protection Agency (USEPA). This protocol employs systematic planning and dynamic work strategies as described in [20]. According to the authors, the TRIAD approach manages decision uncertainty to increase confidence that surveys are made correctly and cost-effectively. This methodology also implies that all activities are carried out on site by senior professionals and with stakeholder participation.

The geophysical survey employed a shielded antenna of 500 MHz and unshielded Rough Terrain Antennas (RTA's) of 50 and 100 MHz. The 500 MHz shielded antenna was used to delineate the breach (Fig. 5a), the phreatic surface and the degree of compaction. The 100 MHz RTA antenna was employed to delineate the boundary of the dam body, its foundation and its cutoff (Fig. 5b). Finally, the 50 MHz RTA antenna was used in delineating the foundation layer and in determining the conformation of the rocky strata (Fig. 5c). GPR acquisitions employed the common-offset method. The antennas have fixed geometry (spacing and orientation) and perform measurements at fixed intervals triggered by the odometer seen in Fig. 5. The purpose of the application of this methodology was to map reflexive sub-surfaces. In this case, the observed changes in the amplitude of reflection and delay time indicated a possible change of material. Data processing started using the background removal filter to minimize the environment noise through the application of a horizontal spatial high pass filter. Then, the bandpass filtering routine was used to increase the signal/noise ratio removing unwanted/outlying frequencies. After that, the time adjustment routine was employed to set the direct wave arrival time. Additionally, the amplitude correction routine was used to accomplish

Fig. 5 GPR antennas: **a** 500 MHz shielded, **b** 100 MHz RTA, and **c** 50 MHz RTA

gain control accommodating for amplitude loss with depth. Finally, the topography routine was employed for correcting the data for the surface topography along the profile.

GPR data were validated by comparison with light percussion drilled borehole logs for the stand pipe piezometers installed on the dam body and its foundation. The survey took place at the downstream slope and the dam crest. Figure 6 show the locations of the traverses along which data were acquired.

3 Results and Discussion

According to Annan and Cosway [21], electromagnetic waves propagate with a normal incidence to the interfaces between materials possessing different dielectric permittivity. In this situation, the magnitude of the reflection coefficient (rGPR) is defined as the ratio of the electric field transmitted by the incident, according to:

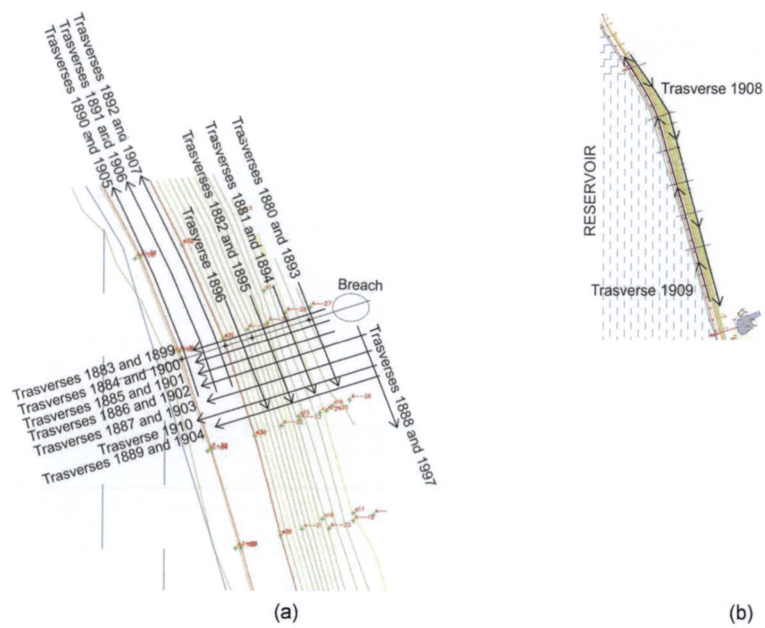

Fig. 6 Layout of the survey: **a** downstream slope, and **b** dam crest

$$r_{GPR} = \frac{\sqrt{k_1} - \sqrt{k_2}}{\sqrt{k_1} + \sqrt{k_2}} \quad (1)$$

where k1 and k2 correspond to the dielectric permittivity of the different materials. Based on this premise, the interpretation of GPR survey sought to find boundaries across which there were mappable contrasts in dielectric permittivity. This would include the contacts between the different materials that make up the dam body and foundation as well as the breach. During interpretation, GPR traces were analyzed, as shown in Fig. 7. It was anticipated that interfaces of interest would be manifested as coherent reflections.

GPR surveys using the 500 MHz shielded antenna were extremely useful in delineating the position and the extent of the breach. According to Di Prinzio et al. [10], GPR is a very efficient tool for mapping shallow targets. The authors also pointed out that the manageability and the non-invasiveness of GPR are particularly suitable to locate voids that are difficult and costly to assess using other intrusive methods. One could

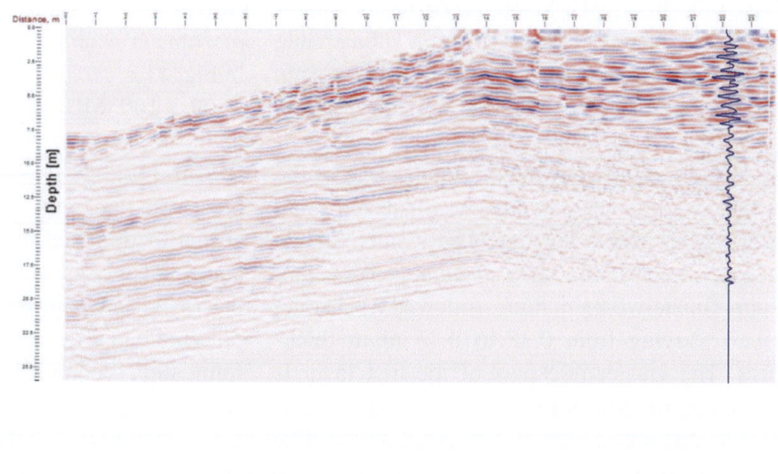

Fig. 7 Interpretation of a processed radargram

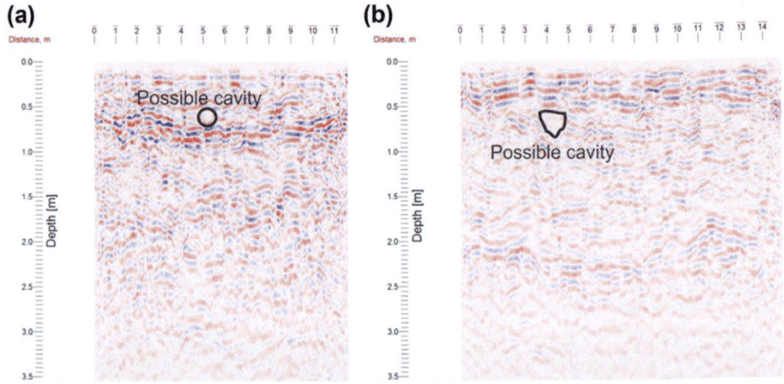

Fig. 8 Interpreted radargrams from traverses 1880 and 1881

Fig. 9 Submersible action camera coupled to a semi flexible plastic pipe used to validate GPR interpretations

see from the interpreted radargrams obtained from sections 1880 and 1881, presented in Fig. 8, that the breach is located under the filter toe and has dimensions about 0.50 m in diameter. Based on this the GPR interpretations, corroborated from images captured from a submersible action camera coupled to a semi flexible plastic pipe as seen in Fig. 9, the schematic diagram presented in Fig. 10 was generated.

Variations in the degree of compaction of the dam body underneath its crest was analyzed using a 500 MHz shield antenna. The results, shown in Fig. 11, present a layer of homogeneous soil approximately 0.85 m thick, followed by distinct layers varying from 0.35 to 0.45 m in thickness. The greater thickness of the first layer is caused by the load imposed by heavy duty vehicles loaded with grains that traveled on the dam crest continuously over the last two decades. It was also observed from Fig. 3.5 that the phreatic surface was positioned approximately 2.60 m from the crest surface by the time the survey was carried out. This information was validated by a piezometer reading which position was indicated in Fig. 11.

The 100 MHz RTA antenna was useful to delineate the dam, and its foundation as can be seen in Fig. 3.6. It was observed that the cutoff presents a thickness of approximately 2.5 m and, at that section, and the height above ground was approximately 7.8 m (i.e., more than 1 m higher that it is stated in the as-designed). The model presented in Fig. 12 also shows that the dam foundation is composed by two distinct materials. When comparing GPR results with the information contained on light percussion rig borehole

Fig. 10 Schematic diagram describing the location of the breach; WL = Water level

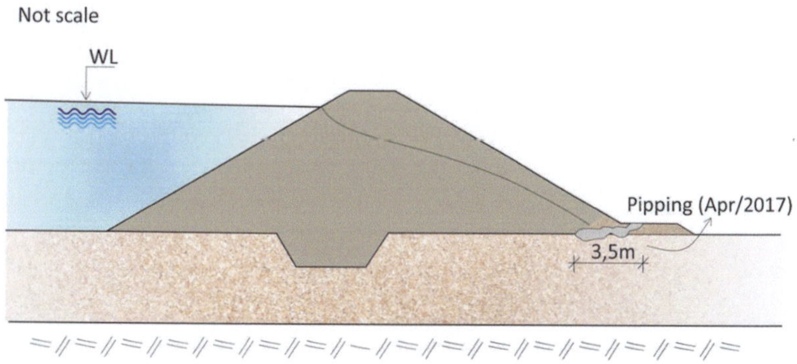

Fig. 11 Degree of compaction obtained from the 500 MHz shielded antenna

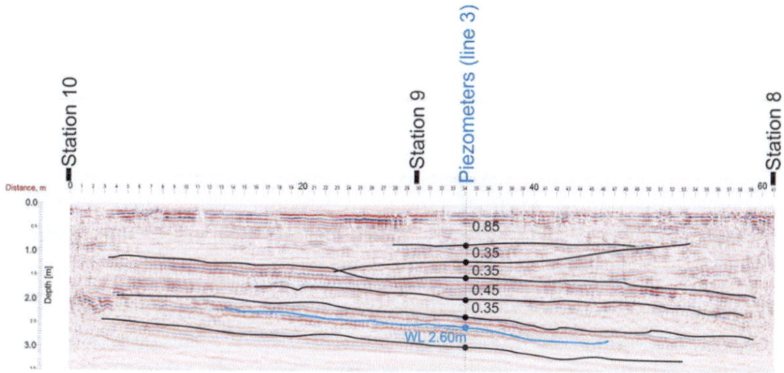

Fig. 12 GPR cross section validated by information of light percussion rig borehole logs

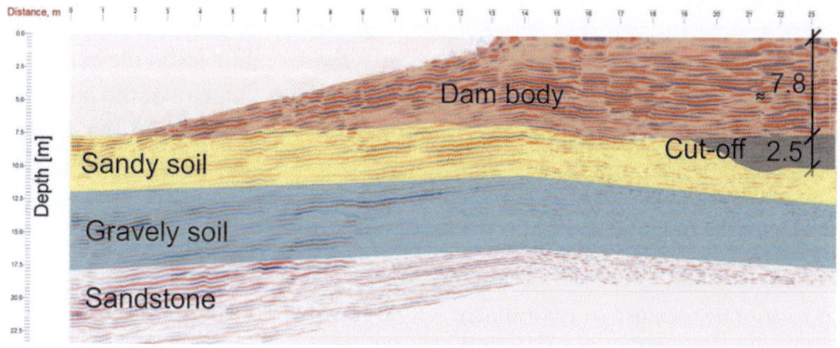

logs produced by the installation of piezometers it was found that these materials consisted by a sandy soil layer positioned above a gravelly soil layer. The boreholes also found the bedrock to be sandstone.

The 100 MHz RTA antenna was also employed to delineate the position of the phreatic surface within the dam body as can be seen in Fig. 13. This information could be useful in the future dam assessment performance since it will enable the determination of the hydraulic heads in different points within the dam body necessary to a future numerical flow modelling. In addition, the position of the phreatic surface will also enable the determination of the pore pressure ratio (ru), a parameter that represents the global pore water pressure condition in a slope stability analysis by the limit equilibrium method.

The lack of maintenance might well be considered the main cause for the breach since it was observed during the GPR survey that the toe filter was clogged on its entire length. However, it was not clear why the breach occurred at that place. To develop a plausible explanation, a GPR survey

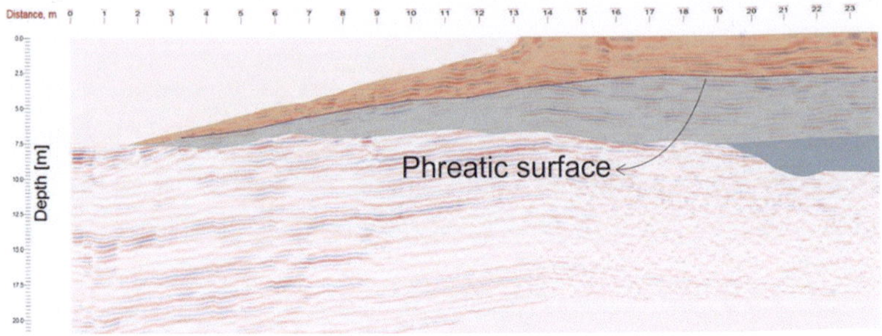

Fig. 13 Phreatic surface

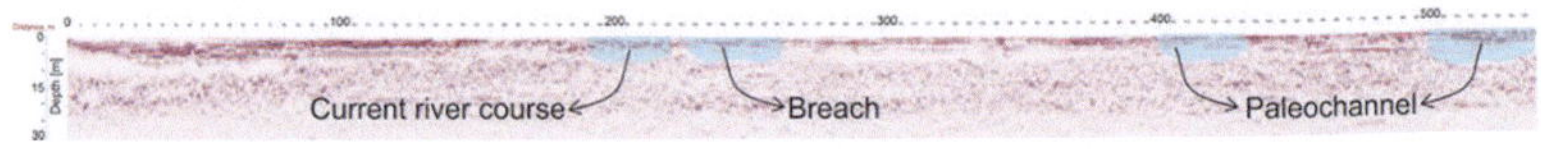

Fig. 14 Position of the interpreted paleochannels

using a 50 MHz RTA antenna was conducted along traverses parallel to the dam's downstream toe. In the radargrams shown in Fig. 14 a series of depressions at the level of contact between the sandy and the gravelly layers were identified. These depressions suggest the presence of paleochannels in the dam's foundation. Paleochannels are important morphological features that provide insight into the drainage systems that have occurred in the past. In addition, Martinez-Moreno also suggest that paleochannels provide key information about leakages. Apparently, these morphological features indicated were the past and the current river paths. The radargram also indicate the position of a paleochannel where small breaches appeared in the past.

The data from the GPR scan was extremely helpful in creating the mitigation techniques used to restrict the release of impounded water. In such cases, the cavity is generally filled with gravel material and a geotextile is used as a separating and filtering surface between the dam body and the filler material. In addition, it was necessary to unclog the toe filter in all its extent prior to the intervention in order to minimize the possible rise of pore pressure during the cavity filling process. Figure 15a shows that a tractor with a water reservoir and a high-capacity pump was employed to remove the fine-grained soil that were clogging the toe filter. The washing process was preceded by the removal of the thin layer that was deposited over the years above the coarse material that constitutes the toe filter (Fig. 15b). A backhoe tractor, shown in Fig. 15c, was employed to dig a ditch besides the tow filter in order to drain the water used in the washing process. The breach shown in Fig. 15d was enlarged by the backhoe tractor in order to be filled with geotextile and gravel. A flow meter coupled on an irrigation piper was then installed to monitor the release of impounded water (Fig. 15e).

The conclusion of the mitigation process had to wait until the end of the season irrigation activities in the farm. When irrigation was concluded, it was possible to lower the water reservoir level and verify the existence of a breach in the upstream slope over the paleochannel indicated by the GPR survey. Similar mitigation measures took place, and they were quite successful as Fig. 16 shows. When the next flood event on the River Balsas took place the water level reached its design level, Fig. 16a, and the excess water was discharged through its spillway (Fig. 16b). There was no release of impounded water at the monitoring station (site of the repaired breach) as Fig. 16c shows.

4 Conclusions

GPR is a useful technique for imaging and assessing for the characteristics and problems of small earth-filled dams. The systematic planning and dynamic work strategies applied for the survey, carried out by senior professionals and with stakeholder participation, enabled authors to characterize River Balsas Dam rapidly and cost-effectively. During the survey, it was possible to

Fig. 15 Mitigation process: **a** agricultural tractor; **b** manual unclogging; **c** drainage trench; **d** breach; **e** cavity filled with gravel material; and **f** flow monitoring

Fig. 16 River Balsas dam first flood event after mitigation measures took place: **a** water reservoir; **b** discharge at the spillway and **c** monitoring station

acquire a detailed knowledge of the height above ground, the structural height, the position of the phreatic surface, the thickness of each foundation material and the position of bedrock. GPR survey was also very useful to delineate the extension of the breach that was affecting the dam's performance since water-filled voids provide an excellent dielectric constant contrast. The survey also showed the importance of paleochannels in providing key information about dam leakages. In sum, it was possible to develop an as-built that not only included the dam's drawings but also kept relevant geotechnical and geological information which might affect River Balsas Dam performance. In addition, GPR profiles provided information that made it possible to mitigate the dam breach in a fast and cost-effective way. However, it should be emphasized that GPR analysis needs expertise and modelling skills. It is very important to validate its findings by confronting them with pieces of information obtained from boring and drilling exploratory holes.

Acknowledgements The authors wish to express their gratitude to Agronol Agro Industrial S/A for the support provided during the GPR survey. The authors also thank EXCEED-SWINDON that provided the support that made it possible to conduct this research.

References

1. Ferreira LG, Fernandez LE, Sano EE, Field C, Sousa SB, Arantes AE, Araújo FM. Biophysical properties of cultivated pastures in the Brazilian Savanna Biome: an analysis in the spatial-temporal domains based on ground and satellite data. Remote Sens. 2013;5(1):307–26.
2. Spring J. Soy boom devours Brazil's tropical savanna. 2018. file:///Users/araruna/Documents/paper%20Barragem%20Agronol/referências/Appetite%20for%20Destruction:%20Brazil's%20soy%20boom%20devours%20tropical%20savanna.webarchive. Assessed 30 Mar 2023.
3. ASWCC (Arkansas Soil and Water Conservation Commission). Inspection and maintenance manual for owner inspection. Little Rock, AR;2002.
4. Trimbath K. New regulations mandate dam inspections in Massachusetts. Civ Eng. 2005;75(12):29.
5. Sari M, Seren A, Alemdag S. Identification of geological structures of Kirklartepe dam by GPR and ERT methods, vol. 1. SRA Academic Publishing; 2021. p. 105–18. ISBN 978-625-7148-11-5.
6. Sjodahl P, Dahlin T, Johansson S. Embankment dam seepage evaluation from resistivity monitoring data. Near Surface Geophys. 2009; 463–474.
7. Bedrosian PA, Burton BL, Powers MH, Minsley BJ, Phillips JD, Hunter LE. Geophysical investigations of geology and structure at the Martis Creek Dam, Truckee, California. J Appl Geophys. 2012;77:7–20.
8. Cardarelli E, Cercato M, De Donno G. Characterization of an earth-filled dam through the combined use of electrical resistivity tomography, P- and SH-wave seismic tomography and surface wave data. J Appl Geophys. 2014;106:87–95.
9. Sentenac P, Benes V, Budinsky V, Keenan H, Baron R. Post flooding damage assessment of earth dams and historical reservoirs using non-invasive geophysical techniques. J Appl Geophys. 2017;146:138–48.
10. Sentenac P, Benes V, Helen Keenan H. Reservoir assessment using non-invasive geophysical techniques. Environ Earth Sci. 2018; 77:293.
11. Sari M, Seren A, Alemdag S. Determination of geological structures by geophysical and geotechnical techniques in Kırklartepe Dam Site Turkey. J Appl Phys. 2020;182: 104174.
12. Alemdag S, Sari M, Seren A. "Determination of rock quality designation RQD in metamorphic rocks: a case study: Bayburt-Kırklartepe Dam. Bull Eng Geol Env. 2022;81:214.
13. Di Prinzio M, Bittelli M, Castellarin A, Pisa PR. Application of GPR to the monitoring of River Embankments. J Appl Geophys. 2010;71:53–61.
14. Busato L, Boaga J, Peruzzo L, Himi M, Cola S, Bersan S, Cassiani G. Combined geophysical surveys for the characterization of a reconstructed river embankment. Eng Geol. 2016;211:74–84.
15. Anchuela OP, Frongia P, Di Gregorio F, Casa Sainz AM, Juan AP. Internal characterization of embankment dams using ground penetrating radar (GPR) and thermographic analysis: a case study of the Medau Zirimilis Dam (Sardinia, Italy). Eng Geol. 2018;237:129–39.
16. Annan AP. Ground penetrating radar (GPR) principles. In: Jol RM, editor. Ground penetrating radar: theory and principles. Netherlands: Elsevier; 2009. p. 4–40.
17. Cerveira CF. Memória Descritiva da Barragem do Açude do Rio das Balsas;1994. p. 21
18. ASTM D2487. Standard practice for classification of soils for engineering purposes (Unified Soil Classification System). ASTM International, West Conshohocken, PA;2017. https://doi.org/10.1520/C0033-03, www.astm.org.
19. Crumbling DM, Griffith J, Powell DM. Improving decision quality: making the case for adopting next generation site characterization practices. Remediat J Environ Cleanup Costs Technol Tech. 2003; 13(2):91–111.
20. Annan AP, Cosway SW. Ground penetrating radar survey design. In: Proceedings of the symposium on the application of geophysics to engineering and environmental problems, vol. 2. 1992. p. 329–52.
21. Martinez-Moreno FJ, Delgado-Ramos F, Galindo-Zaldívar J, Martín-Rosales W, López-Chicano M, González-Castillo L. Identification of leakage and potential areas for internal erosion combining ERT and IP techniques at the Negratín Dam left abutment (Granada, Southern Spain). Eng Geol. 2018;240(5):74–80.

Sustainable Management of the Loumbila Dam Water Resource in Burkina Faso for Market Gardening: Diagnostic and Prospects, a Case Study

Yacouba Sanou and Francois Sawadogo

Abstract

Nowadays, water resources face huge problems such as climatic variability, natural disasters and human activities. Thus, demographic growth not only leads to strong pressure on water resources but also conflicts of use and problems of water management in order to produce vegetables. This work aims to assess strengthen the participatory and sustainable management of the Loumbila Dam water resource in order to optimize efficiency in its management in irrigation and vegetables production. A qualitative and quantitative approach is used. It consists of data collection, interviews with stakeholders, field trips, analysis and processing of data obtained using software (Kobocollecte, Word, and Sphynx). Data collection revealed that the main water users are market gardeners, breeders, arborists, fishermen, hoteliers, construction, and public works companies. It also appears that 61% of these users do not know the structures insole in the management of the dam. In addition, the analysis of the roadmap with the Sustainable Development Analysis Grid (SDAG) showed that the social, ecological and governance dimensions are well taken into account in the current management system with a satisfactory situation, while the dimensions economic, cultural, and ethical are weakly integrated. These analyses enabled us to suggest ways of improving the network, including the education of network users and their periodic training, and to make recommendations for the implementation of a management framework by shore.

Keywords

Agriculture · Burkina Faso · Dam water management · Market gardening · Sustainable management · Water resource

1 Introduction

Burkina Faso, located in the Sahelo-Sahelian strip, has an area of 274,200 km^2 and a population increasing from 14,017,262 inhabitants in 2006 to 20,870,060 in 2019 with a growth rate of 3.1% [1, 2]. Surface water and groundwater are the main sources of freshwater supply in the country. These water sources are seriously threatened by population growth and climatic hazards [3].

Y. Sanou (✉)
Laboratory of Analytical, Environmental and Bio-Organic Chemistry, University Joseph KI-ZERBO, Ouagadougou, Burkina Faso
e-mail: prosper_yacson@yahoo.fr; yacouba.sanou@ujkz.bf

F. Sawadogo
Environment and Sustainable Development Master Program, New Dawn University, Ouagadougou, Burkina Faso

M. Bahadir et al. (eds.), *Agricultural Water Management*, Water and Wastewater Management,
https://doi.org/10.1007/978-3-031-96134-2_9

In fact, the country had 2,100 surface water reservoirs, of which 380 reservoirs are permanent with a storage capacity of 6 billion m^3, which corresponds to 75% of the annual contributions of three river basins [3]. These proportions reflect a low mobilization of river water in the country. Water resources mobilized are used for irrigation, supplying cities with drinking water, producing hydroelectric energy, supplying domestic needs, livestock, mining, etc. The major challenge remains meeting the needs of water users, taking into account the delivery capacity of the various sources of supply. To deal with this situation, Burkina Faso decided to set up an integrated water resources management program in 2001. The objective of which is to propose a scheme for the integrated management of the country's water resources, adapted to the national context, in accordance with the orientations defined by the Burkina Faso Government, and also respecting the international principles in terms of sustainable and ecologically rational management of water resources [3]. Despite the advent of this program, difficulties in the management of water resources are still observed in certain localities such as the commune of Loumbila, where conflicts of use, non-compliance with regulations, and the phenomenon of silting have increased during the last years. To overcome these problems, the administrative and local authorities and Civil Society Organizations (C.S.O) implemented a roadmap for the sustainable management of the dam in 2012. Unfortunately, the illegal occupation of the public water domain, the construction of infrastructure, works and facilities without authorization or declaration, as well as the collection and rejection by agribusiness and construction companies persist.

The general objective of this work was to contribute to the participatory and sustainable management of the Loumbila Dam by (i) identifying the mode of organization of actors in the management of water resources; (ii) analyzing the degree to which the pillars of sustainable development are taken into account in the current management method; and (iii) describing the terms of occupation around the Loumbila water reservoir.

2 Materials and Methodology

2.1 Presentation of Study Area

The study area was located in the Loumbila municipality (12°31′5.39" N, 1°22′8.39" W). It is one of the seven (7) municipalities in the province of Oubritenga, located at 25 km from Ouagadougou city. The department of Loumbila covers an area of 362 km^2 with a density of 65 inhabitants/km^2 and has thirty-one (31) villages [4]. The municipality of Loumbila has a large water reservoir built on the Massili, a watershed of the Nakanbé River. This study relates to this water reservoir. The hydrographic network of the municipality of Loumbila is organized around the Massili watershed. It is made up of dams and other water reservoirs with a tropical rainfall regime and highly dependent on the spatio-temporal distribution of precipitation. The Loumbila Dam is the largest water reservoir in this hydrographic network. It has a capacity of 42,200,000 m^3 and contributes significantly to supplying the city of Ouagadougou with drinking water (Fig. 1). The hydrographic regime of the municipality of Loumbila is average, and the most important water reservoirs besides the Loumbila Dam are the dams of Nomgana, Goué, Ramitenga, Goundry and Nongstenga, and the "boulis" in Kogniga, in Bangrin. In addition, the municipality has about a hundred (100) boreholes distributed in all villages, some of which have a seasonal flow, and about twenty (20) large-diameter wells, only about ten (10) of which have a permanent flow [4].

2.2 Methodology

A method with several tools and structured in several phases was used for the collection of field data, the processing and analysis of the data obtained.

2.2.1 Interview with Stakeholders

The interviews with the stakeholders aimed to understand, on the one hand, their levels of involvement in the management of water

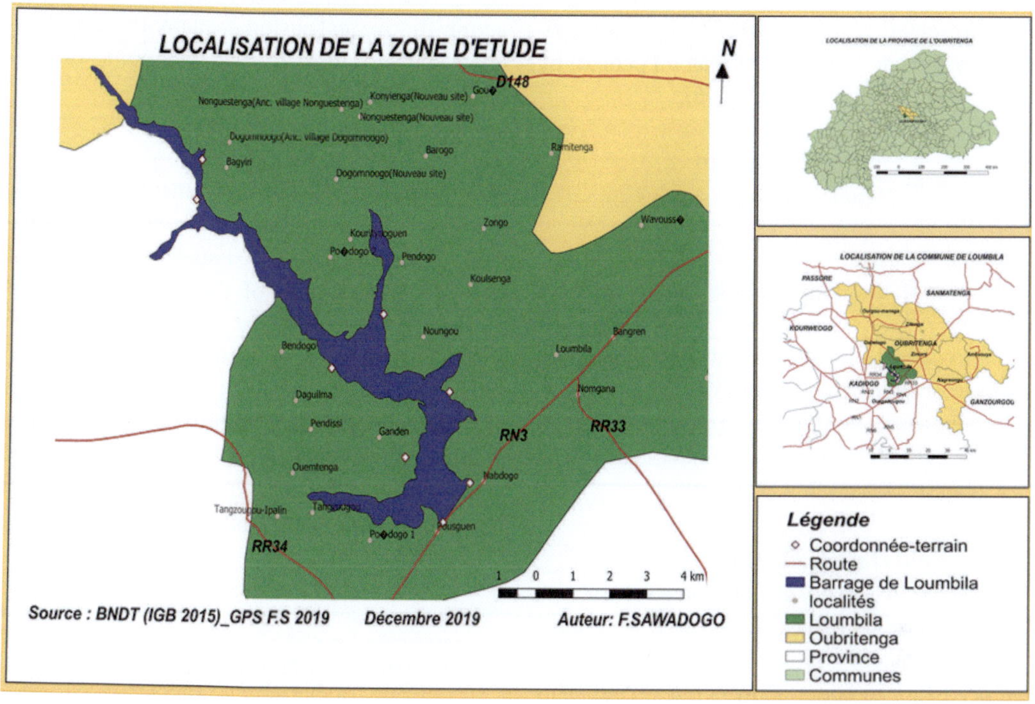

Fig. 1 Geolocalization of the study area [5]

resources, and, on the other hand, the relationships they have with each other. For the realization of these interviews, directors and heads of service, representatives of associations, Non-Governmental Organizations, customary authorities of the region, the province and the commune of Loumbila, who intervene in the management, some waters have been met.

2.2.2 Choice of Samples Surveyed

A random survey was conducted to get their views on the current state of management of the water reservoir as well as to collect their points of view on the establishment of a management framework by shore. The stakeholders involved in water management in the commune of Loumbila are diverse. They bring together administrative authorities, associations, non-governmental organizations and customary authorities. The administrative authorities chosen based on their involvement in water management are the Ministry of the Environment, Green Economy and Climate Change (MEEVCC), the Ministry of Water and Sanitation (MEA), the Ministry of Agriculture and Hydro-agricultural Developments (MAAH) and their decentralized services at the level of the central plateau, Oubritenga and Loumbila. The choice of associations was made based on their links with the use of water and their proximity to our field of intervention. To this end, the associations of the thirteen (13) villages closest to the dam have been identified and ranged. There are thirty-nine (39) associations recognized by authorities that carry out activities related to the use of water. An exhaustive sampling was carried out. There are five (05) Non-Governmental Organizations, which operate in the municipality of Loumbila and have a link with the management or use of water. In addition, investigations concerned the customary authorities, who were selected based on their geographical positions around the dam. The user survey was carried out among 98 water users, made up of market gardeners, fishermen, breeders, arborists, entrepreneurs. Indeed, given the high number of users, the survey randomly was conducted. To do this, the data from the census carried out by the Massili LWC (2,686 users), allowed to have an idea of the minimum size of the representative sample, on which the investigation was made (size greater than 93 people) by applying the following formula [6]:

$$n = \frac{Z^2 N p(1 - p)}{Z^2 p(1 - p) + (N - 1) y^2} \quad (1)$$

n: sample size;

N: size of the target population (number of users);

p: expected proportion of a population response or actual proportion (50%);

z: value associated with the confidence interval, 95% z =1.96 for the confidence interval;

y: margin of sampling error (10%).

2.2.3 Field Data Collection

Data collection is a very essential step for research work, because it makes it possible to collect from stakeholders and in the field the information necessary to better address our theme. This step occurred after the identification of the stakeholders, and the data were collected with different stakeholders. The methodology consisted of structured and semi-structured interviews, interviews and surveys. Interviews with administrative authorities (directors, department heads, and coordinators) and NGOs were conducted using a guide. The technique used to select the related associations was employed previously by Tahirou et al. [7], and it is a group interview with managers, individual meetings using questionnaires with open, semi-open and closed questions. Concerning the customary authorities, the method used was interviews including open questions with the village chiefs, land chiefs and their notables. The Global Positioning System (Garmin GPS64) was used for the geolocation of certain areas around the dam.

2.2.4 Data Processing

(i) Survey data

The data from the questionnaires and interviews were subjected to univariate statistical analysis using SPHYNX V5 to quantify the degree of stakeholder involvement in management. It was first carried out exhaustively on associations, administrative and customary authorities and then on a sample of users to have their points of view on management by shore.

(ii) Roadmap analysis data

The roadmap for the implementation of the recommendations resulting from the discussion workshop on the issues of sustainable use and management of the Loumbila dam is a management tool that defines clear action plans, their staggering in the time and the identification of the resources necessary for their accomplishments. Its general objective was to preserve the Loumbila dam from sedimentation and to enable the structure to fulfill its initial purpose of supplying drinking water to the city of Ouagadougou.

3 Results and Discussion

3.1 Results

3.1.1 Stakeholders in Dam Management

The actors in the management of the Loumbila dam are multiple and diversified. Indeed, they come from the administrative authorities (state structures, state companies, the town hall, the GIP), the customary authorities, the Civil Society Organizations (Non-Governmental Organizations, Associations).

(i) For the administrative authorities, the Ministry of Water and Sanitation (MEA), the Ministry of Agriculture and Hydro-agricultural Developments (MAAH), the Ministry of Environment and Climate Change (MEEVCC), the Ministry of Animal and Fisheries Resources (MRAH) as well as their decentralized structures (Regional Directorates, Provincial Directorates, and departmental Services) have been identified. At the level of the administrative authorities, the heads of the Nakanbé Water Agency (NWA), the National Office for Water and Sanitation (NOWS), the Local

Fig. 2 Performance of social dimension themes (*Source* Study data [8])

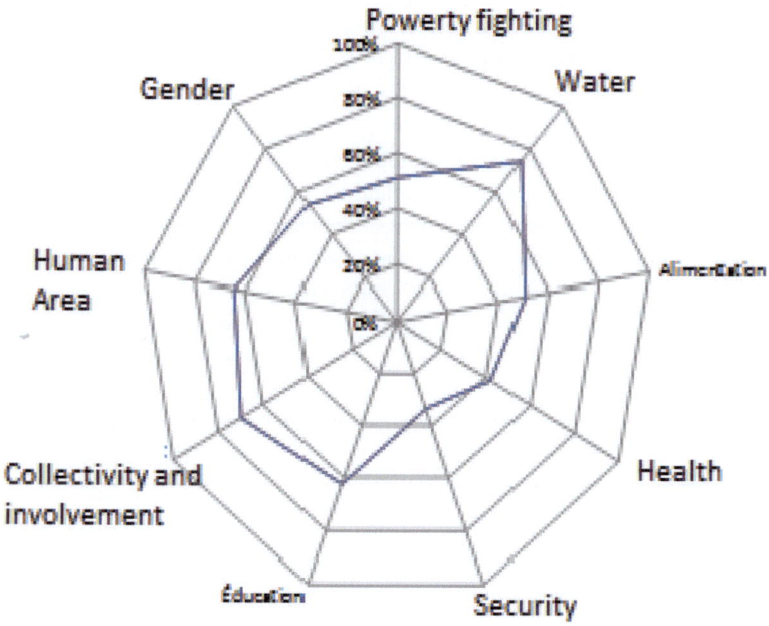

Water Committee (LWC), and the Loumbila Town Hall have been associated to the exchanges.

(ii) For the customary authorities, the village, and land chiefs of Nabdogo, Pousghin, Noungou and Gadin, all localities located near the water reservoir have been met.

(iii) For Civil Society Organizations, the structures encountered are thirty-nine (39) in number for associations and five (5) for NGOs.

3.1.2 Analysis of the Dimensions of Sustainable Development of the Roadmap

Sustainable development includes many dimensions such as social, ecological, economic, governance, ethical and cultural. So, each of dimensions was analyzed.

Social dimension

The social dimension aims to meet social needs and individual and collective aspirations, health and well-being needs, and quality of life needs. It is based on the principles of access to knowledge, health, security and the fight against poverty. This dimension has forty-one (41) objectives, grouped into nine (9) themes (fight against poverty, water, food, health, security, education, community and involvement, human settlements and gender). The results of the analysis showed that the social dimension in the roadmap has an average performance of 60%. Figure 2 shows the performance of the different themes. It appears that the level of consideration of the social dimension in the current roadmap is acceptable. However, topics such as health and security perform below average (50%). The various actors will have to pay more attention to access for all to a quality life in a perspective of intergenerational and intragenerational equity through the application of improvement paths.

From the analysis of the social dimension, the following areas for improvement could contribute to improving performance:

- Start the resettlement and support process with the most vulnerable;
- Take into account the most deprived and vulnerable groups in the implementation of activities.

- Carry out periodic training in health, safety, and the environment for users and waste treatment, while avoiding the production of greenhouse gases;
- Take into account all actors without discrimination, and suggest that the town hall become more involved in the management of the dam.
- Give equal opportunities to men and women in the management of public assets, while paying more attention to women and girls in decision-making.

Ecological dimension

The ecological dimension aims to meet the needs for the quality of the ecosystem and the sustainability of resources, as well as the redefinition of the relationship between human and nature. Its principle is based on the protection of biodiversity, preservation, conservation and management of the environment and environmental services. It has six (6) themes (ecosystems, biodiversity, resources, outputs, land use, and climate change) and twenty-seven objectives. Results of the analysis showed that the ecological dimension in the roadmap has an average performance of 64%. Table 1 shows the performance of the different themes. We find that the themes of the ecological dimension have an acceptable average performance. However, measures to combat climate change are not well taken into account in the roadmap. The various actors will further preserve the biophysical and human environment through the application of improvement tracks to ensure the dam sustainability.

Table 1 Performance of ecological dimension themes (*Source* study data)

Theme	Average weight	Average performance (%)
Ecosystem	2.3	62
Biodiversity	2.31	62
Resources	2.8	72
Extracts	2.6	73
Territory uses	2.0	50
Climate change	1.6	46

The avenues for improvement that contribute to improving the performance of the ecological dimension are as follows:

- Identify ecosystems, propose conservation measures and extend reforestation beyond the banks while restoring degraded soils;
- Identify species with symbolic value and then protect them;
- Encourage crops that produce less waste;
- Regulate the terms of use and provide for conflict management mechanisms;
- Prohibit excessive logging and help vulnerable populations adapt to climate change;
- Identify natural units and provide landscape protection measures.

Economic dimension

This dimension aims to meet the material needs of individuals and communities, as well as their financial empowerment. It echoes the principles of economic efficiency, internalization of social and environmental costs, responsible production and consumption. The economic dimension is divided into eight (8) themes (responsible production, responsible consumption, economic viability, work, wealth and prosperity, energy, entrepreneurship, economic models) and twenty-eight (28) objectives. Its average performance is 53%. Results are shown in Fig. 3, where it appears that there are differences between the average performances of the different themes (37% to 74%). This explains an imbalance between the themes vis-à-vis their importance. The application of improvement tracks will make it possible to balance the different themes and then to achieve sustainable management of the Loumbila dam.

From the analysis of the economic dimensions, the following areas for improvement could help improve the performance of this dimension of the roadmap:

- Develop a network enabling producers to interact with consumers;
- Use mechanisms aimed at limiting the return on capital;

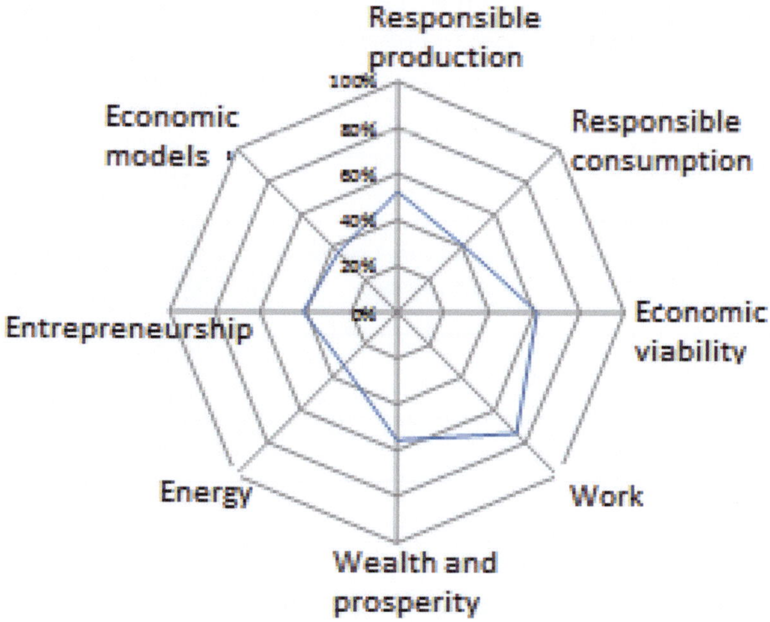

Fig. 3 Performance of economic dimension themes (*Source* Study data [8])

- Support users with inputs, production tools;
- Innovate in production and fishing in order to generate more foreign exchange;
- Include training campaigns in entrepreneurship;
- Support producers in the creation of financing cooperatives.

Cultural dimension

The cultural dimension aims to meet the needs for affirmation, expression, protection and enhancement of the diversity of cultural traits. It is based on the principles of protection of cultural heritage and diversity. This dimension has fifteen (15) objectives, grouped into four (4) themes such as transmission of cultural heritage, cultural and artistic practices, cultural diversity, and contribution of culture to development. The analysis shows that the cultural dimension is not well taken into account in the roadmap, because it only has an average performance of 44%. From the results in Fig. 4, it appears that for the themes of cultural and artistic practices, there are no actions planned or implemented in the current management plan of the dam. Similarly, the contribution of culture to development, good cohesion and support for decision-making is not well defined.

From the analysis of the cultural dimension, the following areas for improvement could contribute to improving its performance:

- Identify cultural sites near the dam and propose protection measures;
- Take into account local and past knowledge in planning.
- Promote cultural diversity and raise awareness among the population to understand changes in behavior;
- Avoid discrimination between natives and non-natives during capacity building measures;
- Set up a program of education, training, exchange in the cultural field and promote interculturality;
- Ensure integration of cultural factors into economic thinking on development;
- Give a good place to the customary society in the management of the dam.

Ethical dimension

The ethical dimension in the development of the roadmap aims to meet the needs of fairness,

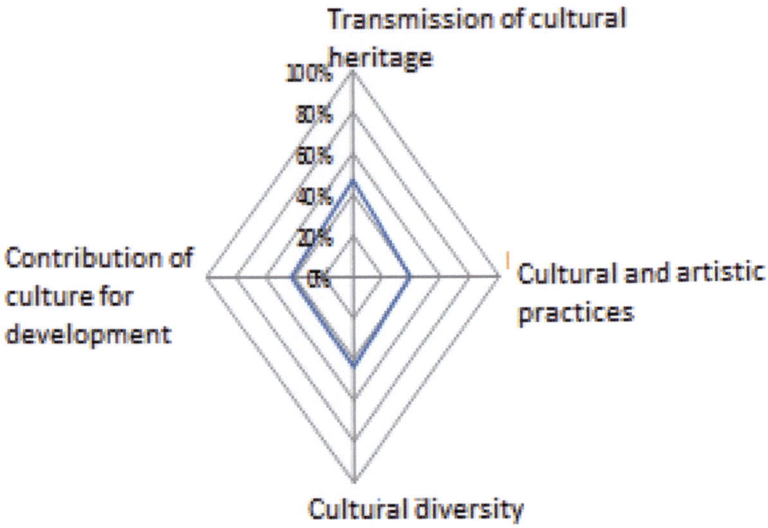

Fig. 4 Performance of cultural dimension themes (*Sources* Study data [8])

Table 2 Performance of ethical dimension themes (*Source* study data)

Theme	Average weight	Average performance (%)
Responsibility	2.5	59
Peace	2.5	50
Kindness	2.5	60
Sharing	2.7	63
Ethical approach	2.5	59

consistency and identification with common values. This dimension includes five themes (responsibility, peace, benevolence, sharing, and ethical approach) and 21 objectives. The results of the analysis showed that the ethical dimension in the roadmap has an average performance of 58%. Table 2 shows the performance of the different themes. It appears that the principles of responsibility, peace and solidarity are slightly taken into account in the current roadmap. The actors must further integrate the common values and enhance them in order to increase the performance of the ethical approach and to achieve the objectives of sustainable management of the dam.

After analyzing the ethical dimension, the avenues for improvement that will contribute to improving its performance are as follows:

- Make accountability reports, make them available and emphasize civic education;
- Fight against social exclusion and set up consultation frameworks to take individual concerns into account;
- Set up a local conflict management committee;
- Involve customary chiefs in conflict resolution and provide fair judicial treatment of cases.
- Promote gender equity, access to public services for all and develop team spirit to the detriment of individualism;
- Improve local governance and decentralization, the transfer of skills and resources to the decentralized level (Local Water Committee, management framework by shore, etc.);
- Reinforce values such as joking relationships, banish behavior deemed immoral by certain sections of society that tend to divide;
- Promote a peaceful political environment.

Governance dimension

The governance dimension aims to meet the needs of participation, democracy and transparency as well as institutional efficiency. It echoes the principles of subsidiarity, sound management, risk prevention and precaution. This dimension has eight (8) themes: institutions, instruments and processes, participation

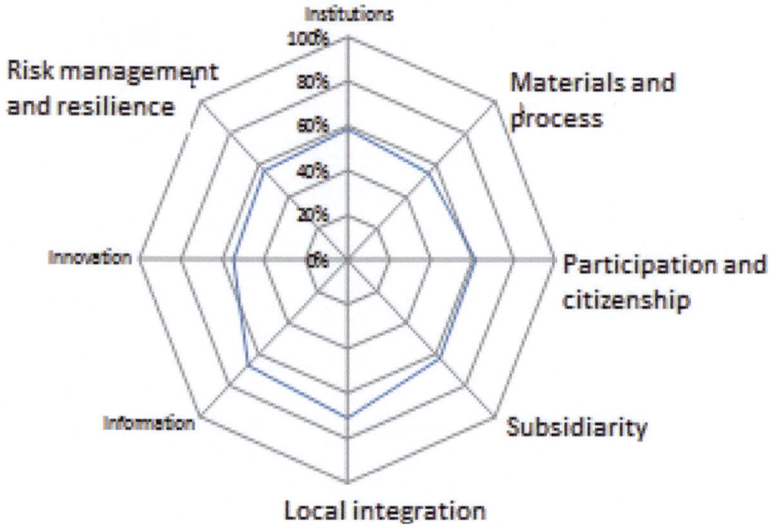

Fig. 5 Performance of governance dimension themes (*Source* study data [8])

and citizenship, subsidiarity, local integration, information, innovation, risk management and resilience, and thirty-four (34) objectives. The results of the analysis showed that the governance dimension has an average performance of 61%. Figure 5 shows the performance of the different themes. We observed that all the themes have a performance greater than 50% with strong local integration (71%). This demonstrates the commitment of all stakeholders to respect rules, good practices and institutions in the management of the dam.

The avenues for improvement that could contribute to improving the performance of the governance dimension are, among others:

- Establish transparent, accessible and impartial jurisdictions with a control monitoring mechanism to minimize corruption;
- Promote group work, take decisions in a participatory manner, and multiply workshops on the sustainable management of the dam;
- Involve and empower all actors with forecasts of consultation mechanisms before the implementation of any project;
- Organize citizen awareness campaigns and build the capacity of associations;
- Promote the potential around water retention in order to encourage research, create partnerships with research institutes.
- Organize workshops on the challenges of climate change and encourage projects involved in adapting to climate challenges;
- Encourage users to evolve in a network while prioritizing innovation.

Global results concerning sustainable development

The general analysis of the dimensions of sustainable development of the roadmap shows that the social, ecological and governance dimensions are taken into account with a satisfactory situation (60% to 79%). However, strong actions remain to be carried out for the ethical, economic and cultural dimensions, which are poorly taken into account with a situation that could be improved (40% to 59%), as indicated in Fig. 6. It appears that the current management model places more emphasis on the ecological dimension through the protection of the water reservoir and its biodiversity. Besides, a variation between the average performances of different dimensions (44% to 64%) was noticed. This

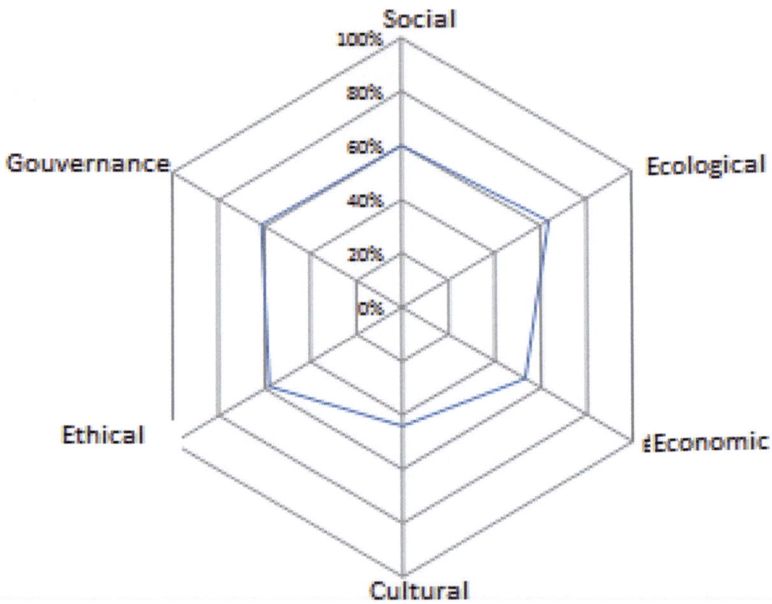

Fig. 6 Performance of the dimensions for sustainable development (*Source* study data [8])

should challenge the various actors to work more for the increase and the achievement of a balance between the dimensions in order to guarantee the sustainability of Loumbila dam.

3.1.3 Assessment of Current Occupations Around the Dam

The Loumbila dam was confronted in recent years with multiple uses, while its primary purpose was to supply drinking water to the City of Ouagadougou. Presently, the banks are occupied by the following users, given in Table 3.

In addition to these user groups, there are farmers, building and public works companies that have temporary uses.

The market gardener culture

The indigenous populations and foreigners practice market gardening more around the dam. They occupy a large part of the banks, and the use of water is done in several ways:

- Direct withdrawal of water from the lake basin using watering cans, when water is still abundant;
- Construction of wells to use water using sumps for watering vegetables;
- Opening of the trenches, allowing the water to be led from the body of water to the cultivation plots;

Table 3 Repartition of users around the dam in their activities sectors (*Source* study data)

N°	Users	Number	Percentage (%)
1	Market gardeners	2323	86.48
2	Sinners	219	8.15
3	Foresters	128	4.77
4	Hotel managers	8	0.30
5	Breeders	7	0.26
6	Drinking water supply	1	0.04
	Total	2686	100

- Pumping water, using machines from the lake to the cultivation plot.

Fishing

Fishing is one of the most practiced activities on the Loumbila dam. The main actors are foreigners. Fishing activity is spread throughout the year. Surveys done in November (2019) identified two groups of actors in this activity: fishermen and dealers. Fishermen use canoes and nets in the water to search for toxic matter and their spoils are sold wholesale to dealers. The latter in turn resell in detail along the national road n°3. We noted the capture of fry and the organizational shortcomings of these actors.

Tourism and hospitality

Tourist and hotel activities have developed around the Loumbila dam. Some of these activities have a close connection with water. Indeed, we note the creation of artificial beaches, recreational activities like Jet skiing, Pedalo riding, Sport fishing and Canoe trips by the promoters of these establishments.

Breeding

Livestock farming occupies an important place among the activities carried out around the dam. The presence of the water reservoir and the proximity of the area to Ouagadougou facilitate the breeding activities. The different practices are poultry farming, fish farming, fattening, etc. The use of water by breeders is done in an archaic way without adequate tools.

3.1.4 Percentage and Distribution of Water Users of Dam

The sample of 98 users covered by our survey enabled us to present the results in Table 4.

There are construction and public works companies that temporarily withdraw water for their activities. Figure 7 depicts the distribution of water users according to their activities in the dam area.

During our user surveys, we also asked for their opinions on the implementation of a management framework by shore. Results showed that 83.67% of surveyed users gave a favorable agreement for the establishment of a management framework by shore. Furthermore, they suggest that the leaders of these management frameworks be dynamic and driven by a spirit of unity and accountability. Indeed, it emerges from the results that 82 users (i.e., 83.67%) answered "yes" for the establishment of a management framework by shore. Only 16 gave an unfavorable opinion. Most of these justify the fact that the current associations involved in the management of the dam have not given satisfactory results. Some users claim not to have knowledge of the structures that manage the Loumbila dam. In addition, the survey revealed that 61.22% of users do not know the structures involved in the management of the Dam, only 24.49% of them recognized the Local Water Committee (LWC) and 11.22% the Nakanbe Water Agency.

Table 4 Users and their numbers from the survey (*Source* study data)

Users	Users number	Percentage (%)
Market gardeners	65	66.33
Fishermen	4	4.08
Breeders	15	15.31
Arborists	6	6.12
Contractors	3	3.06
Other users	5	5.10
Total	98	100

Our surveys also reveal that the majority of occupied land is acquired by inheritance. The owners claim that the market gardening practices used date back to their grandparents, hence there is a need to adopt modern practices and cultures in order to optimize the yields and quality of the market garden products. Indeed, most recognize the efforts made by the authorities in terms of training and awareness raising on good practices but prefer the old methods for lack of financial means. Figure 8 showed a distribution of users in relation to the methods of obtaining plots around the dam. We note four methods of obtaining land around the dam (inheritance, loan, rental and purchase). The land purchased is not numerous and concerns entrepreneurs who close

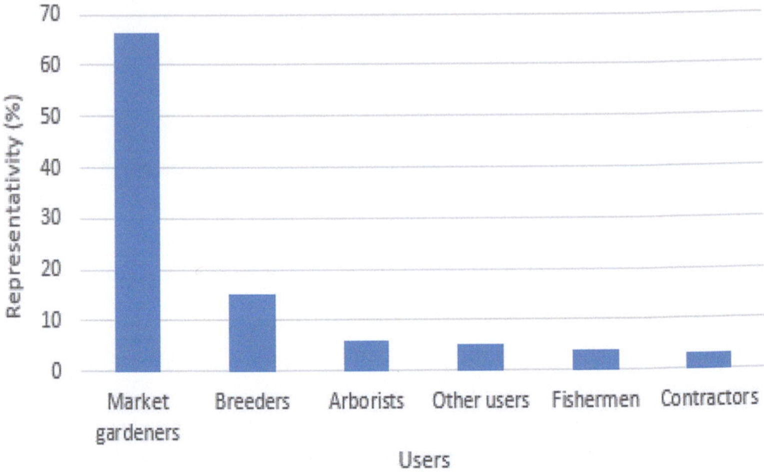

Fig. 7 Distribution of users according to their professions (*Source* study data)

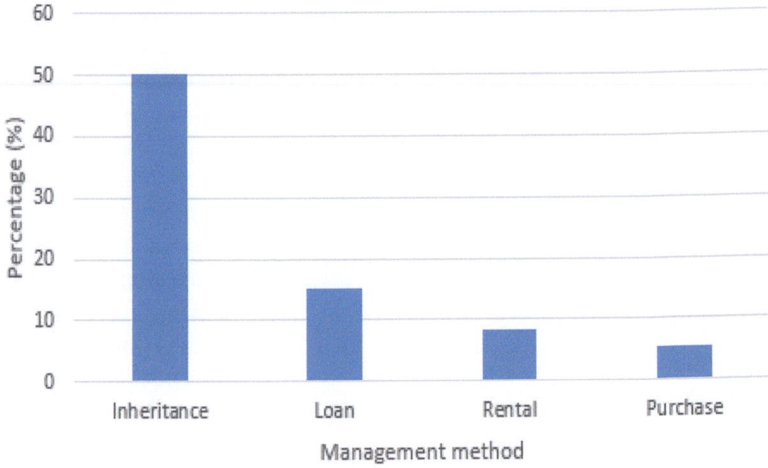

Fig. 8 Percentage of each management method in the dam area (*Source* study data)

it and carry out various activities. Regarding the inherited land, some are in the bed of the dam and the right-of-way due to the increase in the dyke. Consequently, these lands are invaded by water in the rainy season and by crops in the dry season.

3.1.5 Use of Water from Loumbila Dam for Market Gardening: Agricultural Activities

Agriculture remains Burkina Faso's primary economic activity despite unfavorable agro-climatic conditions. It occupies more than 80% of the country's population [9]. In Burkina Faso, there are two types of crops in relation to the season: seasonal crops and off-season crops. Indeed, throughout the country, farmers practice field work between May and October, a period corresponding to the rainy season. Off-season crops, practiced between November and May, concern localities that have benefited from water reservoirs or rivers. In the municipality of Loumbila, the farming system is still traditional, semi-intensive and not very productive. Seasonal migrations and the interest given to secondary activities by young people are the cause of the regression of the agricultural workforce. The different productions encountered in the commune are cereals, cash crops, market gardening, and tubers. Vegetable production is the most practiced one, because it covers the whole year, with more enthusiasm in the dry season. The main vegetable crops are potato, tomato, cabbage, lettuce, onion, carrot, eggplant, green bean, etc. Three main types of farming practices coexist in the municipality: small family farms, more or less intensive farms around the developed perimeters and large agro-pastoral farms around the Loumbila dam [10]. The strong growth in the number of producers testifies to the importance of agriculture around the dam. Due to the quantity of water being limited, the pressure on the water resource becomes more and more important.

The Loumbila Dam was built in 1947 with the aim of supplying drinking water to the City of Ouagadougou. Nowadays, several other activities are carried out around this work. Indeed, we note the occupation of the banks and the basin of the water reservoir by market gardeners with the consequence of siltation, which causes a reduction in the storage capacity of the structure; the uncontrolled use of chemicals, land speculation around the dam, illegal water extraction for civil engineering works, agricultural activities and livestock farms [11]. The Local Water Committee (LWC) of Massili Nord listed 2,686 water users including 2,323 market gardeners, 128 foresters, 07 breeders, 219 fishermen, 08 hotels and a drinking water supply, which is National Office of Water and Sanitation (NOWS). So many activities today threaten the survival of the Loumbila dam in terms of both the quantity of water and its quality. Already in 2012, a workshop had allowed the various actors to lead the reflection for the sustainability of the dam. The Loumbila dam is the summary of all the problems of management of water bodies in our area of expertise. It is under strong anthropogenic pressure. These include, i.e., the occupation of the banks and the basin by market gardeners, the uncontrolled use of chemicals leading to gradual pollution of the water reservoir and the disappearance of fishing potential, land speculation around the reservoir and in the basin of the dam, and the clandestine withdrawal of water for civil engineering works, agricultural and livestock activities [12]. The immediate consequence is siltation, which causes the reduction of the storage capacity of the dam and can lead to the total disappearance of the dam. The market gardeners say they are aware that their activities have certain negative consequences on the dam but insist on saying that they have no other solutions for the quest for their sustenance.

3.2 Discussion

The actors in the management of the Loumbila dam are the administrative authorities, customary authorities and civil society organizations. Each group of actors has its intervention technique for the preservation of water in quality and quantity as well as the ecosystem of the dam dominated by agriculture, tourism and fishing. Overall, management is more focused on implementing the recommendations of the roadmap for the sustainable management of the dam. Our investigations have shown that the administrative actors are not highly visible at the local level (24.49% LWC and 11.22% NWA), leading to omissions of certain very important concerns. In the general dynamic of sustainable development and particularly the sustainable management of water resources, the system of integration through participation at the local level had to be essential. Diop [13] demonstrated that "participatory management is the preferred means of regulation, which should make it possible to achieve the success of integrated and concerted multi-use management". Indeed, we believe that the creation of management frameworks by shore is an ideal management model to strengthen the notion of participation and access to information while empowering local populations in water management. This is confirmed by the national water policy, which through the principle of subsidiarity requires that issues that can be resolved locally must be decided and managed locally [14]. Nowadays, several management tools are used for better water management, but most of them are applied at a higher level thus leading to global resolutions of the difficulties of the water sector without real consideration local concerns. Our study has indicated that the establishment of management frameworks by shore justifies the fact that achieving genuine participatory and sustainable management of water reservoirs necessarily involves the development of local tools adapted to local realities.

In addition, the management of natural resources near the Loumbila dam must be done with great transparency.

Some local actors have pointed out that obtaining financing for the development of new sites can solve the difficulties linked to land speculation. The actors we approached raised financial difficulties in the implementation of their activities. However, an awareness of the issues around water management can provide solutions to certain difficulties related to bad behavior, ignorance of degrading and polluting practices of water retention. The roadmap is the reference tool that guides the implementation of activities within the framework of the management of the dam. Data analysis showed that the cultural, ethical and economic dimensions are not taken into account in this tool, which allows us to affirm that in accordance with the principles of sustainable management, obeying the requirements of sustainable development, where a balance is needed between dimensions, this roadmap has limits. This assertion is based on the research report of the national grouping of regional environmental councils of Quebec on the conceptual and historical evolution of sustainable development, which states that "sustainable development is a dynamic whole, fueled by three fields of actions, which each bring an element of progress, of transformation: the economy, the social dimension and the environment" [15]. Indeed, the trend toward balanced management will make it possible to definitively resolve the difficulties observed in the management of water resources, because it will take into account current and future concerns while promoting broader participation. Hence, the main objective of Integrated Water Resources Management is to achieve a balance between the use of water, its protection and its conservation in order to guarantee the water sustainability and characteristics. For sustainable management, it is necessary to carry out actions that associate and empower local actors. Our results show that at the local level the recognition of the efforts carried out by the central authorities of water management remains poorly perceived. In reality, some users demand the removal of signs prohibiting all activities inside the easement strip, because they were not consulted before the installation. It emerged that representatives of associations participate to the workshops, and take some decision-making in the dam management. However, our user survey showed that 86.73% of respondents do not belong to associations. To do this, it will be more conducive to use communication and participation tools that will inform and reach the maximum number of users. Water users who are recognized by the law are supported and work directly with Nakanbe Water agency and Local Water Committee. The latter want more participation from their local representatives, who are not beholden, and put more forward their interests during decision-making. Despite the conflicts related to the management of water from the Loumbila dam, the actors are working to find more sustainable solutions to ensure the quantity and quality of water. The real problem is related to funding and to the lack of knowledge by some users of good practices in the management of water reservoirs.

4 Conclusions

Research conducted on the sustainable management of the Loumbila Dam has shown the presence of several actors in water management. These actors from the administrative and customary authorities and civil society organizations have specific missions in the maintenance of the dam and the preservation of the quality and quantity of water. The administrative authorities ensure the development and execution of policies related to the management of water resources. They support associations and users in promoting their activities through awareness-raising, training and support advice. Customary authorities and guarantors of tradition in local communities have rules and principles that fall within the framework of natural resource management. Their involvement in water management allows local knowledge and know-how to be taken into account in order to achieve more sustainable results. Civil society organizations are strongly represented by associations, which provide citizen monitoring and act as interlocutors between central structures and users. These associations are in several areas related to the use of water and work for the smooth running of their activities. Despite this strong involvement of the actors in the management, we noted a lack of transparency in the actions of certain actors, which is a source of land speculation near the dam. We also noted the low participation of users in decision-making. Our surveys and literary reviews indicated that the establishment of management frameworks was more than necessary to involve the populations more in the management of the public goods. The analysis of the roadmap revealed that the social, ecological and governance dimensions are well taken into account, while efforts remain to be made for the cultural, ethical and economic dimension. The shortcomings noted at the level of the actors as well as the low consideration of certain dimensions confirm our overall hypothesis. Hence, our recommendation is to create a management framework by shore of the Loumbila dam in order to revitalize the management of this water resource.

Acknowledgements Authors would like to thank DAAD and Exceed- Swindon project for their technical support and enabling to publish this study.

References

1. Institut National de la Statistique et de la Démographie, INSD. Recensement Général de la Population et de l'Habitat. Rapport 2006, 01 BP 374 Ouagadougou 01; 2006. p. 52.
2. Institut National de la Statistique et de la Démographie, INSD. Projections démographiques de 2007 à 2020 par région et province. Avenue de l'Indépendance, 01 BP 374 Ouagadougou 01;2009. p. 10.
3. Direction Générale de l'Hydraulique, DGH. État des lieux des ressources en eau du Burkina Faso et de leur cadre de gestion. Ministère de l'Environnement et de l'Eau, SG, DGH, Burkina Faso;2001. p. 14p.
4. Mairie de Loumbila. Plan Communal de Développement. Mairie de Loumbila, DREP Plateau Central;2017, p. 13.
5. Institut Géographique du Burkina. Rapport sur la cartographie des ressources en eau dans la région du centre. Ouagadougou, Burkina Faso;2015.
6. Groupe de Recherche et d'Echange Technologique, GRET. Calcul de la taille d'un échantillon pour une enquête. Fiches techniques et méthodologiques, Éditions du Gret, 45 bis avenue de la Belle Gabrielle, 94736 Nogent-sur-Marne Cedex, France;2017. p. 713–5.
7. Tahirou S, Zerbo P, Sanou Y, Pare S, Nassirou Ado M. Perceptions Paysannes Des Aléas Climatiques Sur La Production Du Riz Sur Le Périmètre Irrigué De Saga Dans La Vallée Du Fleuve Niger. Eur Sci J. 2021; 18(3):217. https://doi.org/10.19044/esj.2022.v18n3p217.
8. Sawadogo F. Mémoire de Master 2 en management de l'environnement et du développement durable, Université Aube Nouvelle/Ouagadougou, Burkina Faso; 2020. https://iscosafrique.com/2020/11/04/soutenance-de-memoires/.
9. Programme Alimentaire Mondial des Nations Unies, PAM. Analyse Globale de la Vulnérabilité, de la Sécurité Alimentaire et de la Nutrition. Siège social: Via C.G. Viola 68, Parco de Medici, 00148, Rome, Italie;2014. p. 34.
10. Ouédraogo O. Identification des systèmes de productions agricoles intègres existants et propositions de meilleures pratiques pour la commune de Loumbila. Mémoire de master: Université de Ouagadougou, UFR/SH; 2013. p. 04–7.
11. Benon/Yatassaye FN. Bilan de la gestion à mi-parcours du barrage de Loumbila, région du Plateau Central, Burkina Faso;2018. https://www.radarsburkina.net/index.php/fr/environnement/345-envasement-du-barrage-de-loumbila-il-fallait-arreter-les-exploitations-connexes-pour-satisfaire-a-la-priorite-absolue-que-constitue-l-approvisionnement-en-eau-potable-des-populations-et-sauver-loumbila-d-une-catastrophe-ecologique-nana-fatoumata-benon-yatassaye.
12. Executive Direction of Nakanbe Water Agency. Report on assessment of water resource management in Loumbila dam, Burkina Faso;2019.
13. Diop P. épouse Ndao. Vers une stratégie de gestion participative multi-usages de la ressource en eau dans le delta du Fleuve Sénégal: processus de décision et outil de régulation autour du lac de Guires.T hèse de doctorat de l'Université Cheikh Anta Diop de Dakar et de l'Université Paris-Saclay;2017. p. 28–47.
14. Ministère de l'Environnement et du Cadre de Vie. Politique Nationale en matière d'Environnement (PNE). Burkina Faso;2005. p. 30.
15. Regroupement National des Conseils Régionaux de l'Environnement du Québec. Rapport de recherche et d'évolution conceptuelle et historique du développement durable, Atelier d'aménagement, d'urbanisme et d'environnement;1998. p. 1–6.

Determination of Rainwater Harvesting Potential in Greenhouses: A Study in Bursa Uludag University

Busra Yayli and Ilker Kilic

Abstract

Sustainable water management has become more critical in recent years due to climate change and water scarcity, increasing its impact. The rainwater harvesting method is an old, simple, and environmentally friendly method for collecting and using rainwater for areas that have or will face water shortages soon. Rainwater harvesting is an essential alternative to ensure future food supply in agricultural production where water needs are essential. This study aimed to theoretically determine the collectible rainwater potential and to make a cost analysis in the greenhouses of Bursa Uludag University, Faculty of Agriculture, Agricultural Practice Center. In addition, its usability in production in the greenhouse was evaluated. According to the study results, the amount of rainwater collected on the greenhouse roofs examined is 229,612, 272,176, and 79,870 m^3 for G1, G2, and G3, respectively. Among the crops in greenhouses, lettuce has the highest water consumption. Since greenhouse-numbered G3 has a larger area and more production is made, there is a need for more water consumption. Rainwater harvested periodically in the greenhouse can cover 35.7% of crop production in the first and 35% in the second periods. Annual harvested rainwater in the greenhouse provides 35.3% of the water consumption of the crop. The cost of rainwater systems designed for greenhouses is approximately 11,742 \$, and the system's payback period is ten years.

Keywords

Bursa · Greenhouse · Rainwater harvesting · Water scarcity

1 Introduction

Drought indicates its effect ever-increasingly, so the global water crisis is inevitable. Usable water resources are decreasing or disappearing, so the most crucial trigger of sustainable life. Convenient management of available water is vitally important for the sustainable use of water for various purposes.

Countries are among the 'countries shortage of water', if the amount of usable water per capita is 1000–3000 m^3 per year. Türkiye, which has a usable water amount per capita of 1500–1600 m^3/year, is a country that suffers from water shortages [1]. It is predicted that the population of Türkiye will reach 100 million in 2030, and the

B. Yayli · I. Kilic (✉)
Department of Biosystems Engineering, Faculty of Agriculture, Bursa Uludag University, 16059 Bursa, Türkiye
e-mail: ikilic@uludag.edu.tr

B. Yayli
e-mail: busrayayli@uludag.edu.tr

M. Bahadir et al. (eds.), *Agricultural Water Management*, Water and Wastewater Management,
https://doi.org/10.1007/978-3-031-96134-2_10

amount of potable water per capita will decrease to 1120 m^3 [2]. Türkiye will inevitably become a water-poor country due to adverse situations such as the increasing drought problem, unconscious use of water resources, population growth, pollution of water resources, and increased urbanization. Therefore, alternative methods are becoming more critical in supplying the water needed.

60–90% of water stock is used in agriculture depending on the climate and the economic development of the region [3–5]. In agricultural production, where irrigation is essential, alternative methods become imperative for future food supply. Rainwater harvesting is an ancient method from B.C., often used for drinking water or agricultural applications. Rainwater can be collected by appropriate methods and used in agriculture for irrigation in regions with drought. Rainwater harvesting is defined as collecting and accumulating rainwater for various purposes such as domestic, industry, agriculture, and the environment by different methods and reducing surface run-off discharged into the sewer system [1, 6–8]. Harvested rainwater can be used as domestic or drinking water according to its quality. Necessary treatment processes should be provided according to their use as drinking water.

Rainwater harvesting provides a reliable supply of water, where surface and underground water resources are insufficient or where it is economically difficult to provide clean water. There are different techniques of rainwater harvesting. However, the most common ones are grouped as roof systems (micro systems), farm systems (micro systems), valley bed systems (macro systems), and non-valley systems (macro systems) [9, 10]. However, roofs are generally considered rainwater collection areas. The roof has three main components: rainwater harvesting collection area, transport system, and storage facilities [11]. The materials used in the roofs and the functional areas of the roofs affect the water quality, and the collected rainwater generally needs to meet the drinking water quality. With rainfall, urban aerosols, volatile particles, and gases in the atmosphere are mixed with rainwater, the pollutants on the roof surface are washed off, causing the water to become polluted [12]. Since the cost of treating rainwater as drinking water will be high, rainwater harvesting is one of the most popular alternative sources in activities considered domestic water [13].

A significant part of the plant water needs in plant production in the field is provided by precipitation. However, since a closed production area is created in greenhouses, the plants' water needs cannot be supplied by rainwater and irrigation is required [14]. In cases where the water source is far from the greenhouse or the water supply is limited, the water must be transported through pipes to the greenhouse. Considering such difficulties, rainwater harvesting is a popular method in greenhouse operations. Roof areas are considered collection areas in studies conducted with rainwater harvesting in greenhouses [11, 15, 16].

This paper determined the rainwater harvesting potentials of greenhouses belonging to Bursa Uludag University, Faculty of Agriculture. In addition, the rainwater system design and the cost were calculated. The depreciation period of the collected rainwater harvesting system and the usability of the water in the greenhouse were evaluated.

2 Materials and Methods

In this study, the rainwater collection potential of the greenhouses of Bursa Uludag University, Faculty of Agriculture in Türkiye and the irrigation need of the production in the greenhouses were calculated, and the system was designed. The satellite image of the location was obtained from Google Earth and is given in Fig. 1. Although there are four greenhouses, the greenhouse number G2 is not considered in the study, because it is theoretically difficult to calculate rainwater due to its arc roof. G1 and G3 are integrated greenhouses; G1 consists of two greenhouses, G3 consists of four, and G4 consists of one greenhouse. Photographs of the greenhouses taken by the authors are given in Fig. 2.

This study calculated the rainwater capacity separately for each greenhouse with Eq. (1). According to the equation, the roof area of the

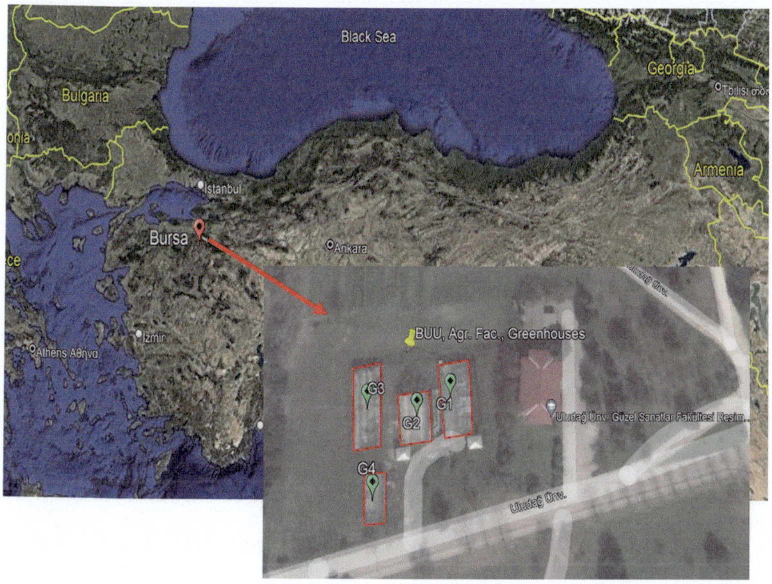

Fig. 1 Study area

Fig. 2 Images of greenhouses

greenhouses, the amount of rain, the roof coefficient, and the filter efficiency coefficient parameters were used.

$$\text{RCW} = \text{RA} * \text{RQ} * \text{RC} * \text{FEC} \quad (1)$$

where RCW: Rainwater capacity

RA*:* Roof area of the greenhouse (m^2)

RQ: Total annual rainfall quantity (mm)

RC: The roof coefficient means that all the rain falling on the roof cannot be recycled. Germany has pioneered collecting and using rainwater with the DIN 1989 standard. The DIN 1989 German Standards defines the roof coefficient as 0.8 [17].

FEC: The filter efficiency coefficient is the efficiency coefficient of the first filter passed to separate the rainwater obtained from the roof from the visible solids. It is a coefficient given by calculating that some water cannot pass through here. The DIN 1989 German Standards defines this coefficient as 0.9 [17].

Plant Production

Cucumbers, beans, and lettuce were produced in monitored greenhouses for one year. It has been taken into account that cucumbers will be planted twice, in spring and autumn. It has been also taken into account that the bean planting will be done in July–September and September–November. Lettuce cultivation is planned for between November–January in the first period and January-March in the second period. Table 1 gives information about the stage of growing the products.

Calculation of Crop Water Demand

The study calculated plant water consumption of lettuce, bean, and cucumber products planned to be produced throughout the year. While calculating the crop water consumption, it was calculated with Eqs. (2) and (3).

$$AET_c = ET_o * k_c \quad (2)$$

AET_c = Actual crop evapotranspiration (mm/day)

ET_o = Reference evapotranspiration (mm/day)

k_c = Crop coefficients

$$CWR_d = AET(1 + I_i) * A_{\text{crop}}/AG \quad (3)$$

CWR_d = daily crop water requirement (mm/day)

I_i = between 0.03 and 0.1 loss factor for the drip irrigation system

A_{crop}/AG = 0.9 for vegetables and flowers on ground beds.

3 Results and Discussion

Considering that production cannot be made in the whole area in greenhouses, it has been regarded that crop production can be made in 2/3 of the floor areas. Accordingly, the production areas in greenhouses are 300, 355, and 104 m^2 for G1, G3, and G4, respectively. The harvest amounts of the crops to be obtained in total production periods and crop production per unit area (m^2) are given in

Table 1 Product information planned to be grown

Planting time	Cucumber	Bean	Lettuce
1. Period			
Sowing seeds	25 March 2023	11 July 2023	15 November 2023
Planting seedlings	10 April 2023	26 July2023	1 October 2023
Last harvest	15 May 2023	10 September 2023	16 January 2023
2. Period			
Sowing seeds	15 May 2023	11 September 2023	17 January 2024
Planting seedlings	1 June 2023	26 September 2023	2 February 2024
Last harvest	10 July 2023	9 November 2023	20 March 2024

Table 2 Total harvest amounts of the crop intended to be grown in greenhouses

Total harvest quantity in greenhouses			
	G1	G3	G4
Cucumber (ton)	5.1	6.1	1.8
Bean (kg)	166	197	58
Lettuce (number)	2913	3453	1013
Total harvest quantity per unit area in greenhouses			
	G1	G3	G4
Cucumber (ton)	0.017	0.017	0.017
Bean (kg)	0.56	0.56	0.56
Lettuce (number)	9.72	9.72	9.72

Table 3 Coefficient, evapotranspiration and water consumption values of the products (mm/day) [18]

	Lettuce	Bean	Cucumber
$K_{c\ ini}$	0.7	0.5	0.6
$K_{c\ mid}$	1	1.05	1
$K_{c\ end}$	0.95	0.9	0.75
$AET_{c\ ini}$	1.59	1.14	1.36
$AET_{c\ mid}$	2.27	2.38	2.27
$AET_{c\ end}$	2.16	2.04	1.70
$CWR_{d\ ini}$	1.52	1.09	1.31
$CWR_{d\ mid}$	2.18	2.28	2.18
$CWR_{d\ end}$	2.07	1.96	1.63

İni: initial stage (sowing seeds), mid: mid-season stage (planting seedlings), end: end of the late season stage (last harvest)

Table 2. According to Table 2, the harvest amount per unit area is the same for the same products in all greenhouses (G1, G3, G4).

The FAO Cropwat 8.0 software was used to calculate the total evaporation (ET_o) values in the FAO Penman–Monteith method. This software used temperature, humidity, wind, and the number of sunny days. The coefficient values (k_c) of single time-averaged products for use with ET_o were taken from the FAO website [18] (Table 3).

In the study, the growing period was given in 3 stages. Cultivation periods vary between 52 and 68 days. In Table 4, the total water consumption of the products according to the 1st and 2nd Periods are given. Lettuce was the product that consumed the most water per unit area.

Harvested Rainwater Capacity

The rainiest seasons are spring and winter in Bursa. The average annual precipitation is 709.5 mm, and the months with the most rainfall are December and January. January has the highest number of rainy days, with 14.95 days. Average monthly precipitation data 1928–2021 from the General Directorate of Meteorology website was used (Table 5) [19]. Although hourly or daily rainfall data give more accurate results in the daily water balance model, the lack of reliability in most countries is one of the most significant limitations [20, 21]. It is seen that most precipitation falls in December and January.

The greenhouse roof areas examined in the study were calculated by measuring directly with the help of a laser meter. Roof areas of the

Table 4 Total crop water consumption according to periods

	G1	G3	G4	Total (m^3/m^2)
1. Period				
Lettuce	0.12	0.12	0.12	0.36
Bean	0.11	0.11	0.11	0.32
Cucumber	0.09	0.09	0.09	0.27
2. Period				
Lettuce	0.13	0.13	0.13	0.39
Bean	0.12	0.12	0.12	0.35
Cucumber	0.10	0.10	0.10	0.29

Table 5 Average rainfall amount of Bursa 1928–2021 [19]

Month	Rainfall (mm)
January	89.4
February	75.7
March	69.9
April	61.8
May	50.9
June	34.8
July	22.3
August	18.3
September	43.7
October	66.5
November	76.3
December	99.9
Total	**709.5**

Table 6 The average amount of rainfall collected on roofs (m^3)

Month	Amount of rainfall collected		
	Greenhouse 1	Greenhouse 3	Greenhouse 4
January	28,932	34,295	10,064
February	24,498	29,040	8,522
March	22,621	26,815	7,869
April	20,000	23,707	6,957
May	16,473	19,526	5,730
June	11,262	13,350	3,918
July	7,217	8,555	2,510
August	5,922	7,020	2,060
September	14,142	16,764	4,919
October	21,521	25,510	7,486
November	24,693	29,570	8,589
December	32,330	38,323	11,246
Total	**229,612**	**272,176**	**79,870**

examined greenhouses are 449, 533, and 156 m^2 for G1, G3, and G4, respectively. The amount of rainwater falling on the total roof areas is given in Table 6. The amount of rainfall collected in the greenhouses is 229,612, 272,176, and 79,870 m^3 for G1, G3, and G4, respectively. It was observed that the highest amount of rainwater collected on the roofs was in December and January. The harvested rainwater for Greenhouse 1 and 3 is 511 m^3 per unit area (m^2). For Greenhouse 4, it is 512 m^3/m^2.

Table 7 shows the total water consumption of the crops produced in greenhouses according to Periods 1 and 2, and the water needs per unit area. The highest amount of water collected per unit area in all greenhouses in both periods was for lettuce. Boyacı and Kartal [15] reported the total rainfall amount as 388.3 L/m^2 and the harvested rainwater as 349.7 L/m^2 in their study in Kırsehir. According to the heating status in greenhouses, the collected rainwater supplied 47.7% of the water consumption in the heated greenhouses and 61.4% in the unheated greenhouses. A study on rainwater harvesting and meeting water consumption in poultry houses stated that the region's annual precipitation was 733.6 L/m^2, and the need could not be met under all conditions. However, it has been said that it can be met with rainwater harvesting in the evaporative scattering systems applied in the henhouse [12].

Table 7 Rainwater harvesting potential of the greenhouses in this study (m^3)

	Greenhouse 1	Greenhouse 3	Greenhouse 4	Total (m^3)	Total (L/m^2)
1. Period					
Lettuce	64.2	71.2	20.9	156.4	191
Bean	15.3	17.7	5.4	38.4	47
Cucumber	32.8	38.8	11.4	83.0	101
2. Period					
Lettuce	56.1	66.5	19.5	142.1	173
Bean	38.4	46.4	3.6	88.4	120
Cucumber	21.5	26.0	7.6	55.1	67

Rainwater Harvesting System Design

The storage structures to be planned should be of sufficient size in order to collect rainwater. The greenhouses examined in the study were designed with the AUTOCAD programme to build the system underground through a rainwater collection tank and a reuse flow chart (Fig. 3). In the system design, considering that collecting can be done with gutters on the edges of the arc-roofed greenhouse, these greenhouses were also planned by including them in the system. The rainwater falling on the roof will be filled into the underground tank by direct water gravity through the gutters. The size of the gutter is designed according to the flow during the heaviest rain. The system has a filter to clean the water from the rain storage tank. In addition, considering that the amount of precipitation is high in some periods, an indicator has been added to show the amount of water in the tank. In this case, there will be no excess water in the tank, and safety will be created against the pressure load that may occur in case of overpressure.

System Depreciation Period

The material and cost calculation for the design of the rainwater system planned in the greenhouses is given in Table 8. If rainwater is used as drinking water in the harvesting method, the water must be disinfected with chlorination, filtered, or boiled [9, 22, 23]. In the study, since the rainwater stored in the system design is not considered to be used as drinking water, it is not expected to be of pure quality in terms of content. The filters are used to treat some pollutants and bird droppings. It is thought that the wastes, which are divided into smaller pieces, can show the effect of fertilization with irrigation on the plants produced.

With the help of the rainwater system to be installed, water savings were achieved in the greenhouses throughout the year. On average, 35.3% of the water needs required for annual production in greenhouses can be supplied, and 1145 $ is saved annually. The cost required for installing the system is 11,745 $, and the system will pay off in 10 years and three months.

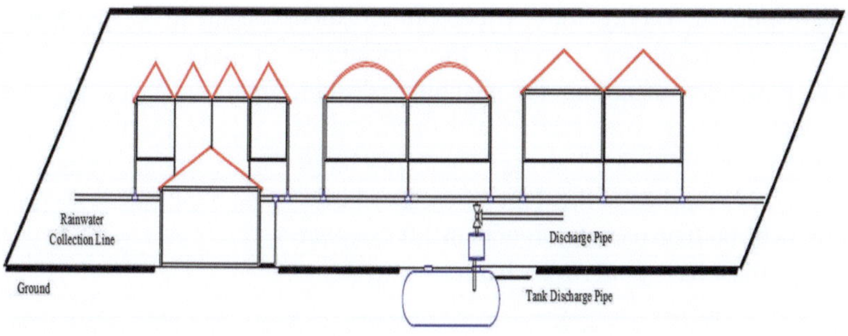

Fig. 3 Rainwater collection system design

Table 8 Materials and costs used in the system

Number	Materials	Piece	Cost ($)
1	100 m^3 underground polyester water tank	1	11,290
2	Elbow pipe	9	13.5
3	T pipe	5	4.5
4	11 Ø pipe	100 m	63.3
5	Three-way valve	1	90.3
6	Filter	1	286

Table 9 Depreciation time of the system installed in greenhouses

	1. Period	2. Period	Total
Amount of water collected (m^3/m^2)	0.34	0.36	0.70
Water consumption (m^3/m^2)	0.95	1.29	1.03
Return rate (%)	35.7	35	35.3
Amount of savings ($)	547	598	1145
Cost ($)			11,745
Payback period (year)			10.25

Table 9 shows the savings made annually from greenhouses and the depreciation period.

4 Conclusions

Factors such as global warming, climate change and increasing population cause direct and indirect harmful effects on usable water resources. Sustainable water resources are needed as an alternative to rapidly polluted and consumed water resources. Advanced rainwater harvesting management is emerging as a sustainable method that can be transferred to future generations.

The Rainwater Collection, Storage and Discharge Systems regulation for rainwater harvesting in Turkey took effect on 23 June 2017. This regulation regulates the procedures and principles regarding the planning, design, project planning, construction and operation of stormwater collection, storage and discharge systems. According to the Regulation, new buildings to be built on parcels larger than 2,000 square meters must have a rainwater harvesting system.

Significant reasons such as global warming causing severe water shortages, the rapidly disappearing water resources, and high-water supply costs necessitate turning to economical and practical alternative methods such as rainwater harvesting. Protecting potential water resources and creating new water resources with the widespread use of alternative methods will support the preservation of ecological balance and sustainable development.

Agricultural production is one of the largest consumers of clean water. Creating new resources will benefit agriculture regarding the depletion of clean water resources. In this respect, rainwater harvesting may be essential for a new clean water source. The fact that rainwater harvested from the available rainfall can provide the water demand all year round without the need for high capital investments makes this method more attractive. Besides, rainwater harvesting system reduces the water footprint of the facilities or regions used.

References

1. Kılıç MY, Abuş MN. Rain water harvesting in a garden house sample. Int J Agric Wildlife Sci. 2018;4(2):209–15.
2. WWF. World Wide Fund for Nature, Su döngüsünü iyileştirmek için: yağmur suyu hasadı. 2020. https://wwftr.awsassets.panda.org/downloads/ysh_web_ekim_2020_1.pdf. Accessed 22 Aug 2022.
3. Velasco-Muñoz JF, Aznar-Sánchez JA, Belmonte-Ureña LJ, Román-Sánchez IM. Sustainable water use in agriculture: a review of worldwide research. Sustainability. 2018;10(4):1084. https://doi.org/10.3390/su10041084.
4. Velasco-Muñoz JF, Aznar-Sánchez JA, Batlles-delaFuente A, Fidelibus MD. Rainwater harvesting for agricultural irrigation: an analysis of global research. Water. 2019;11(7):1320. https://doi.org/10.3390/w11071320.
5. Adeyemi O, Grove I, Peets S, Norton T. Advanced monitoring and management systems for improving sustainability in precision irrigation. Sustainability. 2017;9(3):353. https://doi.org/10.3390/su9030353.
6. Andualem TG, Hagos YG, Teka AH. Rainwater harvesting potential assessment for non-potable use in urban areas. Sustain Water Resour Manag. 2020;6(6):1–8.
7. Üstün GE, Can T, Küçük G. Rainwater harvesting in buildings. Uludağ Univ J Fac Eng 2020;25(3):1593–10. https://doi.org/10.17482/uumfd.765561.
8. Kılıç İ, Yaylı B. A research on the use of rainwater in the faculty of agriculture of Uludag University in Bursa: rainwater harvesting. In: 6th international conference on sustainable agriculture and environment. ISBN 978-605-184-194-6, October 3–5, 2019, Konya;2019. https://www.researchgate.net/profile/Mithat-Direk/publication/337677566_6th_International_Conference_on_Sustainable_Agriculture_and_Environment/links/5f50f7d6458515e96d274e2c/6th-International-Conference-on-Sustainable-Agriculture-and-Environment.pdf#page=70.
9. Yetik AK, Şen B. Importance and techniques of water harvesting systems. Turk J Agri-Food Sci Technol. 2020;8(1):46–53. https://doi.org/10.24925/turjaf.v8isp1.46-53.3952.
10. Hacısalihoğlu S. Contribution of water harvesting practices to the sustainability of rural areas: Bursa-Karacabey case study. Osmaniye Korkut Ata Univ J Inst Sci Technol. 2022;5(2):767–82. https://doi.org/10.47495/okufbed.1001531.
11. Singh KG, Sharda R, Singh A. Harvesting rainwater from greenhouse rooftop for crop production. Agric Res J. 2019;56(3):493–502. https://doi.org/10.5958/2395-146X.2019.00077.2.
12. Çaylı A. ınvestigation of the meeting poultry house water requirement from rainwater harvesting: the case of Kahramanmaraş. J Agric Nat. 2021;24(5):1048–58.
13. Hammes G, Ghisi E, Padilha Thives L. Water end-uses and rainwater harvesting: a case study in Brazil. Urban Water J. 2020;17(2):177–83. https://doi.org/10.1080/1573062X.2020.1748663.
14. Baytorun AN, Zaimoğlu Z, Ünlü M. Determination of harvesting and storage capacity of rain water in greenhouse establishments. Turk J Agric—Food Sci Technol. 2019;7(1):22–9. https://doi.org/10.24925/turjaf.v7i1.22-29.2011.
15. Boyacı S, Kartal S. Rainwater harvesting on greenhouse roof and use in irrigation. Int J Res-Granthaalayah. 2019;7:93–100. https://doi.org/10.5281/zenodo.2580547.
16. Londra PA, Gkolfinopoulou P, Mponou A, Theocharis AT. Effect of rainfall regime on rainwater harvesting tank sizing for greenhouse irrigation use. Hydrology. 2022;9(7):122. https://doi.org/10.3390/hydrology9070122.
17. DIN. Regenwassernutzungsanlagen. 1989, German: Deutsches Institut Normung DIN;1989.
18. FAO. Crop evapotranspiration—Guidelines for computing crop water requirements—Chapter 6—ET_c—Single crop coefficient (K_c). 1998. https://www.fao.org/3/X0490E/x0490e0b.htm. Erişim Tarihi: 19 Sept 2022.
19. MGM. Turkısh state meteorological service. 2022. https://mgm.gov.tr/veridegerlendirme/il-ve-ilceler-istatistik.aspx?k=undefined&m=BURSA. Accessed 19 Aug 2022
20. Matos C, Santos C, Pereira S, Bentes I, Imteaz M. Rainwater storage tank sizing: case study of a commercial building. Int J Sustain Built Environ. 2013;2(2):109–18.
21. Lizárraga-Mendiola L, Vázquez-Rodríguez G, Blanco-Piñón A, Rangel-Martínez Y, González-Sandoval M. Estimating the rainwater potential per household in an urban area: case study in Central Mexico. Water. 2015;7(9):4622–37.
22. Pamuk Mengü G, Akkuzu E. Global water crisis and water harvesting techniques. J Adnan Menderes Univ Agric Fac. 2008;5(2):75–85.
23. Kılıç MY, Adalı S. Rainwater harvesting in a shopping mall example. Bursa Uludağ Univ J Fac Eng. 2022;27(1):29–40.

Historical Change of Water and Agriculture Practices and Regulatory Frameworks in Yemen

Musaed Aklan, Abdulwahab Al-Mujahed, and Raidan Bassah

Abstract

Yemen has a long and rich history of relying on indigenous rainwater harvesting (RWH) and agricultural systems. Until 50 years ago, Yemen had relied on and adapted entirely to the available and renewable water resources by using indigenous water harvesting systems. In the absence of lakes, rivers, and perennial surface water bodies, indigenous water systems had supported the livelihoods of local people, and helped to satisfy their water demands for different uses, domestic, agricultural, and livestock. After the 1970s, with the introduction of deep tube wells and rigs to Yemen, people started relying on groundwater for all their uses. Today, most groundwater basins in Yemen suffer from over-abstraction. Irrigated farming, particularly for qat production, uses more than 90% of the annually abstracted water. This resulted in groundwater levels declining at high unsustainable rates of as much as six meters per year. Yemen today is one of the most water-scarce countries in the world. Therefore, this chapter aims at introducing and documenting the widely used indigenous water systems and exploring their importance for bridging the gap between supply and demand nationwide, and for satisfying water demands for local communities. In addition, the chapter investigates the legal framework and necessary regulations that encourage water users to go back to these indigenous practices and to relieve pressure on groundwater resources.

Keywords

Groundwater · Indigenous practices · Renewable resources · Water-food scarcity

1 Introduction

Yemen is facing a growing number of political, economic, security, and environmental challenges. However, the most important current vulnerability in Yemen is the dwindling reserves of the main natural sources of oil and water [1]. The United Nations Development Program's (UNDP's) index of human development for 2021/2022 ranked Yemen as 183rd out of 191 countries, placing it in the lowest human development category [2]. Though the agriculture sector consumes around 90% of available water sources, Yemen imports most of its food. Yemen

M. Aklan (✉)
Sana'a Center for Strategic Studies, Sana'a, Yemen
e-mail: m.aklan@sanaacenter.org

A. Al-Mujahed
Water and Sanitation Consultant for Yemen Social Fund for Development, Sana'a, Yemen

R. Bassah
Geospatial Data Analyst, Sana'a Center for Strategic Studies, Sana'a, Yemen

M. Bahadir et al. (eds.), *Agricultural Water Management*, Water and Wastewater Management,
https://doi.org/10.1007/978-3-031-96134-2_11

faces an acute water crisis and poverty, and the economic situation of the country has limited the expensive options, such as desalination and the transmission of water from the coastal areas, and/or abstraction and transferring of water from promising basins to the water-stressed highland basins located at 2000–3000 m above sea level [3].

Traditionally, Yemenis relied entirely on renewable natural resources such as locally sourced traditional water harvesting systems, and they developed their own norms and regulations to manage this renewable water source. Rainwater supported all domestic, agricultural, and industrial needs. Following the 1970s introduction of deep tube wells encouraging expanded farming, reliance grew on groundwater extracted from deep drilling without oversight. Unfortunately, Yemenis did not develop traditional norms for managing groundwater, simply because it is not a traditional source, and the successive governments failed to enforce the water law. Consequently, unprecedented agricultural expansion stressed aquifers as the population also concentrated in highland cities. While overpumping for crops, Qat, and illegal drilling drive scarcity, broader influences include weak governance and climate change exacerbating stress. Yemen now depends largely on non-renewable groundwater declining 2–6 m annually to meet all demands, with population growth, environmentally unfriendly urbanization, and expanded irrigation negatively affecting reserves. Traditional norms prove unsustainable amid governance and infrastructure deficiencies require revised management strategies prioritizing renewable sources and controlling withdrawals.

The ancient people did a lot more than manage life resources, they built and managed the natural resources, and many ancient water management systems are still up to today as an amenity for everyone to enjoy [4]. Different experiences and archeological discoveries worldwide attest to these traditional techniques, especially in arid and semi-arid areas, including Yemen. Different practices and systems were developed by Yemeni people in the domestic, livestock, and agricultural domains [5].

Therefore, maximizing the benefits and minimizing the risks of rainwater in such arid and semi-arid regions, using efficient and cost-effective Rainwater Harvesting (RWH) methods, helped to achieve a sustainable society. Indigenous RWH methods, which are being abandoned, have been extensively practiced in arid and semi-arid areas. This chapter will discuss these used systems in Yemen and address the expected impact of upscaling and revitalizing such indigenous RWH systems on mitigating water crises including drought and flood damage. In addition, the chapter investigates the regulatory framework and dispute resolution. It paves the way for the possibility of returning and ways to benefit from historical techniques for harvesting renewable rainwater and relieve pressure on depleted groundwater resources.

2 Water Situation in Yemen

Without lakes, rivers, or permanent surface water bodies, Yemen is one of the most water-scarce countries in the world. The main source of water is rainfall and rapidly declining groundwater. The per capita share of available water in Yemen is in steady decline due to the limited water resources, population growth, poor groundwater governance, and the expansion of irrigated areas [6]. The annual water availability per capita was 162 m^3 in 1992 and dropped down to 74 m^3 in 2021 [7], which is about one percent of the annual global average of 5,925 m^3 per capita, or 13 percent of the annual Middle East and North Africa (MENA) average of 554 m^3 per capita. Available water per capita in Yemen is expected to drop to 55 m^3 by 2030 [6]. This is nearly 20 times below the internationally recognized water poverty line of 1000 m^3 [8].

The water crisis is taking on catastrophic proportions in several places throughout Yemen, as aquifers are depleting fast and limited opportunities exist to refill the groundwater reserves. Yemen has 14 groundwater basins with around 100,000 boreholes (Fig. 1) [9].

The groundwater levels have fallen significantly by 2–6 m/year. For example, the average

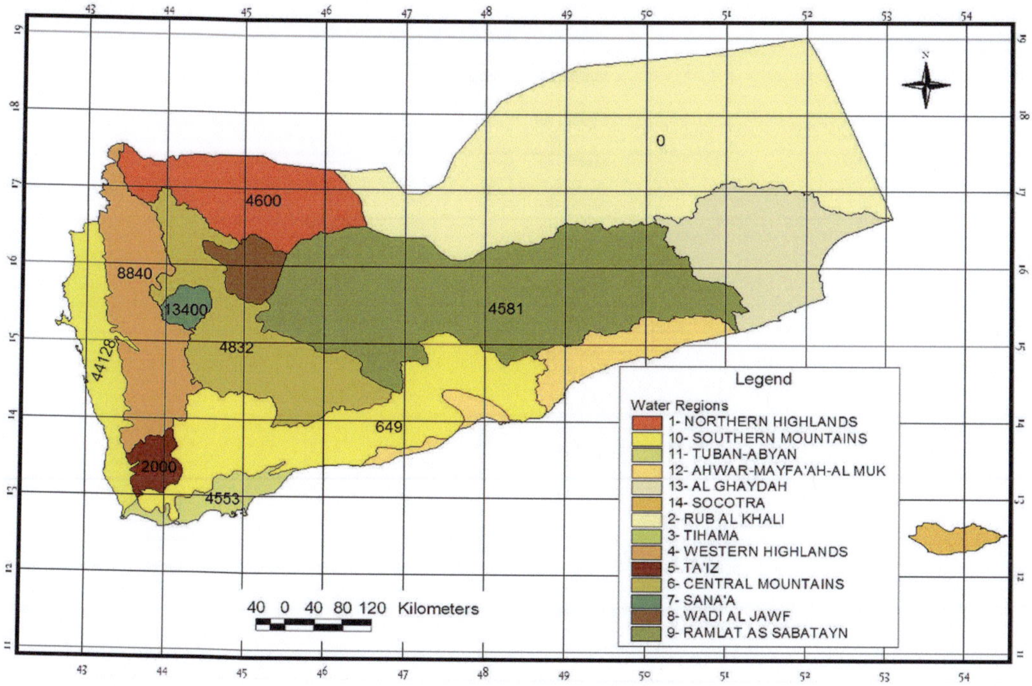

Fig. 1 Water basins and number of water wells in each basin [9]

water table in Sana'a Basin dropped from 30 m in the 1970s to more than 200 m in 2012, and as low as 1200 m in some places [6] Highlands including Sana'a basin, where the rapidly growing capital city of Sana'a is situated, use the biggest share of abstracted groundwater for high-value crop irrigation such as qat and grapes. Groundwater extraction exceeds the recharge rates for all basins in Yemen. In some basins, the abstraction is more than three times the recharge rate, and groundwater levels are dropping as much as six meters per year (Fig. 2) [10].

Today, groundwater is the main source for different uses in the country. Irrigated farming, particularly qat production, uses around 90% of the annually abstracted water and is therefore, the main culprit in the depletion of water tables. As a result, increased salinity of underground water reserves is observed, rendering irrigated fields less productive. The irrigated area in Yemen is rapidly increasing and the rain-fed area is decreasing. Despite the expansion of irrigated agricultural lands by around 10 times during the last five decades, the total agricultural lands are declining. There are currently about 1,100,000 hectares of total irrigated and rain-fed agriculture areas, around 400,000 hectares below the total agricultural lands in the 1970s. This indicates a high rate of desertification and a decline in historical water and agriculture traditions [11]. Mismanagement, climate change, high population growth, rapid environmentally ignorant urbanization[1] and growing needs all contribute to the acceleration of the deteriorating groundwater resources [12].

Water scarcity begins, when water demand exceeds available renewable resources [13], and Yemen has already passed this turning point in the mid-1980s [6]. Since the 1900s, Yemen has been using more water—mostly for agriculture—than its renewable water resources. In 2010, water use was about 4.0 BCM at which time the country's renewable water sources were estimated at 2.5 BCM. The 1.5 BCM shortfall was met by water pumped from non-renewable fossil groundwater

[1]Urbanization resulting in sealing the ground surface with pavement and concrete structures without any attention to compensating the losses in the permeability of the top ground surface.

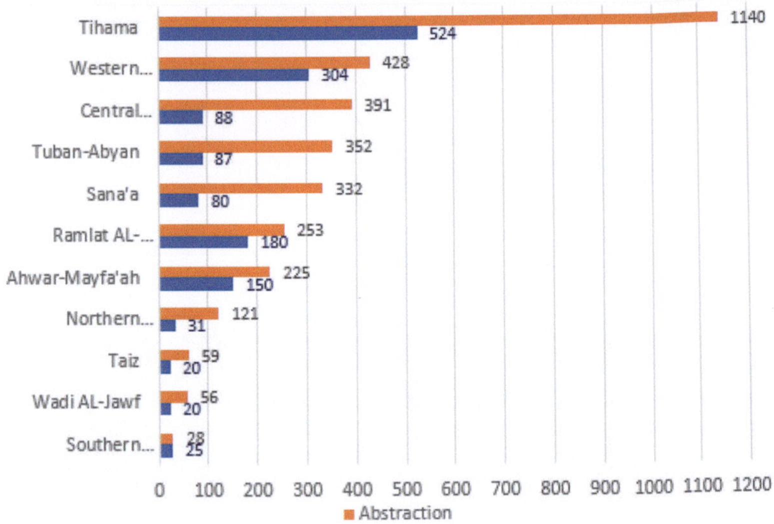

Fig. 2 Groundwater recharge and abstractions (MCM/Y) [10]

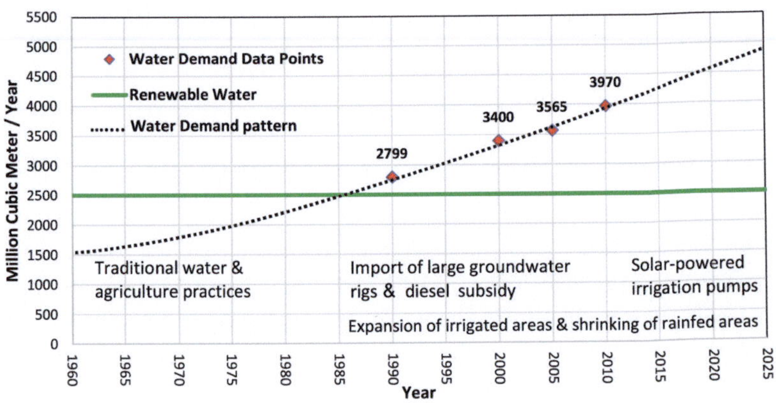

Fig. 3 Water uses & Water Balance (MCM) after [6, 14, 15]

aquifers. The country's aridity, limited water resources, mismanagement, and overexploitation of water resources all contribute to continued water insecurity in Yemen [14–16]. It is expected that the water use will exceed 4.5 BCM by 2025 and that will increase the gap between renewable water sources and demand to more than 2 BCM. Figure 3 summarizes Water Uses & Water Balance in Yemen since 1960.

3 Climate Change and Natural Hazards

Yemen is currently ranked 22nd among the countries most vulnerable to climate change and 171st out of 181 on the Notre Dame Global Adaptation Index (ND-GAIN) in 2022. It is also 12th among the least prepared countries to adapt to climate change. Despite contributing <0.03% of global greenhouse gas emissions, Yemen is one of the least benefitted from global climate funds. Rapid population growth, unplanned urbanization, inadequate environmental regulations, and the concentration of vulnerable populations in flood high-risk areas have exacerbated the impact of these disasters. The ongoing conflict, nearing a decade, has further intensified the humanitarian crisis [17, 18].

Over the past two decades, Yemen has experienced an increase in natural disasters linked to extreme weather events such as cyclones and floods. Only during the last 8 years of ongoing conflict, Yemen has experienced seven major floods and five cyclones resulting in severe flooding, directly affecting food security. In

2022, 63% of the population (approximately 19 million people) could not meet their daily food needs, up from 53% (16.2 million people) in 2021. Crops, particularly staple grains such as sorghum, wheat, and barley, which are crucial for subsistence farming, are highly susceptible to extreme weather. Extreme weather conditions have resulted in the loss of homes, agricultural lands, irrigation systems, livelihoods, and even lives. They have forced approximately 10% of internally displaced people to migrate. Floods, the most significant climate threat in Yemen, can strip the topsoil, depleting essential nutrients and uprooting shallow-rooted seedlings, significantly affecting water security and agricultural productivity [19].

Historically, Yemen managed drought, flash floods, and surface irrigation through community-driven systems rooted in local knowledge. These systems included agricultural terraces, decentralized reservoirs, small dams along flood paths, and spate irrigation that utilized floodwater for agriculture, mitigating floods' impacts and reducing drought and erosion. Given this context, there is a significant need to revive and to fund historical techniques, and to leverage various climate funding sources. These traditional methods are crucial for mitigating climate impacts and enhancing climate adaptation.

4 Historical Water and Agriculture Practices and Regulatory Frameworks

4.1 Historical Water Systems

Ancient Yemenis lived in harmony with nature for thousands of years. Yemeni rainwater Harvesting (RWH) systems and diversion structures were among the most famous water practices in the world. Many ancient RWH systems and impressive structures in Yemen, such as Ma'rib dam and Saharij tanks in Aden are still visible as important archaeological ruins [20, 21]. Yemen is rich in indigenous water systems such as agriculture mountain terraces and spate irrigation systems, because it is believed that the water scarcity problems in Yemen led earlier generations to develop and to employ this great variety of water harvesting techniques. The remains of indigenous water harvesting structures, including dams, reservoirs, delivery, and diversion structures demonstrated the wide variety of water harvesting systems that Yemenis have used to manage water across various agro-ecological areas [22]. Archaeological missions have excavated several flood-control structures in Yemen: at Juban (Albeida), Raiboon (Hadhramout), Aljanad (Taiz), and many other areas. These systems were in use between 800 BC and 300 AD. They served as flood-breakers, sediment traps, and water storage or supply sources [20].

Traditionally, springs, hand-dug wells, *Birak* (Fig. 4), and traditional ponds that collect runoff were the main resources of domestic water. The domestic water supply system in Aden, called Saharij, is one of the oldest water harvesting techniques dating back to the 1st millennium BC and was discovered in the nineteenth century [20]. Figure 4 depicts two cisterns rehabilitated by the Social Fund for Development. The cisterns are secured with a protective fence to prevent children and animals from falling, and each cistern is equipped with a hand pump for convenient access to water.

On the other hand, the historical Ma'rib dam and its irrigation system located in the historical city of Ma'rib is a fine example of controlled irrigation [24]. It stayed in operation for thousands of years before it collapsed around the year 610 AD [25].

4.2 Historical Agriculture Practices

Throughout Yemen's history, the agricultural sector has heavily relied on traditional methods of irrigating systems and livestock. Manual cultivation techniques, with the use of animal traction, have been fundamental in shaping the country's agricultural practices [26]. The Yemeni farmer's close connection with livestock has played a crucial role, contributing significantly to agricultural production income [27]. Livestock

Fig. 4 Roofless RWH system of Birak, in Hejrat Waraf village, Almasane, Amran governorate [23]

has not only served as a source of food and income but also contributed to the overall agricultural ecosystem through manure and labor. Moreover, traditional irrigation systems, including spate irrigation, which relies on the controlled release of floodwater, and water harvesting on terraced fields, have been employed for centuries, highlighting the ingenuity of Yemeni farmers in managing water resources for sustainable agriculture [28]. These historical agricultural methods have not only sustained Yemen's agricultural productivity but also preserved the country's cultural heritage and resilience in the face of evolving challenges.

Ever since the 1960s, Yemen's domestic food production has plummeted sharply with a massive increase in water employed for agriculture. Historically, Yemen was extremely independent, producing over 90% of its staple grains. But decades of Rural workers moved to the cities or migrated abroad, occasional conflict, new emerged water/food policies and changing incentives in markets including remittances support for consumer economy, and imports subsidy by the government, especially cheap, subsidized wheat, steered agricultural attention away from food crops and drove out domestic product from the market. As Yemen became more food import-dependent, it eroded the "economic relevance" of agrarian society, which neglected and disincentivized local grain producers, and augmented the economic domination of powerful import sector conglomerates [29].

The reliance on food imports kept increasing with Yemen today importing over 90% of its food (Fig. 5). At the same time, water supplies, which irrigate more than 90% of agriculture activities, have increasingly been diverted to high-water-crop cash crops like qat instead of food crops [30]. Expanded road networks made transport easier, created the conditions for greater qat production and marketing making it dominates the domestic Yemeni agricultural economy and ecology accounting for 40% of all agricultural water consumption without contributing towards food security. Qat's huge profit margins (around 80%) make it 10–20 times more profitable for producers than "most competing crops" [29, 31]. This environmental misallocation, combined with climatic stress, adverse rural investment, declining soil fertility, and cumulative land degradation, has made Yemen more dependent on food imports to meet its widening food deficit.

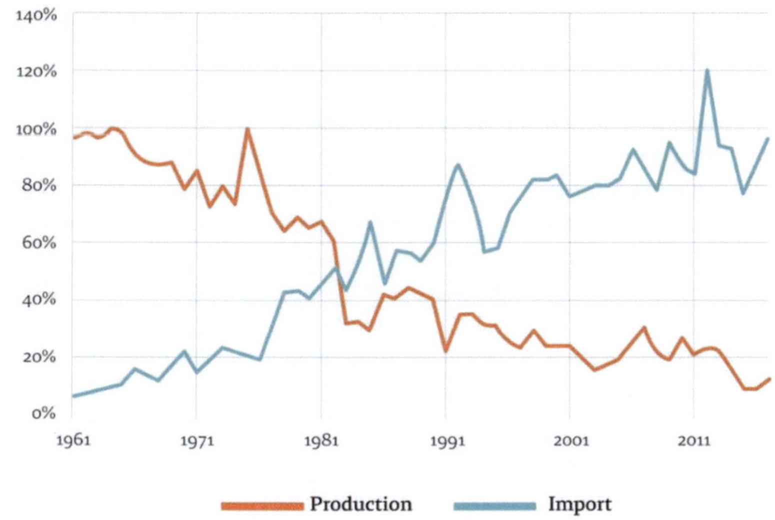

Fig. 5 Grain Domestic production and import shares of total supply quantity (%) [29]

4.3 Management Methods and Water Allocation

In Yemen, different traditional water management agreements or records (known locally as *Marqum*[2]) were widely practiced: one type of *Marqum* agreement regulates the use of natural resources and clarifies the specific rights and duties of each concerned party. Agreements to settle and to regulate existing conflicts is another type of *Marqum* that is still practiced in Yemen. The *Marqum* is signed by the concerned people or by their representatives and then endorsed by a trusted mediator (*wakil, shaikh*). Socioeconomic issues and the full engagement of all locals were always fundamental aspects to be considered. These kinds of indigenous agreements differ from region to region and their content depends on local factors.

Historically, and until 1950, Yemenis have adapted to nature and have managed their available land and water resources by following elaborate indigenous rules and cultural norms that are appropriate for the region and its constraints. These include negotiated systems and rules for the development and management of terraces, spate irrigation, shallow groundwater and other land and water resources. Nevertheless, frequent disputes have arisen, of which some were resolved by negotiation according to traditional rules, mediation and arbitration, but in some cases, the opposing parties resorted to the force of arms [32].

Each Yemeni community has its own rules of water management that have evolved over millennia. Some of the current water rules—so-called *'urf*—were documented in medieval times. Although they have existed since the pre-Islamic era, they are, in general, consistent with Islamic Shari'a. Dozens of old signed rules are documented, with the spate water distribution system alone having 18 different rules. Ancient local water principles that are generally accepted throughout Yemen include the following [33]:

- Upstream riparian users have priority: "*al 'ala fa al 'al*a."[3]
- Hand dug wells should be a certain distance from each other, maintaining a "protection zone," or *haram*.

[2]**Marqum**: is a local word for a written down agreement. After reaching a consensus on it and its wording, it is endorsed with signatures or seals of two or more parties that commit themselves to the regulations and rules stated in that agreement or traditionally called Marqum.

[3]***al 'ala fa al 'ala*** called (الاعلى فالاعلى) and it stipulates that land upstream has the priority of water use than areas which are situated lower, all the way to the lowest level of the stream.

- No one may deprive a person of drinking water—this constitutes the right to quench thirst.
- For springs and other sources that have a stable flow, water rights are based on time, for example in hours or days.
- Dividing the sloping watersheds based on the agricultural land in the bottom. Each agricultural land in the bottom takes the corresponding slopes. The area of the runoff slope sometimes reaches 20 times the area of the agricultural land.

In many cases, these general rules have place-specific sub-rules for different water sources or systems that vary across the country. For instance, the first aforementioned role of upstream riparian has priority or what is called upstream first, has different implementation rules in different areas [32]:

- Water is released from the higher field when that field has enough water, for example when the water in the field is up to knee height.
- With flood irrigation, irrigation begins with the last field left previously unwatered, as is the case in Wadi Zabid, Hodeidah.
- Downstream diversion structures may only receive water when the upstream structure is full or has overflowed.
- During the rainy seasons, the spate is divided between riparian users on a daily basis, starting upstream.

These practices are more sustainable, but they still have equity issues. Although some ancient historical principles affect water rights, they are still acceptable throughout the country, in particular:

- Water is *Mubah*, or "the property of no one."
- Every landowner has the right to dig under their land to abstract groundwater.
- There are no clear individual rights to water, nor is there a regulatory system that can limit water use.

4.4 Dispute Resolution

The unpredictability of floods, gradual changes in wadi morphology, and differences of opinion or access to information between upstream and downstream rights holders can create misunderstandings and disputes. Historically, local irrigation overseers have supervised the distribution of water, but the sheikhs (tribal leaders) judged disputes. To this day, in tribal areas, sheikh (head of the tribe) or aqil (clan head) plays a pivotal role in settling disputes over water. In the past, a sheikh's ability to mediate disputes depended primarily on his social status and knowledge of customary rules.

In recent decades, this has changed, with many sheikhs joining the country's political and economic elite. In tribal areas, some sheikhs have become elected officials, so their wealth, political influence and power provide the basis of their authority to decide and to settle disputes. However, these modern sources of power create the risk of a conflict of interests, because some of these sheikhs have become important commercial farmers and may be the largest water users. In some cases, they have taken advantage of their tribal status to dig more wells and to build large land holdings [32].

5 Classification of Historical Water Systems

The below sections detail different traditional water harvesting systems still practiced in Yemen.

5.1 Rooftop Water Harvesting Systems

Rooftop RWH systems (Fig. 6) were very common in Yemen, particularly in the rural and mountain areas, where there were no networks for tap water or any water source except rainwater. In such areas, rooftop systems are a feasible option to provide water for domestic purposes. Alternative water sources are often

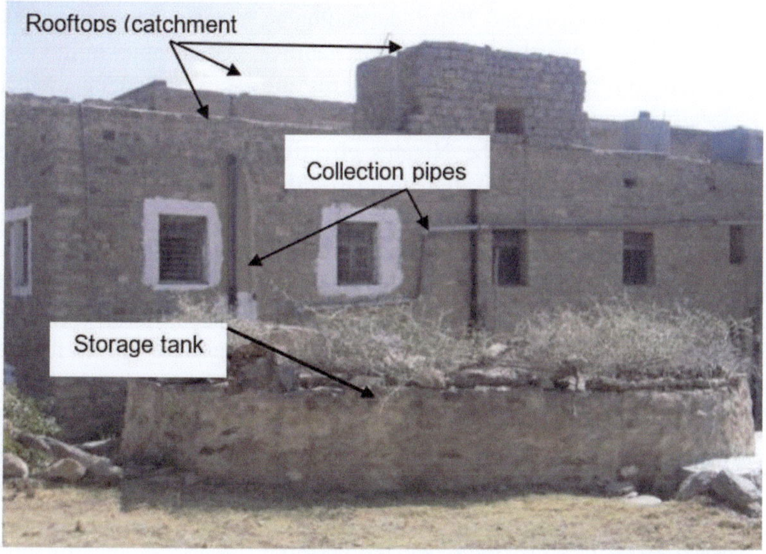

Fig. 6 Rooftop rainwater harvesting system-Taiz [31]

located far away downstream of *Wadis* and using them requires effort and time. However, during the long dry periods, and when the rooftop RWH cistern is dried up, women and children are forced to fetch the water from far away sources [22]. The rooftop RWH cisterns are called locally Seqayat (s. seqayah), they may have circular, rectangular or irregular shapes and their capacity ranges between 10–50 m^3 depending on the roof area and rainfall.

– *Siqayat*, with the singular form being "*Siqaya*," (Fig. 7) is another storage technique typically implemented at the household level for rainwater collection. Rainwater is gathered from nearby clean catchment areas and/or roofs.

Fig. 7 Segaya in impervious soil (Negab) in Ahras village/Bakil/Razeh/Sa'dah [23]

These *siqayat* are categorized into two types, contingent upon the geological conditions of the location.

I. Stone Masonry-Built *Siqaya*: In regions where earth and rock formations are prevalent, the Siqaya is excavated and constructed using stone masonry for the walls and the floor of the cistern. In some areas, such as Raima region, stone arches are utilized to cover the *Siqaya*, while in other regions, reinforced concrete or steel beams and sheets are employed for this purpose. The capacity of the *Siqaya* in this type typically ranges from 20 to 50 m^3.
II. *Siqaya* in Impermeable Soil: In certain parts of Yemen feature geological formations consisting of a top layer of stone with a depth of <1.0 m, beneath which lies a layer of clay soil. The local population, through experience, has recognized the impermeability of clay soil. Consequently, they create an opening through the top stone layer and excavate the clay soil to create a storage capacity, relying completely on the impermeability of clay soil,

ranging between 10 and 20 m^3, depending on the depth of the clay soil and the stability of the top stone layer. These cisterns are locally referred to as "*Nogob*" in singular form "*Nigab*." The key to the Nigab's effectiveness lies in limiting excavation depth to within the depth of clay layer and never extending beyond it. Extending the excavation beyond would render the *Nigab* incapable of storing water. *Nigab* does not involve masonry work; they are primarily the result of excavation. Any masonry work, if present, is limited to raising the opening above ground level to protect it from flooding and traffic. Overflow from the *Nigab* occurs through the opening, and people typically employ a tool to divert excess runoff water away from the *Nigab*, when it becomes full.

5.2 Sloping Land Terraces

Traditionally, RWH has been an appropriate solution to capture rainwater for food production in Yemen. Until 50 years ago, people in the highland areas were relying heavily on rain-fed agriculture on mountain terraces that gave Yemen its unique landscape [34]. The terraces in Yemen have existed since at least 2000 years ago; they are the most spectacular and oldest indigenous water harvesting systems [34]. With a great labor input, millions of terraced fields have been built on steep mountain slopes of the Yemeni highlands [20]. The terraces draw their water from uncultivated areas, rocky surfaces, terraces (one to another) and from *wadi* streams [35]. The terraces are supported by stone masonry retaining walls, built in a way that allows the water to soak through to the shallow soil and when saturated pass in a laminar flow to the next terraces till it reaches the *wadis* downstream [22].

The agricultural terraces of Yemen are considered one of the wonders of the world [4]. Terracing is a good technique for land conservation as it functions as both a soil and water conservation structure [32]. It provides a suitable environment for cultivation of a wide range of arable crops. The modernization of agricultural systems started in the 1960 and 1970s, especially in the valleys and coastal zones, which are accessible for machines. Combined with political instability and other problems led to the abandonment of many of the terraced hillsides, resulting in collapsed walls and soil erosion [4]. Figure 8 shows a section of terraced hillsides of the Haraz Mountains, located to the west of the Yemeni capital of Sana'a. These terraces used to be green with vegetation, coffee bushes, trees, wheat and fruit orchards. However, the

Fig. 8 The Terraced Hillsides of the Haraz Mountains, Sana'a [38]

Fig. 9 Rehabilitation process of Terraced Hillsides [38]

hillsides are now brown and dusty [36]. Modern technologies have degraded the systems, and newer approaches are needed today to correct the imbalances brought about by more recent developments [36, 37].

Despite several initiatives to revive these technologies (Fig. 9), they are not sufficient.

5.3 Flood Water Harvesting (Spate Irrigation)

Historically, Yemenis depend on rain-fed farming activities (as the case of highlands' terraces) and irrigated agriculture from floods along *wadi* courses and coastal plains, as is the case in Tihama, and eastern and southern parts of the country [22]. It is believed that this so-called spate irrigation started in Yemen, when the Neolithic period became more arid. It has been practiced for around 5000 years [39].

Almost 70% of agricultural areas in the coastal and foothill areas depend on spate irrigation for both irrigation and livestock [20]. Approximately 0.1 million hectares are irrigated by flood water, primarily for sorghum and wheat. In Yemen, the use of spate irrigation is declining because of the competition posed by imported, subsidized wheat. Urbanization and new standards of living are also factors that negatively affect the production of local grains of millet and sorghum. Since the 1970s, groundwater wells have been installed in the spate irrigation areas. In Shabwah Governorate, where spate irrigation systems are widely used, approximately 20% of the farms owners have wells. In the irrigated area of Wadi Tuban there are around 1900 operational wells and 2200 wells in Wadi Zabid. The new groundwater source aimed to reduce risk and to guarantee high revenues. However, due to over-pumping and the continued drop of groundwater levels, saline intrusion poses a risk for sustained high productivity [39].

The rules for implementing flood irrigation vary by region and according to available technology. For example, in flood irrigation, it is stated that a farmer has the right to take water until the depth reaches their knees. The indicator that water should be diverted to the next farmer's field is then given by the first farmer, who fires their gun into the air. While this may sound intimidating, it is a simple and convenient method of communication during rain and floods. The result of this method is a uniform distribution of water among users in the valley [32].

Spate irrigation systems are still important in Yemen today. These systems represent about 20% of the irrigated area in Yemen (Table 1) [40].

Consisting of small dams located on the wadi stream, these structures divert floodwater from the streams to farmers' fields during or after heavy rains and are usually temporary (Fig. 10) [32]. The seasonal flooding (*sayl*) in *wadis* in coastal areas, foothills, and springs flow (*ghayl*) in highlands were the main traditional water sources for irrigation in Yemen [41].

Table 1 Spate irrigation in Arab countries, 1989–1995 [40]

Country	Date (year)	Spate irrigation (ha)	Total irrigation (ha)	Spate (%)
Algeria	1992	110,000	555,500	19.8
Morocco	1989	165,000	1,258,200	13.1
Sudan	1995	46,200	1,946,200	2.4
Tunisia	1991	30,000	385,000	8.8
Yemen Rep	1994	98,320	481,520	20.4

Fig. 10 Spate system in Yemen [36, 40]

5.4 Open Catchment Storage Structures

There is a wide variety of open catchment RWH storage structures, which vary in name, size, shape, construction materials, locations, functions and purpose [20, 34]:

- *Berak/Mawajel/Jawabi*

Berak (singular *Birkah*), *Mawajel* (singular *Majel*), and *Jawabi* (singular *Jabia*) are three commonly used names for community-level open rainwater harvesting cisterns (Fig. 11). The first two names, Berak and Mawajel, are prevalent in the upper highlands, while the latter, Jawabi, is commonly found in the northeastern region of Yemen, particularly in Shabwah and Hadhramout. These cisterns can be quite deep, exceeding 8.0 m below ground level, with a storage capacity that can reach or even exceed 1000 m^3.

In ancient times, travel involved riding animals and often took days or weeks to journey from town to town due to the rugged terrain of Yemen. To support these journeys, ancient rainwater harvesting cisterns were constructed along travel routes. At the middle or end of each day's travel, travelers and their animals could find water to quench their thirst.

The construction of these cisterns involves specific techniques:

Floor: The floor is meticulously paved with stone.

Walls: The walls are constructed using stone masonry, ensuring durability and stability.

Joint Filling: In ancient times, joints were filled with Qadhad, a traditional sealing material. In recent days, cement mortar is often used for this purpose.

Surface Finish: Alternatively, the entire floor and interior wall surfaces may be plastered completely with one of the aforementioned materials, enhancing water tightness.

The water collection process begins with open slopes of rocks and soil serving as catchment areas. Collected rainwater flows through open stone channels and passes through a settling basin, locally referred to as "Mashannah," before entering the cistern. This settlement step helps clarify the water and reduce the accumulation of silt in the cistern. The harvested water of Birak is collected from the roofs and the surrounding area. There are different shapes and sizes of

Fig. 11 Different types of Berak/Mawajel/Jawabi (Additional circular hole in the middle of the bottom which is used as a sediment trap) [38]

Birak. *Biraks* are also built outside mosques for outdoor and livestock purposes. The stored water in these cisterns serves multiple purposes, including domestic use and livestock watering. Furthermore, after the introduction of Islam to Yemen, it became customary for every mosque to have at least one such cistern, primarily for prayer and ablution purposes (*wodoo).*

Another type is handmade cisterns that are excavated by hand-drilling, using iron bars in rocky areas and their harvested water is used for drinking. They are built in a cone shape approximately 2 m underground (Fig. 12).

Fig. 12 Rain water harvesting [38]

- *Kerfan*

Kerfan (singular *Karif*), (Fig. 13) are sizable open ponds extensively employed in Hadhramaut. They are excavated in clayey soil and rely entirely on the impermeable nature of clay to retain water. Therefore, no masonry work is required, except at the inlet and outlet (overflow) points to safeguard against water erosion. The fundamental principle here closely resembles that of Nogob, emphasizing the avoidance of complete excavation through the clay soil to prevent water leakage. Consequently, the depth of a Karif depends on the thickness of the underlying clay layer, with storage capacities often exceeding 10,000 m^3 or more. However, the substantial surface area results in significant water losses through evaporation, particularly in the hot climate prevalent in Hadhramaut. The stored water primarily serves domestic, and livestock needs.

Fig. 13 Two Kerfans in Hadhramout [23]

Fig. 14 Jarf [23]

- *Aljeroof*

Aljeroof (singular *Jarf*) (Fig. 14) is used in areas where the top ground surface consists of rocky formations with soil-filled gaps between two rock layers, it is relatively easy to excavate and to construct rainwater storage cisterns. The width of this soil-filled gap controls the shape of the storage cistern, so its width does not exceed few meters, while the length may easily exceed 10 m. However, while they are in use, these systems are relatively uncommon.

Fig. 15 Al-Kohoof [23]

- *Al-Kohoof*

Those that are normally excavated in permeable rocks to collect infiltrated rainwater through the rock layer (Fig. 15). The lower part of the excavation is sealed to store water, and the upper part is left without sealing to allow water infiltration. They are not directly collecting runoff but infiltrated water, hence the water is much cleaner.

5.5 Check Dams

Check dams (Fig. 16) are made mainly from stones. Check dams are small, sometimes temporary dams constructed across the wadies or waterways to counteract erosion and reduce water flow velocity (flood control). It is an effective system in groundwater recharge. It delays (extends) the duration of flow over the wadi and hence increases infiltration opportunity. Check dams are also a good tool to store and/or to divert water to nearby lands through canals/spate irrigation systems. Their water can be used for domestic and agricultural activities.

Check dams have more advantages than big/gravity dams. Table 2 shows a comparison between Beryan gravity dam and Bahman check dams implemented in Sana'a, Yemen.

Check dams have the advantage of shorter construction time, smaller capital costs (cost effective), and shorter gestation period with early benefits and better development of the rural areas. In other words, they are easy to construct, to rebuild and to maintain as they do not rely on advanced technologies. Local materials and traditional knowledge are enough to build such systems. Thereby, they can also be applied in more rural and less advanced communities. The implementation of check dams will not typically displace people or communities, nor destroy natural resources.

Check dams remove fine suspended materials through settlement, providing clear water to the downstream part. This advantage in the downstream could be the only disadvantage of check dams in the upstream areas (location of check dams), where these sediments accumulate and might reduce the infiltration rate. However, this disadvantage can be an advantage, if properly maintained by removing these sediments and material, and restoring the infiltration capacity of

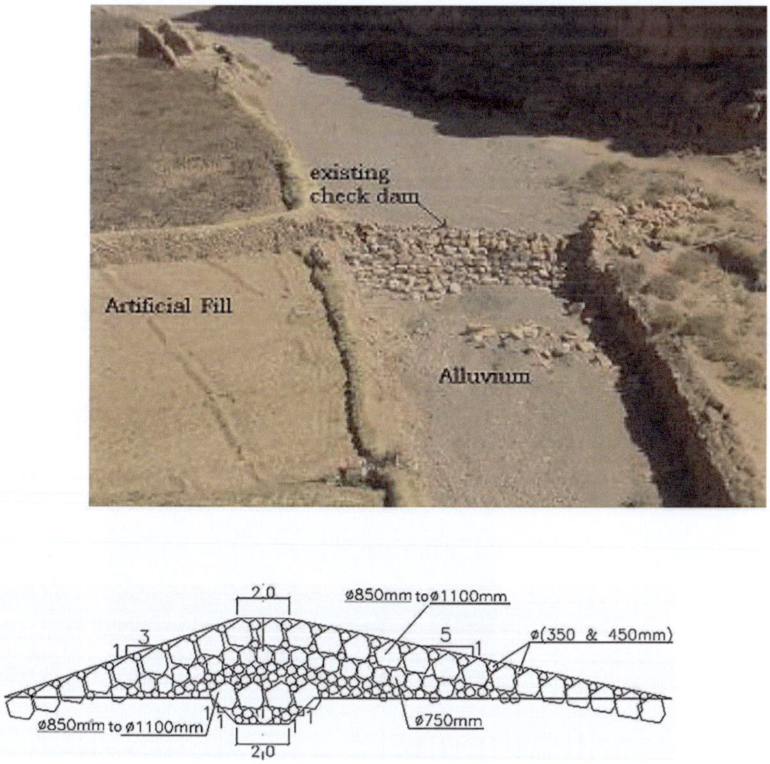

Fig. 16 Check dam and sketch [42]

Table 2 Comparison of the newly constructed Bahman check dams and Beryan gravity dam [43]

Assessed parameters	Bahman check dam	Beryan gravity dam
Average annual runoff, m^3	130,875	62,186
Reservoir capacity, m^3	149,171	145,000
Average annual recharge, m^3	118,252	55,034
Total recharge due to the structure, m^3	122,701	57,320

the wadi floor. The removed material can be used as top-dressing for agricultural soil.

5.6 Dams

The history of water dams in Yemen dates to ancient times, with one of the most notable examples being the Ma'rib Dam, constructed around the eighth century BC. Numerous ancient dams were constructed across Yemen, and their remnants are still visible today. In many cases, only the tops of the dam walls are exposed, while the reservoirs have been filled with sediment carried by runoff. Most historical dams are relatively small, built on wadi channels, and were typically diversion dams for irrigation purposes rather than storage dams.

6 Construction Materials

Al Qadhad

Qadhad, a binding material that consists of natural volcanic glass and hydrated lime, $Ca(OH)_2$, has been used in the construction of all historical water harvesting structures, flood control structures, reservoirs, dams and ponds, in Yemen for thousands of years. Qadhad has also been used as a plastering material for finishing different structures. Qadhad is a strong and durable binding, and cladding construction material. In addition, it is highly resistant to erosion. It has survived severe climatic conditions for thousands of years.

Qadhad can still be found in many ancient buildings, water and agriculture infrastructures, and monuments like the cisterns of Tawila, which are more than 1000 years old. Qadhad can also be seen in the northern and southern ends of the ancient Ma'rib Dam, which is more than 3000 years old. Two ancient water harvesting systems, built with Qadhad, are shown in Fig. 17.

Method of Preparation

The chemical equations, given below, show the different compounds used in the production of Qadhad and their interactions, were not known at all to Qadhad producers. Qadhad production by ancient Yemenis was traditionally based on acquired practices and experiences by people over time.

Qadhad preparation goes through several stages, starting with the preparation of lime and

Fig. 17 Two ancient water harvesting systems, built with Qadhad [38]

Fig. 18 One of traditional method of Qadhad preparation [38]

the collection of volcanic stone. Quick lime (CaO) is prepared by burning small-sized limestone ($CaCO_3$) rocks at high temperatures, up to 1000 °C. Some of the furnaces used for this purpose still exist. The chemical reaction for quick-lime production is given below, Eq. (1).

$$CaCO_3 \rightarrow CaO + CO_2 \quad (1)$$

Hydrated lime $Ca(OH)_2$, locally known as Norah, is produced by adding water to the quicklime and leaving it for two days to ferment, Eq. (2).

$$CaO + H_2O \rightleftharpoons Ca(OH)_2 \quad (2)$$

The volcanic stone or pumice stone, which is also called alhishah, is usually extracted from volcanic caves. The produced Norah is mixed well with crushed pieces of very small volcanic stone (habesh stone or tuff) at a ratio of 2:3 (2 Norah to 3 stones) and water. The mixture is left for two days to a maximum of one week. The mixture may be crushed during fermentation to ensure homogeneity.

Camels have been used for crushing. Qadhad mixture is placed inside a circular channel, where a camel pulls a heavy, high-resistance wheel, attached to a central pivot and capable of rotating. A wooden or iron pole is fixed at one end in the center of the circle, while the other end (the one extending from the wheel outside the circle) is tied with a chain or rope to the camel's back (Fig. 18).

Characteristics of Al Qadhad

The load-bearing capacity of Qadhad ranges between 220 and 340 kg/cm^2 with an average density of approximately 2000 kg/m^3, whilst the load-bearing capacity of ordinary concrete ranges between 250 and 350 kg/cm^2 with a density of 2200–2300 kg/m^3. Despite the similarity in the load bearing capacities, Qadhad has proven to be more resistant to various climatic conditions over decades [38].

7 The Most Important RWH Symbolic Systems for Domestic Use and Irrigation

The sections below describe the two most important historical examples: The Saharij of Tawila (rainwater harvesting cisterns) used for domestic uses and Ma'rib dam used for irrigation.

- *Saharij of Tawila*

In south Yemen, Tawila Reservoirs (Fig. 19) [31], known locally as *saharij* (s. *sihrij*) located

Fig. 19 Middle part view for Tawila RWH Reservoirs (Saharij), Aden [35]

in Crater District, Aden, are very famous indigenous RWH systems. They date back to more than 1000 BC. The main purpose of this system was to supply water to the coastal city of Aden. The system of Tawila channels rainwater runoff from the nearby upstream rock mountains. It comprises of nine reservoirs. These reservoirs are connected to a water distribution system, which includes other reservoirs around the city. The nine tanks act as flood-breakers and sediment media. The total storage capacity of the system is estimated to be 90,000 m^3. This complex system was discovered in 1854, when it was completely covered with debris from the surrounding hills [20, 34].

- *Ma'rib Dam*

The ancient Dam of Ma'rib (also called Sabaein Dam) and its integral irrigation system is considered among the best controlled irrigation systems in the Ancient World [24]. This dam was built 400 m above sea level and collects the runoff water from a 10,000 km^2 watershed of Wadi Saba/Alsud or Abida [34]. More than 9600 ha of the two south and north oases are irrigated by this great historic dam. The dating analysis of ^{14}C confirms that the earliest irrigation systems in Wadi Danah probably date from the late 4th millennium or early 3rd millennium BC. Silt deposition and the declining importance of the dam as a station for convoys are thought to be the main causes leading to the deterioration of its hydraulic system [24].

The Ma'rib Dam was built by the Sabaean civilization and played a crucial role in the region's agriculture and prosperity. It facilitated irrigation in the arid landscape, allowing the development of sophisticated agricultural practices and supporting the growth of a powerful and wealthy society. Over the centuries, the old Ma'rib Dam was maintained, rebuilt, and expanded by various rulers, showcasing the advanced hydrological knowledge and engineering skills of ancient Yemeni civilizations.

The historic Ma'rib Dam is central to Yemeni identity [34]. It is believed that climate change during the 4th millennium BC led to the construction of this irrigation system in the *wadis* bordering Ramlat as-Sabatayn.[4] The long duration of this diversion (not storage) dam proves its success. The well-known exodus of most local inhabitants to other parts of the Arabian Peninsula was the result of a succession of disastrous events, the last of which is described in the Holy Qur'an (Q.34:15). This catastrophic event is considered to have taken

[4]Ramlat as-Sab'atayn: South Arabian Sand Desert.

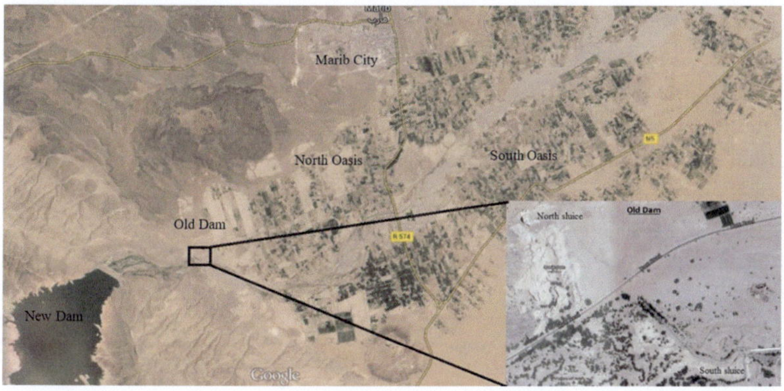

Fig. 20 The location of old ancient Ma'rib Dam and the rebuilt New Dam. Ma'rib. Yemen

place at the beginning of the seventh century AD. With the support of the Abu Dhabi Fund, the new dam, located 2 km upstream from the old dam, was constructed by the Yemeni government and opened officially in 1986 [44] (Fig. 20).

8 Indigenous RWH Systems are Declining

Rainwater harvesting has been practiced since ancient times. However, contemporary people do not rely on water harvesting as they did five decades ago. The area using RWH to support rain-fed agriculture is in steady decline. Although today's water problems have increased interest in indigenous RWH, these systems are still declining despite revival efforts [45–48].

Rainwater harvesting has the potential to supplement surface water and groundwater abstraction. Many indigenous water harvesting systems are still in use, albeit to a lesser extent than in the past. Indigenous RWH using khooshabs[5] in northern Baloochestan in Pakistan, generates up to 90% of the incomes of local families [49]. In the 48 villages of the Banni grasslands of Kachchh, India, indigenous RWH systems developed centuries ago are still the main source of water for more than 100,000 head of livestock [50]. The *Tabias*[6] system in Tunisia protects the entire Kairouan Plain against flooding, recharges groundwater, retains sediments and controls soil erosion [51]. More examples of positive effects of still-working indigenous water harvesting systems are listed in [46].

Despite challenges, indigenous RWH systems are now being reconsidered as an option for improving the social development and stability of local communities [51]. The success of traditional RWH systems over thousands of years has encouraged communities to establish these systems once again and in several new areas. The revived indigenous water harvesting systems have a highly positive impact on water availability, crop production, flood mitigation and groundwater recharge.

9 Conclusions and Recommendations

Yemen is rich in Indigenous RWH and agriculture systems, and this research documents the main Indigenous systems and demonstrates their significance. Although they are in a state of decline, the still-working systems have a great impact and contribute to alleviating water crises and protecting livelihoods. Even with the instability situations and disruption of other water sources, traditional techniques remain an important resort for many people who revived them during times of hardship. From this standpoint, documenting and studying these techniques and working to revive and utilize them is considered vitally important.

[5] ***Khooshabs*** are macro-catchment harvesting systems used to divert flash floodwater to nearby irrigation lands.

[6] ***Tabias*** are one- or two-meter dikes or stone enclosures on low sloping ground that collect runoff and sediments and form the irrigated terraced *wadi* system of Jessour.

What is certain, and tangible is that the water and food crises are increasing, and indigenous RWH can improve water and food availability. In turn, this can combat water scarcity and groundwater depletion while ensuring a sustainable water supply in dry Yemen, where depleted groundwater is the main water source for all uses, and desalination projects are beyond the country's economic capacity.

Incorporating indigenous RWH systems in government policies and strategies will improve access to water and reduce the stress on groundwater. Socio-economic and cultural forces including local context and habits are important factors and inevitably play a role in the relative impacts of RWH programs. The government should thus review water policies and develop a mechanism to encourage the adoption of RWH while taking into consideration the characteristics of each region in terms of water resources, applicability and socio-economic and cultural constraints. The use of simple indigenous water harvesting technologies by local communities is mostly cost-effective and has been widely adopted in the past. However, research on cost–benefit analysis is always essential. The incentive methodology for rainwater harvesting can be put forward by considering the groundwater level and renewability of each region in the country and making legal regulation accordingly.

References

1. Juneau T. Yemen and the arab spring: elite struggles, state collapse and regional security. Orbis. 2013;57:408–23. https://doi.org/10.1016/j.orbis.2013.05.004.
2. UN Yemen Country Results Report; 2022. https://yemen.un.org/en/228267-un-yemen-country-results-report-2022#:~:text=Yemen%20is%20ranked%20183%20out,of%200.455%20in%202021%2F2022.
3. Al-Asbahi QYA. Water resources information in Yemen. In: U. N. Intersecretariat working group on environment statistics (IWG-ENV), Work Session on water statistics Vienna June 20–22, 2005.
4. Guttmann-Bond E. Sustainability out of the past: how archaeology can save the planet. World Archaeol. 2010;42:355–66. https://doi.org/10.1080/00438243.2010.497377.
5. Baquhaizel S, Saeed I, Ghouth M. Traditional irrigation systems and methods of RWH in Yemen (Hadramout and Shabwah governorates). 1996. http://www.hydrology.nl/images/docs/dutch/yemen/Traditional_irrigation_systems_water_harvesting.pdf.
6. Aklan MM, Lackner H. Solar-powered irrigation in Yemen: opportunities, challenges and policies. Rethinking Yemen's Eco. 2021;1–25. https://devchampions.org/publications/policy-brief/Solar-Powered-Irrigation-in-Yemen/.
7. FAO. Aquastat "Country Statistics" tool. 2021. https://data.apps.fao.org/aquastat/?lang=en.
8. Falkenmark M, Lundqvist J, Widstrand C. Macro-scale water scarcity requires micro-scale approaches. Nat Res Forum. 1989;13(4):258–67. https://doi.org/10.1111/j.1477-8947.1989.tb00348.x.
9. NWRA. Water Resources Management Regions. 2009. http://www.nwrayemen.org/index.asp?SeldN=62.
10. World Bank. Yemen—assessing the impacts of climate change and variability on the water and agricultural sectors and the policy implications. 2010.
11. Acacia Water. Water availability in Yemen: Literature review of the current and future water resources and water demand in Yemen. 2021. https://www.undp.org/sites/g/files/zskgke326/files/migration/ye/Water-Availability-Study-in-Yemen.pdf.
12. Haidera M, Alhakimi S, Noaman A, Al Kebsi A, Noaman A, Fencl A, Dougherty B, Swartz C. Water scarcity and climate change adaptation for Yemen's vulnerable communities. Local Environ. 2011;16(5):473–88. https://doi.org/10.1080/13549839.2011.565465.
13. Pontius J, McIntosh A. Water scarcity. In: Environmental problem solving in an age of climate change. Springer textbooks in earth sciences, geography and environment. Cham: Springer;2024. https://doi.org/10.1007/978-3-031-48762-0_8.
14. FAO. Yemen Famine Prevention Plan, January–June 2019. http://www.fao.org/3/CA3134EN/ca3134en.pdf.
15. World Bank. Future impact of climate change visible now in Yemen. 2014.
16. MWE. Water sector in Yemen: resources and services. 2019. https://www.unescwa.org/sites/default/files/event/materials/2._lnob_view_form_yemen-al-sharjabi-mowe_0.pdf.
17. Aklan M. Cyclone Tej Undescores Yemen's Climate Crisis Ahead of COP 28—the Yemen review, September and October 2023—Sana'a Center for Strategic Studies. 2023a. https://sanaacenter.org/the-yemen-review/sept-oct-2023/21252.
18. Aklan M. Extreme weather and the role of early warning systems in Yemen. 2023b. https://sanaacenter.org/publications/main-publications/20542.
19. Aklan M. The impact of flooding on agricultural communities in Yemen. 2024. https://sanaacenter.org/publications/main-publications/22143.
20. Bamatraf A. Water harvesting and conservation systems in Yemen. In: FAO proceedings of the expert

consultation about "Water Harvesting for Improved Agricultural Production" FAO Expert consultation Cairo, 21–25 November 1993, Rome, Italy, Water Reports (FAO). 1020–1203, no. 3;1994. pp. 169–9.
21. Harrower MJ. Is the hydraulic hypothesis dead yet? Irrigation and social change in ancient Yemen. World Archaeol. 2009;41:58–72. https://doi.org/10.1080/00438240802668354.
22. Edmunds W, Cardona C. UNESCO, G-Wadi meeting on water harvesting. BGS Final Report, Aleppo Syria. 2006. http://gwadi.org/sites/gwadi.org/files/Aleppo.pdf.
23. SFD. SFD photographic catalogue, social fund for development, Republic of Yemen. 2024.
24. Francaviglia VM. Dating the ancient Dam of Ma'rib (Yemen). J Archaeol Sci. 2000;27:645–53. https://doi.org/10.1006/jasc.1999.0518.
25. Weiss C, Gerlach I. Qadhat a comparative study of ancient and new lime mortar from the Ma'rib province/Yemen. Archaeol Anthropol Sci. 2009;1:87–94. https://doi.org/10.1007/s12520-009-0007-2.
26. Mohammed F, Al-Shamiry S. Agricultural mechanization development in Yemen (1970–2003). Int J Progressive Sci Technol (IJPSAT). 2020;22(1):351–7. http://ijpsat.ijsht-journals.org.
27. Alasbahi RH, Groot MJ. Ethnoveterinary uses of certain yemeni plants: a review of the scientific evidence. Planta Med. 2022;88(03/04):237–53. https://doi.org/10.1055/a-1612-4556.
28. Al-sabai A, Neszmélyi GI. The challenges and actual questions of the agriculture in Yemen. Studia Mundi – Economica. 2019,6(1).116–4. https://doi.org/10.18531/Studia.Mundi.2019.06.01.116-124.
29. Baldauf, M., Aklan, M. Cultivating an equitable recovery: A political ecology approach to agricultural interventions in Yemen—Sana'a Center for Strategic Studies. 2023. https://sanaacenter.org/publications/main-publications/21634
30. CEOBS. Yemen's agriculture in distress. 2020. https://ceobs.org/yemens-agriculture-indistress/
31. Glass, N. The water crisis in Yemen: Causes, consequences and solutions. 2010. http://www.yemenwater.org/wpcontent/uploads/2013/03/Global_Majority_e_Journal_1-1_Glass.pdf
32. World Bank. Water in the Arab world: management perspectives and innovations. The World Bank, Middle East and North Africa (MNA) Region 1818 H St., NW, Washington, DC. 2009.
33. Al-Eryani M, al-Amrani M. Social and organizational aspects of the operation and maintenance of spate irrigation systems in Yemen. 1998.
34. Oweis A, Prinz D, Hachum AY. Rain water harvesting for agriculture in the dry areas. CRC Press;2012. ISBN 978-0-415-62114-4.
35. Taher T, Saleh SA. Guidelines for roof top RWH systems in the Republic of Yemen. 2010. https://www.bebuffered.com/downloads/Rain%20Water%20Harvesting%20Manual%20-%20WEC-1.pdf.
36. Adeel Z, Schuster B, Bigas H. What makes traditional technologies tick? A review of traditional approaches for water management in Drylands, Desertification series. Hamilton, ON: The United Nations University;2008. ISBN 92-808-6003-8.
37. Kenyon P. Kenyon P. Water, food shortages squeeze Yemen. Washington, DC: NPR: National Public Radio: News & Analysis, World, US, Music & Arts (3 June);2008. http://www.npr.org/templates/story/story.php?storyId=90328214.
38. Aklan M. Revitalizing indigenous water harvesting systems to mitigate drought and flashflood damage-Yemen. In: WatIF 2016 conference. Ontario, Canada: Queen's University;2017. https://www.researchgate.net/publication/385627300_Revitalizing_Indigenous_Water_Harvesting_Systems_to_Mitigate_Drought_and_Flashflood_Damage-_Yemen#fullTextFileContent.
39. Lawrence P, Van Steenbergen F. Improving community spate irrigation. Report OD, 154. 2005. http://r4d.dfid.gov.uk/PDF/Outputs/Water/R8065-OD154.pdf.
40. Perry CJ, Bucknall J. Water resource assessment in the Arab world: new analytical tools for new challenges. 2008.https://www.researchgate.net/publication/283691380.
41. Varisco DM. Sayl and Ghayl: the ecology of water allocation in Yemen. Hum Ecol. 1983;11:365–83. https://doi.org/10.1007/BF00892245.
42. Stanley Consultants. Bahman Check Dams, Sana'a Basin Water Management Project, Project Coordination Unit, Sana'a, Republic of Yemen, Final Report. 2006.
43. Alderwish AM. Integrated water management for small catchments in arid mountainous region, Yemen. Int J Energy, Environ Econ. 2011;19(6):699–715. https://www.scopus.com/inward/record.uri?eid=2-s2.0-79958087290&partnerID=40&md5=eadcddc4a5ed82d7197fe91427bbe6aa.
44. Brunner U. The Great Dam and the Sabean Oasis of Ma'rib. Irrig Drain Syst. 2000;14:167–82. https://doi.org/10.1023/A:1026583213688.
45. Aklan M, de Fraiture C, Hayde L, Moharam M. Why indigenous water systems are declining and how to revive them: a rough set analysis. J Arid Environ. 2022. https://doi.org/10.1016/j.jaridenv.2022.104765.
46. Aklan M, de Fraiture C, Hayde L. Why we should revitalize indigenous water harvesting systems: lessons learned. Int Soil Water Conserv Res. 2024. https://doi.org/10.1016/j.iswcr.2024.05.004.
47. Ferrand E, Cecunjanin F. Potential of rainwater harvesting in a thirsty world: a survey of ancient and traditional rainwater harvesting applications. Geogr Compass. 2014;8:395–413. https://doi.org/10.1111/gec3.12135.
48. Chowdhury K, Behera B. Is declining groundwater levels linked with the discontinuity of traditional water harvesting systems (tank irrigation)? Empirical evidence from West Bengal, India. Groundw Sustain Dev. 2018;7:185–94.
49. Hosseini-Marandi H. Investigation, identification and evaluation of the traditional methods of utilizing floodwater in Sistan-Baluchestan province (in Farsi with

English abstract). S/N 84/139. Tehran, I.R. Iran: Soil Conservation and Management Institute;2005. 80 p.
50. Machiwal D, Kumar S, Sharma GK, Jat SR, Dayal D. Studying an indigenous rainwater harvesting system in Banni grassland of Kachchh, India. Indian J Tradit Knowl. 2018;17:559–68.
51. Berndtsson R, Jebari S, Hashemi H, Wessels J. Traditional irrigation techniques in MENA with a focus on Tunisia. Hydrol Sci J. 2016;61:1346–57.
52. Oweis A, Bruggeman A. Indigenous water-harvesting systems in West Asia and North Africa. Aleppo, Syria: ICARDA; 2004. VI+173 pp. En. ISBN: 92-9127-147X. http://www.researchgate.net/profile/Theib_Oweis/publication/266382589_Indigenous_water_harvesting_Systems_in_West_Asia_and_North_Africa/links/542e7d8d0cf29bbc126f25eb.pdf.

Alternative Water Resources in Agriculture

Reclaimed Domestic Wastewater Reuse in Irrigation: Circular Economy Approach

Lina Abu-Ghunmi and Diana Abu-Ghunmi

Abstract

Water is a vital element for every life on this planet and an essential contributor to economic development. The current unsustainable economic growth, which relies on linear economy (LE), is threatening the planet's valuable water resources through depletion and contamination. Therefore, this chapter envisages reclaimed domestic wastewater reuse in irrigation based on Circular Economy (CE) approach. Accordingly, this chapter begins with an overview of the concept of CE for water and wastewater sectors. Then it introduces the Life Cycle Sustainability Assessment (LCSA) as a practical tool for evaluating the successful implementation of CE model. Afterward, the chapter examines the compatibility of CE concept with current standards and guidelines for reclaimed domestic wastewater to identify the potential transition of the current linear water and wastewater sector toward CE. Finally, the chapter presents a case study on water sector in Jordan. The case study implements mass balance, which is one of CE essential tools, to identify the gaps in the current water budget and to underline the **rethink**-based search for alternative opportunities to fill the gaps. One of the identified alternatives based on the rethink search is **reclaim** and **reuse** of domestic wastewater**,** with focus on reclaimed domestic wastewater **valorization** as irrigation water.

Keywords

Life cycle sustainability assessment · Reduce · Rethink · Wastewater standards

1 Introduction

Water, which is a natural renewable resource regenerated through water cycle, is not only an important resource for life but is also essential for human existence and for economic and social development [1]. Water is a main production input in various economic sectors: domestic, industrial, and energy, as well as agricultural [2]. Hence, water enters the economy for numerous uses; after use, it returns to the water cycle in the form of (1) vapour via evaporation/evapotranspiration, or (2) wastewater: domestic, industrial or agricultural. The rapid increase in the world population and the severe impacts of climate change have increased water demand by all economic sectors,

L. Abu-Ghunmi (✉)
Water Energy and Environment Centre, The University of Jordan, Amman, Jordan
e-mail: lina.abughunmi@ju.edu.jo

D. Abu-Ghunmi
Finance Department, Business School, The University of Jordan, Amman, Jordan

M. Bahadir et al. (eds.), *Agricultural Water Management*, Water and Wastewater Management,
https://doi.org/10.1007/978-3-031-96134-2_12

and the competition among these sectors for water. Furthermore, pressure on water resources has further been exacerbated by the prevailing linear economy (LE), which couples economic growth and resource consumption via take-make-dispose model. Accordingly, LE models are not only threatening water resources' depletion but also threatening the water quality. This is an inevitable consequence of LE, which encourages unsustainable water use, and hence increases wastewater generation that has already exceeded the environment's assimilation capacity.

Consequently, a sustainable economic model to sustain water resources is required as never before. One approach is circular economy (CE), which in contrast to LE decouples economic growth from resources consumption via take-make-reuse model. Thus, CE aims at sustaining economic development by reducing resource consumption, generating zero-waste and regenerating nature [3, 4]. As a result, CE boundaries cover the entire supply-chain, locally, nationally, and globally, through which sustainability aspects: technological, environmental, social, economic, institutional and regulatory are integrated and overarched by proper design. The three principles of CE are (i) elimination of waste and pollution, (ii) materials and products circulation, and (iii) nature regeneration [4], which are translated into reduce, reuse, recycle, recover, among other Rs, and on top of that, is rethink [1, 5]. In order to achieve transition to CE, material, energy, and money balances are the analytical tools used to identify problems/gaps in the current business practices, and then a bottom-up approach is applied for value propositioning to develop and to create non-conventional opportunities, i.e., CE-based business models.

Designing circular economy business models is required across all economic sectors, with the domestic sector being at the top of the list. Worldwide, the agriculture sector is responsible for the highest water use ratio (70%), followed by the domestic sector with a use ratio of 11%, and then the industrial sector, the share of which is 19% [2]. Knowing that the domestic sector has the second highest share of water use, this brings the need to envision domestic wastewater as an imperative renewable resource. Therefore, rapid technological development in treatment/reclamation of domestic wastewater has taken place since the early twentieth century. Furthermore, due to the variation in domestic wastewater qualities, treatment technologies, and operational conditions, country-specific quality standards for treated domestic wastewater reuse have been developed [6–9]. The increasing global water crisis brings reclaimed domestic wastewater to the scene as an alternative to fresh water, especially in the agricultural sector. So, the use of reclaimed wastewater in agriculture is integrated in the modified reclaimed wastewater standards, to include quality and guideline for use in irrigation. Therefore, reclaimed domestic wastewater standards, guidelines and regulations override LE approach, i.e., take-make-dispose, by CE approach.

Accordingly, this chapter discusses reclaimed domestic wastewater reuse in irrigation from the CE perspective. Hence, this chapter covers: (1) CE approach pillars; reduce, reuse, recycle and recover, among the other Rs, which are overarched by the core of the CE "rethinking"; (2) implementation of Circular Economy; (3) reclaimed domestic wastewater standards and guidelines; and (4) a case-study from Jordan.

2 Circular Economy for Water and Wastewater Sector

Overexploiting the finite natural resources, including water, by coupling economic growth and resource consumption under the current LE model has accelerated resources depletion and threatened human development [10, 11]. Consequently, a new approach to economic growth underpinned by sustainability has become a must. For that reason, CE concept, that promotes economic growth by decoupling economic growth from resource consumption, has emerged [10, 11]. In fact, CE is an approach to sustainable economic development that integrates society, economy, and environment. CE is a nature-based economic approach that attempts to regenerate the nature by imitating natural circular processes, i.e., water cycle, food cycle, human life cycle, etc., as well as conserving mass and energy to achieve sustainability. As a result, regeneration is a key pillar of CE model manifested by basically 4 Rs; reduce, reuse, recycle and recover [11]. Smol [1] has added, rethink and reclaim, to these four Rs, to become six Rs that make the CE model for water and wastewater sectors.

Rethink overarches the other Rs by promoting out of box thinking that aims at sustainable waste valorisation. **Reduction** in the water sector aims at minimizing water use, thus eliminating wastewater generation and carbon emissions by returning all reclaimed wastewater components into the economic cycle. Therefore, reducing water use by all water consuming sectors is promoted worldwide. The reduction implementation plan includes activities that consider several sustainability dimensions: (1) technical by improving water network infrastructure to reduce water losses, as well as by providing equipment and appliances that enhance reduction in water use; (2) social by increasing awareness to reduce water use via behavior changes and adoption of water-saving appliances, as well as by minimizing illegal water use; (3) economic by minimizing government subsidies for the water sector, which is a major step toward water use rationalization; (4) institutional by empowering implementation of regulation, which will have an impact on reducing illegal water use; (5) environmental by promoting natural resources conservation, which includes preventing degradation and pollution of water resources, encouraging water use reduction, and consequently reducing wastewater generation. **Reuse** in the water sector aims at expanding the lifespan of used-water (wastewater) for non-potable applications. Thus, reuse implicitly entails reducing the use of freshwater resources, and the valorization of wastewater for different applications. Wastewater for reuse is mainly originated from domestic and industrial sources, and is valorized to be reused in irrigation and industry. Furthermore, reuse, in most cases, embeds **reclamation** of wastewater to meet the reuse quality requirements. Accordingly, reuse implementation plan considers several aspects of sustainability: (1) technical, which includes providing infrastructure to collect, to transport, to reclaim and to store all generated wastewaters, in addition to the use of wastewater treatment technologies that help meeting the reclaimed wastewater quality standards for reuse; furthermore, implementing proper irrigation technologies, such as drip irrigation, for safe application of reclaimed wastewater. (2) Social, which includes promoting and increasing awareness about the advantages of the reuse of reclaimed wastewater in the agricultural and industrial sectors, such as reducing, if not eliminating, fertilization cost. (3) Economic, which includes using pricing mechanisms for fresh water that captures the true value of this resource to encourage users to replace, when applicable, fresh water supply with reclaimed water; in addition to internalising the positive externalities associated with the reclamation and reuse of domestic wastewater. (4) Institutional, which includes providing standards and guidelines that control the implementation of safe reuse of reclaimed wastewater. (5) Environmental, which includes promoting reclaimed wastewater reuse to protect environmental compartments water, soil and air, and consequently, to protect human health. "Reuse" and "Reclamation" are the main Rs that are implemented in the following case study. **Recycle** in the water sector aims at extracting the water component of used-water (wastewater) for potable applications [1]. Water extraction from wastewater, in most cases, is an intensive and high energy consuming process, in which wastewater is **reclaimed** to produce potable water. Water extraction from wastewater is associated with by-products that consist basically of nutrients and/or organic matter. These by-products are the essential elements that are extracted through the **recovery** process in the water sector [1] with the aim of recapturing nutrients and energy from used water (wastewater). Nutrient and energy recapturing, similar to water extraction, are intensive and high energy consuming processes, in which wastewater is **reclaimed** to recover nutrients and energy.

3 Implementation of Circular Economy

For a successful implementation of CE business model or solution, an effective assessment tool is required to quantify the environmental, social and economic costs and benefits associated with such model [12]. Accordingly, the CE business model of wastewater needs to be evaluated to justify the transition of the wastewater industry to circular economy. This evaluation can be achieved by using **L**ife **C**ycle **S**ustainability **A**pproach (LCSA).

In fact, LCSA is an effective tool in assessing the benefits and costs associated with alternative CE business models, compared to each other and compared to their LE counterparts [13]. LCSA is a comprehensive methodology that is used to assess the three dimensions of sustainability (i) environmental, (ii) economic and (iii) social along the "life cycle stages" of products in decision-making process to achieve sustainable solutions [14]. LCSA integrates the **e**nvironmental **L**ife **C**ycle **A**ssessment (eLCA) that examines impacts on the environment; **L**ife **C**ycle **C**osting (LCC), which estimates the financial and internalised environmental costs and benefits; and **s**ocial **L**ife **C**ycle **A**ssessment (sLCA) that assesses the social impact [14]. The scheme for LCSA is [14, 15]

$$\text{LCSA} = \text{eLCA} + \text{LCC} + \text{sLCA}$$

However, to use this scheme, system boundaries should be consistent for the three methods, including life cycle inventory, which should be identical for the three parts of LCSA [15].

The first component of LCSA is eLCA. eLCA entails four steps: (i) goal and scope definition, including system boundaries and functional unit; (ii) life cycle inventory that identifies inputs and output at every single stage of the product life cycle; (iii) life cycle impact assessment, which converts inputs and outputs into impact categories (midpoint and endpoint); and finally (iv) interpretation [14–16].

The second component of LCSA is LCC, which pre-exists eLCA [17]. LCC considers all costs and benefits associated with the use of domestic reclaimed wastewater for irrigation over its life cycle including external costs and benefits that are expected to be privatised or internalised [11, 14, 17]. In particular, LCC focuses on the financial costs and benefits associated with the environmental impacts of the product over its entire life cycle [14, 16, 17]. While environmental LCC extends the financial LCC to cover costs borne by other than the private actors, it does not convert environmental damage to monetary values, as the focus is only on real cash flows that can be internalised [16]. Similar to eLCA, LCC involves four phases: (i) goal and scope including functional unit definitions; (ii) inventory costs; (iii) cost aggregation by categories; and finally (iv) interpretation [14]. Nevertheless, an important consideration in LCC calculations is the selection of the discount rate, which is a function of the goal and scope; it is argued that social discount rate is consistent with environmental protection [17]. Furthermore, low social discount rate (k) is consistent with "intergenerational fairness" [16, 18]

$$\text{LLC} = \sum_{t=1}^{T} \frac{\text{Net Cash Flows}_t}{(1+k)^t}$$

where Net Cash Flows$_t$ includes financial and internalised environmental costs, and benefits, and the latter are represented by a negative sign [16]. This definition of LCC, i.e., Net Cash Flows$_t$ refers to environmental LCC that incorporates, in addition to the conventional (financial) private benefits and costs, external benefits and costs that are internalised or privatised [14, 16].

The last component of LCSA is sLCA. sLCA measures all positive and negative social impacts of using reclaimed wastewater in irrigation along its entire life cycle and hence captures aspects that have both direct and indirect impact on stakeholders [14]. Similar to eLCA, sLCA involves four phases: (i) goal and scope definition, (ii) inventory development, (iii) impact assessment and (iv) interpretation [14, 16]. However, one of the major differences between the two measures is that eLCA captures the social impact only, if it is a result of an environmental impact, while sLCA accounts for all social impacts [16].

4 Reclaimed Domestic Wastewater Standards and Guidelines

According to the current Linear Economic (LE) model, domestic wastewater is a source of contamination (pollution) that should be safely disposed of to avoid the associated risks to humans and the environment. Therefore, in the LE model, the standard business for domestic wastewater is reclamation, which resulted in developing and adopting of "standards/guidelines/regulations for reclaimed domestic wastewater" worldwide. These standards/guidelines/regulations are either international ones, i.e., WHO guidelines [6], or national ones such as the Jordanian standards [7], the EU regulations [8], and the US-EPA guidelines [9].

LE model-based standards for reclaimed domestic wastewater basically requires: (1) treatment of raw domestic wastewater to improve its quality, (2) meeting the standard quality characteristics of reclaimed wastewaters for safe disposal and/or reuse, (3) guidelines and recommendations for the implementation to reduce the risks associated with disposal and reuse, and (4) application of quality monitoring system, by monitoring entities, to ensure upholding of the standard quality for disposal and reuse. Reclaimed domestic wastewater standards provide further details on each of the above requirements, as discussed below.

According to the Jordanian standards for reclaimed domestic wastewater [7], raw domestic wastewater treatment includes mechanical and natural treatment technologies. Mechanical technologies include activated sludge, trickling filter and rotating biological contactors; and the natural technologies include stabilization ponds, and aerated lagoons. In addition, the standard quality characteristics of reclaimed domestic wastewater (Table 1) embeds groups of physical, chemical and biological parameters, which are identified as indicators of the impacts of reclaimed domestic wastewater on human health and environment. The maximum allowable limit for each parameter varies according to reuse/disposal categories as shown in Table 1. Guidelines and recommendations include criteria for selecting discharge location, irrigation system, irrigation time, crop selection and crop harvesting, as well as harvesting protocol. The quality monitoring system includes parameters to be tested and the testing frequency, in addition to the actions to be taken, when quality standards are violated.

In contrast, EU regulations [8] for reclaimed wastewater restricts the reuse for irrigation only. This standard requires that raw wastewater be treated to the secondary level, which includes conventional activated sludge and trickling filter technologies, in addition to disinfection. Furthermore, filtration is required only for irrigation of category A crops. Please note that category A in EU standards refer to crops, while this category is mentioned in the Jordanian standards only for non-crop plantings like in parks, stadium green, and road side. Suitability of reclaimed wastewater for reuse is examined against five parameters (Table 1), which are also part of Jordanian standard [7] parameters (Table 1). Parameters' values, similar to Jordanian standards, vary according to irrigation categories; however, they are more stringent (Table 1). The more stringent values of EU regulations, i.e., category A, are attributed to the use of reclaimed wastewater for irrigating vegetable crops that are prohibited in Jordanian standards. EU regulations specifies irrigation system for each irrigation category, which is the only difference that distinguishes categories B & C. Adding to that and similar to Jordanian standards, quality monitoring and risk management systems are essential part of EU regulations [8].

The CE model, in contrast to LE, recognizes domestic wastewaters as resources [11, 19] that consist of a mixture of raw materials that should be valorized and sustainably exploited by returning them to the economy. This mixture of raw materials includes water as the major element, organic and inorganic matters, in addition to different types of organisms. Reuse and reclaim, among other Rs, are key pillars of CE model for water

Table 1 Jordanian standards [7] and EU regulations [8] for reclaimed (domestic) wastewater

Parameters	Jordanian standards for reclaimed domestic wastewater [7]						EU regulation for water reuse [8]			
	Discharge to Wadis, streams and water courses	Artificial discharge to aquifer	Reclaimed wastewater for irrigation							
				A	B	C	A	B (All irrigation system)	C (Drip or similar irrigation system)	D
			Cut flowers	Parks, stadiums & side roads inside cities	Fruit trees, roads outside cities & green landscaping	Industrial crops, forage crops & forest trees	All food crops consumed raw, where the edible part is in direct contact with reclaimed water, and root crops consumed raw. All irrigation system	Food crops consumed raw, where the edible part is produced above ground and is not in direct contact with reclaimed water, processed food crops and non-food crops including crops used to feed milk- or meat-producing animals. All irrigation system	Industrial, energy and seeded crops. All irrigation system	
BOD5 (ppm)	≤60	≤15	≤15	≤30	≤100	≤200	≤10	≤35		
COD (ppm)	≤150	≤50	≤50	≤100	≤200	≤300				
DO (ppm)	>1	>2	>2	>2						
TSS (ppm)	≤60	≤50	≤15	≤50	≤100	≤10	≤35			
pH	6 to 9									
Turbidity (NTU)		≤10	≤5	≤10		≤5				
NO3-N (ppm)	≤20	≤7	≤16							
NH4 (ppm)		≤4								
T-N (ppm)	≤70	≤20	≤70							
E. coli (MPN/100 ml)	≤1000	<1.1	<1.1	≤100	≤1000		≤10	≤100	≤1,000	≤10,000

(continued)

Table 1 (continued)

Parameters	Jordanian standards for reclaimed domestic wastewater [7]						EU regulation for water reuse [8]			
	Discharge to Wadis, streams and water courses	Artificial discharge to aquifer	Reclaimed wastewater for irrigation							
				A	B	C	A	B (All irrigation system)	C (Drip or similar irrigation system)	D
			Cut flowers	Parks, stadiums & side roads inside cities	Fruit trees, roads outside cities & green landscaping	Industrial crops, forage crops & forest trees	All food crops consumed raw, where the edible part is in direct contact with reclaimed water, and root crops consumed raw. All irrigation system	Food crops consumed raw, where the edible part is produced above ground and is not in direct contact with reclaimed water, processed food crops and non-food crops including crops used to feed milk- or meat-producing animals. All irrigation system	Industrial, energy and seeded crops. All irrigation system	
Intestinal Helminths Eggs (eggs /L)	≤1		≤1							
Fat, Oil and Grease (FOG) (ppm)	≤8	≤2	≤8							
Methylene Blue Active Substances (MBAS) (ppm)	≤25	≤15	≤100							
Phenol (ppm)	≤0.02									
TDS (ppm)	≤1,500									
PO4-P (ppm)	≤5	≤10								
Cl (ppm)	≤500	≤350	≤500							
SO_4 (ppm)	≤500	≤300	≤500							
HCO_3 (ppm)	≤400									
SAR	≤6	9								

and wastewater [1]. Accordingly, it is clear that the current LE-based standard models of reclaimed domestic wastewater go beyond the LE of "take-make-dispose" to achieve "reuse" and "reclaim" that are essential Rs (elements) of CE model. Thus, the current standard models of reclaimed domestic wastewater such as Jordanian standards and EU regulations, among others, are embedding aspects of CE concept, which is an important step to facilitate and accelerate the transition of water and wastewater sectors toward CE. However, while the current LE-based standards for reclaimed water are a good starting point, they should be updated to completely embrace the concept of CE for reclaimed domestic wastewater.

5 Case Study

Domestic wastewater is a non-conventional renewable water resource, and is a function of the increasing population, assuming the availability of water resources and infrastructure, among other factors. The increase in global annual population and climate change are key drivers of water use by all economic sectors, including domestic and agricultural. Most of the water used by the domestic sector is discharged as polluted water, which is well known as "domestic wastewater", and it should be reclaimed by regulations [7, 8]. Therefore, this case study targets reclaimed domestic wastewater reuse in irrigation focusing on Jordan. The case study uses the CE framework, which recognizes wastewater as a non-conventional renewable resource that should be sustainably valorized. This contrasts with LE that recognizes wastewater as waste that should be safely disposed of, rather than a resource. Accordingly, first the case study presents water situation in Jordan to highlight the importance of implementing reclaimed wastewater reuse policy and challenges in Jordan's water sector. Second, it applies an essential CE tool, which is the mass balance, for detailed analysis of Jordan water budget for the year 2022. The outcomes of mass balance do not only identify the gaps but also motivate the **rethink** search for opportunities to fill these gaps. Third, it sheds light on the opportunity of reclaiming and reusing of domestic wastewater with the focus on **valorization** of reclaimed wastewater as irrigation water.

5.1 Water Situation in Jordan

The average annual per capita fresh water share in Jordan is 100 m^3, based on 1133 MCM (million cubic meter) water budget and 11,302,000 populations in 2022 [20, 21]. This share is much lower than the annual 500 m^3/cap global threshold for severe water scarcity [22]. The very low per capita share of water in Jordan is a consequence of severe shortage, which has persisted for decades. This was shown in the annual water budget for the year 2022, in which "renewable conventional" water resources contribute only 50% to water use. The rest of the budget is covered by "non-renewable" (34%), and "non-conventional" resources (16%), 15.7 and 0.3% of the latter are contributions from reclaimed domestic wastewater and water desalination, respectively (Fig. 1). It is clear that reclaimed domestic wastewater is an essential component of the water budget, of which 98.6% is used in irrigation, while 1.4% is used in industrial applications [20].

The severe water shortage in Jordan is basically attributed to natural water scarcity, population growth [23] and semi-arid to arid climate, in addition to poor water management practices. Water scarcity is a consequence of limited renewable surface and groundwater resources, which are estimated at 378 and 213 MCM/yr, respectively, that come from national and international (transboundary) resources, respectively [20]. The semi-arid to arid climate of the country is another contributor to water shortage that is exacerbated by the impact of climate change [22]. Region instability, accompanied by the influx of refugees to the country, is a major contributing factor to the increasing population in the last decades [22]. Therefore, this study aims at analyzing the water budget to identify the gaps and to shed light on potential solutions that are aligned with CE concept.

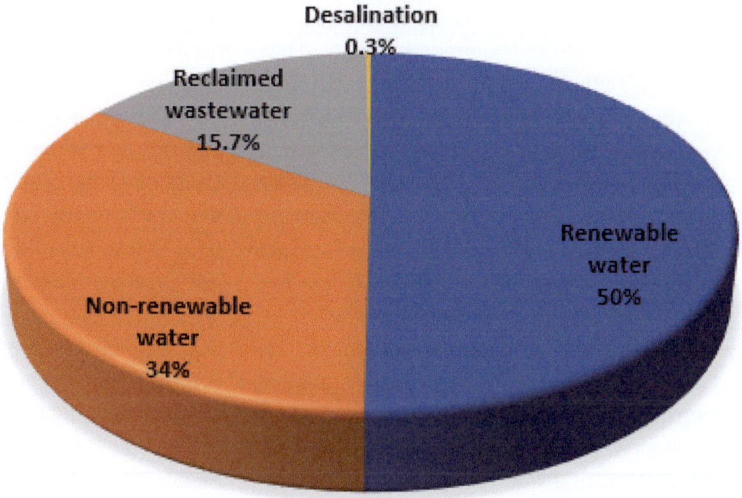

Fig. 1 Water resources contribution to 2022 water budget [20]

5.2 Water Budget Analysis

One-third of Jordan's water budget for the year 2022 came from non-renewable resources, i.e., non-renewable groundwater and over-abstracted renewable groundwater. This raises the question of the sustainability of the current water budget and urges for rethinking innovative solutions to sustainably meet the demand. Therefore, a detailed water mass balance of supply and use has been conducted, based on [20] data, to search for potential opportunities within the concept of CE.

Supply side Jordan received in 2022 a 6,623 MCM of fresh water from three conventional resources, which are 93.5% from precipitation, 3.2% from international water, and 3.3% from basic discharge. However, only 9.3% of this received water is collected, that is 615 MCM and forms the only annual fresh water supply in the water budget. 97% of the uncollected water (i.e., 91.7%), is attributed to evaporation and evapotranspiration, while 3% is attributed to uncontrolled floods. Furthermore, only 6% of the precipitation is collected, of which 2% as run-off stored in open dams and 4% infiltrates to recharge groundwater. Rethinking in water supply side points to further investment in water infrastructure, such as flood control, rainwater harvesting and artificial recharge, which are already recognized by the government of Jordan and highlighted in its national water strategy 2016–2025 [24]. So, it is possible to increase the supply side and to move toward a sustainable water budget. But what about the potential of use-side in solving the problem?

Use-side of the water budget in 2020 was mostly shared by two main sectors, which were domestic 45.5% and irrigation 50.5%, while the other 4% were shared by industry 3% and livestock 1% (Fig. 2). Two striking points on use-side are noticed: (1) On the one hand, water used by the domestic sector shows daily per capita share of 126 L, while the real share was 65 L [20]. This indicates about 48.5% of the water supplied for domestic use was in fact non-revenue

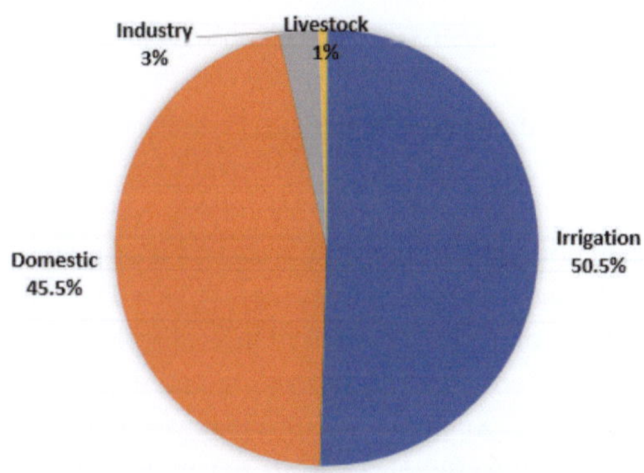

Fig. 2 Water budget distribution among the economic sectors for the year 2022 [20]

water, which is mainly divided into two main categories: physical and commercial losses. Thus, better management of non-revenue water saves up to 22.3% of the total water use, which was 1133 MCM in the year 2022, and hence the total water budget becomes 881 MCM. (2) On the other hand, 66% of the water that was actually used by the domestic sector (which was 51.5% of the total water supplied) is reclaimed. Therefore, collecting the uncollected domestic wastewater and reusing the total amount in irrigation, i.e., 268 MCM saves an extra 8% of the total water demand, making the total saving 30.3% of the water budget. Consequently, the total water budget becomes 790 MCM. **Rethinking** the water use-side indicates that proper management of the domestic water sector focusing on reducing non-revenue water and reusing the uncollected wastewater saves up to 30.3% of the total water budget. This saving results in an annual water use of 790 MCM instead of the current use of 1133 MCM, which consequently reduces the reliance on non-renewable water by a striking 91% and moves the water industry to a circular model of water management.

5.3 Wastewater Sector in Jordan

Domestic wastewaters in Jordan are managed by one of two systems: (1) centralized wastewater system and (2) de-centralized wastewater system [22]. Centralized wastewater system is the scope of this study, because it collects 66.2% of Jordan domestic wastewater and is already included in the water budget as reclaimed domestic wastewater.

Centralized wastewater system serves communities of more than 5000 inhabitants and consists of central sewage network that collects wastewater from domestic sources and transfers it to centralized wastewater treatment plants (CWWTP). The number of CWWTP in Jordan is 33 plants, 30 of which are currently operating and distributed across the kingdom as following: 11 in the northern governorates, 11 in the middle governorates, and 8 in the southern governorates [22, 25, 26]. CWWTPs in 2020/2021 reclaimed 177 MCM of domestic wastewater [25, 26], which is broken downs as follows: 12% from northern governorates, and 83 and 6% from the middle and the southern governorates, respectively [22, 25, 26]. The variation in the volumes of reclaimed wastewater across the governorates is attributed to the variation in the population in these governorates, with 29% of the population living in the north, 63% live in the middle and 8% live in the south [27]. Having shown that 177 MCM of domestic wastewater is reclaimed by the CWWTPs [25, 26], and the per capita share of domestic water is 65 L/d, the estimated percentage of population connected to sewerage system is 69%. This 69% is close to the 67% reported by [25], which is measured as the percentage of population registered to sewerage services divided by population registered to drinking water services.

Reclamation of domestic wastewater by CWWTPs, which represents a major pillar in CE, rely on different technologies. 21 CWWTPs use activated sludge as the main technology. Other technologies are also applied along with activated sludge, or as a standalone that include biological filter, aerated lagoon, natural basin or membrane biological reactor (MBR) [22, 25, 26]. Data obtained from these references are used to generate Table 2, which shows that CWWTP influent design capacity varies according to the served population. It can be seen that out of the 30 operating CWWTPs, one CWWTP serves more than 5.3 million people, receives 71% of the collected wastewater via sewerage network, and discharges 317,000 m^3/d reclaimed wastewater. The rest of collected wastewater, i.e., 29% is reclaimed by the other 29 CWWTPs, which collectively serve 2.1 million of the people connected to sewages. The 29 CWWTPs serve communities of that range between 5000 and 250,000 people and discharge between 7000 and 52,000 m^3/d of reclaimed wastewater, respectively (Table 2). Even though reclamation concept of CE is widely implemented, to achieve full circularity of wastewater, the CWWTPs should focus on sustainable technologies and/or operation conditions that are more consistent with CE that have been emphasized by [28]. For example, nature-based technology, such as stabilization ponds and natural lagoons are more aligned with

Table 2 CWWTPs in Jordan numbers, served population, influents and effluents

CWWTPs		Population[a]		Influent m^3/d^a		Effluent m^3/d^a	
Served population	Number	Average	Total	Average	Total	Average	Total
>5000–25,000	8	14,226 (7,784)	113,808	925	7400	856	6848
>25,000–50,000	6	37,021 (7,774)	222,126	2406	14,438	1918	11,508
>50,000–100,000	6	62,522 (9,393)	375,132	4064	24,384	3927	23,562
>100,000–150,000	5	121,391 (9,138)	606,955	7890	39,450	7288	36,440
>150,000–250,000	4	204,426 (13,979)	817,704	13,288	53,152	12,945	51,780
>5,300,000	1	5,302,424	5,302,424	344,658	344,658	316,986	316,986
Total	30		7,438,149		483,482		447,124

[a]Values were calculated by the authors using data from [22, 25, 26]; (in brackets) standard deviation

the principles of CE than the currently applied mechanical technologies. Further investigation of the discharged reclaimed domestic wastewater is presented in the next sections.

5.3.1 Reuse of Reclaimed Domestic Wastewater in Jordan

CWWTPs locations and reclaimed domestic wastewater quantity and quality are critical factors in identifying reuse opportunities. Jordanian standards [7] valorize reclaimed domestic wastewater as irrigation water, which is used by [22] for indirect and direct irrigation. Reviewing Jordanian standards shows two potential reuses of reclaimed domestic wastewater in indirect irrigation that are (1) discharge to streams, wadies and water bodies, and (2) artificial recharge of aquifers. While direct irrigation category refers to non-vegetable irrigations that are subcategorized into four groups as detailed in Table 1. Furthermore, Jordanian standards have made it clear that the authorized use of reclaimed domestic wastewater is also subject to meeting certain requirements related to taking preventive actions to protect drinking water resources, human health and the environment.

Indirect irrigation indicates that reclaimed domestic wastewaters are received by water bodies, surface and/or groundwater, which are used only for irrigation. This reclaimed waters that are received by water bodies are mixed with its water content, and the resulting blended water is used only in irrigation.

Direct irrigation using reclaimed domestic wastewater is meant for non-vegetable crops, which are classified into four subcategories (Table 1): (1) cut flowers, (2) road-side plants inside cities, parks and stadiums, (3) fruit-trees, road-side plants outside cities and green landscaping, and (4) industrial crops, field crops and forest trees. According to Jordanian standards, these categories prohibit sprinkler irrigation, except for irrigation of golf courses during night; warning signs indicating irrigation by reclaimed wastewater should be posted in green landscaping. Jordanian standards' guidelines go beyond regulating irrigation using reclaimed wastewater; it rather extends to cover harvesting, it stipulates that irrigation should be stopped for two weeks before harvesting, in addition, fallen fruits should be discarded. However, and despite the importance of having storage spaces for reclaimed wastewater to meet irrigation needs, in terms of both time and volume, this aspect is overlocked by Jordanian standards.

Furthermore, Jordanian standards do not include industrial reuses of reclaimed wastewater. However, it kept the door open for other potential reuses providing that research is conducted to assess the reuse impact on human health, environment and water courses.

5.4 Quantity of Reclaimed Domestic Wastewater

The CWWTPs discharge of reclaimed domestic wastewater in Jordan is 88.2% reused in irrigation, while 2.1% is used in industry (Fig. 3). The rest amounting to 9.7% is lost through evaporation.

Indirect irrigation uses 70.8% of the total reclaimed domestic water, while direct irrigation represents 17.4%, which takes place in the vicinity of CWWTPs (Fig. 3). This 17.4% used in direct irrigation is broken down as following: 16% for forage crops, olive trees, fruitful trees and ornamental nursery; 1.3% for landscaping, and 0.1% for forestry irrigation. Reclaimed wastewater that exceeds the irrigation use in the vicinity of some CWWTPs is discharged to streams and wadies to be received by dams. Direct irrigation takes place in the highlands, and the irrigation is managed through contracts between farmers and Water Authority of Jordan (WAJ) that allocate the amount of water in proportion to irrigated land [22].

70.8% of the reclaimed wastewater for indirect irrigation is shared by (i) King-Talal Dam that receives 68.1%, (ii) Wadi Shoaib, that receives 2.1%, and (iii) Kfreen Dam that receives 0.6% (Fig. 3). The dam water becomes blended water and is used for irrigation in Jordan Valley, which makes the food basket of Jordan.

5.5 Quality of Reclaimed Domestic Wastewater

The quality monitoring system of JS (893/2021) for reclaimed domestic wastewater requires instantaneous sample collection from the CWWTPS's effluents at point of discharges. The frequency of sample collections, which is dependent on the quality parameter to be tested, is either quarterly or annually.

The quality characteristics of reclaimed domestic wastewater in Table 3 were reported by [29, 30]. On the one hand, the table shows that the examined quality parameters of "tertiary treated" reclaimed wastewater from a CWWTP in southern Jordan comply with all the reuse options that are identified by JS (893/2021) in Table 1. Thus, this "tertiary treated" reclaimed domestic wastewater can be used not only in direct irrigation for categories A, B and C, but also can be discharged to wadi, streams and water courses to be blended with its water content for indirect irrigation.

On the other hand, the table shows that the examined quality parameters of "secondary

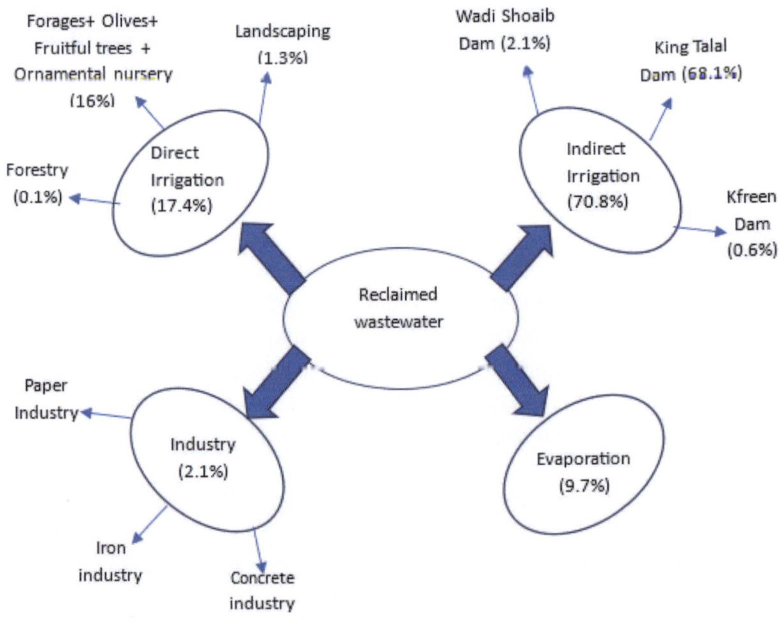

Fig. 3 Reclaimed domestic wastewater in Jordan—disposal and reuses. Authors calculations based on data from [22]

Table 3 Quality of CWWTPs reclaimed domestic wastewaters from Southern and Northern Jordan obtained from [29] to [30] respectively

Parameter		Southern Jordan/CWWTP [29]	Northern Jordan/CWWTP [30]
pH		7.35 (0.1)	
DO	mg/L	6.15 (0.37)	
BOD5 (f)	mg/L	1.08 (0.46)	25–76
COD	mg/L		45–145
TSS	mg/L	3.6 (0.12)	38–174
NO3	mg/L	1.04 (0.23)	0.6–10
NH3	mg/L		0.1–42
T – N	mg/L	1.35 (0.22)	12–52
PO4-P	mg/L		8–17
E. coli	MPN/100 mL	0	2.4E1-2.41E2

treated" reclaimed wastewater from a CWWTP in northern Jordan comply, except for the TSS and PO_4 for some analyzed samples, with the reuse standards of categories B, C and discharge to wadi. However, according to [30] upgrading/ modernization this WWTP will improve the quality to comply with the standards for direct and indirect irrigation.

Rethinking wastewater as a renewable non-conventional water resource should be the key in the search for reuse opportunities of reclaimed wastewater, such as in fishery agro-industry. Another promising effective use of reclaimed domestic wastewater is in hydroponic farming. Furthermore, restoration of ecological system of the degraded system is another option. Finally, water-consuming industries, such as textile industry and other related industries, should be targeted as potential reuse scenarios of reclaimed domestic wastewater.

6 Conclusions

Economy transition from linearity to circularity is a must for sustainable economic development. This transition aims at decoupling economic growth from resource consumptions. The economy major sectors domestic, agricultural and industry are water-dependent, and natural water resources are finite. Accordingly, decoupling the growth in the economic sectors from the consumption of natural water resources is inevitable to sustain economic growth and development. In this context, this chapter investigated the potential transition of water sector from the current linear economy model toward the circular economy model.

Therefore, it shows the importance of the CE six Rs: reduce, reuse, recycle, recover, rethink and reclaim to achieve the transition in wastewater and water management to the CE model. It further illustrates that in order to operationalize the CE models in water industry LCSA is an important assessment tool to evaluate the economic, environmental and social impacts of such transition.

Having established the way forward to sustain water resources, the chapter moves to discuss the current regulations and standards that govern the current practices of reclaimed wastewater reuse in agriculture to identify the gap between the current business model and the targeted model, i.e., CE model.

The chapter concludes with a case study that applies mass balance as CE tool and reports its outcomes. The outcomes of mass balance have identified a number of gaps in water management that prevent this sector from achieving circularity. The identified gaps include; investment in water infrastructure, reduction of nonrevenue water, full collection of domestic wastewater, reuse of total reclaimed wastewater, and upgrade of wastewater treatment technologies. These are critical issues in the current water management that should be targeted by any future plan to accelerate moving the water sector toward circularity. The results of the mass balance have further emphasized the importance of rethinking (one of the CE six Rs in water industry) to find the right pathway to circularity.

References

1. Smol M. Circular economy approach in the water and wastewater sector. Circ Econ Sustain. 2022;(2):1–16. Elsevier Inc.
2. FAO, AQUASTAT—FAO's global information system on water and agriculture. https://www.fao.org/aquastat/en/overview/methodology/water-use. Accessed 31 Dec 2023.
3. Hislop H, Hill J. Reinventing the wheel: a circular economy for resource security, green alliance. 2011. https://green-alliance.org.uk/wp-content/uploads/2021/11/Reinventing-the-wheel.pdf. Accessed 31 Dec 2020.
4. Ellen MacArthur Foundation. https://www.ellenmacarthurfoundation.org/topics/circular-economy-introduction/overview. Accessed 07 Apr 2024
5. Elisha OD. Moving beyond take-make-dispose to take-make use for sustainable economy. Int J Sci Res Educ. 2020;13(3):497–516.
6. WHO. WHO Guidelines for the safe use of wastewater, excreta and greywater. 2006.
7. JS 893. Jordanian standards for water: reclaimed domestic wastewater. Amman, Jordan: Jordanian Standards and Metrological Organization;2021.
8. EU 741. Regulation (EU) 2020/741 of the European parliament and of the council of 25 May 2020 on minimum requirements for water reuse (Text with EEA relevance). Off J Eur Union. L 177/32;2020.
9. EPA/600/R-12/618, USAID. Guidelines for water reuse. EPA. EPHCA, NRMMCA, AHMC;2012.
10. COM; Communication from the Commission to the European Parliament, the Council, the European Economic and Social Committee and the Committee of the Regions Towards a thriving data-driven economy;2014. https://eur-lex.europa.eu/legal-content/EN/TXT/?uri=celex%3A52014DC0442. Accessed 31 Dec 2023.
11. Abu-Ghunmi D, Abu-Ghunmi L, Kayal B, Bion A. Circular economy and the opportunity cost of not 'closing the loop' of water industry: the case of Jordan. J Clean Prod. 2016;131:228–36.
12. Padilla-Riveraa A, Carmo B, Arcesec G, Merveille N. Social circular economy indicators: selection through Fuzzy Delphi method. Sustain Prod Consum. 2021;26:101–10.
13. Ingrao C, Arcidiacono C, Siracusa V, Niero M, Traverso M. Life cycle sustainability analysis of resource recovery from waste management systems in a circular economy perspective Key Findings from This Special Issue. Resources. 2021;10(4):32.
14. UNEP. Towards a life cycle sustainability assessment: making informed choices on products. UNEP/SETAC Life Cycle Initiative. 2011. https://wedocs.unep.org/bitstream/handle/20.500.11822/8001/UNEP_LifecycleInit_Dec_FINAL.pdf?sequence=3&amp%3BisAllowed. Accessed 31 Dec 2023.
15. Kloepffer W. Life cycle sustainability assessment of products: (with Comments by Helias A. Udo de Haes, p. 95). Int J Life Cycle Assess. 2008;13:89–95.
16. Hoogmartens R, Van Passel S, Van Acker K, Dubois M. Bridging the gap between LCA, LCC and CBA as sustainability assessment tools. Environ Impact Assess Rev. 2014;48:27–33.
17. Swarr TE, Hunkeler D, Klöpffer W, Pesonen HL, Ciroth A, Brent AC, Pagan R. Environmental life-cycle costing: a code of practice. Int J Life Cycle Assess. 2011;16:389–91.
18. Howarth RB. Discount rates and sustainable development. Ecol Model. 1996;92(2–3):263–70.
19. Werner C, Fall PA, Schlick J, Mang H. Reasons for principles of ecological sanitation. Inecosan—Closing the loop. In: Proceedings of 2nd international symposium, Lubeck, Germany, 7–11 April 2003. Eschborn, Germany: Deutsche Gesellschaft für Technische Zusammenarbeit (GTZ) GmbH;2003. pp. 23–30.
20. MWI. Water budget, 2022. Ministry of water and irrigation. 2022.
21. DoS. Statistical year book of Jordan 2022. Department of Statistics. 2022.
22. Breulmann M, Muller RA, Al-Subeh A, van Afferden M. Reuse of treated wastewater and biosolids in Jordan—National evaluation. Published by the Helmholtz centre for environmental research-UFZ with support from the Ministry of Water and Irrigation, Amman-Leipzig;2019.
23. Abu-Ghunmi LN. Characterization and treatment of gray water; options for (re)use. Ph.D thesis, Wageningen University, Wageningen, The Netherlands;2009.
24. NWS. National water strategy 2016–2025. Ministry of Water and Irrigation;2016.
25. MWI. Jordan water sector: facts and figures 2020. Annual water report 2022. Ministry of Water and Irrigation-Water authority—Jordan Valley authority;2020a.
26. MWI. Water yearly book: water year 2019–2020. Ministry of Water and Irrigation;2020b.
27. DoS. Statistical year book of Jordan 2021. Department of Statistics;2021.
28. Kayal B, Abu-Ghunmi D, Abu-Ghunmi L, Archenti A, Larkin C, Corbet S. An economic index for measuring

firm's circularity: the case of water industry. J Behav Exp Financ. 2019;21(3):123–9.
29. Al-Momani R, Rasheed M. Treated wastewater uses in irrigation conserve water in world's second water poorest country, Jordan. Fresenius Environ Bull. 2016;25(6):2225–31.
30. Myszograj S, Qteishat O, Sadecka Z, Jedrczak A, Suchowska-Kisielewicz M. Possibilites of reuse of treated wastewater for irrigation purpose in the northern Jordan Valley. Environ Prot Eng. 2014;40(2):17–31.

Determination of the Amount of Treated Wastewater Potential for Irrigation Purposes: A Case Study of Türkiye

Irem Karahan, Serhat Gunaydin, Onur Altun, Yusuf Basaran, Yakup Karaaslan, Afire Sever, and Ilker Angin

Abstract

By considering the amount of usable water per capita, Türkiye is considered a country that suffers from water stress. Therefore, Türkiye must set up procedures based on plans for the effective and planned use of its current water resources to deal with a potential future problem of water scarcity. The option of reusing treated wastewater is one of the most significant techniques for achieving the most effective use of water. A large amount of water is used in agriculture, and the burden on freshwater resources can be reduced through the use of treated wastewater for irrigation purposes. This study aimed to determine the amount of treated wastewater potential for irrigation purposes in Türkiye. To determine water reuse alternatives, a project entitled "Evaluation of Used Water Reuse Alternatives" was conducted by the Republic of Türkiye, Ministry of Agriculture and Forestry, General Directorate of Water Management between the years 2017 and 2019. Used water from wastewater treatment facilities in Türkiye has been assessed for the use in irrigation as part of the project. In this paper, decision-making steps for water reuse in Türkiye are given as a flowchart, and methods for determination of the total amount of the treated wastewater potential for irrigation purposes are explained. As a result of the study, the basin basis used water potential in Türkiye is reported.

Keywords

Irrigation water · Reclaimed water · Wastewater reuse · Water reuse · Türkiye

1 Introduction

With numerous and far-reaching effects, global warming and climate change present a matter of significant societal concern. Changes in precipitation patterns, rise in the frequency and number of extreme weather events, changes in ecosystems, and an increase in climatically associated natural disasters including floods, storms, hurricanes, and drought are some of these effects [1]. These challenges are expected to strongly affect the agricultural sector. Drought in water resources might drastically reduce or completely destroy crop yields, which would result in significant economic losses for countries' agricultural sectors.

I. Karahan · S. Gunaydin · O. Altun · Y. Basaran · Y. Karaaslan · A. Sever
Ministry of Agriculture and Forestry, General Directorate of Water Management, Ankara, Türkiye

I. Angin (✉)
Faculty of Agriculture, Department of Agricultural Structures and Irrigation, Atatürk University, Erzurum, Türkiye
e-mail: iangin@atauni.edu.tr

M. Bahadir et al. (eds.), *Agricultural Water Management*, Water and Wastewater Management,
https://doi.org/10.1007/978-3-031-96134-2_13

Globally, 70% of freshwater is used by agriculture. These rates are 91.2% for South Asia, 85.3% for Middle East and North Africa, 80.7% for Sub-Saharan Africa, 72.1% for Latin America and the Caribbean, 71.8% for East Asia and Pacific, and 35.7% for Europe and Central Asia [2]. One of the most important components of adapting to climate change is managing water resources, which includes reusing the nutrients, energy, and, of course, water that are present in municipal wastewater. The local economy can frequently profit from suitable reuse made possible by wastewater treatment plants (WWTPs). For instance, reusing wastewater effluent from treated WWTPs for agricultural uses can boost crop yields due to its characteristics, which is rich in organic matter, and macro and micro-nutrients. Wastewater is reported to improve plant growth, to reduce fertilizer application rates, and to increase the productivity of poor fertility soils when used as an irrigation source [3].

Reusing water often referred to as water recycling or reclamation is a process that recovers water from several sources, treats it, and then uses it for irrigation, industrial activities, environmental restoration, potable water supply, and groundwater replenishment. Reusing water can improve water security, sustainability, and resilience by offering alternatives to the current water supply [4]. In times of water scarcity, using treated wastewater to irrigate plants is an alternative for irrigation, while it also enhances soil nutrients and productivity without disturbing soil structure [5]. Additionally, by avoiding effluent discharge into water bodies, the use of treated wastewater in irrigation could maintain freshwater supplies and lessen water pollution [6].

The intended use or discharge location, and the applicable regulatory framework would affect collection and treatment design. Compared to freshwater sources, treated wastewater may be predicted to have higher concentrations of pathogens, organic contaminants, heavy metals, and/or hazardous compounds [7]. If not effectively controlled, these chemicals and diseases could not only endanger the health of farmers and consumers but also disturb soil structure. As a result, rules for the reclaimed water from WWTPs are usually stricter than the use of freshwater resources. In terms of quality, it has been determined that even if the used water is purified, it will generally need to be subjected to additional treatment to bring it to the appropriate criteria in terms of the relevant usage area.

Although water reuse is successfully implemented widely in many European Union (EU) Member States and developed countries, especially in the USA (California), Australia, and Singapore, its full potential is not being used even in the most productive countries. Limited awareness of the potential benefits among stakeholders and the general public, and the lack of a supportive and coherent framework for water reuse have been identified as the two main barriers preventing this practice from spreading wider in the EU.

Currently, around one billion m^3 of treated urban wastewater is reused in EU countries per year; this accounts for around 2.4% of the amount of treated municipal wastewater and less than 0.5% of annual EU freshwater withdrawals. Both Member States in the south, such as Spain, Italy, Greece, Malta, and Cyprus, as well as in the north, such as Belgium, Germany, and the United Kingdom, already have several initiatives in the reuse of water for irrigation, industrial uses, and replenishment of the aquifer. Cyprus and Malta currently reuse more than 90 and 60% of their wastewater, respectively, while Greece, Italy, and Spain reuse around 5–12% of their wastewater [8].

Water scarcity has become an increasingly vital issue in Türkiye due to expanding population, urbanization, industrialization, climate change, and global warming. While the annual amount of usable water per capita in Türkiye was approximately 1346 m^3 in 2020, this value is expected to decrease to 1120 m^3 in 2040, and 1069 m^3 in 2050. Türkiye will face a high risk of water scarcity shortly soon. The latest UN climate report singles out the Mediterranean Basin as one of the world's climate hotspots that will bear a disproportionate brunt of global warming [9]. Türkiye is forecast to be more severely affected by global warming than the countries in Europe. Therefore, precautions shall be taken to guard against this danger. It is targeted to increase the reuse rate of treated wastewater in Türkiye to 5% in 2023 and 15% in 2030 [10].

It is necessary to develop practices based on strategies related to the efficient and planned use of existing water resources in the fight against possible water scarcity situations in the future. The option of reusing wastewater is considered one of the most important methods in the efficient use of water. The aim of this study is to complete the evaluation of the reuse of treated wastewater for irrigation in Türkiye because reusing treated effluents gives consumers a consistent supply and helps ensure quality from an environmental and sanitary aspect.

Therefore, this study was carried out to determine the amount of used water potential in all urban and domestic wastewater treatment plants in Türkiye with a capacity of $\geq$2000 m^3/day for irrigation purposes, and quality analysis are done for selected WWTPs.

2 Materials and Methods

2.1 Decision Making Steps for Wastewater Reuse

To determine water reuse alternatives, a project entitled "Evaluation of Used Water Reuse Alternatives" was conducted by the Republic of Türkiye, Ministry of Agriculture and Forestry, General Directorate of Water Management between the years 2017 and 2019. Within the scope of the project, wastewater treated in WWTPs has been evaluated for agricultural irrigation, landscape irrigation, industrial use, groundwater feeding, and environmental use throughout Türkiye.

At first, the current status of WWTPs is identified and an inventory study is done by categorizing treatment plants according to their treatment type and capacity. In addition, planned WWTPs are identified for inventory study. Revision needs of WWTPs are determined after technical analysis. As a result of the inventory study, the total capacity of WWTPs across Türkiye is calculated.

In the second stage, water quality analysis were done to determine the quality of WWTPs effluents. For the aim of the reuse of treated wastewater, total irrigation area calculation and crop pattern identification of irrigation areas were made. The water demand of the areas is determined and the required water quality for crop pattern analysis was completed. Final usage is important for the use of treated wastewater (e.g., direct irrigation, storage, etc.). Therefore, reservoir field requirement and water need per hectare are calculated, and the supply–demand relation is studied in detail. Within the scope of the project, all reports reflecting the water problems of the basins, such as basin master plans, basin protection action plans, river basin management plans, special provision studies, wetland reports, and climate reports were examined. WWTPs in each basin and users, who have the potential to create demand for effluent water from WWTPs, were identified and matched. The water quality determined during the matching was taken into consideration.

To determine the feasibility of the study, a cost analysis was made. An evaluation of the water quality for the proposed reuse area was done, and if the quality is found unsuitable for the intended use, revisions that should be made in the facilities were determined. If there is a need for revision, revision costs are analyzed, and post-treatment cost analysis of WWTPs, where the post-treatment required for the crop pattern, is calculated. The operating cost of WWTPs is also analyzed. Investments to be made outside the facility were determined, and the transfer and storage cost of reusable treated wastewater is calculated. The decision-making steps for wastewater reuse is shown in Fig. 1.

2.2 Wastewater Treatment Plants in Türkiye

As a result of the inventory study conducted for the complete determination of the WWTPs to be considered as a used water source, a total of 614 facilities with a project flow rate of 2000 m^3/day

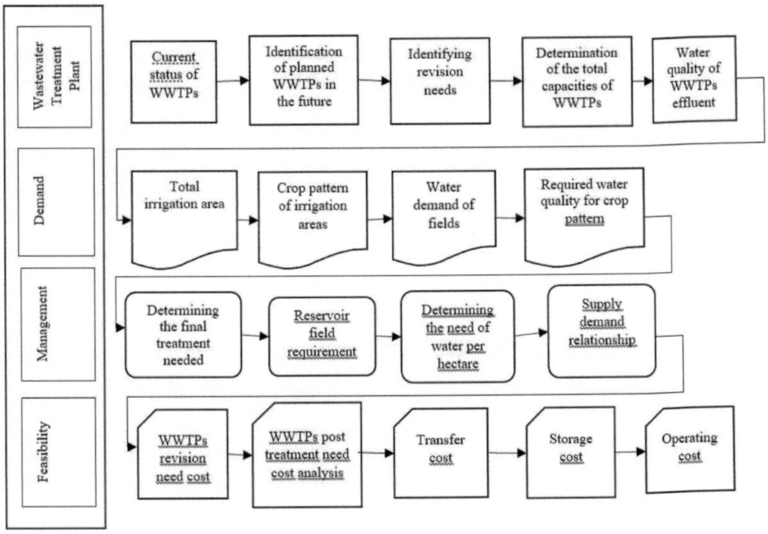

Fig. 1 Decision making flowchart for irrigation with treated wastewater

and above were identified. Although the project flow rates of the four facilities are below 2000 m^3/day, they were also included in the project, because they are the best alternatives for reuse in the particular area of the basins. With the inclusion of these facilities in the inventory, a project inventory consisting of 618 facilities has emerged. Among the facilities to be evaluated, some facilities are in the planning, projecting, and construction phase. The number of facilities according to the operating status is given in Table 1.

The distribution of WWTPs according to the project flow rates are presented in Table 2. To determine the flow rates of some WWTPs, whose project flow could not be determined, a calculation was made over the population considering the daily wastewater generation amount per person as 200 L. The Wastewater Action Plan of the Ministry of Environment and Urbanization was used for the determination of the population.

The process distributions of the WWTPs currently under operation are given in Table 3. Advanced biological treatment is carried out in a total of 186 WWTPs, and 25 of them, the water is discharged to the sea by the deep-sea discharge method. Deep sea discharge is also carried out in 29 of the biological treatment plants. After advanced biological and biological treatment, the most applied method is physical treatment, and 33 of the 34 WWTPs that apply physical treatment carry out deep sea discharge.

Table 1 Operating status of WWTPs

Number	Status	Number of WWTPs
1	Operating	389
2	Under construction	37
3	Projecting	88
4	Planning	89
5	Revision	3
6	Off	1
7	Idle (not activated yet)	11
Total		618

Table 2 Distribution of facilities according to project flow rates

Number	Project flow (m^3/day)	Number of WWTPs
1	<2000	4
2	=2000	17
3	>2000–5000≤	187
4	>5000–10,000≤	134
5	>10,000–50,000≤	194
6	>50,000–100,000≤	42
7	>100,000	40
Total		618

Table 3 Process selections of the operating WWTPs

Number	Treatment type	Number of WWTPs
1	Advanced biological treatment (A^2O, Bardenpho, UCT, etc.)	161
2	Advanced biological treatment + DSD[a]	25
3	Biological treatment	121
4	Biological treatment + DSD	29
5	Package biological treatment	1
6	Physical treatment + Biological treatment + DSD	1
7	Physical treatment	1
8	Physical treatment + DSD	33
9	Chemical treatment + Biological treatment	1
10	Electrocoagulation	1
11	Electrocoagulation + Biological treatment	1
12	Electrocoagulation + DSD	1
13	Stabilization pool	12
14	Natural treatment	1
Total		389

[a]DSD: Deep Sea Discharge

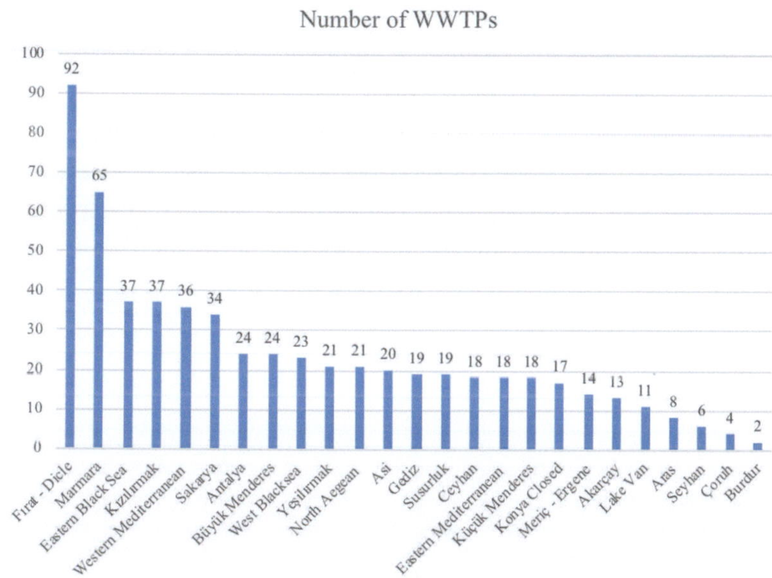

Fig. 2 Distribution of wastewater treatment plants by basin

618 WWTPs in the inventory are not homogeneously distributed over 25 basins of Türkiye. The distribution of the facilities, evaluated within the scope of the project according to the basins, is shown in Fig. 2. Accordingly, the Euphrates-Tigris Basin leads with 89 facilities, and the Marmara Basin follows it with 67 facilities. While the least WWTP is located in the Burdur Basin with two facilities, the second least facility is in the Çoruh Basin with four facilities.

Within the scope of the project, the evaluation of WWTPs was also done on basin basis. As a

Table 4 Process distributions of the WWTPs included in the water quality analysis

Number	Treatment process	Number of WWTPs
1	Advanced biological treatment	150
2	Biological treatment	85
3	Physical treatment	1
4	Stabilization pool	6
Total		**242**

result of the evaluations, the amount of wastewater treated in 618 facilities in Türkiye is determined as 20.3 million m^3/day. This accounts for approximately 7.5 billion m^3/year of wastewater that is treated and lost by being discharged to the receiving environment. The wastewater treated in some of the facilities was not considered as an alternative water source due to the location of the facilities and the absence of any reuse alternatives around them. It has been determined that it is not possible to evaluate the wastewater treated in 55 of 618 facilities in the inventory in this context. It was calculated that the amount of treated wastewater discharged from these facilities is 3.2 million m^3/day in total, which corresponds to approximately 1.2 billion m^3 per year. In this case, it was determined that the total WWTP output amount determined to be reusable throughout Türkiye is approximately 6.2 billion m^3/year.

2.3 Water Quality Analyses of WWTPs

To determine the treated water quality of WWTPs across Türkiye, analyses of the effluent water quality were carried out at 242 WWTPs. The large-capacity facilities with reuse potential are prioritized thereby.

The number of treatment plants selected applying advanced biological treatment and biological treatment is 150 and 85, respectively. To have an idea about the effluent quality and to determine the removal efficiency in the stabilization ponds, six stabilization ponds were also included in this program. 37 of the examined facilities are currently performing deep sea discharge. Process distributions of the WWTPs included in the water quality analyses are given in Table 4.

The distribution of WWTPs across the basins, for which effluent water quality analyses were made, is shown in Fig. 3. While, the highest number of water quality analyses was carried out in the Marmara Basin, no samples were taken from the wastewater treatment plant outlet of Aras Basin because its reuse potential is extremely difficult due to the topography.

The samples were collected as 2-h composite samples with an auto-sampler and stored at 4 °C until analyses. Characteristics of effluents were determined using the methods proposed by the APHA, AWWA, and WPCF [11].

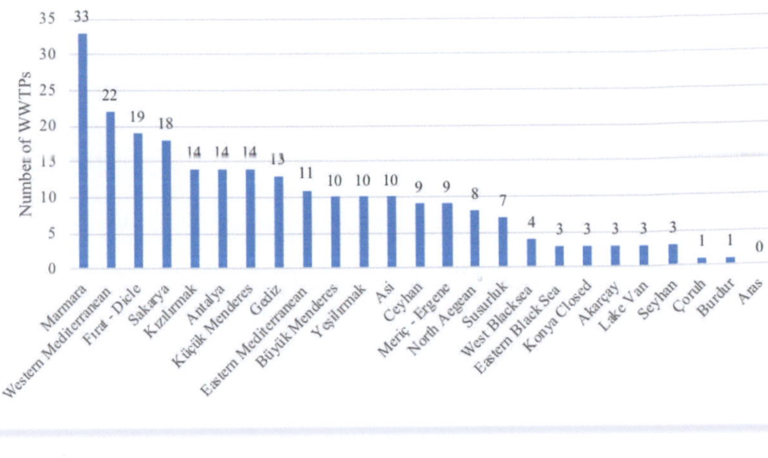

Fig. 3 The distribution of wastewater treatment plants across the basins, for which effluent water quality analyses were made. Revise x-axis: Euphrates-Tigris

Table 5 Minimum, maximum, and average values of analyzed effluents from selected WWTPs

Parameters	Unit	Min	Max	Average	Number of analysis
Temperature	°C	15.2	30.5	25.4	51
pH	–	6.97	9.16	7.78	242
Turbidity	NTU	0.11	43.9	1.22	242
Electrical conductivity	μS/cm	152	28,800	1663	242
SAR	–	0.82	28.5	4.67	241
Total Suspended Solids (TSS)	mg/L	10	752	22	242
Total Dissolved Solid (TDS)	mg/L	72	7500	796	242
Total pesticide	mg/L	0.001	0.001	0.001	242
Total hardness	mg/L	10	1732	353	242
Biological Oxygen Demand (BOD_5)	mg/L	3	58.6	7.0	242
Total Organic Carbon (TOC)	mg/L	5	61.6	9.51	242
Total nitrogen	mg/L	0.47	114	16.8	242
Nitrate nitrogen	mg/L	0.1	38.8	6.5	242
Nitrite nitrogen	mg/L	0.1	4.56	0.20	242
Total Kjeldahl Nitrogen (TKN)	mg/L	0.22	114	10.0	242
Total phosphorus	mg/L	0.03	15.7	2.35	242
Aluminum	mg/L	0.02	4.63	0.08	242
Antimony	mg/L	0.0005	0.0355	0.0009	242
Arsenic	mg/L	0.0005	0.4028	0.0065	242
Copper	mg/L	0.001	0.1435	0.0067	242
Beryllium	mg/L	0.0005	0.0005	0.0005	242
Boron	mg/L	0.02	8.75	0.32	242
Mercury	mg/L	0.0001	0.0095	0.0002	242
Zinc	mg/L	0.005	1.38	0.051	242
Iron	mg/L	0.005	8.56	0.210	242
Fluoride	mg/L	0.1	3.98	0.18	242
Cadmium	mg/L	0.0005	0.0076	0.0005	242
Chloride	mg/L	13.04	6521	414	242
Cobalt	mg/L	0.0005	0.0028	0.0006	242
Chromium	mg/L	0.001	0.0411	0.0022	242
Lead	mg/L	0.0005	0.0404	0.0009	242
Lithium	mg/L	0.001	0.9333	0.0214	242
Manganese	mg/L	0.0005	0.352	0.055	242
Molybdenum	mg/L	0.0005	0.0689	0.0016	242
Nickel	mg/L	0.005	0.0699	0.0072	242
Selenium	mg/L	0.0005	0.0061	0.0006	242
Sodium	mg/L	13.76	2724	231	242
Sulfate	mg/L	15.93	5009	154	242
Vanadium	mg/L	0.0005	0.0252	0.0023	242
Escherichia coli	CFU/100 mL	0	400,000	13,377	242
Fecal coliform	CFU/100 mL	38	660,000	24,059	241

3 Results and Discussion

3.1 Effluent Water Quality of WWTPs

Table 5 summarizes the examination results of selected WWTPs. The values in this table were evaluated according to their suitability as irrigation water. The acquired effluent quality was determined in terms of irrigation water using the Regulation on Technical Procedures for WWTPs, which serves as Türkiye's internal irrigation water standards [12]. As per this regulation, the average values show that effluents can be classified as none/slight to moderate restriction on use. The mean value of treated effluent pH was determined as 7.78. The electrical conductivity (EC) of samples varied between 152 and 28,800 μS/cm, with an average of 1663 μS/cm. The possible reasons for the high EC values are (i) unauthorized industrial wastewater discharges to the facility, and (ii) sea water infiltration into the sewer system. High EC values will not only reduce the treatment efficiency by inhibiting the bacteria responsible for treatment in biological treatment plants but also may cause soil salinization and thus soil degradation. To use treated wastewater as an irrigation supply caution should be taken to reduce the EC. This shows that there is slightly to moderate restriction on use in terms of EC. According to Türkiye's internal irrigation water standards, sodium adsorption ratio (SAR) and EC data should be combined, when determining the infiltration or permeability of treated wastewater into the soil. Since its SAR value is in the range of 3–6 and its EC higher than 1200 μS/cm, it was categorized as no restriction on use for irrigation. In other words, when the EC and SAR values were considered together, it is evident that the effluent may not induce any problem to the infiltration rate.

Results obtained have also shown that many facilities have problems in the removal of nutrients nitrogen and phosphorus. This situation may cause eutrophication if wastewater is stored. When examined independent of the processes, it is observed that removal of the N and P can be done, but discharge standards are not always met. Although there is no problem in terms of average concentration, it is important to examine the maximum concentration and to make the necessary arrangements for nutrient removal in the treatment plants for the protection of the receiving environments. Apart from this, it is foreseen that the effluent of WWTPs with high nutrient concentrations can be used to meet the fertilizer needs of plants. To eliminate the negative effects of N and P, fertilizer application rates should be recalculated when using wastewater as an irrigation source. Irrigation with wastewater should be done with consideration of the N and P demands of the soils and plants.

When evaluated in terms of specific ion toxicity, the average value of Na^+ (relative tolerance of crops to foliar injury from saline water changes by irrigation method, therefore when evaluating Na^+, SAR is used for surface irrigation methods, and me/L for sprinkler irrigation) indicated that treated effluent has slight to moderate restriction on use both in surface and sprinkler irrigation. However, Cl^- values indicated that there is severe restriction of use in surface and slight to moderate restriction on use in sprinkler irrigation. Boron concentration of effluents varied between 0.02 and 8.75 mg/L, with an average of 0.32 mg/L. On average, there is no restriction of use in terms of boron concentration. However, attention should be paid to effluents that have higher concentrations. Heavy metals, which could accumulate in the soil following irrigation, may reach toxic levels to the cultivated crops and the customer. Results obtained have shown that the heavy metal removal efficiencies of the WWTPs are high, and the metal concentrations are below critical limits. For more information about critical limits, please refer to Table Annex 7.2 of the Regulation on Technical Procedures for WWTPs. To maintain sustainable use of treated wastewater, heavy metals in effluents and soils should be tested regularly.

Microbiological evaluations have shown that an adequate fecal coliform and E. coli removal could not be reached even in some facilities with disinfection units. This problem can be solved by improving the operation conditions, especially in facilities with a disinfection unit at the exit. Since the effluents of many WWTPs in Türkiye are used for agricultural irrigation purposes in an unplanned manner, proper removal of microbiological indicators is important for the protection of public health and the environment.

Considering the total pesticide concentrations, no concentration above the detection limit of 0.001 mg/L was detected in any of the samples examined. This shows that there will not be a problem in terms of pesticides in the reuse of treated wastewater.

3.2 Basin Basis Wastewater Potential

Depending on their hydrological characteristics, Türkiye is classified into 25 basins that are given in Fig. 4. Within the scope of this project, wastewater treated in WWTPs has been evaluated for agricultural irrigation, landscape irrigation, industrial use, groundwater feeding, and environmental use throughout Türkiye. Therefore, it should be noted that the amount of wastewater used for irrigation can differ from the actual amount, because the considered basin may have much more suitable option than irrigation.

Marmara Basin

There are 65 domestic and urban WWTPs with a capacity of 2000 m^3/day in the Marmara Basin. The amount of treated wastewater in the Marmara Basin is determined as 7.4 million m^3/day, which corresponds to 2.7 billion m^3/year. Evaluations made have shown that it is not possible to evaluate the effluents of 13 WWTPs with a total project flow rate of 3.4 million m^3/day in any reuse application. For this reason, the amount of reusable treated wastewater in the basin was determined as 4 million m^3/day, which was calculated as 1.5 billion m^3/year. Because of the evaluations made, 257 million m^3/year of treated wastewater in the basin has been suggested to be used for irrigation purposes.

Susurluk Basin

There are 19 domestic and urban WWTPs with a capacity of 2000 m^3/day in the Susurluk Basin. The amount of treated wastewater in this basin is determined as 710,543 m^3/day, which corresponds to 259 million m^3/year. Evaluations made have shown that it is not possible to evaluate the effluent of one of the WWTPs with a project flow of 2000 m^3/day in any reuse application. For this reason, the amount of reusable treated wastewater in the basin was determined as 0.71 million m^3/day, which was calculated to be 259 million m^3/year. Calculations made have shown that 90 million m^3/year treated wastewater in the basin can be used for irrigation purposes.

North Aegean Basin

There are 21 domestic and urban WWTPs with a capacity of 2000 m^3/day in the North Aegean Basin. The amount of treated wastewater in this

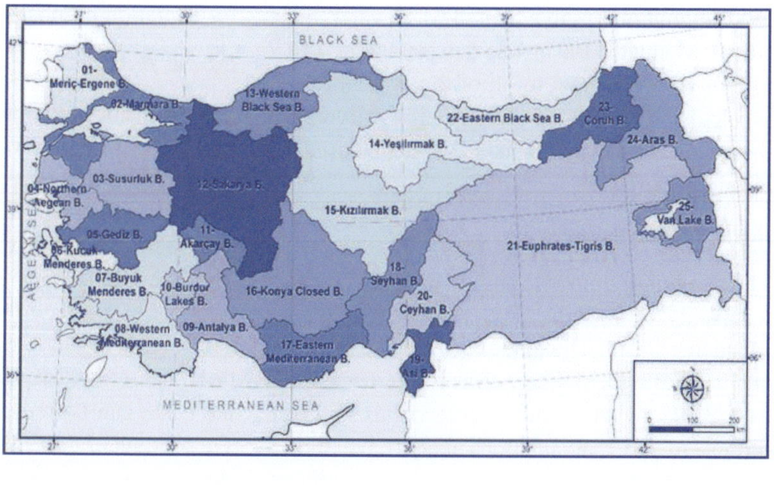

Fig. 4 Basins of Türkiye [13]

basin is determined as 239,969 m^3/day, which corresponds to 87.6 million m^3/year. Evaluations conducted have shown that it is not possible to evaluate one of the WWTPs effluent with a project flow of 30,000 m^3/day in any reuse application. Thus, the amount of reusable treated wastewater in the North Aegean Basin was determined as 0.21 million m^3/day, which corresponds to 76.6 million m^3/year. The amount of treated wastewater that can be used for irrigation purposes was determined as 43.6 million m^3/year.

Gediz Basin

The number of domestic and urban WWTPs with a capacity of 2000 m^3/day is 19 for the Gediz Basin. The amount of treated wastewater in this basin is determined as 871,918 m^3/day, which corresponds to 318 million m^3/year. The total amount of reusable treated wastewater in this basin was determined as 318 million m^3/year. Evaluations made have shown that 112 million m^3/year treated wastewater in the basin can be used for irrigation purposes.

Küçük Menderes Basin

There are 18 domestic and urban WWTPs with a capacity of 2000 m^3/day in the Küçük Menderes Basin. The amount of treated wastewater in the Küçük Menderes Basin was determined as 314.349 m^3/day, which corresponds to 114.7 million m^3/year. According to the evaluations made, it can be stated that it is possible to evaluate the effluent of each wastewater treatment plant in at least one reuse application. Therefore, the total amount of reusable treated wastewater in the Küçük Menderes Basin was determined as 114.7 million m^3/year. As a result of the evaluations, it is calculated that 19.7% (22.6 million m^3/year) of the treated wastewater in the basin can be used for irrigation purposes.

Büyük Menderes Basin

The number of domestic and urban WWTPs with a capacity of 2000 m^3/day is 24 for the Büyük Menderes Basin. The amount of treated wastewater in the Büyük Menderes Basin was determined as 483,873 m^3/day, which corresponds to 176.6 million m^3/year. Evaluations made have shown that it is possible to evaluate the effluent of each wastewater treatment plant in at least one reuse application. Therefore, all of the treated wastewater in the basin can be reusable. The amount of treated wastewater that can be used for irrigation purposes was determined as 75 million m^3/year.

Western Mediterranean Basin

There are 36 domestic and urban WWTPs with a capacity of 2000 m^3/day in the Western Mediterranean Basin. The amount of treated wastewater in the Western Mediterranean Basin was calculated as 341,591 m^3/day, which corresponds to 124.7 million m^3/year. Evaluations made have shown that it is not possible to evaluate five WWTPs effluents in any reuse application. Therefore, the total amount of reusable treated wastewater in the basin was determined as 112.7 million m^3/year. As a result of the evaluations, it is calculated that 32% (36 million m^3/year) of the treated wastewater in the basin can be used for irrigation purposes.

Antalya Basin

The number of domestic and urban WWTPs with a capacity of 2000 m^3/day is 24 for the Antalya basin. The total amount of treated wastewater from these 24 WWTPs was calculated as 961,852 m^3/day, which corresponds to 351 million m^3/year. Evaluations made have shown that it is not possible to evaluate one WWTP effluent in any reuse application. The total amount of reusable treated wastewater in the Antalya basin was determined as 349 million m^3/year. 177.7 million m^3/year treated wastewater in the basin can be used for irrigation purposes.

Burdur Basin

Among the basins, Burdur Basin has the minimum number of WWTPs with a capacity of 2000 m^3/day (only 2 WWTPs). Therefore, the amount of treated wastewater in this basin was determined as 37,138 m^3/day, which corresponds to 13.6 million m^3/year. Evaluations made have shown that it is possible to evaluate the effluent of each wastewater treatment plant in at least one reuse application. Therefore, all of the treated wastewater in the basin can be reusable for different purposes. The amount of treated wastewater that can be used for irrigation purposes was determined as 835,850 m^3/year.

Akarçay Basin

There are 13 domestic and urban WWTPs with a capacity of 2000 m^3/day in the Akarçay Basin. The amount of treated wastewater in this basin was determined as 108,532 m^3/day, which corresponds to 39.6 million m^3/year. Evaluations made have shown that it is possible to evaluate the effluent of each wastewater treatment plant in at least one reuse application. Thus, all of the treated wastewater in the basin can be reusable for different purposes. As a result of the evaluations, it is calculated that 20.3% (8 million m^3/year) of the treated wastewater in the basin can be used for irrigation purposes.

Sakarya Basin

The number of domestic and urban wastewater treatment plants with a capacity of 2000 m^3/day is 34 for the Sakarya basin. The amount of treated wastewater in this basin was calculated as 1.6 million m^3/day, which corresponds to 572 million m^3/year. According to the evaluations made, it is seen that the effluent of each wastewater treatment plant can be used in at least one reuse application. Therefore, the total amount of reusable treated wastewater in the basin was actual to the total discharged amount. As a result of the evaluations made, it is calculated that 261 million m^3/year treated wastewater in the basin can be used for irrigation purposes and the rest amount for different purposes.

West Black Sea Basin

There are 23 domestic and urban WWTPs with a capacity of 2000 m^3/day in the West Blacksea Basin. The amount of treated wastewater in this basin was determined as 313,401 m^3/day, which corresponds to 114 million m^3/year. The evaluations made have shown that it is not possible to evaluate nine WWTPs' effluents in any reuse application. Thus, the total amount of reusable treated wastewater in the basin was determined as 76 million m^3/year. As a result of the evaluations made, it is estimated that 14.6 m^3/year of treated wastewater can be used for irrigation purposes throughout the basin.

Yeşilırmak Basin

The number of domestic and urban WWTPs with a capacity of 2000 m^3/day is 21 for the Yeşilırmak basin. The amount of treated wastewater in this basin was determined as 331,781 m^3/day, which corresponds to 121 million m^3/year. According to the evaluations made, it is seen that the effluent of each wastewater treatment plant can be used in at least one reuse application. Therefore, the total amount of reusable treated wastewater for different purposes in the basin was actual to the total discharged amount (121 million m^3/year). As a result of the evaluations, it is calculated that 53.1% (64.3 million m^3/year) of the treated wastewater in the basin can be used for irrigation purposes.

Kızılırmak Basin

There are 37 domestic and urban WWTPs with a capacity of 2000 m^3/day in the Kızılırmak Basin. The amount of treated wastewater in the Kızılırmak Basin was determined as 531,496 m^3/day, which corresponds to 194 million m^3/year. Evaluations made have shown that it is not possible to evaluate four WWTPs' effluents in any reuse application. Therefore, the total amount of reusable treated wastewater in the basin was determined as 184.9 million m^3/year. Calculations made have shown that 110 million m^3/year (59.5% of total amount) treated wastewater in the basin can be used for irrigation purposes.

Konya Closed Basin

Among the basins, the Konya Closed Basin is the one that suffers from lack of irrigation water most. Therefore, any drop of water is important for this basin. There are 17 domestic and urban WWTPs with a capacity of 2000 m^3/day in the Konya Closed Basin. The amount of treated wastewater in this basin was determined as 489,144 m^3/day, which corresponds to 178.5 million m^3/year. According to the evaluations made, it is seen that the effluent of each wastewater treatment plant can be used in at least one reuse application. Therefore, the total amount of reusable treated wastewater for different purposes in the basin was actual to the total discharged amount (178.5 million m^3/year). The amount of treated wastewater that can be used for irrigation purposes was determined as 154 million m^3/year (86.2% of total discharge).

Eastern Mediterranean Basin

The number of domestic and urban WWTPs with a capacity of 2000 m^3/day is 18 for the Eastern Mediterranean Basin. The amount of treated wastewater in this basin was determined as 491,866 m^3/day, which corresponds to 179.5 million m^3/year. Conducted evaluations have shown that it is not possible to evaluate one WWTP effluent in any reuse application. Therefore, the ratio of reusable treated wastewater to the discharged wastewater was calculated as 99% (177.8 million m^3/year). This shows that effluent of each wastewater treatment plant can be used at least in one reuse application. As a result of the evaluations made it is calculated that 61.5 million m^3/year treated wastewater can be used for irrigation purposes throughout the basin.

Seyhan Basin

There are six domestic and urban WWTPs with a capacity of 2000 m^3/day in the Seyhan Basin. The amount of treated wastewater in the Seyhan Basin was determined as 366,100 m^3/day, which corresponds to 133.6 million m^3/year. According to the evaluations made, it is seen that the effluent of just one WWTP cannot be used in any reuse application. The total amount of reusable treated wastewater for different purposes in the basin was determined as 132.9 million m^3/year. As a result of the evaluations, it is calculated that 32.9% (43.7 million m^3/year) of the treated wastewater in the basin can be used for irrigation purposes.

Ceyhan Basin

The number of domestic and urban WWTPs with a capacity of 2000 m^3/day in the Ceyhan Basin is 18. The amount of treated wastewater in this basin was determined as 361,242 m^3/day, which corresponds to 132 million m^3/year. Evaluations made have shown that the effluent of each wastewater treatment plant can be used in at least one reuse application. Therefore, the total amount of reusable treated wastewater for different purposes in the basin was actual to the total discharged amount (132 million m^3/year). As a result of the evaluations, it is calculated that 44.8 million m^3/year of treated wastewater can be used for irrigation purposes in the basin.

Eastern Black Sea Basin

The number of domestic and urban WWTPs with a capacity of 2000 m^3/day is 37 for the Eastern Black Sea Basin. The total amount of treated wastewater in this basin was determined as 591,600 m^3/day, which corresponds to 216 million m^3/year. Evaluations made have shown that it is not possible to evaluate 27 WWTPs' effluents in any reuse application, due to the topography of the region. Therefore, the total amount of reusable treated wastewater in the basin was determined as 56 million m^3/year. Only, 1.5 million m^3/year treated wastewater can be used for irrigation purposes.

Çoruh Basin

There are four domestic and urban WWTPs with a capacity of 2000 m^3/day in the Çoruh Basin. The amount of treated wastewater in this basin was determined as 21,696 m^3/day, which corresponds to 7.9 million m^3/year. Evaluations conducted have shown that it is not possible to evaluate two WWTPs' effluents in any reuse application, due to the topography of the region. The total amount of reusable treated wastewater in the basin was determined as 4.2 million m^3/year. It is calculated that 59.3% (2.5 million m^3/year) of the treated wastewater in the basin can be used for irrigation purposes.

Lake Van Basin

There are 11 domestic and urban WWTPs with a capacity of 2000 m^3/day in the Lake Van Basin. The amount of treated wastewater in this basin was determined as 292,248 m^3/day, which corresponds to 106.7 million m^3/year. Evaluations made have shown that the effluent of each wastewater treatment plant can be used in at least one reuse application. Therefore, the total amount of reusable treated wastewater for different purposes in the basin was actual to the total discharged amount (106.7 million m^3/year). The amount of treated wastewater that can be used for irrigation purposes was determined as 21 million m^3/year (19.7% of total discharge).

Due to lack of data, the remaining four basins (Meriç-Ergene, Aras, Asi, and Euphrates-Tigris) were not evaluated.

3.3 Cost Analysis

Cost analyses were evaluated under two headings as *in-facility* and *outside-facility* costs. Results obtained in *in-facility cost* calculations were made in a way to include survey, project, and engineering services. Although transmission routes are created following lines such as the main road, and the creeks, and ignoring significant elevation changes, differences during the implementation phase should be taken into account in the calculation of *outside-facility* costs. Therefore, the costs of the land that will be needed for the transmission, pumping center, storage area, and proposed processes are not taken into account in the cost analysis. Cost calculations also do not include land expropriation costs.

Cost analysis included operating and maintenance costs as well as total investment costs. Equipment replacement costs are considered as a part of operating costs. Cost calculations include depreciation accounts and interest expenses, and the annual discount rate is accepted as 10%.

The following formulas were used in calculating the costs.

Total Cost:

$$M_T = M_1 + M_2$$

M_1 = Initial investment cost, and M_2 = Present value of annual operating and maintenance costs,

$$M_2 = A\frac{(1+i)^n}{ix(1+i)^n}$$

A: Annual Operation and maintenance costs, i: Discount rate, and n: Project life (year).

Present Value of Treated Wastewater (M_G):

$$M_G = VX\frac{(1+i)^n - 1}{ix(1+i)^n}$$

V: Volume of water treated in one year, and X: Cost of 1 m^3 of water.

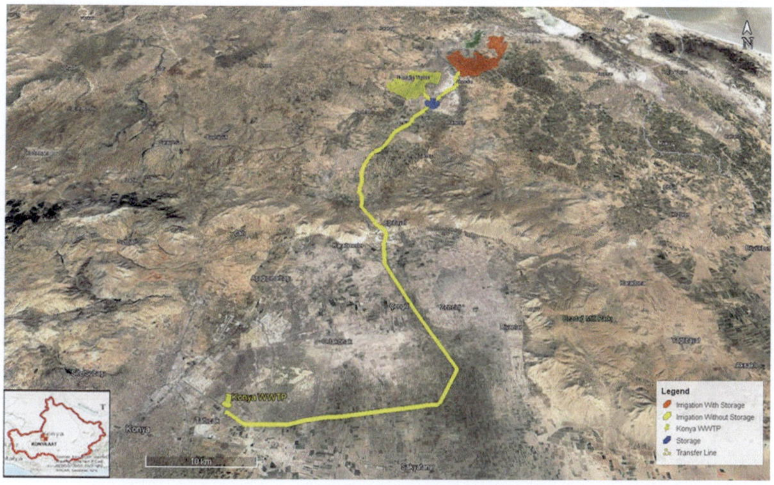

Fig. 5 Recommended reuse areas for Konya WWTP

Konya Wastewater Treatment Plant Example

Water treated in Konya WWTP is discharged into the main discharge channel and reaches Tuz Lake after joining with Keçili Canal. Keçili Canal was built to be Konya's rainwater drainage channel, to protect the city from floods and overflows. Konya WWTP discharges its treated water to this channel with special permission. Recommended reuse areas for Konya WWTP is shown in Fig. 5.

Increasing the WWTP capacity from 200,000 to 400,000 m^3/day: 75 million Euro, construction of advanced treatment processes for irrigation water quality: 15 million Euro, construction of storage pond: 10 million Euro, irrigation infrastructure of 15,600 ha (1st stage): 63 million Euro, irrigation infrastructure of 22,714 ha (2nd stage): 20 million Euro, irrigation infrastructure of 36,670 ha (3rd stage): 47 million Euro. As a total, irrigation of 75,285 ha costs approximately 230 million Euro.

Further data about cost analysis is available upon permission of General Directorate of Water Management of Türkiye.

4 Conclusions

In order to cope with a potential future water scarcity problem, Türkiye must establish practices based on strategies connected to the efficient and planned use of existing water resources. One of the most significant strategies for making optimal use of water is the option of reusing treated wastewater. In this study, reuse potential of effluents of WWTPs in Türkiye is calculated. The quantity of WWTPs output that has been found to be reusable over all of Türkiye is 6.2 billion m^3/year. Quality of the treated wastewater is important for irrigation. The acquired effluent quality determined in terms of irrigation water using Table Annex 7.2 of the Regulation on Technical Procedures for WWTPs, showed that most of the effluents of WWTPs can be used in irrigation with none/slight to moderate restriction on use. However, it should be noted that each WWTP is an individual, and the determination of its effluent characteristics is important for reuse.

Acknowledgements This study was prepared according to the results of the project entitled "Evaluation of Used Water Reuse Alternatives" conducted by the Republic of Türkiye, Ministry of Agriculture and Forestry, General Directorate of Water Management between the years 2017 and 2019.

References

1. Günaydin S. Modelling the potential benefits of irrigation with nutrient-rich treated municipal wastewater effluent for growth of loblolly pine (Master Thesis). The University of Alabama. 2021.
2. World Bank. Water in agriculture. 2017. https://blogs.worldbank.org/opendata/chart-globally-70-freshwater-used-agriculture. Accessed 15 Aug 2022
3. Angin I, Yaganoglu AV, Turan M. Effects of long-term wastewater irrigation on soil properties. J Sustain Agric. 2005;26(3):31–42.
4. US EPA. Basic information about water reuse announcements and schedules. 2019. https://www.epa.gov/waterreuse/basic-information-about-water-reuse.
5. Urbano VR, Mendonça TG, Bastos RG, Souza CF. Effects of treated wastewater irrigation on soil properties and lettuce yield. Agric Water Manag. 2017;181:108–15. https://doi.org/10.1016/j.agwat.2016.12.001.
6. Ofori S, Puškáčová A, Růžičková I, Wanner J. Treated wastewater reuse for irrigation: Pros and cons. Sci Total Environ. 2021;760:144026. https://doi.org/10.1016/j.scitotenv.2020.144026.
7. Yi L, Jiao W, Chen X, Chen W. An overview of reclaimed water reuse in China. J Environ Sci. 2011;23(10):1585–93.
8. JRC Publications Repository. Water reuse in Europe-Relevant guidelines, needs for and barriers to innovation. 2022. https://publications.jrc.ec.europa.eu/repository/handle/JRC92582. Accessed 22 July 2022.
9. Hockenos P. As the climate bakes, turkey faces a future without water. 2021. https://e360.yale.edu/features/as-the-climate-bakes-turkey-faces-a-future-without-water. Accessed 20 July 2022.
10. Ministry of Environmental Urbanization and Climate Change. Aritilmiş Atiksularin Yeniden Kullanim Oraninda Yüzde 4 Olan Yil Sonu Hedefi Aşildi. 2022. https://csb.gov.tr/aritilmis-atiksularin-yeniden-kullanim-oraninda-yuzde-4-olan-yil-sonu-hedefi-asildi-bakanlik-faaliyetleri-34168.
11. Greenberg AE, Trussell RR, Clesceri LS. Standard Methods for the Examination of Water and Wastewater. Washington: APHA, AWWA, WPCF;1985.

12. Anonymous. Atıksu Arıtma Tesisleri Teknik Usuller Tebliği. 2010. https://www.mevzuat.gov.tr/mevzuat?MevzuatNo=13873&MevzuatTur=9&MevzuatTertip=5.
13. Ministry of Forestry and Water Affairs. Strategy of National Watershed Management Report. 2013.

Agricultural Irrigation with Reclaimed Municipal Wastewater: Environmental Implications

Fatma Bedük, Senar Aydin, Mehmet Emin Aydin, and Müfit Bahadir

Abstract

Wastewater irrigation seems an inevitable way of sustainable agricultural activities because of water scarcity. While this method offers many opportunities, like conservation of clean water sources for domestic consumption, better crop yield, lower fertilizers need, there are also some drawbacks, such as accumulation of pollutants in soil, translocation of the pollutants from soil to crops, and hygienic problems in crops caused by irrigation methods that result in consumer health risk. While accumulation and translocation of heavy metals in soil and cultivated crops are well documented, fate of all synthetic organic pollutants and microplastics (MPs) in the soil ecosystem is not yet clear. Although wastewater irrigation raises concerns about sustainability of agriculture and human health, preserving domestic wastewater from industrial discharges, successful wastewater treatment methods, and selection of true irrigation methods may ensure safe and profitable application of wastewater irrigation. The aim of this review is to point at potential contaminants resourced from insufficiently treated wastewater. For this purpose, a literature review was conducted about heavy metals, synthetic organic pollutants, and MPs sourced contamination of soils from wastewater irrigation.

Keywords

Heavy metals · Microplastics · Pharmaceuticals · Synthetic organic pollutants · Municipal wastewater · Irrigation

1 Introduction

More than two thirds of freshwater is used in the agriculture sector. The increasing need for water to sustain agricultural activities and the decrease of precipitation in some parts of the world as a result of global warming have resulted in agricultural drought. Excessive abstraction of salty groundwater for agricultural irrigation threats soil fertility [1]. While higher quality water is used for domestic needs, lower quality water is preferred to be used for irrigation purposes. Especially in arid and semi-arid regions, and

F. Bedük (✉) · S. Aydin
Engineering Faculty, Environmental Engineering Department, Necmettin Erbakan University, Konya, Türkiye
e-mail: fabeduk@erbakan.edu.tr

M. E. Aydin
Engineering Faculty, Civil Engineering Department, Necmettin Erbakan University, Konya, Türkiye

M. Bahadir
Institute of Environmental and Sustainable Chemistry, Technical University of Braunschweig, Braunschweig, Germany

M. Bahadir et al. (eds.), *Agricultural Water Management*, Water and Wastewater Management,
https://doi.org/10.1007/978-3-031-96134-2_14

in agricultural areas, where most of the water resources are consumed, the utilizing of treated or untreated wastewater for irrigation purposes is considered as a solution for the sustainability of agriculture. Approximately 6 million ha of agricultural lands all over the world has been irrigated with wastewater [2]. It seems sure that water shortage is going to be a common problem of more countries in the course of time, and wastewater irrigation will continue to be evaluated as a solution for sustainable agricultural activities.

Despite there is a positive effect of wastewater on agricultural productivity in the short term, there are many studies on its negative effects on plant and product quality [3–5]. While wastewater contains organic matter and mineral elements useful for growing crops, it also contains toxic substances and persistent organic pollutants (POPs). Pollution of soils with heavy metals and toxic organic compounds because of the use of wastewater for irrigation is an important problem for many countries [6–8]. Legally or illegally, insufficiently treated or untreated wastewater are used for agricultural purposes. When literature studies are examined, it is seen that one of the most important ways that people are exposed to toxic pollutants is the consumption of products grown in soil contaminated with these pollutants [9, 10].

Regulations for discharge or reuse of reclaimed water mainly include conventional parameters, such as chemical oxygen demand (COD), biological oxygen demand (BOD), total nitrogen (TN), total phosphorus (TP), total suspended solids (TSS), heavy metals, and microbiological parameters, etc. While tertiary treatment effluents of wastewater treatment plants (WWTPs) can meet criteria set for conventional parameters, secondary treatment, which is more generally applied, usually does not meet most of these criteria. Effluents of conventional WWTPs still contain excess nitrogen (N) and phosphorus (P), emerging pollutants, mineral salts, pathogens, and some toxic substances. A significant number of emerging pollutants is not monitored regularly in wastewater, since their presence in the environment is not regulated by legislation. While pesticides, polycyclic aromatic hydrocarbons (PAHs), polychlorinated biphenyls (PCBs), nutrients, and heavy metals are regulated contaminants, pharmaceuticals and MPs are candidates for future regulations. Unclear risks to soil and agricultural products, hence human health results in conflicting opinions about regulations. At the same time, there is a need for analytical capacity of laboratories, which is one of the main restrictions affecting regulation of the pollutants.

Contamination of agricultural products can occur in the form of physical effects of irrigation water on the product, and uptake of chemical content from the soil through the roots and accumulation in the crops. The behavior and interaction of pollutants in soil covers various physical, chemical and biological processes occurring in all phases (solid, liquid, gas) in the soil media. These are generally adhesion in and on the soil, infiltration, diffusion and transport by soil solutions, and chemical changes (transformation) in the soil. The type of soil, cultivated plant characteristics, climate and wastewater characteristics affect the soil contamination potential.

The purpose of this review paper is to highlight the effects of wastewater irrigation on soil and agricultural products. For this purpose, first, the effects of physicochemical properties of soil on its contamination potential are presented, and then soil contaminants, such as heavy metals, synthetic organic pollutants, pharmaceuticals, and MPs are reviewed. Factors affecting bioaccumulation, transportation, and sorption–desorption behavior of these contaminants in soil and plants are emphasized.

2 Effects of Physicochemical Properties of Soil on Contamination

While the properties of soil, such as geochemical fractions, cation exchange capacity (CEC), organic matter content, acidity, and alkalinity affect its contamination potential, wastewater irrigation also changes the properties of soil, such as its salinity, organic matter content, and microbial activities [11, 12]. The mobility and retention of environmental pollutants are strongly influenced by sand, silt, and clay contents of the soil. The high clayey structure of the soil prevents the infiltration of pollutants and causes their accumulation in the upper layer.

Irrigation with reclaimed wastewater affects alkalinity and acidity of the soil. As the acidity in the soil increases, firstly the replaceable bases in the soil are replaced by hydrogen (H^+). The displaced bases are taken up by plants or washed away with irrigation water and rainwater in the form of soluble salts. Wastewater causes high electrical conductivity (EC) in the soil due to its increased salinity, caused by dissolved solids and major ions. Wastewater irrigation results in an increase of exchangeable sodium (Na^+) concentration in soil. A review of 21 independent studies about the effects of wastewater irrigation on soil properties demonstrated a significant soil salinity by increasing the concentrations of potassium (90.1%), sodium (59.9%), etc. Soil EC also increased by approximately 57.2% [13].

Ion exchange is the process, in which the solid phase containing clay and organic matter (OM) in the soil absorb anions and cations and replaces them with an equivalent amount of other ions into the soil solution. Cation exchange capacity (CEC) is responsible to retain nutrients on the top soil. Therefore, high CEC means generally that the soil is fertile. CEC is influenced by the binding strength of the absorbed cations. As soil pH increases, H^+ dissociates from the SiOH functional groups of clay minerals, and the resulting negatively charged SiO^- ions combine with appropriate cations. As the amount of clay and organic matter in soil increases, CEC also increases. Clay type, clay amount, organic matter amount, and pH are the factors that affect the CEC of the soil. The increase of the CEC value also increases the buffering feature of the soil. Buffering in soil is defined as the resistance to change its pH. However, higher CEC of the soil result in higher accumulation of the pollutants in the soil and vice versa [14].

Soil is a habitat for algae, bacteria, fungi and other microorganisms. Dominant groups such as algae and cyanobacteria are important for soil fertility and soil health. However, environmental contaminants affect soil microorganisms. Bedük [12] questioned effects of simultaneous occurrence of Cadmium (Cd) and Azithromycin (AZI) on ammonium oxidation bacteria in soil. At environmentally available concentrations of 11 mg/kg Cd and 5 mg/kg AZI, 50% inhibition of potential nitrification rate was determined for the sample soil, which is an indication of risks that are vital for soil fertility.

3 Contamination of Soil and Agricultural Products with Reclaimed Municipal Wastewater

3.1 Heavy Metals

Metal ions in irrigation water might leach to groundwater, be uptaken by the cultivated plants, and be subjected to chemical reactions in the soil [15]. Metal concentration in soil is influenced by the complexation, oxidation/reduction, precipitation/dissolution, and adsorption/ desorption reactions [16]. Uncontrolled industrial discharges are main sources of heavy metals contaminating domestic wastewaters. The transfer of heavy metals from soil to edible plants has been widely studied. It was determined that interactive effects of soil properties, such as soil pH, CEC, and OM content were more effective to control heavy metal transfer from soil to the crop, than the heavy metal concentrations in the soil itself [17]. Mao et al. [18] reported that heavy metal concentrations in soil were lower for higher soil pH, while accumulation of heavy metals by rice shoot was higher for low soil pH. It was determined that heavy metal uptake of plants increased in saline and waterlogged soils than in non-saline and well drained soils [19].

Plant type is another factor affecting the bioaccumulation and tolerance to heavy metals. Heavy metal contents of selected often consumed vegetables and crop plants, irrigated with wastewater in a region in Punjab, Pakistan were explored. Brinjal, wheat, tomato and spinach were determined to bioaccumulate heavy metals to a higher extent than red corn, white corn, luffa, apple gourd, and cabbage, the former exceeding the limits set by WHO/FAO for Cr, Ni, Mn, and Pb [20].

Some plant types are used for phytoremediation of heavy metal contaminated soils due to their high bioaccumulation capability for heavy metals. Therefore, irrigation with heavy metals containing wastewater should be applied with caution for hyperaccumulator species, such as sunflowers (*Helianthus annuus*) and Indian mustard (*Brassica juncea*) [21, 22].

Heavy metals accumulate in different parts of the plant, such as root, shoot, and leaves. Higher heavy metal concentrations in the aerial parts of plants (stem and leaf) than in the root indicate the mobility of these metals. So as to determine the heavy metal storage ability of a plant, the bioaccumulation factor (BF) is calculated by the rate of heavy metal concentration in the root (C_{root}) to heavy metal concentration in the soil (C_{soil}) as BF $= C_{root}/C_{soil}$. A similar calculation can be used for the determination of the translocation factor (TF), by dividing the heavy metal concentration in shoot (C_{shoot}) through the heavy metal concentration in root as TF $= C_{shoot}/C_{root}$. An investigation was carried out for fodder plant Jew mallow (*Corchorus olittorius*) in order to assess the effect of wastewater irrigation on bioaccumulation of heavy metals. It was determined that Jew mallow phyto-accumulated Cd, Ni and Mn in its root, while it translocated Pb, Cu, Cr, Fe, and Zn to its leaves [23]. While the **BF** was higher than 1 for all examined heavy metals, only Pb, Cu, Cr, Fe, and Zn presented **TF** higher than 1. Low levels of TF ensure safe application of wastewater irrigation for edible plants. It was determined that irrigating Citrus trees with wastewater is safe in the means of heavy metals for the Citrus fruit produced, since heavy metals do not translocate from the root to the fruit [24]. Gupta et al. [25] determined the bioaccumulation of Fe, Mn, Pb, Cd, and Cr in different plant parts of three species *Colocasia esculentum, Brassica nigra*, and *Raphanus sativus*. While the total heavy metal accumulation in roots was higher for *Raphanus sativus*, the species *Brassica nigra* accumulated the highest concentration of heavy metals in its shoots.

Aydin et al. [3] determined that wastewater irrigation might not be safe for wheat growth, since the highest concentrations of Pb, Cu, and Zn were determined in the grains, when compared with the leaf and stem of the plants. Untreated domestic and industrial wastewaters of Konya City in Türkiye were conveyed through a main drainage channel and were used downstream for agricultural irrigation for a long time until 2010. It was determined that long-term irrigation of agricultural areas resulted in heavy metal contaminations of both soil and cultivated crops in the region. The highest notable soil contamination was determined for Cd (8.23–11.6 mg/kg). Maximum concentrations in wheat grain samples irrigated with wastewater were found for Pb, Cr, Cu, Cd, Zn, and Ni, respectively (8.44, 1.30, 9.10, n.d, 29.31, and 0.94 mg/kg, respectively). Cd contamination was reported for wheat grain, cultivated in wastewater irrigated area in Tianjin, China. It was found that 54.5 and 18.25% of soil samples were contaminated with Cd and Zn, respectively, while TF was the highest for Zn and Cd, when compared with Cu, Pb, Cr, and Ni. While Zn contamination of wheat grain was reported to be below tolerance limits, Cd was in the range of 0.019–0.117 mg/kg, and one wheat grain sample exceeded tolerance limits, set in China (0.1 mg/kg). A significant positive correlation (correlation coefficient of $r = 0.565$) was determined between Cd contents in the wheat grain and the soil [26].

Co-existing of heavy metals in soil is another factor influencing both accumulation of these compounds in soil and their uptake by plants. Xu et al. [27] performed a research study on the effects of co-existing Pb, Zn, and Cu on Cd uptake of wheat grains. A positive correlation was found for soil Cd content and Pb, Zn, and Cu existence ($r = 0.459$–0.946). Zn decreased the uptake of Cd by wheat to a higher extent, when compared to Pb and Cu.

Some metals like Cu, Zn, and Fe in small amounts are micronutrients, necessary for plant development, while some others have toxic effects even at very low concentrations, like Co, Pb, and Cd. Plants are often exposed to toxic metals that can cause temporary or long-term stress. Stress emerges as decrease in growth, seed germination, plant biomass, root and stem length, etc. [28]. Kalavrouziotis and Drakatos [29] performed a research study on the effects of wastewater irrigation containing essential elements on some Mediterranean plants. It was determined that *Nerium oleander* exhibited strong toxicity toward Cu^{2+}, Mn^{2+}, and Zn^{2+} (up to 1000 mg/L) while *Myoporum sp.*, and *Geranium sp.* showed tolerance toward these ions.

One of the main concerns in irrigation with wastewater is the risk posed to human health by utilization of contaminated foods. Health risk index (HRI) is calculated by the daily intake of metals (DIM) to reference oral dose (Rfd) ratio. Metal content in contaminated plants (mg/kg) (C_m), conversion factor (f) (such as 0.085), daily ingestion of contaminated food (D), and body weight (Bw) are used for DIM calculation (DIM $= C_m * f *$ D/Bw). The average adult and child body weights were determined as 61.6 and 18.6 kg, respectively [30]. Rfd values for Cr, Cu, Zn, Pb, Cd, and Ni were 1.5, 0.04, 0.3, 0.0035, 0.001, and 0.02 mg/kg . day, respectively [31]. HRI value bigger than 1 is found to be unsafe for the consumers [20]. Toxic concentrations of Cd and Pb were determined in *Triticum aestivum* (wheat) and *Zea mays* (maize) grains, cultivated at south Cairo, Egypt, under wastewater irrigation. HRI > 1 for Cd and Pb indicated a consumer health risk for both grain types [32]. Wastewater irrigation of fodder for animals farming has been a wide application in Pakistan. It was determined that heavy metal concentrations of raw milk of buffalo, fed by wastewater irrigated Maize and Brassica plants, indicated a high carcinogenic health risk for consumers [33].

Wastewater irrigation may pose a potential risk for shallow groundwater sources via vertical infiltration. Hydraulic conductivity, porosity, bulk density, and saturated soil water content, etc. are the factors affecting the mobility of heavy metals in the soil. Amiri and Nakhaei [34] reported that Ti and Co have the highest and the lowest mobility in soil, respectively. Yang et al. [35] conducted an investigation on the migration of heavy metals to deep soils. Samples taken from 30 m depth in a wastewater-irrigated region were analyzed for heavy metals. As, Cd, and Pb were determined in high concentrations in deep soil layers. 73 mg/kg as was determined in 29 m soil depth, while 214 mg/kg Pb was determined in 30 m soil depth.

3.2 Synthetic Organic Pollutants

Treated or untreated wastewater used in agricultural irrigation contains many synthetic organic pollutants, such as polycyclic aromatic hydrocarbons (PAHs), pharmaceuticals and their metabolites, xenobiotic chemicals, phenolic compounds, etc. Chlorinated hydrocarbons like aldrin, dieldrin, endrin, chlordane, DDT, heptachlor, mirex, and toxaphene as old pesticides; polychlorinated biphenyls (PCBs) and hexachlorobenzene (HCB) as industrial chemicals; polychlorinated dibenzo-p-dioxins and dibenzofurans (PCDD/PCDF) as combustion by-products of chlorine containing materials; polybrominated flame retardants (BFRs), etc., are summarized as persistent organic pollutants (POPs). These pollutants can accumulate in the soil because of long-term irrigation, can contaminate agricultural products, and be bio-magnified through the food chain. POPs mostly pass through conventional WWTPs, since primary or secondary treatment processes are insufficient to remove these pollutants from wastewater [36]. Besides, wastewater irrigation affects the behavior of pesticides in soil regarding their sorption–desorption and/or degradation patterns [37]. Dissolved organic matter moieties, such as hydroxyl, carbonyl, carboxylic, and phenolic compounds, suspended solids, heavy metals, nutrients, inorganic salts, etc., in wastewater interact with the soil-POPs interactions [38].

PCBs have been used as coolant and insulating fluid for transformers and capacitors, stabilizing additive in flexible PVC coatings of electrical cables, and electronic equipment, cutting oils, flame retardants, hydraulic liquids, seals, adhesives, wood polishes, paints, and dust removal agents. PCBs leak into the nature from the equipment still being used [39]. PCBs are resistant to decompose in nature, and their average half-life in soil and sediment are on average 57 days. PCBs continue to pose a danger by being transported long distances in air, and by accumulating in soil and sediment. There are just limited publications about PCBs accumulation in soil after wastewater irrigation. Higher chlorinated congeners of PCBs, such as PCB 153 (2,2′,4,4′,5,5′-Hexachlorobiphenyl) and PCB 170 (2,2′,3,3′,4,4′,5-Heptachlorobiphenyl) are reported to have high persistency in soil [40, 41].

Important sources of PAHs are industrial processes, such as aluminum and coke production, coal gasification, iron and steel mining, vehicle emissions, and fossil fuels, as well as incomplete combustion of organic materials like wood. PAHs molecules are usually classified according to their ring numbers as 2–3 rings, 4 rings, and 5–6 rings PAH. Low molecular weight PAHs have high volatility and thus higher atmospheric mobility than higher molecular weight PAH, the latter resulting in soil deposition [42].

There are studies revealing POPs pollution in different parts of the world. For instance, in China, wastewater containing both domestic and industrial discharges have long been used for agricultural irrigation. Research was carried out in Shenyang, Liaoning Province in northeastern China, so as to define PCBs and PAHs contamination of soils after 40 years of wastewater irrigation. 14 out of 16 PAHs, listed by the US Environmental Protection Agency (USEPA), were detected in those soil samples. The total PAHs contents were between 950 and 2790 μg/kg. The background concentrations in unpolluted control soils in the same area were determined as 275–317 μg/kg, which demonstrated the effect of wastewater irrigation. Six PCB congeners were determined in the concentration range of 4.4–20.14 μg/kg, and PCB 209 (full-chlorinated decachlorobiphenyl) was determined as the dominant congener. It was emphasized that the sum of PCBs exceeded 8 μg/kg for almost all tested samples [41].

Similar findings were reported for Konya City, in the central part of Türkiye. Long-term irrigation with untreated wastewater until 2010 resulted in not only soil contamination by PAHs, PCBs, pesticides, and heavy metals, but also of crops' contamination, cultivated in the polluted field [3, 5]. Total six regulated PCBs concentration (28, 52, 101, 138, 153, 180) was determined as 1.12–29.71 μg/kg for wastewater irrigated soil samples. While PCB 153 was determined as the highest concentration, PCB 28 ve PCB 52 were the lowest concentrations among the analyzed PCBs. Domination of higher chlorinated PCBs were a common finding for soil analysis, which could be attributed to their higher molecular weight, lower volatility and higher persistency. Total 16 USEPA PAHs were determined in the range of 27.94–125 μg/kg in the same field samples. Naphthalene, phenanthrene, anthracene, fluoranthene, fluorene, and pyrene were determined as the dominant PAH compounds. Cultivated wheats in the field were also analyzed for PAHs and PCBs. Total PAHs in wheat samples were determined in the range of 0.22–0.9 mg/kg, while acenaphthylene was the dominant one among PAHs. Total PCBs in wheat samples was determined in the range of 1.8–14.4 μg/kg.

Contamination of soil with POPs was reported even after using secondary treatment effluents in short term. 18 PCBs, 13 PAHs, and 16 organochlorine pesticides (OCPs) were analyzed in soil samples collected in Nabeul, Tunisia, at the end of the irrigation season. The total PAHs (acenaphthylene, fluorene, phenanthrene, anthracene, pyrene benzo[a]anthracene, chrysene, benzo[b]fluoranthene, benzo[k]fluoranthene, benzo[a]pyrene, dibenzo[a,h]anthracene, indeno[1,2,3-cd]pyrene, and benzo[ghi]perylene), PCBs (28, 52, 77, 81, 101, 105, 114, 118, 123, 126, 138, 153, 156, 157, 167, 169, 180, 189), and OCPs (the highly chlorinated α-HCH, β-HCH, γ-HCH, aldrin, cis-chlordane, 2,4-DDE, trans-chlordane, 4,4-DDE, dieldrin, 2,4-DDD, endrin, 4,4-DDD, 2,4-DDT and 4,4-DDT as well as the monochlorinated herbicides atrazine and alachlor) were 120.01–365.18 μg/kg for PAH; 11.26–21.89 μg/kg for PCB, and 12.49–21.81 μg/kg OCP, respectively. Total PAHs concentration increased three folds at the end of the irrigation season than before the season. Higher concentrations of PAHs in top soil than the lower parts of soil were attributed to deposition of PAHs resourced from suspended solids in treated wastewater besides of atmospheric deposition. When the total PCBs was 11.26 μg/kg before irrigation season, it was 21.89 μg/kg at the end of the season; even PCBs were below detection limits in treated wastewater used for irrigation. The lowest total OCPs was determined as 12.49 μg/kg at the beginning of the irrigation season, while it was 21.81 μg/kg at the end of the season. DDT was predominant among the total OCPs, even it is internationally prohibited for long, which was attributed to its low biodegradability [40].

A human health risk was reported in many studies as a result of the consumption of POPs contaminated agricultural products. Long-term consumer health risk, indication of a cancer risk, has been determined based on the estimated and the acceptable daily intakes. Short-term consumer health risk was correlated with estimated short-term intake and acute reference dose (ARfD) (mg/kg * bw * day). ARfD data are available only for some of the POPs [43]. Bedük et al. [5] reported both short-term and long-term consumer health risk for adults and for children as a result of the consumption of pesticide contaminated wheat, cultivated in wastewater irrigation field.

3.3 Pharmaceuticals

Pharmaceuticals that are designed to medicate humans and animals have potential adverse effects on non-target organisms in the aquatic and terrestrial ecosystems [44]. Analgesic and anti-inflammatory drugs, antibiotics, lipid regulators, beta-blockers, and psychiatric drugs are the most widely used pharmaceuticals classes worldwide. Kapelewska et al. [45] reported that annual global production of pharmaceutical compounds is estimated at approximately 20 million tons. Especially after the Covid 19 pandemic, the consumption of pharmaceutical has increased even more on a global scale. Therefore, they are the most frequently detected compounds among emerging contaminants in the aquatic environment [46–50]. After using pharmaceuticals, they are excreted from the body in the form of parent compound and the metabolites, and reach the wastewater. Wastewater and sludge treatment processes in conventional WWTPs have not been designed to remove micropollutants such as pharmaceuticals. Furthermore, transformation products of pharmaceuticals are generated during the pharmaceutical removal by treatment processes such as biological, photolysis and disinfection processes in WWTPs [51]. Therefore, WWTP effluents that subsequently discharged these chemical substances into the environment are the most important source of pharmaceuticals and their transformation products in the environments [52].

631 pharmaceuticals were determined at measured environmental concentration values in environmental matrices such as groundwater, surface water, tap/drinking water, and soil in 71 countries covering all continents [53]. Pharmaceuticals are highly bioactive, even at low concentrations as ng/L or μg/L detected in the environment, and have adverse effects on non-target organisms. Generally, conventional parameters such as pH, conductivity, total suspended solid, and COD are continuously monitored in municipal WWTPs. Environmental concentration limits of pharmaceuticals have not been established in regulations yet [44, 45, 51].

While analgesic and anti-inflammatory drugs like acetaminophen, ibuprofen, and naproxen were efficiently removed (>80%) in WWTPs including conventional activated sludge systems [49], the removal efficiencies of psychiatric drugs were quite low, for example, 11% for carbamazepine [54], 8% for diazepam [55], 33% for fluoxetine [56], and 48% norfluoxetine [57]. The total antibiotic removal in urban WWTP in Türkiye was 79% in summer and decreased to 36% in winter. The removal efficiency of tetracycline group antibiotics (chlortetracycline, oxytetracyclin, doxycycline) was found to be lower (<18%) [48]. The removal rates of the antibiotics in the treatment plant were determined between 20 and 77% [58]. Papageorgiou et al. [59] reported that the majority of the 55 pharmaceuticals and personal care products cannot efficiently been removed in WWTP in Greece. According to environmental risk assessments, the anti-inflammatory drug (diclofenac) and antibiotics (trimethoprim and ciprofloxacin) exhibited high potential risk (risk quotient (RQ) > 1) for aquatic ecosystem. Verlicchi et al. [60] reported that <10% removal efficiency could be found for diclofenac, ibuprofen, indomethacin, ketoprofen, mefenamic acid, and tramadol in municipal WWTPs with a conventional activated sludge. In addition, 14 pharmaceuticals including antibiotics (erythromycin, ofloxacin, sulfamethoxazole, clarithromycin, amoxicillin, tetracycline and azithromycin), psychiatric drugs (fluoxetine and diazepam), analgesics-anti/inflammatory drugs (ibuprofen and mefenamic acid) and lipid regulators (fenofibric acid, fenofibrate and gemfibrozil), detected in WWTP effluents, exhibited high risk for the environment. Azithromycin and clarithromycin antibiotics and atenolol β-blockers in the effluent from the urban Konya WWTP in winter in Türkiye posed a high risk (RQ > 10) for the aquatic organisms' algae and fish in the receiving environment [48, 61]. While trimethoprim and sulfamethoxazole antibiotics pose high acute ecotoxicological risk in eight WWTPs effluents in Greece, diclofenac and triclosan compounds pose high chronic ecotoxicological risk for the aquatic environment (RQ > 1) [62]. Similarly, Santos et al. [63] determined antibiotics (ciprofloxacin, ofloxacin, sulfamethoxazole, azithromycin and clarithromycin) in WWTP effluents could pose an ecotoxicological risk to algae.

As a result of wastewater treatment in WWTPs, large amounts of sewage sludge are formed as a by-product. The use of stabilized sewage sludge in agricultural areas as soil amendments is a common practice due to its high organic matter and nutrient content [64]. Approximately, half of sewage sludge produced in 28 European Member States, United States and China was applied on agricultural land [65, 66]. Generally, the main removal mechanisms of pharmaceuticals in conventional WWTPs are biodegradation and adsorption. About, 70% of antibiotic compounds in influents are transferred to sewage sludge in the treatment process [67]. Therefore, agricultural use of stabilized sludge may cause environmental problems. Aydin et al. [61] determined 38 selected pharmaceuticals in the range of 280–4898 μg/kg * dw in the anaerobically treated sludge of the urban WWTP in Konya, Türkiye. The annual pharmaceutical mass load discharged into the environment was calculated as 71.6 kg. While five antibiotics (i.e., azithromycin, clarithromycin, erythromycin, sulfamethoxazole, and doxycycline), one analgesic (i.e., acetylsalicylic acid) and one beta-blocker (i.e., atenolol) in the digested sludge posed high risks, azithromycin in the digested sludge amended soils exhibited high short chronic risk. Verlicchi and Zambello [68] determined high environmental risks pose for azithromycin, erythromycin, ofloxacin, roxithromycin, and sulfamethoxazole antibiotics, acetaminophen, ibuprofen, naproxen, and salicylic acid analgesics and anti-inflammatories, and propranolol beta-blocker in the digested sludge. Also, the residual of ciprofloxacin, ofloxacin and tetracycline compounds after sludge applications on soil pose high risk for organisms. In addition, trimethoprim, ciprofloxacin and norfloxacin antibiotics in soils, amended with digested sludge, pose high acute and chronic risk on the environment [69]. Martín et al. [70] determined that the highest ecotoxicological risk in digested sludge was $RQ = 4.4$ for ibuprofen, a non-steroidal anti-inflammatory drug.

Because of the use of recycled wastewater and sludge in agricultural lands, many pollutants other than pharmaceuticals can pass into the soil. Pollutants can also be transported to surface water resources by surface runoff. In addition, the pollutants in the applied wastewater or sludge can be transported to the groundwater through infiltration. Chronic ecotoxicological risks can be expected due to long-term irrigation of soil with wastewater or amendment with sewage sludge. Therefore, wastewater and sludge treatment plants need to be modified with advanced treatment processes in order to remove pharmaceuticals from municipal wastewater. In addition, it is necessary to determine limit values for pharmaceuticals in regulations for the reuse of wastewater in agricultural irrigation and the application of stabilized sludge in agricultural areas.

3.4 Microplastics (MPs)

Plastics, which are high molecular polymers, are widely used in all areas of life because they are cheap and durable. Polyethylene (PE), polypropylene (PP), polyvinyl chloride (PVC), polystyrene (PS) and polyethylene terephthalate (PET) are the plastics commonly used today and most abundant in the environment, representing approximately 90% of global plastic production [71]. Plastics larger than 100 mm are classified as megaplastics, 5–100 mm sizes as macroplastics, 0.330–5 mm as MPs, and plastics smaller than 0.330 mm as nanoplastic [72]. MPs were first defined as small-sized plastic particles in 2004 and are now defined as plastic particles smaller than 5 mm [73]. MPs are categorized into primary and secondary MPs. Primary MPs are manufactured to be <5 mm in size for a specific application such as pellets or microbeads. Secondary MPs are formed as a result of the breakdown of plastics as a result of physical abrasion, thermal and photo degradation, degradation by microorganisms, and hydrolysis in water. MPs contain various chemical additives such as phthalates, bisphenol A, nonylphenols, antioxidants, flame retardants, fillers, stabilizers, plasticizers, and colorants to increase the functionality of plastic products [74].

MPs have become pollutants of interest all over the world in recent years because of causing environmental pollution. Since MPs are difficult to naturally biodegrade in the environment, they have a long lifespan [75]. Many studies show that MPs can be transported over long distances under the influence of external forces such as wind, river and ocean currents. In recent years, MPs have been called as a new class of persistent, bioaccumulative, and toxic (PBT) pollutants [76]. MPs tend to adsorb heavy metals and organic pollutants due to their small size, large surface area and hydrophobic properties. In addition, they are often eaten by organisms in the aquatic environment, mistaking them for food. MPs enriched in organisms through the food chain can eventually reach humans and cause serious health problems [77]. In 2018, 74 different MPs emission sources were detected, and according to studies, paints, tires, plastic pellets, synthetic textiles, geotextiles and detergent capsules were determined as the main emission sources [78]. It is estimated that global plastic use will triple by 2060, if current consumption habits continue. As a result, it is estimated that the amount of plastic waste released into the environment will double, reaching 44 million tons, and the amount of plastic accumulated in water will be more than triple [79].

The sources of MPs s in the environment may be land or marine based. It is estimated that more than 80% of MPs found in the seas are from land-based sources. In particular, WWTPs were considered as major sources of MPs in the environment through sludge and effluent discharges [80]. The presence of primary MPs in wastewater is a result of microbeads used for cosmetics and personal care products such as face wash, shower gels, and toothpaste [81–83]. The most important secondary source of MPs in wastewater is synthetic fibers in washing machine and laundry effluents. For example, in average 18 million synthetic microfibers were released for a load of 6 kg of synthetic fabrics and textiles [84]. High amounts of MPs such as PE and nylon from washing sponges during manual dishwashing are released into the sewerage [85]. Car tires and road marking paints are another important source of MPs in the wastewater. It is estimated that per capita MPs emission ranges from 0.23 to 4.7 kg/year, with a global average of 0.81 kg/year due to mechanical wear and tear of car tires [86]. In cities with combined sewer systems, rainwater transports MPs to wastewater through surface runoff. MPs transported on a regional or even global scale in the atmosphere reach sewage through rainfall [87]. The size, color, shape and polymer type of MPs in the wastewater reaching the WWTP vary depending on the source of the wastewater. In the literature, MPs in fiber and fragment morphology have been predominantly detected in domestic wastewater. The most common polymers determined in influent of WWTPs were PET, PS, PP and PE [88, 89].

MPs found in wastewater negatively affect the efficiency of water and sludge treatment in WWTPs. Zhang and Chen [90] reported that MPs in influents reduce the effluent quality and energy production from sludge. WWTP is designed to remove conventional parameters like suspended solid, chemical oxygen demand, nitrogen and phosphorus from wastewater. Removal rate of MPs in primary, secondary and advanced treatment processes has been reported in the different region of the world. In the literature, MPs removal efficiency by preliminary wastewater treatment that consists of screening, aerated grit chamber and primary sedimentation was determined as 35.3% in Thailand [91], 41.2–47.1% in China [92], 58.6–74.8% in South Korea [93], 69–74% in Morocco [94], and 78% in India [95]. Preliminary treatment is very effective for the removal of MPs. For example, Patil et al. [95] reported 78%, 6.22 and 2.39% for the primary, secondary and tertiary treatment in Bhandewadi sewage treatment plant in India, respectively. Also, Meng et al. [96] determined the removal efficiency of MPs as 62.9–70.4% in primary treatment, 55.6–57.5% in secondary treatment and 34.6–44.9% in tertiary treatment in China. The total removal efficiency of MPs in conventional WWTPs consisting of screening, grit removal, biological treatment and secondary sedimentation was reported as 98% in Austrailia [97], 93.7% in Bursa, Türkiye [98], 84.7–93.0 in İstanbul, Türkiye [99], 90.87% in Iran [100], 90.3% in China [92], 90.3% in Spain [101], and 71% in Karaman, Türkiye [102]. Although WWTPs include similar treatment processes in studies, secondary treatment MPs removal efficiency varies quite widely. It has been reported in several studies that the removal efficiency of MPs in WWTPs including advanced treatment was determined as higher than 97% [93, 103–106].

Although high MPs removal is achieved in WWTPs, MPs are constantly released into the environment through WWTP effluents due to the high wastewater flow rate discharged into the receiving environment. Liu et al. [107] reported that the abundance of the MPs in the effluent ranged from 5.00×10^5 to 1.39×10^{10} MPs/L on a daily rate in 38 WWTPs in 11 countries. Depending on the flow rate of the WWTP, billions of MPs are discharged daily with the effluent. Irrigation with wastewater provides the main input pathways for MPs into the soil. Especially in developing countries, agricultural irrigation with inadequately treated or untreated wastewater containing large amounts of MPs causes significant amounts of MPs to release into terrestrial environment.

MPs retained in the sludge during treatment processes reach the terrestrial environment because of the use of sewage sludge or biosolids as a soil amendment. The highest concentration of MPs in stabilized sludge so far was determined 169,000 MPs/g d.w. in anaerobically digested sludge in Denmark by Vollertsen and Hansen [108]. Harley-Nyang et al. [109] reported average MPs concentration in sewage sludge and biosolids as 208.3 MPs/g d.w., according to the results of a study covering 25 countries. The potential MPs load per year released into the environment through the application of sludge or biosolids to agricultural soils was estimated as 2.09×10^{12} in the USA in 2018, 7.30×10^{12} in China, 8.5×10^{13} in Germany, 1.38×10^{14} in Sweden, 1.03×10^{14} in Spain, 4.82×10^{12} in France, 1.29×10^{15} in England, 1.84×10^{13} in Italy, 1.04×10^{11} in Belgium, and 1.12×10^{11} in Türkiye. Nizzetto et al. [110] estimated that 63,000–430,000 and 44,000–300,000 tons of MPs were put through sludge application to European and North American farmlands, respectively. Corradini et al. [111] found that the number of MPs in the soil increased with increasing sludge rate in fields, where sludge is applied at different rates. As a result of the study investigating the fate of MPs in agricultural soils through sludge applications, it was determined that sludge application significantly increased the MPs content in the soil. Since the transport of MPs in the soil by infiltration and surface water flows is low, the soil can be an important reservoir for MPs in the long term [112]. The European Union Sewage Sludge Directive (86/278/EEC) in Europe regulates the application of sewage sludge to agricultural land [113]. There is no limit on MPs in this directive, which aims to protect the environment and in particular the soil, when sewage sludge is used in agriculture.

MPs existing in the effluent water and stabilized sludge may pose a threat to the aquatic and terrestrial environment. So far, many researchers have detected MPs in agricultural soil amended with sewage sludge and irrigated with wastewater [114]. Zhou et al. [115] detected between 2.2 $\times 10^4$ and 6.9 $\times 10^5$ MPs/kg in soils in Wuhan city in China. Moreover, MPs in soil contain heavy metals (Cd, Cr, Pb, Ag, Cu, Sb, Hg, Fe, and Mn) in different concentrations. Therefore, MPs in soils posed higher risk to soil organisms. MPs in the form of fibers were predominantly detected in agricultural soils. Fiber-type MPs are dominant forms detected in wastewater effluents and sludge, and are carried to wastewater by laundry washing water [116].

Contamination of soil with MPs affects its physical and chemical properties such as soil bulk density, porosity, water holding capacity, and soil aggregate stability [117]. MPs might cause changes in the nutrient cycle in the soil, affecting the dissolved organic matter and enzyme activity therein [118]. When MPs enter the soil, they release toxic and endocrine disrupting substances (bisphenol A). In addition, additives such as plasticizing agents (phthalates) and pollutants (toxic organic and inorganic pollutants) adsorbed to MPs are also released into the soil [119]. MPs in the soil can be ingested by animals in the terrestrial environment as well as in the aquatic environment. Therefore, the chemicals and pollutants contained in MPs can cause various toxicological effects in animals. For example, Cao et al. [120] reported that MPs inhibit the reproduction and growth of earthworms living in the soil and increased their mortality. In addition, MPs might cause some adverse effects on terrestrial plants growing in soil. It was determined by Qi et al. [121] that MPs residues in the soil have negative effects on the vegetative and reproductive growth of wheat. It was observed that when *Lolium perenne* seeds were exposed to MPs in the soil, the germination rates of the seeds decreased, and their shoot heights was negatively affected. In addition, the pH of the soil decreased, and the size distribution of the soil aggregate changed [122]. Therefore, MPs found in soil present potential risks on agricultural sustainability.

4 Conclusions

Findings in the literature reveal that wastewater irrigation has potential risk of contamination for soil and agricultural products. Even heavy metals and some POPs have been taken under control by regulations, many other organic pollutants and MPs have been overlooked. According to the outcomes of this review paper, irrigation with wastewater and application of soil amendments using biosolids, which are the main sources of MPs for agricultural soil, should be reconsidered. In order to reduce the MPs load on the aquatic and terrestrial environment, existing wastewater and sludge treatment processes need to be modified with economical treatment methods to ensure better removal of MPs. So as to sustain a safer approach, reclaimed wastewater quality should be improved and proper irrigation methods selected. Measures should be taken at the source to reduce the pollutants input to municipal WWTPs. It is clear that soil type and choice of crop are critical for risk reduction.

References

1. Bahçeci B, Tarı AF, Bahçeci İ. The effect of different depths of salty groundwater on yield and soil salinity of some pasture crops. J Agric Sci. 2023;29(2):546–54.
2. Sharma L, Sarangthem I, Thangjam R, Sadhukhan R, Oinam N, Yanglem B, Jatav H. Sewage sludge and its health risk assessment: opportunities and challenges. In: Rajput V, Yadav A, Jatav H, Singh S, Minkina T, editors. Sustainable management and utilization of sewage sludge. Cham: Springer;2022. p. 425.
3. Aydin ME, Aydin S, Bedük F, Tor A, Tekinay A, Kolb M, Bahadir M. Effects of long-term irrigation with untreated municipal wastewater on soil properties and crop quality. Environ Sci Pollut Res. 2015;22:19203–12.
4. Bahadir M, Aydin ME, Aydin S, Bedük F, Batarseh M. Wastewater reuse in middle east countries—a review of prospects and challenges. Fresenius Environ Bull. 2016;25(5):1284–304.

5. Bedük F, Aydin ME, Aydin S, Tekinay A, Bahadir M. Pesticide pollution in soil and wheat: risk assessment of contaminated food consumption. Fresenius Environ Bull. 2017;26(3):2330–9.
6. Sharma RK, Agrawal M, Marshall F. Heavy metal contamination of soil and vegetables in suburban areas of Varanasi, India. Ecotoxicol Environ Saf. 2007;66:258–66.
7. Aiello R, Cirelli GL, Consoli S. Effects of reclaimed wastewater irrigation on soil and tomato fruits: a case study in Sicily (Italy). Agric Water Manag. 2007;93:65–72.
8. Samia K, Abdelkarim C, Yassin H, Teruo H. The effect of long-term soil irrigation by wastewater on organic matter, polycyclic aromatic hydrocarbons, and heavy metals evolution: case study of Zaouit Sousse (Tunisia). Arab J Geosci. 2013;6:4337–46.
9. Khan K, Khan H, Lu Y, Ihsanullah I, Nawab J, Khan S, Shah NS, Shamshad I, Maryam A. Evaluation of toxicological risk of foodstuffs contaminated with heavy metals in Swat, Pakistan. Ecotoxicol Environ Saf. 2014;108:224–32.
10. Li Q, Chen Y, Fu HB, Cui ZH, Shi L, Wang L, Liu ZF. Health risk of heavy metals in food crops grown on reclaimed tidal flat soil in the Pearl River Estuary, China. J Hazard Mater. 2012;227–228:148–54.
11. Borisover M, Lordian A, Levy GJ. Water-extractable soil organic matter characterization by chromophoric indicators: effects of soil type and irrigation water quality. Geoderma. 2012;179–180:28–37.
12. Bedük F. Combined effects of cadmium and azithromycin on soil nitrification process. Water. 2023;15:881.
13. Gao Y, Shao G, Wu S, Xiaojun W, Lu J, Cui J. Changes in soil salinity under treated wastewater irrigation: a meta-analysis. Agric Water Manag. 2021;255:106986.
14. Zhang Q, Li Z, Huang B, Luo N, Long L, Huang M, Zhai X, Zeng G. Effect of land use pattern change from paddy soil to vegetable soil on the adsorption-desorption of cadmium by soil aggregates. Environ Sci Pollut Res. 2017;24:2734–43.
15. Cary L, Surdyk N, Psarras G, Kasapakis I, Chartzoulakis K, Sandei L, Guerrot C, Pettenati M, Kloppmann W. Short-term assessment of the dynamics of elements in wastewater irrigated Mediterranean soil and tomato fruits through sequential dissolution and lead isotopic signatures. Agric Water Manag. 2015;155:87–99.
16. Draszawka-Bołzan B. Heavy metals in soils. World News Nat Sci. 2015;2:20–37. EISSN 2543-5426.
17. Zhuang Z, Nino-Savala AG, Mi Z, Wan Y, Su D, Li H, Fangmeier A. Cadmium accumulation in wheat and maize grains from China: interaction of soil properties, novel enrichment models and soil thresholds. Environ Pollut. 2021;275:116623.
18. Mao C, Song Y, Chenc L, Ji J, Li J, Yuan X, Yang Z, Ayoko GA, Frost RL, Theiss F. Human health risks of heavy metals in paddy rice based on transfer characteristics of heavy metals from soil to rice. CATENA. 2019;175:339–434.
19. Zouari M, Bloem E, Souguir D, Schnug E, Hachicha M. The nutritional composition of six plant species after irrigation with treated wastewater and possible hazards by heavy metal accumulation. Environ Sci Pollut Res. 2023;30:97700–11.
20. Atta MI, Zehra SS, Dai D-Q, Ali H, Naveed K, Ali I, Sarwar M, Ali B, Iqbal R, Bawazeer S, Abdel-Hameed UK, Ali I. Amassing of heavy metals in soils, vegetables and crop plants irrigated with wastewater: Health risk assessment of heavy metals in Dera Ghazi Khan, Punjab, Pakistan. Front Plant Sci. 2023;13:1080635.
21. Surucu A, Marif AA, Majid SN, Farooq S, Tahir NA. Effect of different water sources and water availability regimes on heavy metal accumulation in two sunflower species. Carpathian J Earth Environ Sci. 2020;15(2):289–300.
22. Jabeen R, Ahmad A, Iqbal M. Phytoremediation of heavy metals: physiological and molecular mechanisms. Bot Rev. 2009;75:339–64.
23. Ahmed DA, Slima DF. Heavy metal accumulation by *Corchorus olitorius* L. irrigated with wastewater. Environ Sci Pollut Res. 2018;25:14996–5005.
24. Abdel-Shafy HI, El-Khateeb M. Heavy Metals in citrus fruits as affected by primary treated sewage irrigation. Egypt J Chem. 2021;64(1):165–76.
25. Gupta S, Satpati S, Nayek S, Garai D. Effect of wastewater irrigation on vegetables in relation to bioaccumulation of heavy metals and biochemical changes. Environ Monit Assess. 2010;165:169–77.
26. Zeng X, Wang Z, Wang J, Guo J, Chen X, Zhuang J. Health risk assessment of heavy metals via dietary intake of wheat grown in Tianjin sewage irrigation area. Ecotoxicology. 2015;24:2115–24.
27. Xu J, Hu C, Wang M, Zhao Z, Zhao X, Cao L, Lu Y, Cai X. Changeable effects of coexisting heavy metals on transfer of cadmium from soils to wheat grains. J Hazard Mater. 2022;423(B):127182.
28. Cunha Neto ARD, da Silva IG, Calvelli JVB, Martins GEC, Carvalho M, Barbosa S. Toxicity of heavy metals that affect germination, development and cell cycle of *allium cepa* L. Bull Environ Contam Toxicol. 2023;111:22.
29. Kalavrouziotis IK, Drakatos PA. Irrigation of certain Mediterranean plants with heavy metals. Int J Environ Pollut. 2002;18(3):294–300.
30. Huang M, Zhou S, Sun B, Zhao Q. Heavy metals in wheat grain: assessment of potential health risk for inhabitants in Kunshan, China. Sci Total Environ. 2008;405(1):54–61.
31. US, Epa (United States Environmental Protection Agency). Guidelines for the health risk assessment of chemical mixtures. Fed Reg. 1986;51(185):34014–25.
32. Farahat EA, Galal TM, Elawa OE, Hassan LM. Health risk assessment and growth characteristics of

wheat and maize crops irrigated with contaminated wastewater. Environ Monit Assess. 2017;189:535.
33. Iqbal Z, Abbas F, Ibrahim M, Qureshi TI, Gul M, Mahmood A. Human health risk assessment of heavy metals in raw milk of buffalo feeding at wastewater-irrigated agricultural farms in Pakistan. Environ Sci Pollut Res. 2020;27:29567–79.
34. Amiri N, Nakhaei M. Evaluating the potential of treated municipal wastewater reuse in irrigation and groundwater recharge; 5-year contaminant transport modeling. Int J Environ Sci Technol. 2023. https://doi.org/10.1007/s13762-023-05293-x.
35. Yang S, Feng W, Wang S, Chen L, Zheng X, Li X, Zhou D. Farmland heavy metals can migrate to deep soil at a regional scale: a case study on a wastewater-irrigated area in China. Environ Pollut. 2021;281:116977.
36. Szulzyk-Cieplak J. Removal of hardly biodegradable organic compounds from wastewater by means of reagentless methods. J Ecol Eng. 2017;18(5):63–71.
37. Rodríguez-Liébana JA, ElGouzi S, Mingorance MD, Castillo A, Peña A. Irrigation of a Mediterranean soil under field conditions with urban wastewater: effect on pesticide behaviour. Agr Ecosyst Environ. 2014;185:176–85.
38. Peña A, Delgado Moreno L, Rodríguez-Liébana JA. A review of the impact of wastewater on the fate of pesticides in soils: effect of some soil and solution properties. Sci Total Environ. 2020;718:134468.
39. Haddaoui I, Mahjoub O, Mahjoub B, Boujelben A, Bella GD. Occurrence and distribution of PAHs, PCBs, and chlorinated pesticides in Tunisian soil irrigated with treated wastewater. Chemosphere. 2016;146:195–2015.
40. Wang T, Wang Y, Fu J, Wang P, Li Y, Zhang Q, Jiang G. Characteristic accumulation and soil penetration of polychlorinated biphenyls and polybrominated diphenyl ethers in wastewater irrigated farmlands. Chemosphere. 2010;81(8):1045–51.
41. Song YF, Wilke BM, Song XY, Gong P, Zhou QX, Yang GF. Polycyclic aromatic hydrocarbons (PAHs), polychlorinated biphenyls (PCBs) and heavy metals (HMs) as well as their genotoxicity in soil after long-term wastewater irrigation. Chemosphere. 2006;65(10):1859–68.
42. Bedük F, Aydin S, Ulvi A, Aydin ME. Fingerprint of persistent organic pollutants (POPs) in the environment: ecological assessment and human health effects. In: Water and wastewater management, global problems and measures. Springer;2022. pp. 153–177. ISSN 2731-3166.
43. JMPR, Pesticide Residues in Food, Report of the 2006 Joint FAO/WHO Meeting of Experts, Rome, Italy, 3–12 October 2006.
44. Kümmerer K. Pharmaceuticals in the environment: sources, fate, effects and risks. Springer Science & Business Media;2008.
45. Kapelewska J, Kotowska U, Karpińska J, Kowalczuk D, Arciszewska A, Świrydo A. Occurrence, removal, mass loading and environmental risk assessment of emerging organic contaminants in leachates, groundwater and wastewaters. Microchem J. 2018;137:292–301.
46. Wang Y, Liu J, Kang D, Wu C, Wu Y. Removal of pharmaceuticals and personal care products from wastewater using algae-based technologies: a review. Rev Environ Sci Bio/Technol. 2017;16(4):717–35.
47. Aydin S, Bedük F, Ulvi A, Aydin ME. Simple and effective removal of psychiatric pharmaceuticals from wastewater treatment plant effluents by magnetite red mud nanoparticles. Sci Total Environ. 2021;784:147174.
48. Aydin S, Aydin ME, Ulvi A, Kilic H. Antibiotics in hospital effluents: occurrence, contribution to urban wastewater, removal in a wastewater treatment plant, and environmental risk assessment. Environ Sci Pollut Res. 2019;26:544–58.
49. Aydin S, Aydin ME, Ulvi A. Monitoring the release of anti-inflammatory and analgesic pharmaceuticals in the receiving environment. Environ Sci Pollut Res. 2019;26:36887–902.
50. Ulvi A, Aydın S, Aydın ME. Fate of selected pharmaceuticals in hospital and municipal wastewater effluent: occurrence, removal, and environmental risk assessment. Environ Sci Pollut Res. 2022;29:75609–25.
51. Hernández-Tenorio R. Hydroxylated transformation products of pharmaceutical active compounds: generation from processes used in wastewater treatment plants and its environmental monitoring. Chemosphere. 2023;349:140753.
52. Alharbi OA, Jarvis E, Galani A, Thomaidis NS, Nika MC, Chapman DV. Assessment of selected pharmaceuticals in Riyadh wastewater treatment plants, Saudi Arabia: mass loadings, seasonal variations, removal efficiency and environmental risk. Sci Total Environ. 2023;882:163284.
53. Beek TAD, Weber FA, Bergmann A, Hickmann S, Ebert I, Hein A, Küster A. Pharmaceuticals in the environment—global occurrences and perspectives. Environ Toxicol Chem. 2016;35(4):823–35.
54. Santos JL, Aparicio I, Callejon M, Alonso E. Occurrence of pharmaceutically active compounds during 1-year period in wastewater from four wastewater treatment plants in Seville (Spain). J Hazard Mater. 2009;164:1509–16.
55. Suárez S, Ramil M, Omil F, Lema JM. Removal of pharmaceutically active compounds in nitrifying–denitrifying plants. Water Sci Technol. 2005;52:9–14.
56. Radjenović J, Petrović M, Barceló D. Fate and distribution of pharmaceuticals in wastewater and sewage sludge of the conventional activated sludge (CAS) and advanced membrane bioreactor (MBR) treatment. Water Res. 2009;43:831–41.
57. Zorita S, Martensson L, Mathiasson L. Occurrence and removal of pharmaceuticals in municipal sewage treatment system in the south of Sweden. Sci Total Environ. 2009;407:2760–70.

58. Brown K, Kulis J, Thomson B, Chapman T, Mavhinney D. Occurrence of antibiotics in hospital, residential and dairy effluent, municipal wastewater and the Rio Grande in New Mexico. Sci Total Environ. 2006;366:772–83.
59. Papageorgiou M, Kosmo C, Lambropoulou D. Seasonal occurrence, removal, mass loading and environmental risk assessment of 55 pharmaceuticals and personal care products in a municipal wastewater treatment plant in Central Greece. Sci Total Environ. 2016;547–69.
60. Verlicchi P, Al Aukidy M, Zambello E. Occurrence of pharmaceutical compounds in urban wastewater: removal, mass load and environmental risk after a secondary treatment—a review. Sci Total Environ. 2012;429:123–55.
61. Aydin S, Ulvi A, Bedük F, Aydın ME. Pharmaceutical residues in digested sewage sludge: occurrence, seasonal variation and risk assessment for soil. Sci Total Environ. 2022;817:152864.
62. Kosma CI, Lambropoulou DA, Albanisa TA. Investigation of PPCPs in wastewater treatment plants in Greece: occurrence, removal and environmental risk assessment. Sci Total Environ. 2014;466:421–38.
63. Santos L, Gros M, Rodriguez-Mozaz S, Delerue-Matos C, Pena A, Barceló D, Montenegro MC. Contribution of hospital effluents to the load of pharmaceuticals in urban wastewaters: identification of ecologically relevant pharmaceuticals. Sci Total Environ. 2013;461:302–16.
64. Mejías C, Martín J, Santos JL, Aparicio I, Alonso E. Occurrence of pharmaceuticals and their metabolites in sewage sludge and soil: a review on their distribution and environmental risk assessment. Trends Environ Anal Chem. 2021;30:e00125.
65. Buta M, Hubeny J, Zielinski W, Harnisz M, Korzeniewska E. Sewage sludge in agriculture—the effects of selected chemical pollutants and emerging genetic resistance determinants on the quality of soil and crops—a review. Ecotoxicol Environ Saf. 2021;214:112070.
66. Collivignarelli MC, Abbà A, Frattarola A, Miino M, Padovani S, Katsoyiannis I, Torretta V. Legislation for the reuse of biosolids on agricultural land in Europe: overview. Sustainability. 2019;11:6015.
67. Sun C, Li W, Chen Z, Qin W, Wen X. Responses of antibiotics, antibiotic resistance genes, and mobile genetic elements in sewage sludge to thermal hydrolysis pre-treatment and various anaerobic digestion conditions. Environ Int. 2019;133:105156.
68. Verlicchi P, Zambello E. Pharmaceuticals and personal care products in untreated and treated sewage sludge: occurrence and environmental risk in the case of application on soil—a critical review. Sci Total Environ. 2015;538:750–67.
69. Bastos MC, Soubrand M, Guet TL, Floch ÉL, Joussein E, Baudu M, Casellas M. Occurrence, fate and environmental risk assessment of pharmaceutical compounds in soils amended with organic wastes. Geoderma. 2020;375.
70. Martín J, Camacho-Muñoz MD, Santos JL, Aparicio I, Alonso E. Occurrence of pharmaceutical compounds in wastewater and sludge from wastewater treatment plants: removal and ecotoxicological impact of wastewater discharges and sludge disposal. J Hazard Mater. 2012;239–240:40–7.
71. Venghaus D. Microplastics in urban water management. Tech Trans. 2017;137–46.
72. Masura J. Microplastics in puget sound: from 2008 to present. University of Washington Tacoma;2017.
73. Thompson RC, Olsen Y, Mitchell RP, Mitchell RP, Davis A, Rowland SJ, John AWG, McGonigle D, Russell AE. Lost at sea: Where is all the plastic? Science. 2004;304(5672):838–8.
74. Hahladakis JN, Velis CA, Weber R, Iacovidou E, Purnell P. An overview of chemical additives present in plastics: migration, release, fate and environmental impact during their use, disposal and recycling. J Hazard Mater. 2018;344:179–99.
75. Mohanan N, Montazer Z, Sharma PK, Levin DB. Microbial and enzymatic degradation of synthetic plastics. Front Microbiol. 2020;11:580709.
76. Alava JJ, Jahnke A, Bergmann M, Aguirre-Martínez GV, Bendell L, Calle P, Domínguez GA, Faustman EM, Falman J, Kazmiruk TN, Klasios N, Maldonado MT, McMullen K, Moreno-Báez M, Öberg G, Ota Y, Price D, Shim WJ, Tirapé A, Vandenberg JM, Zoveidadianpour Z, Weis J. A call to include plastics in the global environment in the class of persistent, bioaccumulative, and toxic (PBT) pollutants. Environ Sci Technol. 2023;57(22):8185–8.
77. Liu R, Dong Y, Quane G, Zhu H, Xu Y, Elwardany RM. Microplastic pollution in surface water and sediments of Qinghai-Tibet plateau: current status and causes. China Geol. 2021;1:178–84.
78. Brown E, MAcDonald A, Allen S, Allen D. The potential for a plastic recycling facility to release microplastics pollution and possible filtration remediation effectiveness. J Hazard Mater Adv. 2023;10:100309.
79. OECD, Global plastic waste set to almost triple by 2060, says OECD. https://www.oecd.org/environment/global-plastic-waste-set-to-almost-triple-by-2060.htm. Accessed 01 Dec 2023.
80. Reddy AS, Nair AT. The fate of microplastics in wastewater treatment plants: an overview of source and remediation Technologies. Environ Technol Innov. 2022;28:102815.
81. Lei K, Qiao F, Liu Q, Wei Z, Qi H, Cui S, Yue X, Deng Y, An L. Microplastics releasing from personal care and cosmetic products in China. Mar Pollut Bull. 2017;123:122–6.
82. Praveena SM, Shaifuddin SNM, Akizuki S. Exploration of microplastics from personal care and cosmetic products and its estimated emissions to marine environment: an evidence from Malaysia. Mar Pollut Bull. 2018;136:135–40.
83. Madhumitha CT, Karmegam N, Biruntha M, Arun A, Al Kheraif AA, Kim W, Kumar P. Extraction, identification, and environmental risk assessment

of microplastics in commercial toothpaste. Chemosphere. 2022;296:133976.
84. Galvão A, Aleixo M, De Pablo H, Lopes C, Raimundo J. Microplastics in wastewater: microfiber emissions from common household laundry. Environ Sci Pollut Res. 2020;27:26643–9.
85. Lassen C, Hansen SF, Magnusson K, Hartmann NB, Rehne Jensen P, Nielsen TK, Brinch A. Microplastics: occurrence, effects and sources of releases to the environment in Denmark. In: Danish environmental protection agency. 2015. https://www.orbit.dtu.dk. Accessed 01 Dec 2023.
86. Kole PJ, Löhr AJ, Van Belleghem FGAJ, Ragas Ad MJ. Wear and tear of tires: a stealthy source of microplastics in the environment. Int J Environ Res Public Health. 2017;14:1265.
87. Allen S, Allen D, Phoenix VR, Le Roux G, Durántez Jiménez P, Simonneau A, Binet S, Galop D. Atmospheric transport and deposition of microplastics in a remote mountain catchment. Nat Geosci. 2019;12:339–44.
88. Gies EA, LeNoble JL, Noël M, Etemadifar A, Bishay F, Hall ER, Ross PS. Retention of microplastics in a major secondary wastewater treatment plant in Vancouver, Canada. Mar Pollut Bull. 2018;133:553–61.
89. Liu X, Yuan W, Di M, Li Z, Wang J. Transfer and fate of microplastics during the conventional activated sludge process in one wastewater treatment plant of China. Chem Eng J. 2019;362:176–82.
90. Zhang Z, Chen Y. Effects of microplastics on wastewater and sewage sludge treatment and their removal: a review. Chem Eng J. 2020;382:122955.
91. Tadsuwan K, Babel S. Microplastics contamination in a conventional wastewater treatment plant in Thailand. Waste Manage Res. 2021;39(5):1–8.
92. Zhang L, Liu J, Xie Y, Zhong S, Gao P. Occurrence and removal of microplastics from wastewater treatment plants in a typical tourist city in China. J Clean Prod. 2021;291:125968.
93. Kwon HJ, Hidayaturrahman H, Peera SG, Lee TG. Elimination of microplastics at different stages in wastewater treatment plants. Water. 2022;14(15):2404.
94. Hajji S, Ben-Haddad M, Rida AM, De-la-Torre GE, Alla AA. Occurrence, characteristics, and removal of microplastics in wastewater treatment plants located on the Moroccan Atlantic: the case of Agadir metropolis. Sci Total Environ. 2023;862:160815.
95. Patil S, Kamdi P, Chakraborty S, Das S, Bafana A, Krishnamurthi K, Sivanesan S. Characterization and removal of microplastics in a sewage treatment plant from urban Nagpur, India. Environ Monit Assess. 2023;195:47.
96. Meng X, Bao T, Hong L, Wu K. Occurrence characterization and contamination risk evaluation of microplastics in Hefei's urban wastewater treatment plant. Water. 2023;15(4):686.
97. Ziajahromi S, Neale PA, Silveira I, Chua A, Leusch F. An audit of microplastics abundance throughout three Australian wastewater treatment plants. Chemosphere. 2021;263:128294.
98. Üstün GE, Bozdaş K, Can T. Abundance and characteristics of microplastics in an urban wastewater treatment plant in Turkey. Environ Pollut. 2022;310:119890.
99. Vardar S, Onay TT, Demirel B, Kideys AE. Evaluation of microplastics removal efficiency at a wastewater treatment plant discharging to the Sea of Marmara. Environ Pollut. 2021;289:117862.
100. Takdastan A, Niari MH, Babaei A, Dobaradaran S, Jorfi S, Ahmadi M. Occurrence and distribution of microplastics particles and the concentration of di 2-ethyl hexyl phthalate (DEHP) in microplastics and wastewater in the wastewater treatment plant. J Environ Manag. 2021;280:111851.
101. Bayo J, Olmos S, López-Castellanos J. Microplastics in an urban wastewater treatment plant: the influence of physicochemical parameters and environmental factors. Chemosphere. 2020;238:124593.
102. Bilgin M, Yurtsever M, Karadagli F. Microplastic removal by aerated grit chambers versus settling tanks of a municipal wastewater treatment plant. J Water Process Eng. 2020;38:101604.
103. Ben-David EA, Habibi M, Haddad E, Hasanin M, Angel DL, Booth AM, Sabbah I. Microplastics distributions in a domestic wastewater treatment plant: removal efficiency, seasonal variation and influence of sampling technique. Sci Total Environ. 2021;752:141880.
104. Yuan F, Zhao H, Sun H, Zhao J, Sun Y. Abundance, morphology, and removal efficiency of microplastics in two wastewater treatment plants in Nanjing, China. Environ Sci Pollut Res. 2021;28:9327–37.
105. Salmi P, Ryymin K, Karjalainen AK, Mikola A, Uurasjärvi E, Talvitie J. Particle balance and return loops for microplastics in a tertiary-level wastewater treatment plant. Water Sci Technol. 2021;84(1):89–100.
106. Hidayaturrahman H, Lee TG. A study on characteristics of microplastics in wastewater of South Korea: identification, quantification, and fate of microplastics during treatment process. Mar Pollut Bull. 2019;146:696–702.
107. Liu W, Zhang J, Liu H, Guo X, Zhang X, Yao X, Cao Z, Zhang T. A review of the removal of microplastics in global wastewater treatment plants: characteristics and mechanisms. Environ Int. 2021;146:106277.
108. Vollertsen J, Hansen A. Microplastics in Danish wastewater: sources, occurrences and fate. [Online] The Danish Environmental Protection Agency, Copenhagen. 2017. https://vbn.aau.dk/en/publications/microplastic-in-danish-wastewater-sourcesoccurrences-and-fate. Accessed 19 Dec 2023.

109. Harley-Nyang D, Memon FA, Baquero AO, Galloway T. Variation in microplastics concentration, characteristics and distribution in sewage sludge & biosolids around the world. Sci Total Environ. 2023;891:164068.
110. Nizzetto L, Futter M, Langaas S. Are agricultural soils dumps for microplastics of urban origin? Environ Sci Technol. 2016;50:10777–9.
111. Corradini F, Meza P, Eguiluz R, Casado F, Huerta-Lwanga E, Geissen V. Evidence of microplastics accumulation in agricultural soils from sewage sludge disposal. Sci Total Environ. 2019;671:411–20.
112. Schell T, Hurley R, Buenaventura N, Mauri P, Nizzetto L, Rico A, Vighi M. Fate of microplastics in agricultural soils amended with sewage sludge: is surface water runoff a relevant environmental pathway? Environ Pollut. 2022;293:118520.
113. European Union (EU). Council Directive on the protection of the environment, and in particular of the soil, when sewage sludge is used in agriculture (EU Directive 86/278/EEC). Brussels: Council of the EU;1986.
114. Wang C, Zhao J, Xing B. Environmental source, fate, and toxicity of microplastics. J Hazard Mater. 2021;407:124357.
115. Zhou Y, Liu X, Wang J. Characterization of microplastics and the association of heavy metals with microplastics in suburban soil of central China. Sci Total Environ. 2019;694:133798.
116. Yang L, Zhang Y, Kang S, Wang Z, Wu C. Microplastics in soil: a review on methods, occurrence, sources, and potential risk. Sci Total Environ. 2021;780:146546.
117. Ya H, Jiang B, Xing Y, Zhang T, Lv M, Wang X. Recent advances on ecological effects of microplastics on soil environment. Sci Total Environ. 2021;798:149338.
118. Liu H, Yang X, Liu G, Liang C, Xue S, Chen H, Ritsema CJ, Geissen V. Response of soil dissolved organic matter to microplastics addition in Chinese loess soil. Chemosphere. 2017;185:907–17.
119. Bläsing M, Amelung W. Plastics in soil: analytical methods and possible sources. Sci Total Environ. 2018;612:422–35.
120. Cao D, Wang X, Luo X, Liu G, Zheng H. Effects of polystyrene microplastics on the fitness of earthworms in an agricultural soil. IOP Conf Ser: Earth Environ Sci. 2017;61:12148.
121. Qi Y, Yang X, Pelaez A, Huerta Lwanga E, Beriot N, Gertsen H, Garbeva P, Geissen V. Macro- and microplastics in soil-plant system: effects of plastic mulch film residues on wheat (*Triticum aestivum*) growth. Sci Total Environ. 2018;645:1048–56.
122. Boots B, Russell C, Green D. Effects of microplastics in soil ecosystems: above and below ground. Environ Sci Technol. 2019;53:11496–506.

Marginal Water Use and Management in Agriculture

Berkant Odemis and Dursun Buyuktas

Abstract

One of the most important problems experienced in arid and semi-arid areas is the gradual decrease and pollution of water resources. This problem, which arises due to many factors, especially climate change, causes a decrease in agricultural production in many regions. As is known, the most effective application in increasing agricultural production is irrigation. Irrigation alone can increase the yield by 40%. However, in projections on water resources, it is seen that access to irrigation water will become more difficult in the future. As the annual world population increases by an average of 80 million people per year, it is inevitable to increase the freshwater provision in the world by 64 km^3 per year. While only a few countries had water problems in the 1950s, it is stated that by 2050, 66 countries will have moderate or severe water shortages, where 2/3 of the world's population will live. For this reason, it is imperative to operate water resources more appropriately at local, regional, national, and international levels, to store water when it is plentiful, and more importantly, to use non-traditional water resources (saline water, drainage water and domestic-industrial wastewater) besides traditional water resources (surface water, stream and river water, lakes and groundwater). Saline water (or unconventional water sources) includes urban industrial wastewater and drainage water of unconventional quality. It is estimated that 70% of the world's agricultural land will become problematic soils in the next 50 years because of improper operation of freshwater resources and will be irrigated with wastewater (drainage and domestic-industrial wastewater). Currently, saline water is widely used in many parts of the world, especially in countries like Australia, Egypt, India, Israel, Pakistan, the USA and the former USSR. However, since the long-term use of saline water can change the physiochemical properties of the soils, it requires important knowledge of many parameters for its use. These parameters are, i.e., (a) water quality and pollution sources, (b) plant salt tolerance, (c) time the plant is exposed to salt, (d) spatial and temporal distribution and variation of salt in the root zone and soil, (e) differences in resistance of varieties to salt, (f) climatic characteristics of the area where the water is used, and (g) soil properties. There are uncertainties about irrigation scheduling that should be applied in the long-term use of

B. Odemis (✉)
Faculty of Agriculture, Biosystems Engineering, Mustafa Kemal University, Hatay, Türkiye
e-mail: bodemisenator@gmail.com

D. Buyuktas
Faculty of Agriculture, Agricultural Structures and Irrigation, Akdeniz University, Antalya, Türkiye

M. Bahadir et al. (eds.), *Agricultural Water Management*, Water and Wastewater Management,
https://doi.org/10.1007/978-3-031-96134-2_15

saline water, the leaching process to reduce soil salinity, the cyclic irrigation strategy, and the frequency of irrigation to reduce the effect of salt in the soil.

Keywords

Climate change · Marginal waters · Saline soil · Soil management · Water quality

1 Introduction

Water scarcity is one of the most important environmental stress factors affecting agricultural productivity. Approximately 40% of the world is under the influence of an arid and semi-arid climate, where water scarcity is a common problem. The scarcity and irregularity of precipitation that occur as a result of global climate change increases the severity of drought problems in many areas of the world. The general opinion is that the amount of precipitation has decreased but the amount of evaporation and evapotranspiration has increased, especially in subtropical climate areas and the Mediterranean Region. This causes water scarcity and even drought problems in many regions. Currently, the world's average per capita water consumption is 800 m^3 per year. As the world's population increases by an average of 80 million people per year, the need for freshwater will increase by 64 km^3 at the same time to meet human needs. While only a few countries had water problems in the 1950s, it is stated that in 2050, 66 countries will have moderate or severe water shortages, where 2/3 of the world's population will live. According to a more pessimistic forecast, if the trends in water withdrawal under current practices and policies continue, approximately 60% of the world's population will face water stress by 2025 [1]. Water shortages are particularly worrisome in Middle Eastern countries (Table 1). It is inevitable that there will be future social explosions and economic crises in these countries, where the water required to meet food requirements began to decrease years ago. This region is the driest in the world, accounting for only 1% of the world's freshwater resources. However, some countries in other regions have also faced a water shortage recently [2]. Datasets and maps published in recent years show that an increasing number of countries will suffer from water shortages owing to increasing water scarcity [3].

According to estimates, approximately 600 km^3 of water was used annually for agriculture, industry, and domestic use in the early 1900s, this value reached 5000 km^3 in 2000. Although it varies according to the socio-economic conditions of the countries, approximately 70% of the water resources are used in agriculture. While this value is 41% and 79% in high- and middle-income countries, respectively, it increases to 90% in low-income countries. While the amount of water used in agriculture is 96% in Sudan and 74% in Türkiye, this rate is less than 1% in Germany and The Netherlands [4].

Irrigation is the most effective method to increase agricultural production. Fertilization, plant protection practices, plant growth regulators, and plant variety are important factors for increasing agricultural production. However, none of these factors can provide the effect of water alone. Irrigation alone can increase the yield by 40%. For this reason, irrigation is almost mandatory to increase the amount of agricultural products, especially in arid and semi-arid areas. However, projections of water resources show that access to irrigation water will become more difficult in the future.

While an average of 1,091 km^3/yr of water is used to irrigate approximately 255 million ha in the world, it is estimated that this amount will be 1,151 km^3/yr or 1,177 km^3/yr according to the ECHAM4 and HadCM3 models, respectively, in 2070. Modeling studies predict that in 2070, net irrigation water requirement will increase in 12 of the 17 regions, where agricultural production is intense, and the highest increase will be seen in South Asia (Pakistan, India, Bangladesh). In addition, it is stated that the net irrigation requirement will increase by 5.5% in 66% of the irrigated areas by 2070. In a study conducted with two different models (MRI and CCSR) in the Seyhan Plain (Turkey), it was determined that the precipitation amount would decrease by 29.4–34.7%, the

Table 1 Estimates and projections of populations and renewable water resources (RWR) per capita in 2005 and 2030 for countries [a] that are expected to fall below the water scarcity level (1000 m^3 capita/yr) in the year 2030 [3]

Country	Population (*10^3)		RWR (m^3 capita/yr)[b]	
	2005	2030 [c]	2005	2030 [d]
Kuwait	2,671	4,198	7	5
United Ara	3,106	4,056	48	37
Saudi Arab	25,626	43,193	94	56
Libya	5,768	8,123	104	74
Singapore	4,372	4,934	137	122
Jordan	5,750	8,643	157	104
Yemen	21,480	50,584	191	81
Israel	6,685	8,970	254	190
Oman	3,020	5,223	331	191
Algeria	32,877	44,120	435	324
Tunisia	10,042	12,351	458	372
Burundi	7,319	13,652	492	264
Rwanda	8,607	13,453	604	387
Egypt	74,878	109,111	779	534
Burkina Faso	13,798	27,910	906	448
Morocco	31,564	42,505	919	682
Kenya	32,849	41,141	919	734
Lebanon	3,761	4,692	1,170	938
Somalia	10,742	24,407	1,257	553
Malawi	12,572	19,834	1,376	872
Pakistan	161,151	271,600	1,382	820
Syria	18,650	28,750	1,410	915
Eritrea	4,456	7,942	1,414	793
Ethiopia	74,189	127,220	1,483	865

[a] Countries with limited area, population or data (e.g., Bahamas, Qatar, Malta) were not included
[b] RWR includes internal renewable water resources plus or minus the flows of surface water and groundwater entering or leaving the country, respectively
[c] UN medium variant population projections for the year 2030
[d] Based on the assumption that the RWR in the countries in 2030 will remain the same as in 2005

river flows by 37.5%–46.4%, and water storage capacity by 24.7% until 2070. At the same time, it is predicted that the amount of agricultural water use will increase by 13% in citrus fruits, 11% in fruits and cotton, and 32% in vegetables. Similarly, it is estimated that precipitation will decrease by 30–40% until 2070 in Thrace, Aegean, Western and Central Mediterranean, some parts of Southeastern Anatolia and Central Anatolia region of Turkey [5].

The decrease of water amount due to climatic changes (decrease in precipitation, increase in evaporation, etc.), as well as the increase in urban population also cause serious stress on freshwater resources. Therefore, it is necessary to increase either the crop production per hectare or the arable land cultivated. Considering that the increase in the crop yield is mostly obtained from irrigated areas, it is seen as a stronger necessity to increase the floorspace of irrigated agricultural areas. Limited agricultural areas and salinization problems in soils due to irrigation water quality reduce crop production. For this reason, the effective operation of water resources in all areas, where water is used, storage when water is abundant, and more importantly, the use of non-traditional water resources in addition to traditional water resources (surface waters, stream and river waters, lakes and groundwater) is mandatory.

Non-traditional water sources serve as valuable supplements to address water scarcity in regions facing extreme shortages of renewable water resources. These alternative water sources are particularly vital for agricultural and other purposes and are harnessed through specialized processes. Examples of such water sources include rainwater harvesting, collection, treatment, and utilization of wastewater, collection and reuse of agricultural drainage water, extraction of groundwater with varying salt content, and desalination of seawater. Incorporating these non-traditional water sources for irrigation requires the implementation of appropriate strategies to effectively manage soil, water, and crop yield. Careful planning and efficient water management techniques are crucial to optimizing the use of these resources and ensure sustainable agricultural practices. By employing these strategies, regions facing water scarcity can mitigate the impact of limited water availability and foster more resilient agricultural systems [6–8].

2 Properties of Marginal Waters

(a) **Origin**

Marginal waters are waters that do not have the usual qualities and come from the following sources: (1) water produced from domestic, commercial, and industrial uses, (2) water reaching the drainage system through the soil profile as a result of surface runoff and deep percolation in irrigated agriculture; and (3) groundwater from different sources (e.g., leachate from underground saline layers or from areas with seawater intrusion). The quality of such water includes many complex factors and is influenced by many variables. Therefore, it is not convenient to classify traditional water. The water resources mentioned have two basic characteristics; the first one is their high organic matter and microorganism content (e.g., urban wastewater), and the second one is their high salt concentrations (industrial and agricultural wastewater).

There is no generalized definition of the nature of marginal water resources. However, in the general use process, it can be defined as water that has the potential to cause some problems. For example, drainage and municipal wastewater may not be of suitable quality for irrigation because of the high salt content and health hazards. Therefore, the use of such water in irrigation should be carefully planned. In addition, significant differences in its acquisition should be considered. For example, saline (non-traditional) water resources require more rigorous monitoring and complex operating modes than other well-qualified water resources do.

In the last 30–40 years, criteria have been established for the use of unconventional water resources for irrigation worldwide. Examination of the standards regulating the use of wastewater in agricultural production has shown that these vary according to the country. It was stated that each country made arrangements in accordance with its own conditions, and some restrictions were imposed on the use of the aforementioned water. It was emphasized that the limitations in question were due to the long-term negative effects of wastewater on soil, plants and humans in terms of salinity, alkalinity, trace elements, and other pollutants [9].

(b) **Quality**

Evaluation of marginal water resources is more complex than non-saline water, because the ‘quality’ of water depends on its power to change the environment it affects. For example, the poor quality and salty water used for agricultural purposes may not create the expected adverse conditions in light-textured soil and salinity-resistant plants. However, the same quality of water can cause unexpected negative situations in heavy-clay texture soil and plants sensitive to salinity. Therefore, the mutual interactions of the soil, plant, drainage, and climatic conditions of the region are very important in determining the quality of irrigation water in each region. Because of the complex relationship between the factors mentioned, numerous classification methods have been suggested and applied, although an international comprehensive classification that determines the suitability of water for irrigation has not yet been developed. Many of these methods are based on three primary criteria: (1) salinity, which measures the total amount of soluble salts, (2) sodicity, determined by the sodium carbonate and bicarbonate ions, and (3) the toxicity of specific ions such as chloride and boron. These criteria serve as essential indicators in evaluating water quality for irrigation purposes [10].

Salinity is usually expressed in terms of electrical conductivity (EC), and sodicity as a measure of sodium adsorption ratio (SAR). Electrical conductivity is based on the increase in the conduction of the electric current with an increase in the salt concentration of the solution. The sodium adsorption ratio (SAR) varies according to cation density. This ratio reflects the proportional efficiency of sodium ions in cation exchange reactions occurring in soil complexes.

3 Potential of Marginal Water

As a result of improper operation of freshwater resources, 70% of the world's agricultural lands will become problematic soils in the next 50 years and will likely be irrigated with wastewater (drainage and domestic-industrial wastewater). In a meeting held by the World Health Organization (WHO) in 1973, it was stated that wastewater is an application that helps both to conserve water resources and to minimize the harmful effects of irrigation on the environment and ecology; and it was accepted to be used primarily in agricultural irrigation [11]. Today, treated, diluted, or untreated wastewater is used on approximately 2×10^6 ha of land around the world, especially in countries where water is scarce (Table 2). For example, Egypt's total water consumption for 2015 was 76.4×10^9 m^3, of which 8.9×10^9 m^3 belonged to drinking/health use. This amount of freshwater produces approximately 5×10^9 m^3 of wastewater, of which 3.7×10^9 m^3 is treated and reused. At least 75% of the wastewater collected in Egypt is treated, providing significant water savings [12]. Already at the end of the twentieth century, it was determined that 14.35 million m^3 of drainage water was generated in Egypt every year, and the salinity level of 3.5 million m^3 of this was more than 4.0 dS/m. In addition, it is known that 2.37 million m^3 of drainage water in 1988 and 4.45 million m^3 of drainage water in 1992 were used for irrigation purposes [13].

Israel has made significant strides in the utilization of wastewater and treated wastewater for irrigation purposes. Presently, the country purifies an impressive 85% of its wastewater, making it the global leader in this practice. In comparison, Spain comes second with a usage rate of 25%. Israel employs a well-structured approach to manage its treated wastewater. Approximately 10% of the treated wastewater is discharged into rivers, while another portion is utilized to combat forest fires. The remaining 5% is discharged into the sea, efficiently reusing valuable resources. Additionally, Israel has established itself as a pioneer in desalination technology. The country boasts the world's largest seawater desalination plant, despite the process being known for its energy-intensive and expensive process. Nevertheless, the advanced technologies implemented allow Israel to produce an impressive 1,000 L of drinking water per person in just one week, at a cost of 58 cents. The government of Israel has set ambitious targets for the future, aiming to produce 900,000 m^3 of drinking water annually through five major desalination plants within a few years [4]. Furthermore, Israel has demonstrated successful use of groundwater with a salinity of

Table 2 Volume of wastewater generated annually in some countries of Central and West Asia and North Africa

Country	Reporting year	Wastewater volume (× 10^6 m^3/yr)
Algeria	2004	600
Bahrain	1990	45
Egypt	1998	10,012
Iran	2001	3,075
Jordan	2004	76
Kuwait	1994	119
Kyrgyzstan	1995	380
Lebanon	1990	165
Libya	1999	546
Morocco	2002	650
Oman	2000	78
Saudi Arabia	2000	730
Syria	2002	825
Tunisia	2001	240
Turkey	1995	2,400
United Arab Emirates	2000	881
Uzbekistan	2004	170
Yemen	2000	74

3.0 dS/m by implementing pressurized irrigation methods and adhering to a leaching requirement of 25–30%. This approach ensures effective water usage and sustainable agricultural practices in the region.

Numerous studies have been conducted in countries like Tunisia, India, and the USA to investigate the possibilities of using saline/waste water in agriculture. In the Pecos Valley region of the United States, waters with an average salinity of 2,500 mg/L was successfully used on 81,000 ha of farmland for 30 years [14]. It was stated that in the irrigation of the cotton plant in the aforementioned region, water containing iron and gypsum with an electrical conductivity of 8 dS/m is used by the furrow method, and the water of 5 dS/m is used by the sprinkler method at night, early in the morning and in the evening. Alfalfa and other forage crops were grown using irrigation water with a conductivity of 3–5 dS/m with minimal yield loss. The successful use of saline groundwater for irrigation for 10 years in the dry and hot regions of Arizona has been also reported [15].

Irrigation, whether with saline water or fresh water, is a practice that stores salt in the soil. Of the total 14 billion hectares of land in the world, 6.5 billion hectares are located in arid and semi-arid regions, where irrigation is mandatory, and about 1 billion hectares of this is naturally salty soil. In addition to this area, approximately 20% of cultivated areas and 33% of irrigated area are affected by high salinity [16], and saline area increases by 10% each year [17]. Agricultural productivity of 10 ha of land that can be cultivated worldwide is lost every minute, and 3 ha of this is due to salinity. 1.6 million hectares of arable land are lost every year in this way. It is predicted that 45.4 million hectares of the 220 million hectares of irrigated area (about 250 million hectares in some sources) are affected by salt, and 25% of the freshwater resources are polluted to an unusable extent in the same period [18].

4 Problems in the Use of Marginal Waters

The rules to be followed for the use of marginal water vary according to the environment, in which they will be applied. Soil, plant, and natural environmental conditions are significantly affected by marginal water quality, duration of use, and climatic conditions. However, there are four main difficulties in the use of such water in social life: (a) social difficulties (in the public acceptance of wastewater reuse), (b) administrative difficulties (plant selection, irrigation and soil-based practices), (c) identification of potential problems and risks to human health, and (d) environmental threats. While using wastewater (saline water), the characteristics of the environment, where the water will be applied, should be well known from an environmental point of view. Long-term use of saline/wastewater can change the physicochemical properties of soils. Also, the bioaccumulation of pathogens and toxic chemicals are the further disadvantages of wastewater reuse in agriculture.

5 Use of Marginal/Saline Water in Agriculture

5.1 Use in Crop Production: Points to Be Considered and Precautions to Be Taken

The utilization of saline water predominantly leads to the physical and chemical degradation of the soil profile, resulting in adverse effects on the conducive environment within the root zone for plant growth and sustainability. As the salt concentration in the root zone of the plant increases, the speed of movement of water in the soil decreases, and the plant cannot find water when it needs it and gets stressed. In addition, water vapor loss under saline conditions increases the osmotic regulation against the gas diffusion resistance of the leaves and affects the plant water balance. Transpiration is reduced and the plants tend to regulate turgor. However, stomatal closure results in increased resistance owing to the diffusion of CO_2 in the leaf. This situation leads to a slowdown of photosynthesis and development in the plant and decreases in crop yield and quality.

(a) **Salt Tolerance**

Plants react differently to salinity stress. Numerous studies have been conducted on the salinity level of plants during the entire growing season. The results showed that, in general, the salt tolerance of field crops is higher than that of fruit trees and vegetables, and vegetables are the most sensitive plant group to salinity. The response of plants to salinity (and threshold values) under current climate and soil conditions has been revealed in different studies conducted in different parts of the world [19].

The response of crop yield to salinity was expressed by the following two-part function (salt threshold value and slope value indicating yield decrease due to increased salt).

$$\mathrm{Yr} = 100-(\mathrm{ECe}-t) * s$$

In this equation, Yr represents the proportional yield decrease in percentage, ECe is the electrical conductivity of the soil saturation extract, which indicates the salinity level in the soil, t is the threshold salinity value, at which the yield begins to decrease (It signifies the critical point where salinity starts affecting crop productivity), and s represents the slope value, indicating the rate of yield decline with increasing salinity (It quantifies how much the crop yield decreases as salinity levels rise).

This function allows us to assess the impact of soil salinity on crop yield by calculating the percentage reduction in productivity based on the actual electrical conductivity of the soil saturation extract (ECe) in comparison to the threshold salinity (t) and the slope value (s). By knowing these parameters, we can better understand how salinity affects agricultural productivity and make informed decisions to manage soil salinity for sustainable crop production.

Salt tolerance is characterized by both the threshold and slope values. Tolerant plants like cotton, barley, sugar beet, and asparagus are characterized by high thresholds and small slopes, whereas sensitive plants like beans, onions, carrots and strawberries are characterized by low thresholds and large slopes. Soil electrical conductivity here is the average soil salinity during plant growth and development. The type of salt and periodic variation in salt concentration are not included in this function. Under realistic field conditions, the salt concentration varies over time locally and across the profile. Uniformity (equal distribution of salt in soil) is not common.

The salt tolerance of plants is influenced by various factors, including soil type, vegetation, climate conditions, and irrigation management. The response to salinity can be influenced by changes in soil management during the growing season, spatial and temporal variations in salt concentration in the plant root zone, the duration of plant exposure to salt, and the prevailing climatic conditions (temperature, humidity, wind, and evaporation) of the region. Additionally, soil characteristics (clay or sandy soil) and topographic features also play a significant role in determining the plant's ability to tolerate salt stress. Understanding these factors is crucial for developing effective strategies to manage salt-affected soils and promote crop productivity under varying environmental conditions.

(b) **Duration of Exposure to Salinity**

Studies on the determination of the salinity threshold values of plants assume that salinity is at a standard constant level during the growth phase of the plant; therefore, it does not provide information about the response of plants to salinity at different developmental stages from planting to harvest. However, plants may react differently to salinity during each developmental stage. While the sensitivity to salinity is high at the initial stages, the tolerance increases toward the harvest time and the sensitivity decreases (Table 3). This situation sheds light on the periods during which saline or marginal water can be used. Generally, toward harvest time, using saline water instead of available freshwater resources increases water use efficiency. However, it is difficult to determine the salt tolerance at the individual developmental stages. The effect of salinity at any stage of growth makes it difficult to determine its relative sensitivity experimentally, as it can affect the response at later stages. Studies claiming that sensitivity differs during growth periods have shown that there is actually a sensitivity to the length of time the roots are exposed to salinity. For example, in a study examining the relationship between salinity and the growth stage in wheat, the sensitivity of the plant at the seedling stage was determined

Table 3 Salinity response functions as affected by the time and stage of crop growth [21]

Crop	Growth stage[a]	r[b]	Slope[b]	Threshold[b]	50% YD	Source
Sorghum	Vegetative		−8.8	3.3	9.0	[22]
	Reproductive		−7.7	6.3	13.4	
	Maturation		−8.4	9.0	14.9	
Wheat (hard)	Vegetative		−8.0	2.5	8.8	[20]
	Reproductive		−4.8	4.6	15.0	
	Maturation		−7.8	11.6	18.0	
Cowpea	Vegetative		−7.4	0.3	7.0	[23]
	Reproductive		−5.2	1.0	10.6	
	Maturation		−5.5	1.8	10.9	
Maize	102 DBH		−15.8	2.0	5.1	[24]
	81 DBH		−7.8	2.0	8.6	
	52 DBH		−7.4	3.4	10.2	
	TW	0.96	−0.15	204		
Bean	50 DBH		−17.1	2.6	5.5	[25]
	40 DBH		−10.0	2.6	7.6	
	20 DBH		−2.6	1.5	21.0	
	TW	0.96	0.23	87		

[a] Salinization during designated growing period or starting at designated day before harvest (DBH). TW = time-weighted salinity

[b] Correlation coefficient, slope of response curve in % per dS/m for mean EC_e, or % per dS/m/decrease for time-weighted. EC_e, threshold of response curve in dS/m or dS/m/decrease, respectively, and EC_e at which yield was reduced by 50% yield decrement) in dS/m

[20]. Salinity was initiated and maintained for 21 days at four different times, from pre-sowing to 16 days post-sowing, after which the plants were harvested. It was observed that the later the exposure time to salinity, the lower was the reduction in yield. Studies have shown that the seedling stage is the most sensitive time to salinity. However, because soil salinity increases over time, the average salinity level, to which the plant is exposed, should be considered as effective salinity [21, 22].

(c) **Spatial and Temporal Variability of Salinity**

Salinity under field conditions can vary spatially and temporally. Salinity in irrigated agricultural areas varies along the soil profile depending on the amount of irrigation water used. When the amount of leaching water applied in the use of saline water is kept low, the amount of salt increases along the vertical profile and the maximum concentration occurs at the base of the root zone. However, an increase in the leaching fraction or excessive irrigation application initially causes a balanced and homogeneous distribution of salinity in the profile. The variability increases over time. Soil moisture, which is the field capacity (or saturation point) in the first irrigation, decreases with transpiration and evaporation. As the soil moisture in the upper soil layer decreases over time, the upward movement of water begins with capillarity, while the water in the upper layer decreases by evaporation, and salt accumulates on the soil surface. Therefore, the highest salt accumulation in the period between the two irrigations was observed in the upper soil layers and below the root zone under normal conditions (homogeneous soil profile and the same soil texture conditions). Soil salinity in these layers can be two to three times higher than that in the other layers. Roots in the upper layers with the highest root density are more exposed to salinity than those in the other layers. A similar trend was observed under conditions with a shallow water table. The dynamic nature of salinity (EC_e) along the profile causes the roots to be constantly exposed to different salt concentrations. Here, a question should be discussed: (a) is the plant's response to salinity affected by the average salinity between irrigations (or from planting to harvest), or (b) by the weighted average of salinity in the profile (under continuous sampling conditions). Most of the evidence so far indicate that the average salinity with depth integrated over time is the effective salinity affecting the plant. In addition, the electrical conductivity of the soil water (EC_{SW}) or saturation extract (ECe) was found to be more correlated with the yield. As a result, the arithmetic average of the salinity obtained from the samples taken along the root depth can be considered as the effective salinity to determine the salt response of the plant.

(d) **Differences Between Varieties**

Salt stress can have varying effects on different varieties of the same plant species as well as across different plant species. Research on cultivar responses to salinity is predominantly conducted under controlled laboratory conditions, where the use of a single abiotic stress is often the most feasible option. However, this controlled environment might not always reflect the real-world conditions faced by plants in the field. In laboratory settings, attempts have been made to understand the reactions of plants to the most abundant ions in nature, namely Na+ and Cl-. These ions can elicit distinct responses in plants, including leaf burns and eventual defoliation. Moreover, there are differences in how fruit trees and rootstocks respond to salt stress, particularly in their ability to transport chloride to the upper parts of the plants. Overall, the effects of salt stress can vary significantly among different plant varieties and species, and studying these responses under laboratory conditions helps provide valuable insights. However, it is essential to consider the complexities of field conditions where plants might encounter multiple abiotic stress factors simultaneously. This calls for a comprehensive approach in understanding and managing salt stress in agricultural and natural ecosystems.

The salt tolerance of the varieties of field and horticultural plants is not very high. As there are many bred varieties, their yield abilities are tested first. For this reason, there is insufficient information regarding the salt tolerance of cultivars. However, in field crops, like cotton [26–28], soybean [29, 30], corn, rice [31], wheat [32], barley [33], vegetables, tomato [34] and lemon [35], salt tolerances were determined. Although the investigations on this subject are not sufficient, it is inevitable that these studies will increase in the future. The decrease in qualified water resources due to climate change will make the use of waste/saline/drainage water mandatory. This will make it necessary to investigate the resistance of plant species and varieties to salinity, especially by genetic engineering. The genes controlling tolerance to salinity, and how the plant behavior will be under conditions of multiple stresses (temperature, water, disease, etc.), are among the important future topics in the field of plant genetics.

(e) **Climate Characteristics**

Climatic features significantly affect the soil salinity level. Agricultural areas, especially arid, semi-arid, and coastal regions, are highly sensitive to the effects of climate change on soil salinity. Temperature, humidity, and precipitation are the main climatic factors affecting soil salinity. Although not all at the same level, these factors affect soil salinization, and thus, the response of plants to salinity. In this sense, the most critical parameter is temperature. High temperatures increase the stress level of plants due to transpiration or the effect of temperature on the photosynthesis process in the leaf. Increased stress causes changes in the salt response function in salt-sensitive plants. High temperatures also significantly increase the evaporation. High temperature causes rapid salt accumulation in the upper soil layer as a result of capillary rise, especially in areas, where fallow and ground water level are high. In the upper soil layers, where the roots are dense, this causes a decrease in the available water for the plant.

Relative humidity can partially prevent stress in the plants. The damage caused by osmotic stress due to salinity is lower at high relative humidity than in dry air. High temperature and low relative humidity can alter the salinity response of plants, such that the threshold salinity value decreases and the slope increases, making the plant more sensitive to environmental conditions.

Soil salinity is diluted by precipitation, creating a favorable environment for the plant. In saline soils (or areas irrigated with saline water), low precipitation causes salts to be transmitted to the crop root zone layer, whereas high precipitation causes percolation under the root zone. This provides significant advantages for sustainable soil salinity management in areas where drainage is effective.

(F) **Soil Properties**

Soil properties in saline soils or areas irrigated with salt water are at least as important as the salt tolerance abilities of plants. Soil fertility, soil structure, infiltration, hydraulic conductivity, and aeration rate play an active role in soil salinization.

Crop yield loss due to salinity is closely related to soil fertility. Increasing salinity in soil rich in organic matter with high fertility causes more yield loss than in poor soil. Therefore, plants are more sensitive to salinity when the conditions that make the soil fertile are suitable. If soil fertility is low, the additional damage to the increased salt will be very low, as crop productivity will already be low.

The effects of soil texture and structure on salinity occur under the influence of properties that depend entirely on structure.

The infiltration rate, hydraulic conductivity, water holding capacity, and saturated water content affect the salinity and applications required for leaching salts. For example, the success of leaching to remove salinity depends on the soil texture. Clay soil has the highest water holding capacity and sandy soil has the lowest water holding capacity. For the same evapotranspiration rate, sandy soil will lose proportionally more water than clay soil, therefore, the salt concentration in the soil solution will increase faster in sandy soil. In order to reduce the damage caused by increased salt concentrations, the irrigation interval should be kept short as the soil sand content increases.

Leaching is important in the rehabilitation of saline soils, and soil texture must be taken into account for successful leaching. Leaching is the process of replacing the soil solution with incoming irrigation water. The amount of water required for leaching largely depends on the nature of the soil (and therefore its water holding capacity). Because the water holding capacity is high in clay soil and low in sandy soil, the same amount of precipitation or irrigation water will cause deeper leaching in sandy soil than in clay soil. For this reason, the success of medium textured or sandy soils is higher than that of clay soils in reclamation practices. The same is true for the soil reclamation process.

The infiltration rate is a highly variable parameter even in small areas and varies according to the soil texture and structure. The duration of saline water use and the types and concentrations of salts in water affect the infiltration rate. Prolonged use of water with high salt content results in reduced infiltration and increased runoff. Salts act on the reduction of infiltration, as follows: Cations (Na, K, Ca, and Mg) in salt water can easily adhere to the surfaces of negatively charged clay particles. Because of their + 2 valency, Ca and Mg can adhere to clay surfaces more easily than the + 1 valence of Na. However, the abundance of Na in nature causes it to be held on clay surfaces rather than on Ca and Mg. Na, which has a larger hydration diameter than other cations, causes dispersion in clay complexes and formation of an impermeable crust layer on the soil surface. Sodium less than 5 g/L in soil solution reduces dispersion between soil particles and promotes aggregate formation [36].

The easiest way to estimate the possible reduction of infiltration rate is determining the SAR (sodium adsorption ratio) value of the irrigation water. The SAR is defined as the ratio of Na to Ca + Mg and is calculated using the formula $SAR = Na/(\sqrt{}(Ca + Mg)/2)$. A high SAR value indicates a possible decrease of the infiltration rate. Rainfall on soil with a high SAR reduces water penetration (infiltration rate) into the soil, which causes runoff and erosion. The easiest solution to the infiltration problems caused by Na is to apply Ca-rich lime to soils as reclamation agents. Enriching the soil with Ca and Mg, either by adding it to the irrigation water or by scattering it on the soil surface, can improve infiltration by reducing the proportional amount of Na in the clay complexes. The most important factor in these applications is the leaching rate. It is particularly important that leaching be sufficient to remove Na from the medium.

The infiltration rate is the basic soil property used in the design of irrigation systems, whereas hydraulic conductivity is mainly used in the design of drainage systems. The hydraulic conductivity is strongly affected by the composition and concentration of the irrigation water. In particular, the SAR value is effective for hydraulic conductivity. The reduced hydraulic conductivity due to SAR can be compensated for by increasing the total salt concentration. The hydraulic conductivity, when the SAR is 30 and the total salt concentration is 3 mEq/L, can be the same as that when the SAR 10 salt concentration is 0.5 mEq/L.

6 Soil Management in the Use of Saline Water

(a) **Irrigation Schedule**

In the case of saline water used in agricultural irrigation, the most important problem is the accumulation of excessive salt and sodium in the root zone. In this case, an irrigation schedule that provides salt balance in the soil–water system should be applied. An irrigation strategy should be established, in which any irrigation application, whether high or low salt content, will result in minimal salt accumulation. During the irrigation period, "salt" is stored in the soil in many ways, especially in irrigation water, fertilizer applications, pesticides, hormone applications, seepage from neighboring lands and capillarity. Most of the salt comes from irrigation applications. Because irrigation scheduling determines when, how much and by which method irrigation water should be given, it prevents salt accumulation caused by excessive irrigation. The application of water as needed by the plant also reduces the amount of salt to be carried to the upper layers, with the effect of capillarity in areas with high ground water.

(b) **Leaching**

In irrigation with saline water, the plant's response to water and salinity is continuous, and plant growth and development increases with plant water consumption. Because salinity reduces plant growth and development, it also reduces plant water consumption. This necessitates the removal of salt from the root zone. Leaching is the most effective way to remove sal from the root zone. Leaching can be considered the key to the successful application of wastewater or saline water. Although there are many parameters that determine the amount of leaching water, the most important parameter is evapotranspiration. The amount of leaching water is added as an additional amount of water to the amount of evapotranspiration and applied to the soil. In general, the amount of leaching water is calculated 20–40% more than that of evapotranspiration. This ratio generally varies with the amount of water supply available, drainage conditions and plant salt tolerance. The leaching process controls the salt accumulation in the soil in all aspects. The concentration of salts, their diversity, their dissolution amounts, the amount of drainage water, the quality of groundwater, and most importantly, the crop yield, can be controlled by the leaching process. The salt deposition process is closely linked to the evaporation-transpiration process. Salts are left behind and accumulate as water is consumed by plants or evaporates from the soil surface. Under normal conditions without precipitation, the maximum salt accumulation can only be reached in the layer close to the soil surface, where water loss by evaporation is predominant. The fact that the root density of the plants is higher in the layers close to the soil surface also contributes significantly to this accumulation. Although the increase in salt concentration with the water uptake of the roots affects water uptake, this situation changes depending on the concentration and the time the salt stays in the layer. Controlled leaching ensures that this concentration is carried at the base of the root zone. Reducing the amount of leaching water does not increase the salt concentration but causes the salt to remain in the layers close to the surface, resulting in a gradual decrease in rooting depth. Therefore, the maximum level, at which salts accumulate in the soil, depends more on the salt tolerance of the plant than on the leaching fraction.

(c) **Cyclic Irrigation**

The sensitivity of plants to salt during development periods is different. This indicates that different types of water can be used during the growing season. In general, plants are sensitive to salt during the first period of germination and plant development and require the use of high-quality water during these periods. Marginal quality water can only be used during later development periods. This practice has fewer adverse effects on plant growth. In this way, plants have passed their salt-sensitive period before the use of saline water. In cases where high-quality water is available but limited, the use of marginal quality water for irrigation or the alternating (cyclic) use of the plant in different growing periods, can be considered as an important strategy in terms of protecting freshwater resources. In addition, saline and non-saline water can be applied alternately. Although it has been stated that the physical and chemical properties of soil may deteriorate, if the alternating method is used for a long time, it is a more valid approach in terms of plant breeding than the mixing method. As long as the leaves of the plant are resistant to salt, it is more valid to use non-saline water in the first stage (germination and vegetative development periods) and saline water in the other stages of development in cases where both water sources are available.

(d) **Irrigation Frequency**

When saline water is used, the upward transport of salt by capillarity during the period between the two irrigations causes an increase in salt accumulation in the part, where the roots are dense. This increase may cause the salt resistance threshold to exceed several times in one irrigation period and result in yield losses. For this reason, it is recommended to increase irrigation frequency in salt-sensitive plants. As the soil dries between irrigations, not only the matric potential, but also the osmotic potential decreases as the concentration of the soil water solution increases. This situation depends on the shape of the wetting–drying (desorption) curve of soil and plant water consumption. As a result, the pre-irrigation salt concentration is lower at shorter irrigation intervals. If water with low salt content is to be used for irrigation, where frequent irrigation prevents seasonal salinization of the soil, leaching should be performed once or twice a year, or at least at the end of the irrigation season. Although it is generally recommended to increase irrigation frequency when using saline water, there is little data to support this recommendation. Increasing the irrigation frequency can increase the salt concentration in the root zone toward the upper layers. As is known, salinity reduces evapotranspiration, which causes the soil to dry more slowly than under non-saline conditions. Therefore, for the same irrigation interval, the pre-irrigation soil water potential may be lower under non-saline conditions than under saline conditions, resulting in lower yields. In addition, the shorter the irrigation interval, the greater the amount of evaporation in evapotranspiration, resulting in additional water application and an increase in salt load. Different studies recommend and do not recommend increasing the frequency of irrigation when using saline water. For example, it has been reported that increasing irrigation interval provides a significant advantage when saline water is used in Guayule plant. On the other hand, it was stated that there was no advantage in increasing the frequency of irrigation with saline water in kidney bean, citrus, eggplant, corn and grass irrigation, and it was even determined that it had a more damaging effect than a beneficial effect on pepper and barley [21]. As a result, most of the information in the literature shows that increasing the irrigation frequency does not provide any advantage or can even be harmful when irrigation is performed with saline water. However, it should not be forgotten that the damage can be eliminated by leaching.

(e) **Other Factors**

Tillage is very important for increasing permeability and breaking the hard layer formed on the soil surface in soils irrigated with saline water. If the soil is not cultivated according to the technique, a salty layer may form on the soil with the plow base and close to the surface. Because sodium soils tend to form sludge and hard layers, the soil should be kept dry by careful processing, and heavy equipment should not be used frequently.

Deep plowing is important for implementing cultural measures. In sodium soils, deep plowing up to 60 cm loosens aggregates, increases the soil–water holding capacity, and thus helps control salt accumulation when saline water is used. Vegetative yield can be increased by changing the plowing depth every three or four years. Choosing the right type of plow, row and spacing of the plows, depth of the plow and soil moisture content during plow allows good soil cultivation and soil structure development. It may be necessary to plow the soil to a depth of 2.2 m with special tillage tools and break the impermeable layer that prevents water from penetrating deep. It has been stated that the evaporation rate is reduced, and leaching efficiency is increased by 10% by shallow plowing of the soil surface [37].

Organic and green fertilization have two important effects on saline-sodium soils and soils irrigated with waters with high SAR values: (a) increasing soil permeability and (b) releasing carbon dioxide and some organic acids formed during organic decomposition. This phenomenon causes the release of calcium as a result of the decrease of soil pH, and thus, the dissolution of $CaCO_3$ and other minerals. Thus, the exchangeability of Na with Ca and Mg increases and as a result, the exchangeable sodium percentage (ESP) decreases. Green manure has a similar effect as organic manure [13].

Mulching is a practical method to decrease soil moisture losses due to evaporation and, consequently, reduce soil salinization. When soluble salt concentrations are anticipated to be high at the soil surface, mulching with saline water facilitates salt leaching and lowers sodium content. This enables the cultivation of salt-tolerant plants in these conditions.

It has been determined that the evaporation rate of groundwater at depths of 90, 120, and 180 cm from the surface can be approximately 8, 3, and 1 mm/day, respectively, on a fine sandy-loam dry soil surface. Therefore, whenever practicable, mulching should be encouraged to reduce the upward flow of soluble salts [13]. Mulching has been shown to increase soil temperature, delay soil moisture loss, and control weed growth in an area irrigated for 2 years with saline groundwater with a mineral content of 2–5 g/L at a depth of 10–20 m from the soil surface in China. Most importantly, it has been found to prevent increases in soil salinity [38]. It was reported that polyethylene and straw mulch increased soil temperature by approximately 6 °C at 5 cm depth and 4 °C at 10 cm depth, while maintaining soil moisture and reducing evaporation and salt accumulation [39].

Places, where plants are planted in conditions where saline water is applied, are very important especially for surface irrigation methods. Aforementioned process should be carried out in places where salt is the least in order to ensure germination and emergence. The use of slopes for seeds is suggested instead of furrow ridges in furrow irrigation [40]. Farmers in Israel and the USA alternately irrigate the furrows to prevent the effect of irrigation water salinity and sow the seeds on the furrow ridge boundaries to protect the plants from the negative effects of salt [41].

7 The Effects of Climate Change on Soil Salinization

Climate change has a direct impact on agricultural productivity. Elevated atmospheric CO_2 levels (> 400 ppm) contribute to climate change, along with increased air temperatures. Sudden and significant changes in daily, seasonal, and interannual temperatures, alterations in the rainy and dry periods, heavy rainfall events, prolonged droughts, excessive frost, and rising fire hazard levels are all climate change-related factors that influence agricultural production. These factors pose challenges and uncertainties for farmers and can affect crop yields and overall agricultural sustainability.

Climate models predict that terrestrial ecosystems, soil properties, surface water, and stream flows will be significantly affected by climate change [42]. Climate parameters are expected to be primarily affected by precipitation, potential evapotranspiration (ET) and temperature [43]. These parameters are very important for the continuity of agricultural production in irrigated areas. Ironically, arid and semi-arid areas, which account for 40% of the world's food production, require irrigation for maximum efficiency. Therefore, changes in the aforementioned climatic parameters affect all elements of the irrigation phenomenon. Considering the effect of irrigation practices on soil salinization, it can be understood how strong the relationship between climate change and soil salinization is.

There are two types of salinity in soils: primary and secondary salinity. Primary salinity is caused by minerals during soil formation as a result of natural events. For this reason, even if it is not exposed to any activity, there is salt in the soil. Secondary salinity, on the other hand, occurs because of anthropogenic effects at the end of the management of natural resources. The main factors determining secondary salinity are urbanization and agricultural activities. Whether secondary or primary salinity, accumulation involves one process. These processes can be broadly classified as: (a) soil mineral decomposition, (b) salts from irrigation water, (c) salts from wastewater, (d) fertilizer and pesticide applications, (e) saline groundwater intrusions, and (f) salts in water from seawater intrusion in coastal areas. Salts coming from these sources to any area cause different amounts of distribution in land areas owing to the effects of soil, climate, topography, and biological factors.

The element that causes salt accumulation in the soil in an irrigated area is primarily evapotranspiration. Regardless of the quality of water after irrigation, both evaporation and transpiration result in water loss from the soil. The soil moisture, which is initially at field capacity, gradually decreases to the wilting point (dry conditions). In this dynamic process, the water evaporated from the soil surface moves continuously from the lower layers to the upper layers. During this movement, as water evaporates from the soil surface, it leaves the containing salt in the soil. Therefore, salt accumulation in the soil creates a decreasing gradient along the soil profile, especially between the two irrigations (unless there is an extraordinary situation such as precipitation and seepage from neighboring areas). The top layer becomes the saltiest layer of the profile in this process and continues until the next irrigation. The salt concentration between the upper and lower layers can sometimes be 5–10 times (depending on the irrigation interval). Because air temperature is very effective in increasing the amount of evaporation under climate change conditions, increasing the temperature causes both evaporation and evapotranspiration to increase. For example, increasing temperatures in the Eastern Mediterranean Region of Turkey over the last 10 years have caused evapotranspiration to increase by approximately 20%. This situation has caused a significant increase in salinity particularly in areas where the Amik Plain is located.

The presence of ground water also causes salt accumulation in the soil. Evaporation (and transpiration) is effective in transporting salt to the upper layers, especially in arid and semi-arid areas, where a shallow water table (< 2 m) exists. In the absence of a drainage system, salt transport is continuous, and the soil becomes saline after a while. In soils where the soil texture is excessively clayey, the capillary rise will be higher, and there will be more salt movement and accumulation in the upper layer. In undrained areas with low rainfall and low leaching, soil salt accumulation can reach extraordinary proportions. Soil salinization follows a sustainable process, if the rainfall is sufficient and drainage conditions are suitable. For this reason, it is inevitable that salt accumulation in the soil will accelerate in areas where decreasing precipitation amounts cannot provide leaching. This process results in the formation of unproductive saline soil and desertification. On the other hand, heavy rainfall in areas, where soil depth is low and precipitation is high, causes the groundwater level to rise and salty with capillarity in dry periods.

Climatic factors affecting salt storage in the soil cause an increase in salinity and rapid degradation of soil properties with a synergistic effect. Increasing temperature and evaporation, decreasing precipitation and increasing salt accumulation affect the physical and chemical conditions of the soil, reducing seepage (leaching) of water (and salts) from the soil. Soil texture is effective in this process. Especially in clay soils, increased salinity causes soil dispersion and decreases the aeration rate. Sodium in water is more effective than other cations and negatively affects the hydraulic properties of soils (infiltration, hydraulic conductivity, and water holding capacity. Dispersed clay particles hinder water infiltration, causing precipitation and irrigation water to run off the surface instead of entering the soil. In regions with saline groundwater or treated wastewater as the only water source, proper irrigation and drainage systems are essential to prevent excessive salt accumulation in the soil during irrigation.

Salinization in coastal areas is caused by salts carried by humid air from the sea and the utilization of saline groundwater for irrigation. Even without additional applications, approximately 200–300 kg of salt can accumulate per hectare annually through atmospheric salinization. The fluctuating sea level due to rising temperatures also contributes to salinization of agricultural land and well water. In alluvial regions, seawater intrusions may extend up to 300 m. Moreover, excessive use of well water in coastal areas leads to the inflow of saline water from a saline aquifer into a non-saline aquifer [44].

8 Conclusions

Under the recent living conditions changing due to climate change, the use of waste water (saline/drainage/urban and domestic) is inevitable in almost every sector, from drinking water to agricultural irrigation, and from domestic consumption to industrial use. Over the last 50 years, many strategies (deficit irrigation, partial root drying method, etc.) and methods (drip irrigation, sprinkler irrigation, etc.) have been developed to conserve freshwater resources. These methods and strategies can also be applied to the use of saline water. This is an important advantage, and it seems that the use of these waters will continue to increase in arid and semi-arid areas, where irrigation is mandatory. However, the main problem here is how the operation methods should be followed in the use of these waters. In fact, the application of water to the soil, whether saline or non-saline, causes more or less salt accumulation. With a rough calculation, as a result of 5 irrigations (assuming 100 mm of requirement for each irrigation) in an area of 10 ha, where flood irrigation is practiced, approximately 50,000 m^3 of water will be applied in one irrigation season. Assuming the electrical conductivity (salinity) of this water is 0.7 dS/m, approximately 12 tons of salt will accumulate on the same land in one season with only irrigation water. Considering that the water in question is waste/saline/drainage water, it is obvious that the salt accumulation in the soil will be much higher. However, it is possible to reduce this accumulation with several precautions. First of all, instead of traditional and common irrigation practices, producers should adapt to the usage principles of this new water source. Otherwise, in addition to yield reductions due to climate change, yield reduction due to soil salinity will make the situation more critical. Despite possessing sufficient knowledge about using marginal waters, the long-term impacts of droughts on plants and soils remain incompletely understood. Additionally, there is a growing need for scientific criteria to determine the extent to which saline water (sea water) inflows, resulting from excessive aquifer use in coastal regions, can advance into inner areas. Preventing soil salinization and devising suitable crop patterns and irrigation management practices from coastal to inland regions are becoming increasingly critical, necessitating further research and thoughtful consideration.

References

1. Cosgrove WJ, Rijsberman FR. World water vision: making water everybody's business. London: Earthscan Publications Ltd.; 2000.
2. Allan JA. The middle east water question: hydropolitics and the global economy. London, UK: I.B. Tauris & Co., Ltd.; 2001. p. 382.
3. Qadir M, Sharma BR, Bruggeman A, Choukr-Allah R, Karajeh F. Non-conventional water resources and opportunities for water augmentation to achieve food security in water scarce countries. Agric Water Manag. 2007;87(1):2–22.
4. Anonymous. 2020. http://www.Thetower.Org/4305oc-İsrael-Recycles-90-Of-İts-Wastewater-Four-Times-More-Than-Any-Other-Country/.
5. Demir I, Kılıç G, Coşkun M. PRECIS bölgesel iklim modeli ile Türkiye için iklim öngörüleri: HADAMP3 sres A2 senaryosu, Iv. Atmosfer Bilimleri Sempozyumu, Bildiriler Kitabı; 2008. p. 365–373.
6. Voutchkov N. Desalination—water for the next generation. Filtr Sep. 2005;42(2):14–25.
7. Oweis TY, Hachum A, Bruggeman A. Indigenous water-harvesting systems in West Asia and North Africa. In: International center for agricultural research in the dry areas (ICARDA), Aleppo, Syria; 2004. p. 173.
8. Qadir M, Oster JD. Crop and irrigation management strategies for saline–sodic soils and waters aimed at environmentally sustainable agriculture. Sci Total Environ. 2004;323:1–19.
9. Ayers RS, Westcot DW. Water quality for agriculture. FAO Irrigation and Drainage Paper No: 29. 1989. p. 163.
10. Grattan SR, Rhoades JD. Irrigation with saline ground water and drainage water. In: Tanji KK, editor. Agricultural salinity assessment and management, vol. 71. ASCE; 1990, p. 432–449.
11. Kanber R, Kırda C, Tekinel O. Irrigation water quality and salinity problems in irrigation. Çukurova University. University publication No: 21. Adana/Turkey p. 277.
12. Elbana TA, Bakr N, Elbana M. Reuse of treated wastewater in Egypt: challenges and opportunities. In: Negm A, editor. Unconventional water resources and agriculture in Egypt, vol. 75. The Handbook of Environmental Chemistry; 2017. p. 429–453.
13. Rhoades JD, Kandiah A, Mashali AM. The use of saline waters for crop production, vol. 48. FAO; 1992. p. 133.
14. Miyamoto S, Moore J, Stickler C. Overwiev of saline water irrigation in far West Texas in: water today and tomorrow (unpublished); 1984.
15. Dutt GR, Pennington DA, Jr Turner F. Irrigation as solution to salinity problems of river basin. In: French FR, editor. Salinity in water courses and reservoirs. Michigan: Ann Arbor Science. 1984. p. 465–472.
16. Shrivastava P, Kumar R. Soil Salinity: A Serious environmental issue and plant growth promoting bacteria as one of the tools for its alleviation. Saudi J Biol Sci. 2015;22:123–31.
17. Szabolcs I. Soils and salinization. In: Pessarakli M, editor. Plant and crop stress, Part I. New York: Marcel Dekker, Inc. 1994. p. 3–13.
18. Ghassemi F, Jakeman AJ, Nix HA. Salinization of land and water resources. Human, causes, extent, management and case studies. Australia: Centre for Resource and Environmental Studies, The Australian National University. 1995. p. 576.
19. Maas EV. Salt tolerance of plants. Appl Agric Res. 1986;1:12–26.
20. Maas EV, Poss JA. Salt sensitivity of wheat at various growth stages. Irrig Sci. 1989;10:29–40.
21. Shalhevet J. Using water of marginal quality for crop production: major issues. Agric Water Manag. 1994;25:233–69.
22. Maas EV, Poss J, Hoffman GJ. Salinity sensitivity of sorghum at three growth stages. Irrig Sci. 1986;7:1–11.
23. Maas EV, Poss JA. Salt sensitivity of cowpea at various growth stages. Irrig Sci. I989b;10:313–320.
24. Kaddah T, Ghowal SI. Salinity effects on the growth of com at different stages of development. Agron J. 1964;56:214–217
25. Lunin J, Gallatin MH, Batchelder AR. Effect of stage of growth and time of salinization on growth and chemical composition of beans. I. Total salinization in one irrigation. Soil Sci. 1961;91:194–202.
26. Saba N, Haq I, Dahot MU, Hussain M. Responses of different cotton genotypes under saline stressed conditions. Pak J Biotechnol. 2010;7(1–2):51–55. ISSN. 1812-1837.
27. Bibi Z, Khan NU, Latif A, Khan MJ, Manyuan G, Niu Y, Khan QU, Khan MJ, Khan IU, Shaheen S, Sadoza GU. Salt tolerance in upland cotton genotypes to NaCl salinity at early growth stages. J Anim Plant Sci. 2016;26(3):766–75.
28. Qadir M, Shams M. Some agronomic and physiological aspects of salt tolerance in cotton (G. Hirsutum L.). J Agron Crop Sci. 1997;179:101–6.
29. Papiernik SK, Grieve CM, Lesch SM, Yates SR. Effects of salinity, imazethapyr, and chlorimuron application on soybean growth and yield. Commun Soil Sci Plan. 2005;36(7–8):951–67.
30. Bustingorri C, Lavado RS. Soybean growth under stable versus peak salinity. Sci Agric. 2011;68:102–18.
31. Lutts S, Kinet JM, Bouharmont J. Changes in plant response to NaCl during development of rice (*Oryza sativa* L.) varieties differing in salinity resistance. J Exp Bot. 1995;46(12):1843–1852.
32. Farooq S, Farooqe A. The use of cell membrane stability (cms) technique to screen for salt tolerant wheat varieties. J Plant Physiol. 2006;163(6):629–37.
33. Katerji N, Van Hoorn JW, Hamdy A, Mastrorilli M, Fares C, Ceccarelli S, Grando ST. Classification and

salt tolerance analysis of barley varieties. Agric Water Manag. 2006;85(1–2):184–92.
34. Gawad GA, Arslan A, Gaihbe A, Kadouri F. The effects of saline irrigation water management and salt tolerant tomato varieties on sustainable production of tomato in Syria (1999–2002). Agric Water Manag. 2005;78(1–2):39–53.
35. Gimeno V, Syvertsen JP, Nieves M, Simón I, MartÍnez V, Garcia-Sanchez F. The orange varieties as inter-stocks ıncrease the salt tolerance of lemon trees. J Hortic Sci Biotechnol. 2009;84(6):625–31.
36. Xing X, Kang D, Ma X. Differences in loam water retention and shrinkage behavior: effects of various types and concentrations of salt ions. Soil Tillage Res. 2017;167:61–72.
37. Minhas PS, Khosla BK. Leaching of salts as affected by the method of water application and atmospheric evaporativity under shallow and saline water-table conditions. Jour Agric Sci Camb. 1987;109:415–9.
38. Jiang H, Jinsong Y, Huijun Y. Effects of irrigation with saline water on soil salinity and crop yield. Plant Nutr Fertil Sci. 2003;10(6):599–603.
39. Ramakrishna A, Tam HM, Wani SP, Long TD. Effect of mulch on soil temperature, moisture, weed infestation and yield of groundnut in northern Vietnam. Field Crop Res. 2006;95(2–3):115–25.
40. Hamdy A. Saline irrigation management for a sustainable. In: Use advanced short course on sustainable use of non conventional water resources in the Mediterranean Region, Aleppo, Syria; 1998. p. 91–145
41. Oster JD. Irrigation with poor quality water. Agric Water Manag. 1994;25:271–97.
42. Patterson LA, Lutz B, Doyle MW. Climate and direct human contributions to changes in mean annual streamflow in the South Atlantic, USA. Water Resour Res. 2013;49:7278–91.
43. Yeo A. Predicting the interaction between the effects of salinity and climate change on crop plants. Sci Hortic. 1999;78:159–74.
44. FAO. The state of the world's land and water resources for food and agriculture—systems at breaking point. Synthesis report, Rome; 2021. p. 82.

Addressing Climate Change Effects on Freshwater Resources: Case Study on Büyükçekmece Lake Water Budget with Zero Liquid Discharge Prospective

Kemal Güneş, Mehmet Beşiktaş, Irfan Yolcubal, Azime Tezer, Ömer Visali Sarıkaya, Muhammet Burak Bozçelik, Cihangir Aydoner, Selma Ayaz, Mehmet Şimşek, and Danuta Leszczynska

Abstract

Given ongoing climate changes, the need to conserve existing water resources through water/wastewater reuse and reclamation is becoming urgent. The general concept, Zero Liquid Discharge (ZLD), is a realistic option to counteract the increased evapotranspiration and decreased precipitation caused by climate-change-driven temperature variations and overconsumption of original freshwater resources. The studied Büyükçekmece Lake Dam is a reservoir supplying about 17% of Istanbul's population with drinking water. The elements of the shifting water budget for the Büyükçekmece Lake Dam were estimated for the middle and the end of the twenty-first century and compared with the future projections of climate change models for the worst-case scenario. Outcomes from those models anticipated that seasonal temperature variations could surge from the current 2.0 °C to 3.2 °C in mid-century and 4.0 °C to 6.0 °C at the end of the century. According to this scenario, rainfall in Istanbul is predicted to decrease by 12% compared to 1986–2005 rainfall data. According to the 2017 report by the Istanbul Metropolitan Municipality (IBB), the average monthly precipitation is expected to decrease by approximately 5% in winter and up to 33% in summer by the end of the century. Even if there were no significant variations in the annual precipitation, the drop in annual runoff would reach 15% due to increased evapotranspiration. As a result, annual streamflow to the lake would drop about 22% in the mid-century and 56% at the end of the century compared to the period between 1968 and 2005. These estimates revealed that the water capacity of Büyükçekmece Lake Dam would drop to 90 million m^3 in the mid-century and to 45 million m^3 at the end of the century. One

K. Güneş (✉) · M. Beşiktaş · Ö. V. Sarıkaya · M. B. Bozçelik · C. Aydoner · S. Ayaz
Climate Change and Life Sciences, TUBITAK—Marmara Research Center (MRC), Gebze, Kocaeli, Türkiye
e-mail: kemalgunes.tr@gmail.com

I. Yolcubal
Faculty of Engineering, Department of Geology Engineering, Kocaeli University, Izmit, Kocaeli, Türkiye

A. Tezer
Faculty of Architecture, Department of Urban and Regional Planning, Istanbul Technical University, Taksim, Istanbul, Türkiye

M. Şimşek
Faculty of Engineering, Civil Engineering Department, Şırnak University, Şırnak, Türkiye

D. Leszczynska (✉)
Department of Civil and Environmental Engineering, Jackson State University, Jackson, MS, USA
e-mail: danuta.leszczynska@jsums.edu

I. Yolcubal
Faculty of Mines Department of Geology Engineering, Istanbul Technical University, Istanbul, Türkiye

M. Bahadir et al. (eds.), *Agricultural Water Management*, Water and Wastewater Management,
https://doi.org/10.1007/978-3-031-96134-2_16

of the solutions addressing incoming shortages is employing the LZD approach, which focuses on reusing/reclaiming already spent drinking water instead of over- drafting fresh resources.

Keywords

Büyükçekmece Lake · Climate change · Water budget · Water scarcity · Zero Liquid Discharge (ZLD) · Water reuse · Reclamation of treated wastewater

1 Introduction

Fresh water supplies are already under intense pressure because of increased demand for drinking water due to population growth, industrialization, greater demand for irrigation water to feed the expanding population, and changes in land use. In addition, water resources, which are already restricted, also appear to be more fragile due to climate change and temperature fluctuations, triggering irregularities in the amount and severity of rainfall in a given time period.

The RCP (Representative Concentration Pathways) scenarios describe the trajectories for carbon dioxide emissions and subsequent atmospheric concentration from 2000 to 2100. These scenarios are delivered from the RCP 2.6, RCP4.5, RCP 6.0, and RCP 8.5 models (Fig. 1). The RCP 8.5 expresses the worst-case scenarios, indicating that carbon dioxide emissions continue to rise rapidly through the early and mid-parts of the century.

According to various studies of climate scenarios, winter rainfall in the western provinces of Türkiye has been declining for the last 50 years as summer temperatures have increased [2]. Worldwide, the temperature increase is predicted to be 0.85 °C by 2030, 1.33 °C by 2050, and 2.48 °C by 2100 [3]. In addition, according to the Intergovernmental Panel on Climate Change (IPCC) data, human activities are estimated to cause global warming of about 1.0 °C, possibly in the range of 0.8 °C to 1.2 °C above the pre-industrial use levels. It is believed that global warming is likely to reach 1.5 °C between 2030 and 2052, if it continues to increase at the current pace [4].

To address the ongoing climate changes, various countries have opted to develop resilience against drought, highlighted as one of the outcomes of climate change, by considering unconventional water treatments, such as desalination and treating water in certain hotspot areas [5]. In a study conducted in China, the impacts of climate change on irrigation water for grain production were investigated. The study reported that while climate change's influence was moderate from 1960 to 1990, it intensified after 1990, leading to a significant increase of 100 billion m^3 (BCM) in additional irrigation water consumption [6]. According to a study conducted in Iran, assessing the impact of climate change on water resources using the Soil and Water Assessment Tool (SWAT) model and considering representative concentration pathway scenarios (RCP 2.6, RCP 6, RCP 8.5), it was predicted that shortly the surface runoff will increase in spring and fall while decreasing in winter and summer. Additionally, it was projected that evapotranspiration will be highest from 2070 to 2100, causing a decrease in the yield of rain-fed crops [7]. In nearby Pakistan, which shares a geographic proximity to Iran, an increase in streamflow was shown during summer and fall, while a decrease was expected in winter and spring [8]. Since climate change directly alters rainfall patterns, it will considerably affect crop yields in rain-fed irrigation across different world regions. For instance, projections using the "Coupled Models Inter-Comparison Project Phase 5 (CMIP5)" model in the Awash Basin of Ethiopia indicate that water stress affects all sectors in every season [9]. Given that the agriculture sector accounts for 70% of the annual consumption of freshwater globally, it is inevitable that this sector will be the most affected by climate change both presently and in the future. Various studies suggest that irrigation water demand may increase from 4% to 25% by 2050 [10].

Dams are reservoirs that primarily meet agricultural irrigation, industrial, and domestic water needs, while also possessing unique ecosystems. Therefore, the impacts of climate change directly

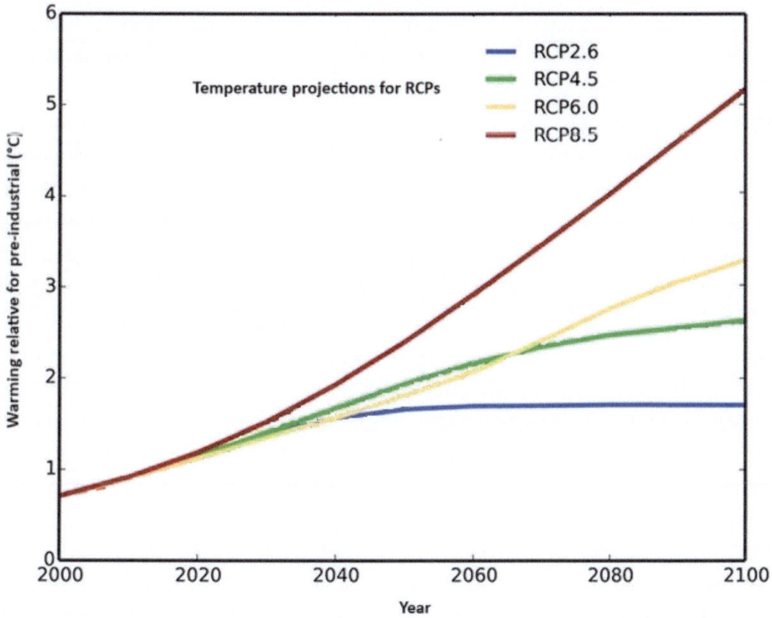

Fig. 1 Global mean temperature projections from a climate model (called MAGICC6) relative to a pre-industrial average (1850–1900) for RCP2.6, RCP4.5, RCP6.0 and RCP8.5 (Adapted from [1])

affect these structures, leading to increased challenges in managing economic, ecological, and environmental issues.

The presented study aims to determine the shifts in the lake water budget because of the ongoing rise in temperature and evapotranspiration and the decline in precipitation based on predicted climatic conditions of Büyükçekmece Lake. The water conservation strategy to counteract foreseeable climate-driven changes proposes the reclamation and reuse of already spent water, which will decrease withdrawal and, at the same time, decrease the load of pollutants entering the Lake with traditionally treated wastewater and not treated runoff from agriculture.

1.1 Global Key Indicators in Climate Change

According to NASA's latest data, when looking at certain climate change indicators on Earth, it is observed that CO_2 levels have reached 422 ppm, global temperatures have risen by 1.4 °C compared to the pre-industrial era, sea levels have increased by approximately 10.1 cm since 1993, ocean heat content has reached 360 zettajoules (10^{21} joules), and methane emissions have shown an upward trend at a level of 1922 ppm. On the other hand, the Arctic Sea ice minimum extent has decreased by 12.2% per decade since 1979, and ice sheets are reported to be losing 406 billion metric tons per year [11]. Similarly, according to findings obtained from the ERA 5 dataset within the Copernicus Climate Change Service (C3S), considering that 2023 was 1.48 °C above pre-industrial levels, it is reported that the annual temperature in 2024 will exceed 1.5 °C above pre-industrial levels and is likely to reach or surpass 1.55 °C [11].

1.2 Climate Change and Europe

A study on the direct and delayed effects of climate change reveals that Europe faced a severe summer drought in 2022 with significant socio-economic consequences. Central and Southern Europe experienced the largest water storage deficit since satellite observations began in 2002, marking the highest levels in the past sixty years. The study highlights that while the effects of global warming on drought continue, the risk of drought is expected to increase further in the future due to intensified human-induced warming [12].

Furthermore, in a study evaluating the current and future water availability in Europe on a regional basis, the countries at greater risk are Spain, Portugal, North Macedonia, Greece, Bulgaria, Albania, France, and Italy. Additionally, countries such as Slovakia, Ireland, Belgium, Luxembourg, Croatia, and Romania are identified as being at mild risk, while Northern Arctic countries like Sweden, Finland, Norway, and Russia exhibit strong water availability, with a slight increase also observed [13]. A study examining the major impacts of climate change on the agricultural sector in Europe's main biogeographical regions (Fig. 2) highlights that temperature increases will be significantly higher, particularly in countries bordering the Mediterranean [14]. Additionally, a decrease in rainfall and an increase in drought are expected. Therefore, it can be said that the Mediterranean basin, where agricultural production is concentrated, may face substantial losses in agricultural yield.

Although similar conditions are expected in Central Europe, but the anticipated risks are not as severe as those in the Mediterranean basin. In other parts of Europe, issues related to extreme rainfall, storms, and flooding emerge as the primary concerns.

1.3 Importance of Climate Change and Water in Türkiye

According to the report *The Impact of Climate Change on Water Resources Project*, which evaluates Türkiye's 25 water basins up to the year 2100, significant increases in average temperatures are expected. The rise in temperatures is predicted to exceed the anticipated levels by 1–2 °C in the Eastern and Southeastern regions of the country. The temperature increase will be more pronounced in the summer months, particularly in the Southern and Western regions. A general decline in total rainfall is projected, except for the Black Sea coast, where rainfall could increase by up to 150 mm. Rainfall decreases are expected to be most severe during the winter months. Significant water shortages are anticipated in the Euphrates-Tigris and Konya Closed Basins, while water resources are generally expected to remain sufficient in the Marmara, Susurluk, Northern Aegean, Western Mediterranean, Western Black Sea, Yeşilırmak, Antalya, Aras, and Lake Van basins [15].

As seen, except for one region, rainfall regimes in most other regions of Türkiye are projected to decline due to climate change. Climate change scenarios also emphasize a 25% reduction in

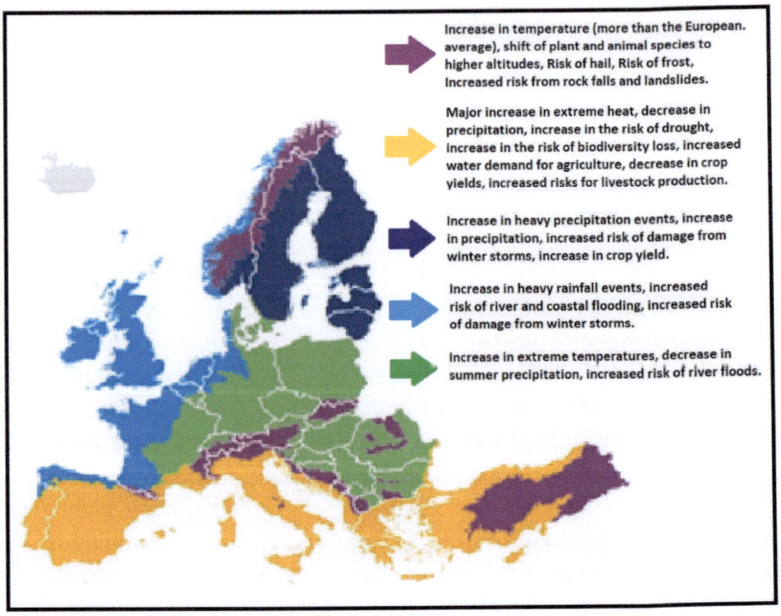

Fig. 2 Major climate change impacts on the agricultural sector for major biogeographic regions in Europe (Adapted from [14])

water resources across Türkiye by 2100 [16]. The decrease in water resources is of particular concern to the agricultural sector. Türkiye, which has a total surface area of 78 million ha, has 24 million ha of irrigable land. Technically and economically, irrigable agricultural land is 8.5 million ha, of which approximately 83.5% of the 8.5 million ha of agricultural land can be irrigated. Additionally, 77% of total water use in Türkiye is allocated to agriculture [17]. Therefore, the negative impacts of climate change on the agricultural sector are expected to intensify in the future. In this context, rain-fed agriculture is likely to be the most adversely affected by unfavorable climate conditions. As a result, reductions in crop yields and losses in biodiversity are among the other anticipated negative outcomes.

To build resilience against the impacts of climate change, improving efficiency in sectoral water use is fundamental. For countries like Türkiye, where the share of water used in agriculture is higher than in other sectors, maximizing efficiency in this sector is becoming increasingly essential. As a result, the construction and deployment of modern irrigation systems, the digitization of agriculture, vertical farming, and similar agricultural technologies are the primary measures to consider. Additionally, channeling treated domestic wastewater to different uses through advanced treatment technologies could help preserve renewable water resources. Separating greywater from domestic wastewater for specific treatment and reuse in toilet flushing, for instance, could save up to 60% of domestic water consumption. Rainwater harvesting and its use, particularly in urban areas, would contribute to both environmental protection and more efficient use of existing water resources. However, to achieve success in these areas, the effective implementation of regulations, guideline documents, and incentive methods is necessary.

In industry, the sectors with the highest water consumption are chemicals, textiles, and food, which account for 39.3%, 19.3%, and 14.5% of industrial water use, respectively [18]. Together, these three sectors consume approximately 73% of the total water used in industry. Therefore, incorporating circular economy tools into these sectors, as well as others, would make industrial water management more sustainable and climate-resilient.

1.4 Study Area

The study was conducted on Büyükçekmece Lake (Fig. 3), located in Istanbul, providing drinking water for 2,600,000 people. Büyükçekmece Lake also provides drinking water for several other districts, such as Büyükçekmece, Beylikduzu, Esenyurt, Avcılar, Başakşehır, Çatalca, and Silivri [19].

Büyükçekmece Lake Basin is approximately 63.2 ha, and most of the land consists of forest and agricultural areas. The drainage area of the lake covers a land of 620 km^2. The total deposition volume of Büyükçekmece Dam is 148,943,000 m^3. According to 2016 data, the lake's average depth was only about 5 m, and it is considered a shallow lake, meaning it is

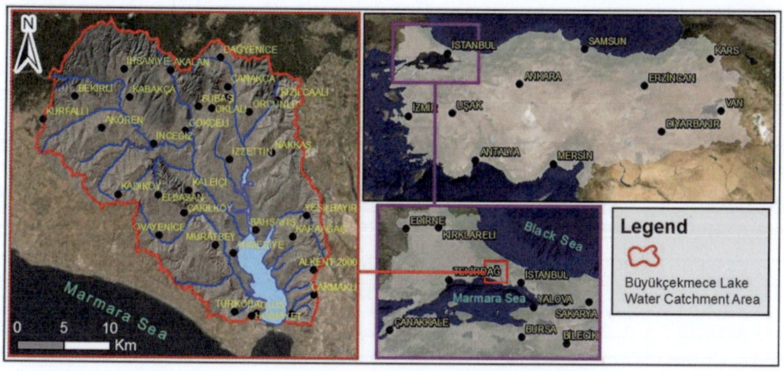

Fig. 3 The study area (Büyükçekmece Lake Basin)

more affected by drought. About 82.2 million m^3/year from the Büyükçekmece Lake Dam is withdrawn to feed the existing drinking water treatment plants. After the construction of Hamzalı Dam, a total of 23.90 million m^3/year of water will be released to Büyükçekmece Dam Lake, and from there, it will be treated at Büyükçekmece drinking water treatment plants and incorporated into the existing Istanbul drinking water systems.

2 Materials and Methods

The objective was to assess the changes in the water budget by considering the increases of temperature and evapotranspiration (ET), and decrease of precipitation under the climatic conditions of Büyükçekmece Lake, assuming no adjustments in current land use.

Long-term data from three meteorological stations (Table 1) around the basin was used to estimate the mean annual precipitation in the basin using the isohyetal method. Annual evapotranspiration was evaluated with the Turc method [20]. Mean annual runoff in the basin was calculated from long-term gauge data (1967–2005) of Hamzadere (Sarısu), Karasu, and Tahtaköprü (Çakıl) streams (Table 2), obtained from State Hydraulic Works of Türkiye [21].

3 Results and Discussion

Analysis of long-term meteorological data revealed that the mean annual temperature in the Büyükçekmece Lake watershed was 14.58 °C, and the mean air temperature in the basin varied from 6.6 °C in winter to 23.4 °C in summer.

Table 2 Stream gauge stations in Büyükçekmece Lake Basin

Stream gauge	Drainage area (km^2)
D02A021/Hamzadere stream (Bahsayıs)	143.00
D02A024/Tahtaköprü stream (Tepecik)	95.75
D02A015/Karasu stream (Incegiz)	174.90

Mean annual precipitation in the Büyükçekmece Lake basin was estimated to be 644.5 mm using the isohyetal method. The extent of rainfall in the basin increased toward the northwest of the Lake (Fig. 4). Mean evapotranspiration (ET) estimated from the Turc method was 385 mm/year. The highest amount (52-59 mm/month) of ET occurred from June through August, while the lowest ET (9.2–12 mm/month) occurred between December and February. The analysis found that evapotranspiration (ET) removes 59 % of annual precipitation in the basin.

The mean annual runoff in the basin was calculated to be 202 mm/year. The highest mean monthly runoff (34.9–44.0 mm/month) occurred between December and March, while the lowest mean runoff (0.8–1.5 mm/month) was observed from July to September. The runoff coefficient in the basin was 0.31, indicating that total runoff in the basin accounted for 31% of the annual precipitation. From the water budget calculation, the mean yearly infiltration in the basin was estimated to be 57.1 mm/year (Tables 3 and 4).

The Büyükçekmece Lake water budget, the mean annual inflow to the lake reservoir between 1968 and 2005, was estimated as 125.42 million m^3 [21]. Based on the mean annual precipitation (644.5 mm) and Class-A evaporation data (385

Table 1 Meteorological Station around Büyükçekmece Lake Basin (masl = meters above sea level)

Latitude	Longitude	Meteorological station	Elevation (masl, m)	Mean annual precipitation (mm)
41.07310 0°	28.239400°	Silivri MGI (1958–1986)	10	506.7
41.045300°	28.590000°	D02M011 (Büyükçekmece Dam) (1967–2015)	10	570.5
41.189638°	28.409818°	D02M039 (Çatalca MGM) (1964–2015)	170	723

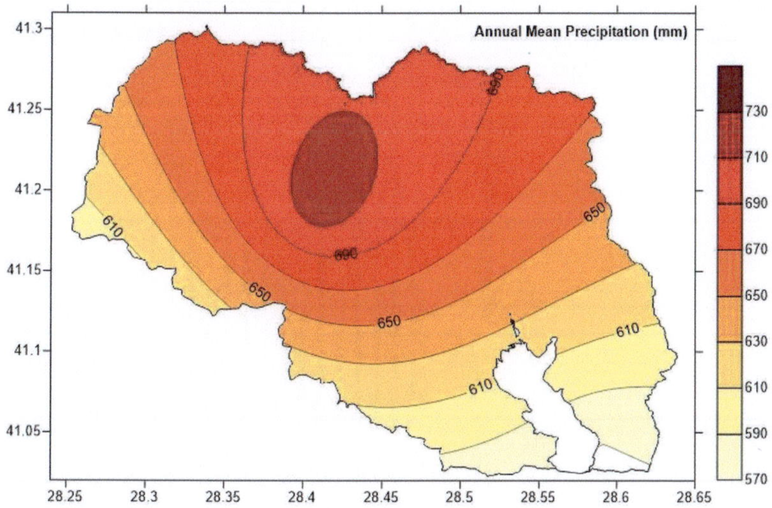

Fig. 4 Isohyetal map of mean annual precipitation in the Büyükçekmece Lake Basin

mm/year) obtained from the Büyükçekmece Dam meteorological station in the same period, the net natural outflow (E-P) from the lake surface was calculated as 6.42 million m^3. Therefore, the mean water capacity of Büyükçekmece Dam Lake was estimated to be 119 million m^3/year (Fig. 5).

The IPCC report 2018 stated that human activities were estimated to have caused about 1 °C of global warming above pre-industrial levels. Therefore, global warming is likely to reach 1.5 °C between 2030 and 2052 [4]. Because climate change is a process with multi-component effects, its actual effects may be less or more severe than anticipated. Although no clear consensus has yet been reached on the impact of climate change, the severity of these effects is likely to increase in the current century.

Different global climate models (CMIP5, CORDEX, World Clim) were applied at different local resolutions to estimate the change in temperature through the twenty-first century, considering various RCP scenarios that describe the trajectories for carbon dioxide emissions and resulting atmospheric concentration from 2000 to 2100. These scenarios were named RCP 2.6, RCP4.5, RCP 6.0, and RCP8.5. Based on RCP8.5 (worst case scenario), concentrations of CO_2 in the atmosphere will accelerate, reaching 950 ppm by 2100, and continue to increase for another 100 years. In comparison, by the RCP4.5 scenarios, the CO_2

Table 3 Mean monthly evapotranspiration in the Büyükçekmece Lake Basin, estimated based on the Turc method [20]

Months	I	II	III	IV	V	VI	VII	VIII	IX	X	XI	XII
Tmax (°C)	9	9.1	11.6	16.5	21.5	26.1	28.8	28.9	25.3	20.1	15	10.9
Tmin (°C)	3.7	3.2	4.8	8.4	12.9	17.3	20.1	20.5	16.9	13.1	8.8	5.6
Tmean (°C)	6.1	5.7	7.7	12	16.9	21.6	24.3	24.4	20.7	16.1	11.5	8
Rs (MJ/ $m^2.d^{-1}$	6.3	9.1	13.2	18.6	22.0	23.6	22.8	20.1	16.3	11.03	7.4	5.6
ET (mm/month)	9.2	11.8	19.8	35.0	48.7	57.7	58.7	52.3	40.6	25.8	15.7	10.3

T *mean*: mean air temperature, Rs: estimated solar radiation, MJ: Megajolule, d: day, ET: reference evapotranspiration

Table 4 Mean monthly runoff in Büyükçekmece Lake Basin (Modified after DSI, [20])

		II	III	IV	V	VI	VII	VIII	IX	X	XI	XII	Total
Runoff (mm/month)	44.0	40.0	37.0	18.4	8.2	3.4	1.1	0.8	1.5	5.3	7.9	34.9	202.3

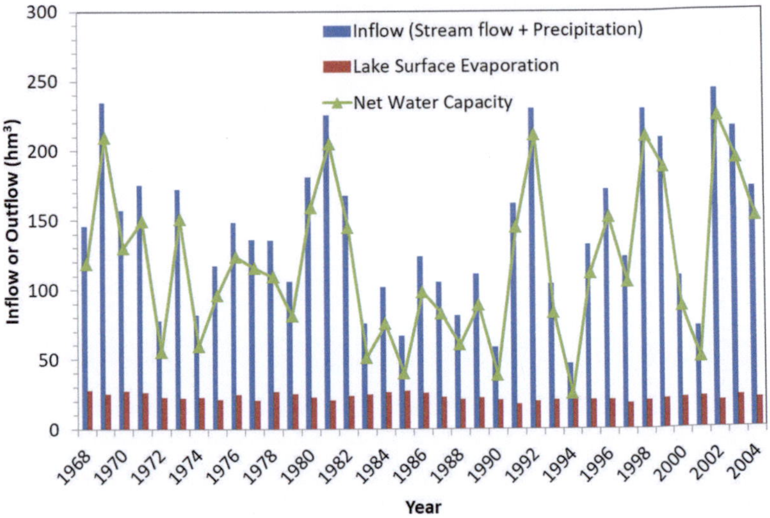

Fig. 5 Water budget elements of Büyükçekmece Lake Dam between 1968 and 2005

concentration will continue to rise to about 520 ppm in 2070 and will continue to increase but more slowly. The latest RCP scenarios are usually employed for local climate modeling studies.

The simulation results of climate change models in the IPCC report show that the mean annual air temperature in Istanbul will rise about 1°C at the end of the century (RCP 2.6, best case scenario), 2.5 °C (RCP 4.5) and 5 °C (RCP 8.5, the worst-case scenario) compared to mean air temperature between 1986 and 2005 [21]. According to the RCP 8.5, the seasonal variation in temperature increase ranged from 2 °C in January (from 1980 to 2012) to 3.2 °C at mid-century, and 4 °C to 6 °C at the end of the century. Simulation of climate models also indicated that annual precipitation variation in İstanbul will show a drop from 2.5 % (RCP4.5) to 12 % (RCP8.5) compared to the period between 1986 and 2005. The mean drop rate in monthly precipitation at the end of the century, based on the RCP 8.5 scenarios, would occur from about 5% in winter to about 33% in summer [22]. The climate change model (EC-Earth) used with higher local resolution (3 km) for Türkiye also showed that for the RCP 8.5 scenarios, the mean annual air temperature in Istanbul would increase by about 1.93 °C in mid-century to 3.71 °C at the end of the century as compared to the period between 1986 and 2005 [14]. Results from the EC-Earth climate change model also suggested that annual precipitation would drop from about 5.8% in the mid-century to 6.4% at the end of the century [22]. Incidentally, based on the RCP 8.5 scenarios, the annual evapotranspiration would increase from 4 to 11% toward the end of the century. In comparison, a yearly runoff would decrease by 19.5% and 30.6% in the mid-century and the end of the century, respectively. According to World Clim data, in the Büyükçekmece Basin, the annual precipitation is projected to decrease by 6.5% between 2041 and 2060, with this decline expected to continue, reaching up to 12% between 2061 and 2080 [22].

General outcomes from different climate change models for Istanbul suggested that—in the worst-case scenario—the negative impact of climate change on the limited water resources will begin to be felt from mid-century and substantially progress toward the end of the century. It is safe to assume that even if there would be no variation in the annual precipitation in Istanbul, a drop in yearly runoff would reach 15% due to an increase in evapotranspiration [22].

Interestingly, the analysis based on the data of Florya meteorological stations, located near the Büyükçekmece Lake Basin and containing more current data sets (1980 to 2015), provides results different from that what was predicted. It

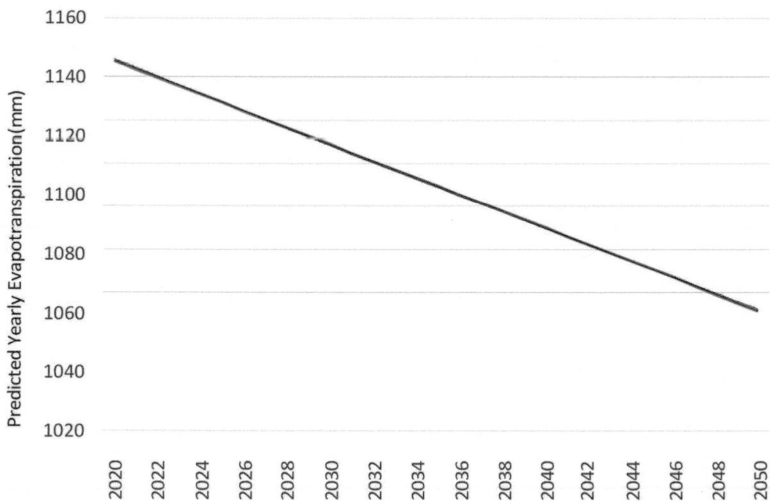

Fig. 6 Estimated ET future trend according to Florya meteorological station data (1980–2012)

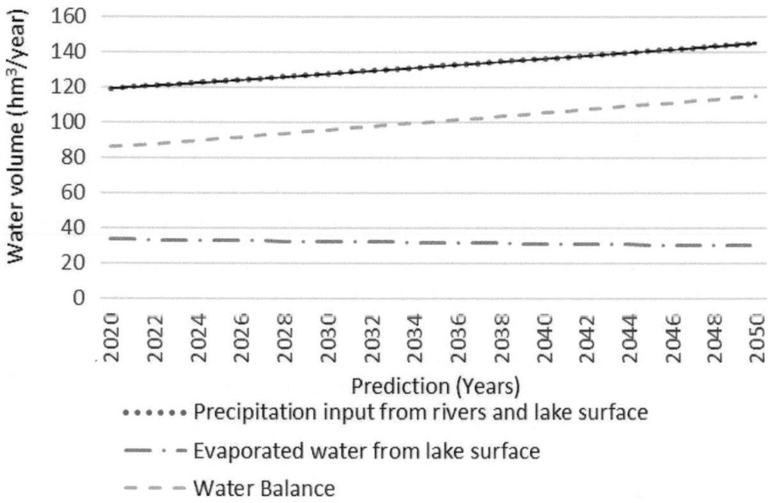

Fig. 7 Büyükçekmece Lake Dam's water budget trend regarding Florya meteorological station data

shows a relative decrease in evapotranspiration (ET) between 1980 and 2012 (Fig. 6).

Based on the predictive models, the ET value is expected to increase with temperature and solar radiation. Still, variation in rainfall has a significant influence. Conversely, the results from the analysis of the lake water budget suggest that if the current conditions continue, the water balance of Büyükçekmece Lake Dam between 2020 and 2050 will be positive (Fig. 7).

However, the positive water budget trend for the Büyükçekmece Lake Dam (based on the Florya meteorological data) will end after 2050 as the temperature is expected to increase further. Consequently, the adverse effects start to occur, as specified in the worst-case scenario models. The negative trends may begin as early as after 2035s.

The future projections of Büyükçekmece water budget elements based on the worst-scenario climate change model RCP 8.5 outcomes are shown in Table 5 for the current, mid-, and end-of-the-century data.

In the assessments of the lake water budget for the upcoming years, it was assumed that current land use remained the same. Future infiltration estimates were done using the coefficient of infiltration determined in this study. Annual runoff was projected by subtracting the sum of yearly evapotranspiration and infiltration from yearly precipitation in the basin. The annual net evaporation

Table 5 Future projection of Büyükçekmece water budget elements considering the outcomes of climate change models based on the RCP8.5 scenario

Parameters	1968–2005	Mid-century	End of century
Annual temperature (°C)	14.6	17.1	19.6
Precipitation (mm/year)	644.5	612.3	567.2
Evapotranspiration (mm/year)	385.0	400.4	427.4
Infiltration (mm/year)	57.1	54.2	50.3
Runoff (mm/year)	202.3	157.6	89.6
Inflow to lake (million m^3/year)	125.4	97.7	55.5
Net evaporation from Lake surface (million m^3/year)	6.4	8.1	10.8
Water capacity of the Lake (million m^3/year)	119.0	89.6	44.7

from the Lake Surface was assessed by subtracting the evaporation value from rainfall over the Lake Surface.

As seen in Table 5, annual streamflow to the lake is predicted to drop about 22% in the mid- century and 56% at the end of the century compared to 1968–2005. These estimates reveal that the water capacity of Büyükçekmece Lake would be about 90 million m^3 in the mid-century and 45 million m^3 at the end of the century. Considering the recent usage of Büyükçekmece Lake and future population growth in Istanbul, water deficiency in the lake water budget will be noticeable starting from the mid-century, even assuming no change in annual precipitation occurred in the basin. In addition, the drop in yearly streamflow would be 7.6% in mid- century to 21% at the end of the century compared to 1968–2005. It means that the annual water capacity of the lake would drop to 90 million m^3 at the end of the century due to an increase in yearly evapotranspiration, with a 5 °C rise in mean annual temperature. All these results suggested that climate change in the twenty-first[t] century would have a prominent impact on the limited water resources of Istanbul.

4 Summary of the Case Study Outcomes

The zones severely affected by climate change are diminishing freshwater resources in the future. These conditions should be considered not only in terms of drinking water scarcity but also in terms of the surrounding ecosystem. Büyükçekmece Lake Dam is a shallow lake that feeds mainly on surface water's inflows, one of the most fragile features of this lake regarding climate change. According to current climate scenarios, the response of the presented lake's budget to climate change in the future will be substantially affected not only by the current water availability. Also, a drop in the lake water capacity will make this water more polluted due to incoming loads of pollution discharged from agriculture and city activities. However, suppose the adjustments (starting today) in the watershed that feeds the lake are completed by 2035, the expected effects of water scarcity will decrease while maintaining its quality and increasing its usability.

4.1 Zero Liquid Discharge Approach for Freshwater Conservation

A city as populous as Istanbul needs more freshwater serving as a drinking water source. There are several immediate measures to enhance the resilience of the Büyükçekmece Dam reservoir to climate change, including (a) minimizing losses and leaks during water delivery from the reservoir to the ultimate consumer; (b) sustainable management of pollution from agricultural and livestock activities in the basin; (c) addressing points in the reservoir basin, where there may have future issues with wastewater management, even presently, there is lack of industry near the immediate basin; and (d) continuously conducting awareness campaigns emphasizing the importance of water management.

Conservation of dwindling water resources due to overuse and climate change could also be achieved through the further development, enhancement, and ultimate implementation of large-scale reusing and reclamation of treated wastewater produced by human activities. The tactic that has recently gained popularity is presently known as a Zero Liquid Discharge (ZLD) approach.

The concept of recycling previously used water has a very long history, dating to the early Bronze Age (3000 BC) in Greece and Crete. Over centuries, domestic wastewater was used for the reclamation of nutrients and reuse in aquaculture (fish ponds) or fertilizing farming fields, and later (twentiethcentury) for industrial applications [23]. The middle of the twentieth century was marked as the starting point of advanced wastewater reclamation, known as direct potable reuse (DPR), by augmenting potable water supplies in the city of Windhoek in Namibia (1968) [24]. The eighties of the twentieth century are recognized as the beginning of big reuse/reclamation projects implemented worldwide. However, the resistance of public opinion and the additional cost of advanced wastewater treatment before reuse kept the progress at a slower-than-deserved pace.

The term Zero Liquid Discharge (ZLD) was initially used for industrial wastewater treatment, as regulations prohibited wastewater release from particular industries. It started with power plants discharging water with increased salinity, which deteriorated receiving water bodies. One of the first cases (1970s) was the power plants discharging to the Colorado River [25]. Currently, the LZD approach is recognized and adapted as an umbrella for the general design of enhanced wastewater management. Depending on the needs of surrounding areas, it provides the custom-made reuse/reclamation of treated wastewater instead of releasing it back into the environment. The additional treatment is always tied to a particular reuse and required standards.

The full implementation of the LZD approach will allow (a) the conservation of freshwater resources, (b) the prevention of the additional load of pollutants discharging as stormwater runoff or treated wastewater to the water bodies, and (c) the immediate functional reuse of treated wastewater instead of simple release to the environment.

The local LZD approach would benefit and protect freshwater in the Büyükçekmece Lake by diverting the treated domestic wastewater from nearby plants to satisfy the needs of the neighboring agricultural irrigation. The immediate benefits will be bi-folded: (a) reducing an extensive withdrawal of fresh water from the lake for lone agricultural purposes, and (b) reducing the total cost of additional fertilizing of irrigated plants/crops, as treated wastewater contains abundant nutrients.

5 Conclusions

It is evident that the impact of climate change on shallow lakes, especially those like Büyükçekmece, will have highly dramatic consequences. In addition to the decrease in precipitation, its irregularity can also lead to adverse conditions. After a prolonged dry period, sudden heavy downpours are very likely to transport significant amounts of sediments into the lakes, causing a rapid reduction in lake volume as well as acute pollution. As is well known, freshwater resources are already crucial today, and considering the effects outlined in climate change scenarios, they will undoubtedly become even more vital in the future. The priority order of usage for these essential resources, as specified by laws, is as follows: human consumption (drinking water), ecosystem, agriculture, and industry. Although this is the priority order for water usage, the agricultural sector is undoubtedly the most critical sector that requires attention. The first reason for this is that the agricultural sector accounts for 70% of global total freshwater consumption, making it imperative to focus conservation efforts primarily on this sector. Secondly, as the global population continues to grow, the increasing demand for food supply will inevitably lead to a greater need for water in the future. This, in turn, highlights the necessity of using water more efficiently in the agricultural sector.

In summary, this study emphasizes the need for maximum attention to the general protection measures outlined below in basins where natural lakes and reservoirs serve as freshwater sources:

- Redefining agricultural production patterns based on water availability rather than specific crops, taking climate change scenarios into account, and ensuring efficient water use for this purpose,
- Avoiding urban planning decisions that would lead to population increases in existing residential areas,
- Preventing the establishment of any industrial sector within lake or reservoir basins,
- Refraining from activities that would reduce the effectiveness of ecosystem services,
- Ensuring that all waste is managed in accordance with relevant regulations and, if possible, transferred outside the basin.

Acknowledgements The authors would like to thank the TUBITAK Marmara Research Center and the General Directorate of the Istanbul Water and Sewerage Administration for their essential contribution to the study.

References

1. UK Department for Environment Food & Rural Affairs, Department for Business, Energy & Industrial Strategy, Met Office Hadley Centre, Environment Agency. UKCP18 Guidance: Representative Concentration Pathways. 2019. https://www.metoffice.gov.uk/binaries/content/assets/metofficegovuk/pdf/research/ukcp/ukcp18-guidance---representative-concentration-pathways.pdf
2. Dalfes N, Karaca M, Sen OL. Climate change scenarios for Turkey: preliminary studies. United Nations Development Programme, Climate Change & Turkey. Effects, Sectoral Analysis, Socio-Economic Dimensions. UNDP. 2007.
3. Harmancioğlu NB, Ozkul S, Fistikoglu O, Barbaros F, Onusluel G, Cetinkaya CP, Dalkilic Y. Possible effects of climate change on the Buyukmenderes and Gediz River basins. United Nations Development Programme, Climate Change & Turkey. Effects, Sectoral Analysis, Socio-Economic Dimensions. UNDP. 2007.
4. IPCC. Summary for Policymakers. In: Global warming of 1.5 °C. 2018.
5. Moroto A-F, Olcina J, Hernandez M. The use of non-conventional water resources as a means of adaptation to drought and climate change in semi-arid regions: South-Eastern Spain. Water. 2019;11:93. https://doi.org/10.3390/w11010093
6. Lu S, Bai X, Li W, Wang N. Impacts of climate change on water resources and grain production. Technol Forecast Soc Chang. 2019;143:76–84. https://doi.org/10.1016/j.techfore.2019.01.015.
7. Shahvari N, Khalilian N, Mosavi SH, Mortazavi SA. Assessing climate change impacts on water resources and crop yield: a Varamin plain basin, Iran case study. Environ Monit Assess. 2019;191:134. https://doi.org/10.1007/s10661-019-7266-x.
8. Mahmood R, Jia S, Babel MS. Potential impacts of climate change on water resources in the Kunhar River Basin, Pakistan. Water. 2016;8:23. https://doi.org/10.3390/w8010023.
9. Taye MT, Dyer E, Hirpa FA, Charles K. Climate change impact on water resources in the awash basin, Ethiopia. Water. 2018;10:1560. https://doi.org/10.3390/w10111560.
10. Xing-Guo M, Shi H, Hui LZ, Xia LS, Jun X. Impacts of climate change on agricultural water resources and adaptation on the North China Plain. Adv Clim Chang Res. 2017;8:93–8. https://doi.org/10.1016/j.accre.2017.05.007.
11. https://science.nasa.gov/climate-change/
12. Bevacqua E, Rakovec O, Schumacher DL, Kumar R, Thober S, Samaniego L, Seneviratne SI, Zscheischler J. Direct and lagged climate change effects intensified the 2022 European drought. Nat Geosci. 2024;17:1100–7.
13. Garrote L, Iglesias A, Granados A. Country-level assessment of future risk of water scarcity in Europe. Proc IAHS. 2018;379:455–62. https://doi.org/10.5194/piahs-379-455-2018.
14. EEA Report/No 04/2019. Climate change adaptation in the agriculture sector in Europe.
15. SYGM. Water Efficiency Strategy Document and Action Plan in the Changing Climate Framework (2023 – 2033). Ministry of Agriculture and Forestry, Türkiye. 2023.
16. SYGM. (2016). Climate Change Impact on Water Resources Project. Ministry of Agriculture and Forestry, Türkiye.
17. DSI, 2023. DSI 2023 Activity Report. Ministry of Agriculture and Forestry, Türkiye.
18. Kavuncu B, Ekmen E, Yaman Ö, Yazan SY, Kanmaz N, Ünver Ü. Industrial water consumption treatment in Türkiye. İleri Mühendislik Çalışmaları ve Teknolojileri Dergisi. 2022;3(1);19–33 (Turkish).
19. ISKI (Istanbul Water and Sewerage Administration) (2016). 1/100000 Scale Environmental Layout Plan and Report (2009) of Istanbul Metropolitan Municipality.

20. Turc L. Water requirements assessment of irrigation, potential evapotranspiration: simplified and updated climatic formula. Ann Agronom. 1961;12:13–49.
21. DSİ. Revised Work plan report of Istanbul Dams. DSI XIV Bolge Mudurlugu (in Turkish); 2005. p. 122.
22. IBB. Istanbul Climate Change Action Plan: Climate Scenarios (In Turkish), Istanbul Metropolitan Municipality, Environmental Protection and Control Department; 2017. p. 146.
23. Water Reuse Europe. https://www.water-reuse-europe.org/
24. Haarhoff J, Van der Merve B. Twenty-five years of wastewater reclamation in Windhoek, Namibia. Water Sci Technol. 1996;33(10–11):25–35. www.sciencedirect.com/science/article/abs/pii/0273122396004039
25. Tong T, Elimelech M. The global rise of zero liquid discharge for wastewater management: drivers, technology and future directions. Environ Sci Technol. 2016; 6846–6855. https://doi.org/10.1021/acs.est.6b01000.

Water Policies and Management—An Outlook

Cost–Benefit Analysis of Subsidization of Water-Efficient On-Farm Irrigation Systems in the Northern Jordan Valley, Jordan

Koichi Noguchi, Kevin Wheeler, and Abbas Al-Omari

Abstract

Achieving cost recovery is a major challenge in subsidizing water-efficient irrigation systems. This study examined the conditions, under which the Jordan Valley Authority in Jordan can recover subsidization costs by raising irrigation water tariffs while maintaining or enhancing farmers' profit. Crop model simulations using the FAO-Aquacrop software, coupled with cost–benefit analysis, showed the degree to which irrigation system improvements enhance farmers' yield and profit. The analysis showed that replacing sprinkler and furrow irrigation systems with water-efficient systems, such as surface drip irrigation and subsurface drip irrigation, improves yield for lemons, tomatoes, and wheat. In addition, the analysis showed that replacing sprinkler irrigation with subsurface drip irrigation allowed for the largest yield gain. It was also shown that the additional yield would enable lemon and tomato farmers to pay higher tariffs, allowing full cost recovery for the irrigation system improvements. For example, in the simulation scenario, where sprinkler irrigation is replaced with surface drip irrigation for tomato production, it was shown that the current water tariff of 0.024 JOD/m^3 can be increased up to 1.62 JOD/m^3, as a consequence of increased yield without reducing farmers' profit (1 JOD = 1.41 USD). For wheat, however, there was no tariff that enables full cost recovery while maintaining farmers' profit. To actualize the possible tariff, increase for lemon and tomato farmers, strictly regulating water extraction among farmers and providing them with sufficient time for adaptation will be necessary.

Keywords

Aquacrop · Cost–benefit analysis · Irrigation water tariff · Irrigation development · Jordan

K. Noguchi (✉)
School of Geography and the Environment, University of Oxford, Oxford, UK
e-mail: kch.noguchi@gmail.com

PwC Consulting LLC, Chiyoda City, Tokyo, Japan

K. Wheeler
Environmental Change Institute, University of Oxford, Oxford, UK

Water Balance Consulting LLC, Boulder, CO, USA

A. Al-Omari
Water, Energy and Environment Center, University of Jordan, Amman, Jordan

M. Bahadir et al. (eds.), *Agricultural Water Management*, Water and Wastewater Management,
https://doi.org/10.1007/978-3-031-96134-2_17

1 Introduction

1.1 Background

The Hashemite Kingdom of Jordan (hereinafter Jordan) has been battling one of the most severe water scarcity situations in the world [1, 2]. In addition to the high potential evaporation rate (e.g., 2000 mm/year near the Red Sea), 92% of Jordan's land receives an annual average precipitation of <200 mm [3, 4]. The water scarcity is predicted to deteriorate; climate change studies of the region project an increase in the potential evapotranspiration rate by 2% and a decrease in the mean annual precipitation by a maximum of 25% by 2100 [3, 5]. The natural and immigration-induced population growth has further accentuated this extreme water scarcity, as Jordan's population increased from 0.9 million in 1960 to 11.2 million in 2022 [6]. Migration from neighboring countries and rural areas to cities has intensified water scarcity, as urban residents consume more utility water [1, 7].

There is a common belief that one of the keys to tackling scarcity of high-quality water in Jordan is irrigation efficiency enhancement in the Jordan Valley (JV) [2]. Over 50% of Jordan's total water demand is for agriculture [8], and approximately 70% of the domestic fruit and vegetable products are produced in the JV [4, 9]. The climatic and geological conditions vary significantly across three geographic regions in the JV: the northern, middle, and southern JV. The northern JV (NJV) receives the benefit of relatively high precipitation, mild temperature, and fertile soil [7]. Southern JV is not naturally favorable for agricultural production [10, 11].

Improving irrigation efficiency requires a shift from inefficient equipment—such as sprinkler and surface irrigation—to water-efficient equipment such as drip irrigation. One challenge of expanding drip irrigation is its capital intensity [12, 13]. Overcoming this challenge requires the Jordan Valley Authority (JVA), the government agency in charge of agricultural water distribution, to subsidize irrigation system improvements. The JVA has historically subsidized irrigation in the JV to support food production associated with population growth. However, the insufficient cost recovery has undermined the financial sustainability of the subsidization; the operating expenses (OpEx) and full cost recovery rate are estimated at 60% [14] and below 20% [4], respectively. These figures demonstrate that a tariff raise is inevitable for the JVA's financial sustainability. This chapter investigates whether such a tariff raise is feasible from the perspectives of both the JVA and farming's profitability; that is, it explores whether a tariff raise that enables the cost recovery for subsidizing water-efficient irrigation systems can be introduced without diminishing farmers' profit.

1.2 Tariff Raise and Its Potential Impacts in the JV

The feasibility of a tariff raise in the JV depends upon two elements: the deployed irrigation method and quality of water provided. A cost–benefit analysis by Tabieh et al. [11] deploying the residual imputation method showed that farmers sprinkler- and drip-irrigating their farmland in the JV can pay of 0.071 and 0.843 JOD/m^3, respectively (1 JOD = 1.41 USD). Furthermore, the farmers are paying up to 0.30 JOD/m^3 for pumping costs of groundwater, implying their willingness to pay more for water provision [15]. Positive incentives such as improved quality of water provided would make it easier for the farmers to accept higher tariffs [12, 14].

Farmers affected by the tariff raise in the JV are likely to take some of the following coping measures, potentially resulting in water-saving [15, 16]: (i) modification of cropping patterns [17], (ii) cropped area change [11, 16], (iii) lease of land [10, 13], (iv) improvement of irrigation equipment and management [13], and (v) adoption of water-efficient irrigation systems [13]. Increase in tariff would affect farmers disproportionately; farmers growing crops with high water requirement would lose their profit the most significantly [12, 14, 18]. In fact, studies strongly suggest tariff raise would limit crops that can be grown feasibly in the JV [16, 17]. However, it should be noted that these studies as well as Tabieh et al. [11] assume that farmers will continue cultivating their land as long as their net profit is positive. This assumption may not be held, as farmers can also monetize their land by leasing it [10, 13]. These predicted farmers' behaviors may lead to water-saving. In fact, it is estimated that a tariff raises to 0.25 JOD/m^3 would lead to a reduction in cultivated areas and thus a reduction in water consumption [16].

However, the abovementioned water-saving may need to be accompanied by additional policy and/or regulatory instruments [10, 13, 19]. The economic incentive to reduce water use may be small unless water can be stored or traded [13]. The financial and logistical burdens of installing advanced irrigation systems and developing new marketing channels for less thirsty crops also hinder cropping pattern change and agricultural water conservation [10]. Unstable water supply may prompt farmers to exhaust available irrigation water regardless of the financial disincentive [19].

Furthermore, several studies have challenged the idea of increasing irrigation efficiency to tackle water scarcity in recent years [20–22]; paradoxically, enhancing irrigation efficiency does not necessarily increase available water on a basin scale. For instance, Ward and Pulido-Velazquez [20] reported that in the Upper Rio Grande Basin, where subsidies promoted adoption of drip irrigation, disseminating drip irrigation enabled higher yields but diminished downstream flows due to reduced return flows. Grafton et al. [21] offered the conceptual explanation for this paradoxical phenomenon. They maintain that higher irrigation efficiency is achieved by reducing evaporation and/or deep drainage. While deep drainage is often treated as non-beneficial water consumption, a portion of deep drainage can be used by other irrigators, when it recharges groundwater or emerges as surface runoff. Hence, increasing irrigation efficiency by reducing deep drainage does not necessarily contribute to water use reduction on a basin scale. Advanced irrigation technologies are considered to enhance irrigation efficiency by suppressing evaporation, which is considered 'unproductive consumption' [22]. Thus, one can argue that introducing this type of water-efficient irrigation system improves basin-scale water balance.

1.3 Research Question and Contribution

There are a few existing studies that have investigated the effect of water pricing on farmers' profits and behaviours (e.g., Molle et al. [13]). These studies investigated how a higher tariff may prompt farmers to take adaptation measures such as adoption of water-efficient irrigation technologies. This article further argues that introducing such technologies may not only cause reduction in water use but also increase farming's profitability, which may in turn enable higher tariffs and thus improve their cost recovery without compromising farmers' profit. This hypothesis was explored by investigating the extent to and conditions, under which the JVA can recover subsidization cost for water-efficient irrigation systems by increasing the water tariff while maintaining profit of farmers, by conducting the cost–benefit analysis coupled with crop model simulations. The effect of the types of in-farm irrigation systems on yields was simulated first. The JVA's and farmers' net profit was then assessed under various irrigation development and tariff scenarios, using the predicted yields, together with the estimated cost of such development.

This study's contribution to agricultural water management literature is threefold. Firstly, it provides further guidance on tariff setting in the JV by identifying feasible tariffs for both farmers and the JVA. Secondly, it sheds light on the effect of improving on-farm irrigation systems on irrigation water tariffs, which existing literature has not thoroughly addressed and will aid in the examination of future irrigation development projects. Thirdly, it provides a versatile cost–benefit analysis framework for on-farm irrigation development.

2 Materials and Methods

This study was carried out in the following two steps. Firstly, the crop production simulation was conducted using the FAO-Aquacrop software (hereinafter Aquacrop) [23] under different irrigation systems to reveal how the on-farm irrigation system type influences the produced biomass and yield per hectare. The obtained yields were then converted to fresh yields using crop water content. Secondly, the cost–benefit analysis of introducing water-efficient irrigation systems was conducted using capital expenditure (CapEx) data obtained from literature and the calculated fresh yields. The cost–benefit analysis was conducted under a range of tariff rates for sensitivity analysis to reveal the tariff rates that satisfy both the JVA's cost recovery and farmers' profit requirements.

2.1 FAO-Aquacrop Crop Production Simulation

Lemons, tomatoes, and wheat were selected for cultivation simulation. These crops have the largest cultivated land area in the NJV in each crop category (fruits, vegetables, and field crops) [24] and data availability for the analysis. The Aquacrop model was applied, which allows for the robust crop production simulation based on water balance with minimal data requirements [25]. In Aquacrop, the yield is calculated based on the yield response factor and evapotranspiration. These values are modified according to the soil and atmospheric conditions, and crop characteristics. Aquacrop's performance under the Mediterranean climate has been proven for citrus [26, 27], tomatoes [28, 29], and wheat [30, 31].

Aquacrop requires appropriate crop characteristics, climate dataset, and initial conditions for simulation. Crop characteristics of lemons, tomatoes, and wheat were obtained from agronomy literature (Table 1 in Supplementary Materials). Data obtained in the JV were used for most of the simulation, however, where such data were absent, data typical in Mediterranean climate conditions were adopted. Climate data recorded in the weather station in Irbid were assigned throughout the simulation (Table 2 in Supplementary Materials). The NJV and Irbid are in the same climatic zone [3]. Instead of values of a specific year, long-term average values between 1970 and 2018 were used to limit the influence of interannual variability. Influence of soil fertility stress and salinity was considered negligible as the soil is relatively fertile and the high precipitation prevents salinization in the NJV [7]. For the initial condition of the soil profile, it was assumed that all water in macropores has drained away due to gravity flow (i.e., the soil profile is initially at the field capacity).

Allocated irrigation water quota in the JV varies by crop category and season (Table 3 in Supplementary Materials). However, farmer surveys by the World Bank [14] revealed that actual water allocation falls short of these quotas. To take this deficiency into account, it was assumed that farmers have access to 80% of their allocated water [14, 19]. Irrigation was set to occur weekly, starting on the first day of the simulation.

In Aquacrop, the effect of irrigation methods on the water balance is incorporated by modifying a fraction of the wetted ground surface area; a total of 100, 80, 30, and 0% of the surface area was used for sprinkler irrigation, furrow irrigation, surface drip irrigation (SDI), subsurface drip irrigation (SSDI) [32]. In the JV, 75, 24%, and <1% of irrigated land are drip-, surface-, and sprinkler-irrigated, respectively [11]. SSDI is rarely seen in Jordan, however, it was included in the analysis because of its high water-use efficiency [33, 34]. Though production factors such as wind speed and relative humidity were left out, the data types used in this study provide a reasonable approximation collectively [23]. Further calculation procedures of Aquacrop can be found in Sect. 6.1 in Supplementary Materials.

Table 1 Crop characteristics data for the Aquacrop simulation

Data type	Values		Data sources
Reference harvest index (HI) (%)	Lemon	50	Raes et al. [32][a]
	Tomato	60	Raes et al. [32][a]
	Wheat	50	Raes et al. [32][a]
Total length of crop cycle (days)	Lemon	334	Raes et al. [32][a]
	Tomato	137	AL-Zu'bi and Al-Kharabsheh [49]
	Wheat	240	Raes et al. [32][a,b]
Planting date	Lemon	January 1st	AL-Zu'bi and Al-Kharabsheh [49]
	Tomato	March 1st	AL-Zu'bi and Al-Kharabsheh [49]
	Wheat	November 1st	Mithen et al. [45]
$L_{ini} + L_{dev}$ (Number of days required for canopy coverage to reach the maximum value of 99%, counting from the planting date) (days)	Lemon	120	AL-Zu'bi and Al-Kharabsheh [49]
	Tomato	71	AL-Zu'bi and Al-Kharabsheh [49]
	Wheat	170	Raes et al. [32][a,b]
L_{mid} (Number of days required for canopy senescence to begin, counting from the end of L_{dev}) (days)	Lemon	123	AL-Zu'bi and Al-Kharabsheh [49]
	Tomato	41	AL-Zu'bi and Al-Kharabsheh [49]
	Wheat	40	Raes et al. [32][a,b]
L_{late} (Number of days required for the crop to be ready for harvest, counting from the end of L_{mid}) (days)	Lemon	91	AL-Zu'bi and Al-Kharabsheh [49]
	Tomato	25	AL-Zu'bi and Al-Kharabsheh [49]
	Wheat	30	Raes et al. [32][a,b]
Seeding rate (plants/ha)	Lemon	2000	Abu-Awwad [42][c]
	Tomato	17,600	Al-Qerem et al. [44]
	Wheat	1,500,000	Al-Sayaydeh et al. [51][d,e]
Canopy size when transplanted or planted (cm²/plant)	Lemon	50	Raes et al. [32][a]
	Tomato	10	Raes et al. [32][a]
	Wheat	5	Al-Sayaydeh et al. [51][e]

[a]General values under the Mediterranean climate were used
[b]Values for winter wheat under the Mediterranean climate was used
[c]While Abu-Awwad [42] reported the seeding rate for lemon trees in the Jordan Valley to be 400 plants/ha, the minimum seeding rate on Aquacrop is set at 2000 plants/ha. It was judged that this modification does not significantly influence the simulation result as the maximum canopy coverage was capped at 99%. This was justified as the predicted yield was in good agreement with the yield observed by Abu-Awwad [42]
[d]Barley and wheat were considered to be planted at the same seeding rate
[e]Following Al-Sayaydeh et al. [51], it was assumed that the seeds are sown so that the plant density of 1,500,000 plants/ha is achieved. Hence, the emergence rate was assumed to be 100%. When seeds are directly sown, Aquacrop automatically sets the initial canopy size of 5 cm^2/plant when planted

Aquacrop returns dry biomass and dry yield as outputs. Previous economics studies that utilized Aquacrop (e.g., D'Odorico et al. [35]) calculated crops' economic values using dry yield. However, such calculation may underestimate crops' economic values significantly, because some crops have high water content, and their market prices are set based on fresh yields' weight [36]. Dry yields predicted by the Aquacrop simulation were hence converted to fresh yields using typical water contents of each crop (Table 4 in Supplementary Materials).

Table 2 Climate data used for the Aquacrop simulation

Data types	Descriptions	Data sources
Monthly precipitation (mm)	Average monthly precipitation recorded in Irbid (north Jordan) between 1989 and 2018	JMD [52]
Reference daily evapotranspiration (mm)	Average daily pan evaporation recorded in the King Hussein Nursery station in Irbid (north Jordan) between 1970 and 2002	Al Mahamid [53]
Maximum temperature (°C)	Average monthly maximum temperature recorded in Irbid (north Jordan) between 1989 and 2018	JMD [52]
Minimum temperature (°C)	Average monthly minimum temperature recorded in Irbid (north Jordan) between 1989 and 2018	JMD [52]
Soil type	Clay	MWI[a]
CO_2 concentration	Representative Concentration Pathway (RCP) 4.5	Van Vuuren et al. [54]

[a]Soil data were obtained from the Ministry of Water and Irrigation of the Hashemite Kingdom of Jordan by one of the authors

Table 3 Allocated water quotas for each crop category [12] (a) and adjusted water quotas used in the Aquacrop simulation (b) for each crop

(a)

Periods	Quotas (m³/ha/day)		
	Vegetables	Citrus	Banana
December 16th to March 15th	10	On-demand but ≤20	On-demand but ≤20
March 16th to March 31st	15	On-demand but ≤20	On-demand but ≤20
April 1st to April 15th	15	20	30
April 16th to April 30th	20	20	30
May 1st to June 15th	20	30	50
June 16th to August 15th	On-demand but ≤10	40	70
August 16th to September 15th	10	40	70
September 16th to October 15th	15	30	50
October 16th to October 31st	20	30	50
November 1st to December 15th	20	On-demand but ≤20	On-demand but ≤20

(b)

Periods	Adjusted quotas (mm/week)		
	Tomato	Citrus	Barley and wheat
December 16th to March 15th	6	11	11
March 16th to March 31st	8	11	11
April 1st to April 15th	8	11	17
April 16th to April 30th	11	11	17
May 1st to June 15th	11	17	28
June 16th to August 15th	6	22	39
August 16th to September 15th	6	22	39
September 16th to October 15th	8	17	28
October 16th to October 31st	11	17	28
November 1st to December 15th	11	11	11

Table 4 Typical water contents of lemons, tomatoes and wheat

Crop types	Water contents (%)	Data sources
Lemons	83	Furr and Taylor [58]
Tomatoes	94	Davies and Hobson [59]
Wheat	5	Nedeva and Nikolova [60]

2.2 Cost–Benefit Analysis and Sensitivity Analysis

Sensitivity analysis of the irrigation water tariff was conducted using the outputs from the Aquacrop simulation. The irrigation water tariff varied between 0.04 and 0.40 JOD/m^3. While this tariff range is significantly higher than the current rate (=0.024 JOD/m^3), studies suggest that the current tariff is underrepresenting the cost and value of irrigation water [16]. In fact, several studies investigated similar tariff ranges (e.g., [17, 18]).

A cost–benefit analysis was conducted by calculating the net present value (NPV) of irrigation development projects under each tariff. NPV was calculated as follows:

$$\mathrm{NPV} = \sum_{t=0}^{n} \frac{B_t}{(1+r)^t} - \sum_{t=0}^{n} \frac{C_t}{(1+r)^t} \quad (1)$$

where t, n, B, C, and r denote year, the number of years in the time horizon, expected future benefits, expected associated costs, and discount rate, respectively. A typical discount rate of 6% [37] was applied over a time horizon of 25 years (i.e., $r = 0.06$, $n = 25$). The time horizon was determined based on the reported lifetime of irrigation systems in the JV [14].

The benefits and costs were calculated using the residual imputation method, which is widely used in irrigation projects' valuation [38]. In this method, the value of all produced outputs is first calculated. The cost of all production inputs except one, the value of which is to be determined, is then deducted from the sum of the outputs' value. The remaining value equals the value of the production input. Considering value of water as a production input and following the formulation of Just et al. [39], the benefits and costs are described as follows:

$$\mathrm{TR} = B + C + B_{\mathrm{qr}} + B_{\mathrm{nw}} \quad (2)$$

where TR, B, C, B_{qr}, and B_{nw} denote, respectively, the total revenue, benefit, cost, value deriving from the limitedness of supply, and non-water related value exceeding the minimum production cost. Here, enhancing the efficiency of on-farm irrigation systems increases yield and hence B by improving the production condition. Assuming the irrigation system does not influence B_{qr} and B_{nw}, inserting the yields calculated by Aquacrop and irrigation water tariffs into Eq. (2) provides benefits and costs for farmers. Inserting the tariffs and CapEx data provide benefits and costs for the JVA.

2.3 JVA's and Farmers' Costs and Benefits

The JVA's cost comprises operating expense (OpEx) and capital expense (CapEx). OpEx of irrigation systems depends on four factors [38]: (i) the amount of water pumped, (ii) the cropping system, (iii) each year's growing conditions, and (iv) the type of crop. In this study, the amount of applied water was consistent across all scenarios, and the results were only compared among the same crop and cropping system. Annual cultivation conditions did not differ as the simulation duration was limited to <1 year. Thus, OpEx was assumed to be constant for all the irrigation systems. The CapEx for each irrigation system was obtained from literature (Table 5 in Supplementary Materials). The values reported in USD were converted to JOD by the rate 1 USD = 0.707 JOD. All the values were adjusted to the values

Table 5 CapEx of each irrigation system by crop type

Irrigation type	Crop type	Reported CapEx (USD/ha)	Source	Surveyed year	Converted CapEx (JOD/ha)
Sprinkler	Lemon	1429	FAO [61]	2004	3468
	Tomato	1429	FAO [61]	2004	3468
	Wheat	1429	FAO [61]	2004	3468
Furrow	Lemon	537.5	Venot et al. [12][a]	2003	1348
	Tomato	880	Venot et al. [12][a]	2003	2208
	Wheat	880	Venot et al. [12][a]	2003	2208
Surface drip	Lemon	1075	Venot et al. [12]	2003	2697
	Tomato	1760	Venot et al. [12]	2003	4415
	Wheat	1760	Venot et al. [12]	2003	4415
SDI	Lemon	2057	Amosson et al. [37]	2002	5243
	Tomato	2057	Amosson et al. [37]	2002	5243
	Wheat	2057	Amosson et al. [37]	2002	5243

[a]Following World Bank [14], the cost of furrow drip irrigation was assumed to be half of surface drip irrigation's CapEx

in 2019, which is the most recent year when all types of used data are publicly available for using the consumer price index (CPI) [40].

The JVA's total benefit was calculated by multiplying the irrigation water tariff and the amount of consumed water. Thus, the JVA obtains additional profit by imposing a higher irrigation water tariff.

Farmers' total additional cost was calculated by multiplying the irrigation water tariff and the amount of water consumed. Thus, a higher irrigation water tariff increases the cost for farmers.

Farmers receive additional profit from the irrigation system improvement if it increases yields. Such predicted additional profit was calculated using the farm gate price of each crop. The farm gate price reflects farmers' perception of the value of crops [35]. The farm gate prices of lemons, tomatoes, and wheat in 2019 were retrieved from DOS [36].

2.4 Key Assumptions and Limitations

Two significant assumptions in this analysis are acknowledged. Firstly, the crop production simulation and cost–benefit analysis did not consider the effect of irrigation water quality. While water from some wells and wastewater treatment plants is reported to be of low-quality [1], it was assumed that its effect on crop production is negligible due to limitation of the data. Secondly, it was assumed that additional yields enabled by more efficient irrigation technologies will be sold without affecting the market price, whether for domestic consumption or export. The potential effect of affordable alternative produce such as imported commodities was not considered. It has been reported that some farmers in the JV see marketing and exporting their products as challenges [10]. However, this assumption was found justifiable because the conducted analysis was on individual farm-level (i.e., small-scale), indicating additional yield can be sold through farmers' existing channels. It may require additional consideration where the obtained result is to be applied to all the JV.

3 Results

3.1 FAO-Aquacrop Crop Production Simulation

The use of SSDI enabled the largest fresh yield, followed by SDI, furrow irrigation, and sprinkler irrigation (Table 6). This trend was constant for all crops. However, these crops exhibited the significantly different magnitude of the yield difference. The effects of irrigation type on lemons and wheat's yields were rather limited. Compared to sprinkler irrigation, water application by SSDI only led to the increase in fresh yields of lemons and wheat by 13.6 and 18.1%, respectively. On the other hand, tomatoes responded more acutely to the improvement of the irrigation methods. The use of SSDI increased tomatoes' fresh yield by 189.8% (13.7–39.7 t/ha * yr) compared to tomatoes' fresh yield obtained under sprinkler irrigation.

3.2 Cost–Benefit Analysis and Sensitivity Analysis

Sprinkler irrigation and furrow irrigation were classified as water-inefficient irrigation systems and SDI and SSDI as water-efficient irrigation systems. Based on this classification, a total of four scenarios for irrigation development: (a) shifts from sprinkler irrigation to SDI, (b) furrow irrigation to SDI, (c) sprinkler irrigation to SSDI, and (d) furrow irrigation to SSDI. These scenarios are in alignment with the national strategy of Jordan, which aims to increase the coverage of drip irrigation [41].((Fig. 1))

Table 6 Estimated fresh yields of lemons, tomatoes, and wheat under different on-farm irrigation systems

	Fresh yield (t/ha*year)		
	Lemons	Tomatoes	Wheat
Sprinkler	11.8	13.7	2.2
Furrow	12.2	19.4	2.3
SDI	13.0	33.3	2.5
SSDI	13.4	39.7	2.6

To further explore the irrigation water tariff range that satisfies both farmers' profits and the JVA's cost recovery requirements, the sensitivity analysis of the irrigation water tariff up to 0.40 JOD/m^3 was conducted for lemons (Fig. 2), tomatoes (Fig. 3), and wheat (Fig. 4). For lemon, the JVA was able to recover the subsidization cost for adopting SDI by increasing the irrigation water tariff to a minimum of 0.04 JOD/m^3 (Scenarios (a) and (b)) and SSDI to 0.08 JOD/m^3 (Scenarios (c) and (d)). Farmers' profit was maintained positive at these tariff rates. In Scenario (a) for instance, lemon farmers receive the additional net profit of 7.6 * 103 (JOD/ha)/25 years at a tariff rate of 0.04 JOD/m^3, and the net profit remained positive until the tariff rate of up to 0.10 JOD/m^3. The tariff rate, at which farming becomes unprofitable, varied significantly between Scenarios (a) and (d) for lemon; the additional net profit turned negative, when the tariff was raised to 0.10, 0.16, and 0.12 JOD/m^3 for Scenarios (b)–(d), respectively. These results show that the irrigation tariff range, in which the government can recover the subsidization cost for irrigation development while maintaining lemon farmers' profits differs by irrigation type. Such ranges were identified as 0.04–0.10, 0.04–0.08, 0.04–0.14, and 0.04–0.10 JOD/m^3, for Scenarios (a)–(d), respectively.

The trend observed in the sensitivity analysis of tomato production (Fig. 3) was similar to that

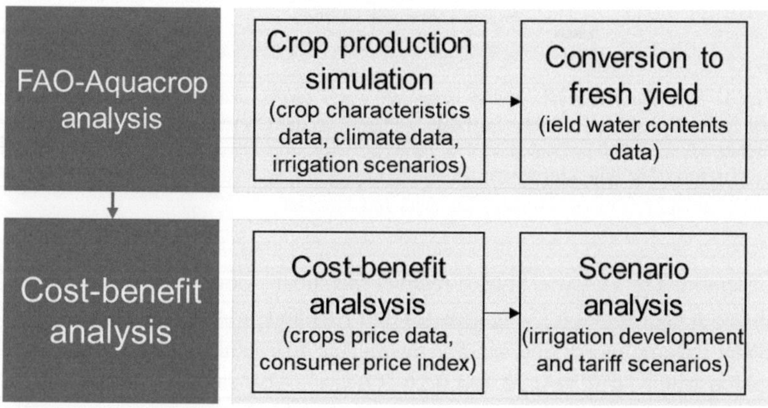

Fig. 1 Schematic diagram of methodology

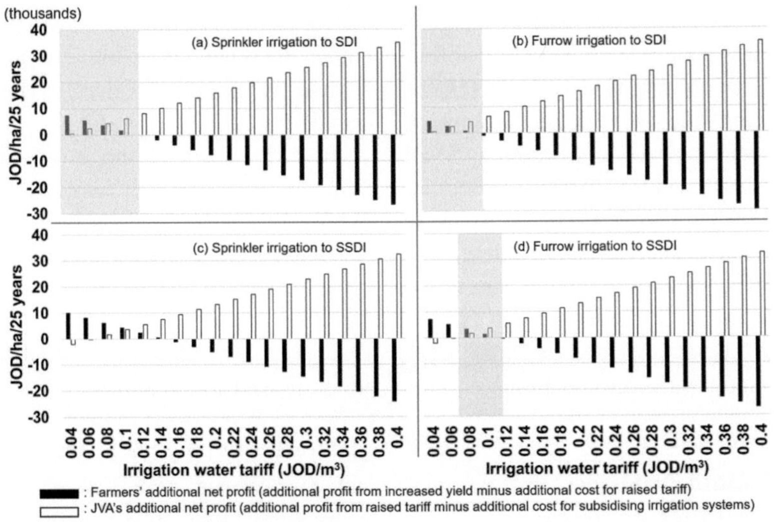

Fig. 2 Sensitivity analysis of the irrigation water tariff for lemons. A total of four on-farm irrigation system update scenarios were considered: **a** sprinkler irrigation to SDI, **b** furrow irrigation to SDI, **c** sprinkler irrigation to SSDI, and **d** furrow irrigation to SSDI. Dark and white bars signify farmers' and the JVA's additional net profit, respectively. The vertically shaded bands represent the tariff ranges that make both the JVA's and farmers' additional net profit positive

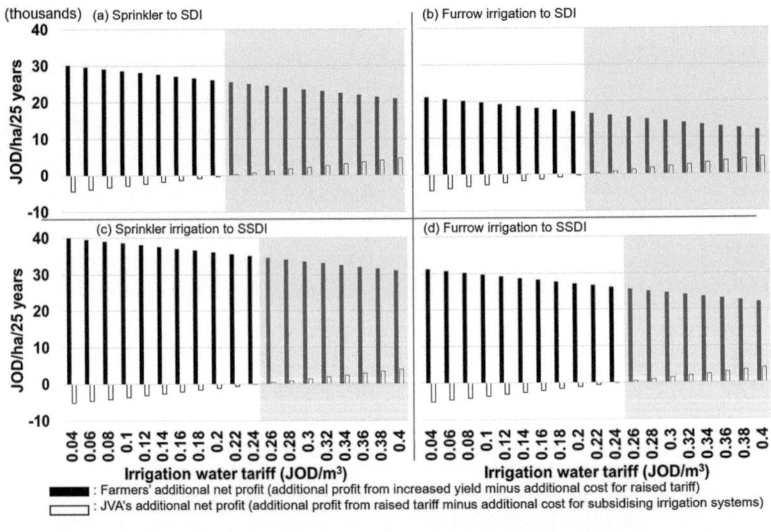

Fig. 3 Sensitivity analysis of the irrigation water tariff for tomatoes. A total of four on-farm irrigation system update scenarios were considered: **a** sprinkler irrigation to SDI, **b** furrow irrigation to SDI, **c** sprinkler irrigation to SSDI, and **d** furrow irrigation to SSDI. Dark and white bars signify farmers' and the JVA's additional net profit, respectively. The vertically shaded bands represent the tariff ranges that make both the JVA's and farmers' additional net profit positive

of lemon. The JVA was able to recover the subsidization cost for SDI (Scenarios (a) and (b)) and SSDI (Scenarios (c) and (d)) by raising the irrigation water tariff to 0.22 and 0.26 JOD/m^3. At both tariff rates, the additional profit for farmers was bigger than their additional cost. The farmers' net profit remained positive in the irrigation tariff range of 0.04–0.40 JOD/m^3. Though not shown

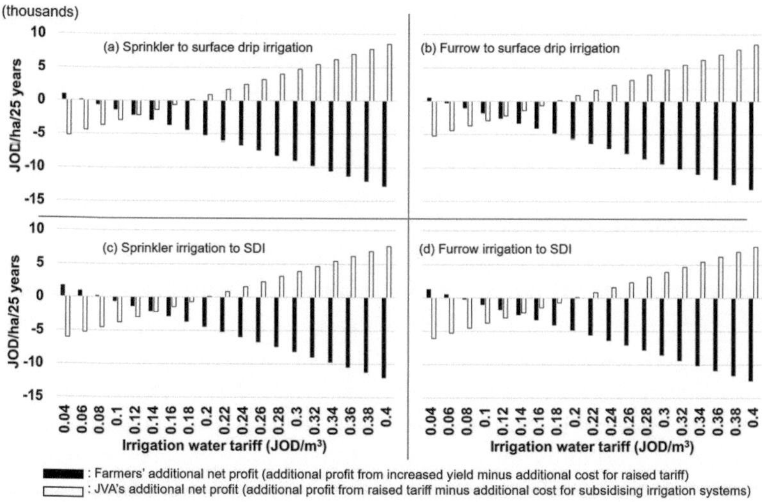

Fig. 4 Sensitivity analysis of the irrigation water tariff for wheat. A total of four on-farm irrigation system update scenarios were considered: **a** sprinkler irrigation to SDI, **b** furrow irrigation to SDI, **c** sprinkler irrigation to SSDI, and **d** furrow irrigation to SSDI. Dark and white bars signify farmers' and the JVA's additional net profit, respectively. The vertically shaded bands represent the tariff ranges that make both the JVA's and farmers' additional net profit positive

in Fig. 3, further calculation identified that the additional net profit for farmers turns negative at tariff rates of 1.24, 0.90, 1.64, and 1.28 JOD/m^3 for Scenarios (a)–(d), respectively. The irrigation tariff range that ensures the profitability of farming and the JVA's cost recovery was determined to be 0.22–1.22, 0.22–0.88, 0.26–1.62, and 0.26–1.26 JOD/m^3 for Scenarios (a)–(d), respectively.

In contrast, the tariff range that maintains both the JVA's and farmer's additional net profit positively did not exist for the wheat production, regardless of the irrigation system update scenario. For instance, in Scenario (c), the additional net profit for farmers remained positive up to a tariff of 0.08 JOD/m^3 and turned negative at a tariff rate of 0.10 JOD/m^3. The JVA's additional net profit did not turn positive until a tariff rate of 0.20 JOD/m^3. This result indicates that the irrigation system improvements for lemon production would place the financial burden on the JVA and/or wheat farmers.

4 Discussion

The objective of this research was to reveal the extent and conditions for the JVA to recover the cost for subsidizing advanced irrigation systems by raising tariffs without harming farmers' profit. The analysis revealed that such cost recovery is possible for high-value crops and that the tariff rate acceptable for farmers depends on the crop type, the types of the irrigation system being replaced and introduced.

4.1 Validation of Aquacrop Simulation

The predicted fresh yields agree reasonably well with literature [14, 42–45]. Abu-Awwad [42] reported a maximum lemon yield of 22.2 and 16.2 t/ha when sufficient water (i.e., meeting the average transpiration rate) and insufficient water (i.e., 75% of the average transpiration rate) in their cultivation experiments, respectively. For tomatoes, past literature report about 60–90 t/ha of the annual yield [43, 44], which is approximately 50–130% of the yield predicted in this study. Mithen et al. [45] reported wheat yields of 3–4 t/ha, when the water application rate was set at 80–120% of wheat's crop water requirement.

Their result may indicate that the Aquacrop simulation in this study underestimated the lemon yield, however, these discrepancies have likely derived from the difference in the production environment. In Abu-Awwad [43], their lemon cultivation experiments were conducted in a controlled environment. For tomatoes and wheat cultivation, the amount of applied water was determined based on tomatoes' potential evaporation rate, likely exceeding the allocated quota [43–45]. Given that actual water supply falls short of the quotas [14], this study's approach may provide more accurate representation of water application and hence yields farmers typically achieve.

Besides, the different response of tomatoes and lemons to introduction of SDI and SSDI can be explained by their root structure. Typically, citrus trees have fibrous roots (i.e., roots are concentrated near the ground surface) [46], while tomato plants are characterized by distinct taproots (i.e., a primary central root extends to a relatively deep soil, which smaller roots sprout from) [47]. It is likely that irrigating deep soil with SDI and SSDI improved the rhizosphere condition of tomatoes significantly, whereas its effect on the shallow root of lemons was relatively limited.

4.2 Physiological Response to Different Irrigation Methods Based on the Aquacrop Model

The difference in the obtained yields can be explained physiologically by stomatal closure and acceleration of senescence due to water stress. In Aquacrop, the yield is calculated as a product of the harvest index (HI) and biomass, which is a product of water productivity and total transpiration Sect. 6.1 in Supplementary Materials). Given that water productivity is constant [25], the yield fluctuation is explained by total transpiration and HI.

In fact, for tomatoes and wheat, cumulative transpiration (T_r) showed a similar trend (Fig. 5) to the yield (Table 6); it was the largest when the crop was subsurface drip-irrigated, followed by SDI, furrow, and sprinkler irrigation. The decrease in transpiration is due to stomata closure from the root's exposure to water stress (activation of water stress coefficient Ks) due to the intensified evaporation.

In contrast to tomatoes and wheat, lemon's cumulative transpiration did not differ significantly by irrigation type, indicating that the yield decline occurred due to modification of HI. In fact, the irrigation type influenced the HIs of lemons and tomatoes significantly. Figure 6 shows HI after the lag phase of biomass production for each crop. After the lemon cultivation initiated on January 1st, HI started differing by irrigation type in mid-September. The difference in the HI values expanded until October 8th, when phenological maturity was reached and remained constant after. Lemon's harvest index reduction is due to the direct effect of stomatal closure. For tomatoes, the HI increase first halted and physiological maturity was reached in early June under sprinkler irrigation, followed by furrow irrigation, SDI and SSDI. The different timings of physiological maturity indicate that the HI increase was inhibited by acceleration of senescence (activation of water stress coefficient for canopy senescence Ks_{sen}).

This dissection of physiological response suggests a potential water management strategy. Reducing water quotas may not harm yields in the periods, when cumulative transpiration did not differ by irrigation type (e.g., January and early February for tomatoes). In these periods, water spared by water-efficient irrigation did not contribute to the yield increase, indicating that water quotas can be reduced while maintaining the yield, if water-efficient irrigation is introduced.

4.3 How Does the Choice of On-Farm Irrigation System Affect Yields?

The Aquacrop simulation showed yields can be increased by replacing conventional on-farm irrigation systems with water-efficient systems like SSDI. Tabieh et al. [11] also reported that farmers, who utilize drip irrigation systems, have the highest ability to pay for water (0.84 JOD/m^3) due to higher yields, followed by those with SDI (0.38 JOD/m^3) and sprinkler irrigation (0.07 JOD/m^3). Their study provided general trends with no differentiation between crop types. This study further advanced their findings by revealing that this trend is constant for lemons, tomatoes, and wheat, and the magnitude of the yield increase differs by crop, tomatoes showing the largest increase. Furthermore, this study revealed that SSDI, an unconventional irrigation method in the JV, enables the highest yield for all crops.

The increases in biomass and yields are attributed to the soil water balance. The fraction of the wetted ground surface is minimal under SSDI, followed by SDI, furrow irrigation, and sprinkler irrigation. The dry ground surface suppresses the evaporation and hence water loss. The remaining irrigation water is either absorbed by the root or lost from the rhizosphere by deep percolation. Deep percolation is limited in the NJV as the dominant soil is fine-textured (clay). Thus, suppressing evaporation by maintaining the ground surface dry improved the root zone condition and enabled more biomass and yield production. It should be noted that, however, providing more water in the root zone over the plant's water requirement does not make a significant contribution to a higher yield and may result in waterlogging [48].

4.4 Extent to and Conditions Under Which Farmers' Profit Can Be Maintained

While the JVA's cost recovery necessitates a tariff raise [16, 17], too steep an increase would undermine farmers' profit and drive some farmers out of business. Poor farmers could be forced to rent out their land [12, 13], expanding the inequality between the rich and the poor. Such a divide in the JV may undermine Jordan's national food security [1]. It is then crucial to ensure that the tariff raise does not undermine farmers' profit.

The result indicates that the extent to which the profitability of farming can be maintained

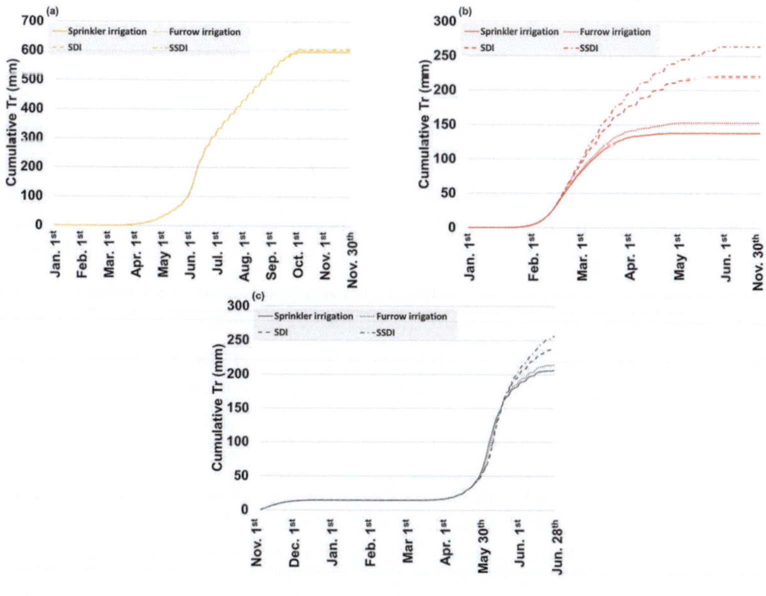

Fig. 5 Timely cumulative transpiration of **a** lemons, **b** tomatoes, and **c** wheat under different irrigation systems in the NJV. The values for each crop's growing season are shown

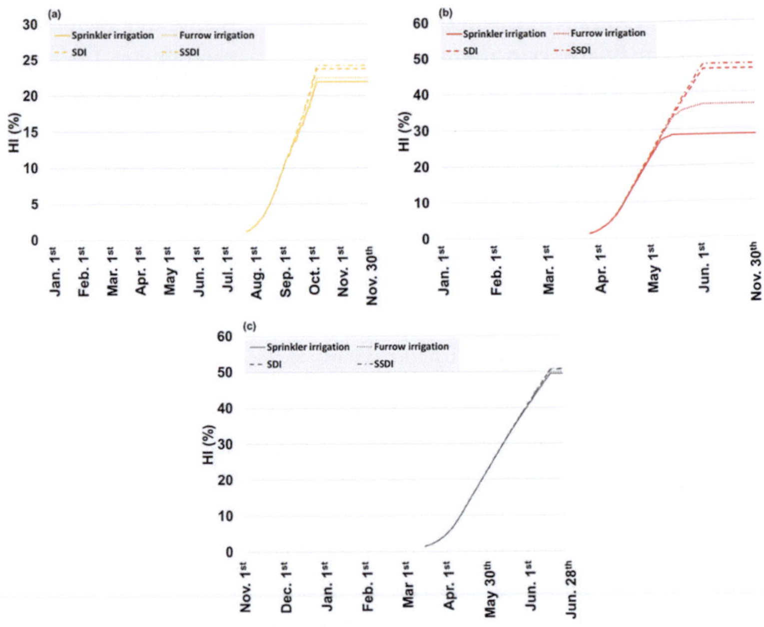

Fig. 6 Timely HI values of **a** lemons, **b** tomatoes, and **c** wheat under different irrigation systems in the NJV. The values from the end of a lag phase to the end of the growing season of each crop are shown

varies by crop. For instance, boosting wheat yields by introducing drip irrigation did not offset the additional water cost, when the tariff was raised to 0.10 JOD/m^3 in any Scenario, while tomatoes' irrigation system replacement offset the tariff increase to the maximum of 1.62 JOD/m^3.

To maintain farmers' profit after a tariff, raise, conventional irrigation systems like sprinkler and furrow irrigation need to be replaced by water-efficient drip irrigation. SSDI allowed for a steeper tariff raise than SDI, as its high efficiency [33] enabled more yield. The extent of the replacement's effect also depended on the existing irrigation system.

The profit increase was the most significant when sprinkler irrigation was replaced, indicating that such replacement should be prioritized over the replacement of furrow irrigation as it enables the largest yield. These findings resonate with Tabieh et al. [11], which revealed that drip irrigation yields a higher net value added by water than surface and sprinkler irrigation. The cost–benefit analysis of this study further advanced the literature on water tariffs in two ways. Firstly, it revealed an overlooked benefit of subsidizing water-efficient irrigation systems for the JVA; it allows for a higher tariff without compromising farmers' profit. Secondly, it showed that SSDI enables higher yields, profits, and hence a more significant tariff increase.

4.5 Extent to and Conditions Under Which the Subsidization Cost of the JVA Can Be Recovered

While ensuring farmers' profit is crucial for Jordan's food security, the irrigation water tariff needs to satisfy financial sustainability of the JVA's operation. The cost–benefit analysis indicates that the extent the JVA can recover the subsidization cost also depends on the crop type. On one hand, the JVA could not recover the subsidization cost for improving irrigation systems for wheat by imposing a higher tariff, as it would diminish farmers' profit. On the other hand, imposing a much higher tariff enabling the cost recovery was possible for high-value crops such as lemons and tomatoes. Tabieh et al. [11] also argued that farmers' ability to pay depends on the crop type they are growing, as the net profitability of water differs by crop. For instance, tomato and orange farmers can pay 1.307 and 0.464 JOD/m^3 for irrigation water, respectively. The finding also agrees with Al-Zu'bi and Al-Kharabsheh [49]. They reported that water productivity (monetary value created by a unit volume of water) differs by crop. Crops with high water productivity like lemons and tomatoes allows for a higher tariff and thus the cost recovery.

Overall, the Aquacrop analysis and cost–benefit analysis showed that the extent and conditions depend on at least three factors: (i) the crop type, (ii) irrigation system being replaced, and (iii) irrigation system being newly introduced. The full cost recovery for the subsidization was possible for high-value crops like lemons and tomatoes without harming farmers' profit, as their increased yields offset farmers' additional cost for the raised tariff, assuming the average weather conditions in the region. The full cost recovery may not be possible when the region experiences extreme weather. In contrast, for a low-value crop like wheat, only the partial cost recovery was possible, as the additional cost outweighed the additional profit. Hence, one of the conditions for the full cost recovery is that high-value crops are cultivated on the irrigated land. This result indicates that different tariff rates should be set for different crops.

It should be noted that the cost–benefit analysis of this study did not account for the indirect or non-economic benefits of irrigation development such as environmental impact. Thus, the estimated benefits for the JVA should be seen as minimum values. Another condition is that the irrigation system improvement leads to an increase in yields. Unless the yield increase gives farmers sufficient profit to offset the additional water cost, it undermines the financial sustainability of their farming. Sprinkler irrigation should first be replaced and then the furrow irrigation replacement should be implemented.

The tariff ranges can also be expanded by incorporating other cost-effective measures, such as use of less capital-intensive irrigation systems and improving billing and collection efficiencies of the tariff. For instance, a ring-shaped emitter for SSDI is known to be less capital intensive and easy to maintain, as it is made from locally available materials [50]. Studies indicate that this type of emitter is effective in cultivation of various high-value crops such as tomatoes [34, 50]. Utilizing such emitters will reduce the CapEx for the JVA, allowing for a lower tariff rate. Besides, billing and collection efficiencies of the irrigation water tariff is currently estimated to be as low as 82 and 75%, respectively [14]. Improving these inefficiencies would enhance the JVA's revenue and reduce the required tariff raise. While this study's cost–benefit analysis was based on typically reported values, exploring such cost-effective options will further aid with the JVA's cost recovery, assisting Jordan's sustainable food production.

4.6 Challenges Around Implementation of Irrigation Development Projects

One challenge to irrigation system development in practice is its implementation. This study did not consider the Jevons paradox, which refers to the phenomenon that enhancing resource-use efficiency may intensify the resource consumption as the per-unit value of the resource is increased [21]. Pérez Blanco [22] reported its occurrence in the water management space in several countries like India, Spain, and the United States. To prevent this phenomenon, although challenging, effective regulation of water extraction need to be put in place [1, 21]. Besides, in improving irrigation systems, it is of paramount importance to install water-efficient irrigation systems before the tariff raise and provide farmers with sufficient time to adapt to the new irrigation system. Subsurface drip irrigation system, though highly water-efficient, is known to be high maintenance [34] and may require farmers to learn its usage. This time lag will prevent farmers from the loss of profitability due to adaptation.

5 Conclusions

One of the keys to water conservation in a water-scarce country like Jordan is efficient irrigation water use. However, heavy subsidization of water-efficient irrigation systems may compromise the JVA's financial sustainability and hinders the adoption of advanced irrigation technologies. This study examined to what extent and under what conditions the government can recover the cost of subsidizing water-efficient irrigation systems while maintaining farmers' profit in the NJV.

The analysis revealed that switching from sprinkler and furrow irrigation to water-efficient SDI and SSDI enables higher yields of lemons, tomatoes, and wheat. The extent of the increase differed by crop; tomatoes showed the largest yield increase, followed by lemons and wheat. The full cost recovery for subsidizing SDI and SSDI systems was shown to be possible for high-value crops like lemons and tomatoes but is difficult for low-value crops like wheat. Successful implementation of such irrigation updates is also subject to other factors, such as regulation of water extraction, minimizing CapEx, improving tariff billing and collection efficiencies. Exploring these factors' influence on water tariffs will further aid in planning sustainable agricultural water management of Jordan.

6 Supplementary Materials

6.1 Calculation Procedures of Aquacrop

Aquacrop follows a production function proposed by Doorenbos and Kassam [55]:

$$\left(1-\frac{Y_a}{Y_x}\right)=K_y\left(1-\frac{\mathrm{ET}_a}{\mathrm{ET}_x}\right) \tag{3}$$

where Y_a [M/L] and Y_x [M/L] denote the actual and maximum yields, and ET_a [L/t] and ET_x [L/t] are the actual and maximum evapotranspiration rates, respectively. These values are modified based on the soil and atmospheric conditions and crop characteristics. K_y [−] is the yield response factor that represents the relationship between the evapotranspiration deficit and yield decline.

The yield (Y) in Aquacrop's production function (Eq. 3) is a product of the biomass and harvest index:

$$\mathrm{Y} = \mathrm{B} * \mathrm{HI} \tag{4}$$

where B [M/L^2] and HI [−] denote the biomass and harvest index [25]. HI is the proportion of the yield over crop biomass and differs by crop type and increases monotonically with phenological stages [56]. B can be calculated as follows:

$$\mathrm{B} = \mathrm{WP}^* \sum T_r \tag{5}$$

where WP [M/L^3] and T_r [L/T] are the water productivity and transpiration, respectively [25]. WP is the proportion of biomass over cumulative transpiration. Timings of biomass production are dictated by the growing degree day (GDD) [T], which is reported to be an accurate indicator of phenological events controlled by received thermal energy [57]. Aquacrop calculates GDD following McMaster and Wilhelm [57]:

$$\mathrm{GDD} = \left[\frac{(T_{\max} - T_{\min})}{2}\right] - T_{\mathrm{base}} \tag{6}$$

if

$$T_{\max} < T_{\mathrm{base}} \tag{6.1}$$

then

$$T_{\max} = T_{\mathrm{base}} \tag{6.2}$$

where $T_{\max}$, $T_{\min}$, and T_{base} [T] denote the maximum, minimum, and base temperatures, respectively. The base temperature is a threshold value below which crop growth halts. GDD is then used to calculate the canopy growth coefficient (CGC), which is the canopy cover growth rate per time (GDD) [25]. Canopy cover (CC) [L^2] is finally calculated using CGC:

$$\mathrm{CC} = \mathrm{CC}_0 e^{\mathrm{CGC}*t} \tag{7}$$

where CC_0 $[L^2]$ and t $[t]$ are the initial canopy size and time. CC is used to separate E $[L/t]$ and T_r in calculating ET_a. As the plant matures, CC starts declining, which can be expressed as follows:

$$CC = CC_x\left[1 - 0.05\left(\exp^{\frac{CDC}{CC_x}t} - 1\right)\right] \quad (8)$$

where CC_x denotes the maximum canopy coverage, and CDC [−] is the canopy decline coefficient.

Water stress inhibits biomass and yield production by suppressing canopy growth, T_r, and HI and by accelerating canopy senescence. Except for HI, these physiological processes are modified by the water stress coefficient, which is a function of the soil water content that takes a value between zero (maximum stress) and one (no stress). CGC, CDC, and T_r under water stress are adjusted as follows:

$$CGC_{adj} = Ks_{exp}{*}CGC \quad (8.1)$$

$$CDC_{adj} = \left(1 - Ks_{sen}^{8}\right){*}CDC \quad (8.2)$$

$$T_r = Ks * T_{r_x} \quad (8.3)$$

where Ks_{exp} and Ks_{sen} [−] are the water stress coefficients of expansive growth and canopy senescence. T_{rx} denotes the maximum transpiration. For T_r, Ks [−] results from stomatal closure induced by low soil water contents (Ks_{sto}) or anaerobiosis (Ks_{aer}) [56]. HI's modification due to water stress is divided into three stages. In the first stage, water stress affects HI positively after the leaf growth threshold (introduced as an adjustable crop parameter a [−] in Raes et al. [56]) has been reached, in which stage water stress inhibits canopy expansion. In the second stage after the stomata inhibition threshold (an adjustable crop parameter b [−]) has been reached, water stress reduces HI as crop transpiration is suppressed due to stomata closure. Finally, after the pollination inhibition threshold (an adjustable crop parameter p [−]) has been reached, the water stress' negative impact is further accentuated by failure of pollination. Further details of Aquacrop's calculation procedure is found in Steduto et al. [25] and Raes et al. [56].

Water loss by evaporation is adjusted by the area of wet ground surface in Aquacrop as follows:

$$E_{adj} = E * f_w \quad (9)$$

where E and E_{adj} $[L/t]$ denote the evaporation, and evaporation adjusted for the irrigation method, respectively. f_w [−] signifies the fraction of wetted surface. Thus, when the wet surface is smaller, water loss by evaporation is suppressed.

Disclaimer The opinions expressed in this article are those of the authors. They do not purport to reflect the opinions or views of PricewaterhouseCoopers or its members.

References

1. Greenwood S. Water insecurity, climate change and governance in the Arab world. Middle East Policy. 2014;21(2):140–6. https://doi.org/10.1111/mepo.12077.
2. Hussein H. Lifting the veil: unpacking the discourse of water scarcity in Jordan. Environ Sci Policy. 2018;89:385–2. https://doi.org/10.1016/j.envsci.2018.09.007.
3. Menzel, L., Teichert, E., Weiss, M. Climate change impact on the water resources of the semi-arid Jordan region. In: Heinonen M. editor. Proceedings of the 3rd international conference on climate and water. Helsinki: Finnish Environment Institute;2007. pp. 320–5.
4. Ministry of Water and Irrigation (MWI). National Water Strategy 2016–2025. 2016.
5. The Hashemite Kingdom of Jordan (THKJ). Jordan's third national communication on climate change. 2014a.
6. World Bank, Population, total. Jordan: World Bank;2024. https://data.worldbank.org/indicator/SP.POP.TOTL?locations=JO.
7. Kool J. The Jordan Valley. In: Kool J, editor. Sustainable development in the Jordan valley. New York: Springer International Publishing; 2016. p. 5–60.
8. Ministry of Water and Irrigation (MWI). Jordan water sector facts and figures 2017. 2017.
9. Al-Omari AS, Al-Karablieh EK, Al-Houri ZM, Salman AZ, Al-Weshah RA. Irrigation water management in the Jordan valley under water scarcity. Fresenius Environ Bull. 2015;24(4):1176–88.

10. Courcier R, Venot J, Molle F, Suleiman L, Jridi A. Historical transformations of the lower Jordan River Basin (in Jordan): changes in water use and projections (1950–2025) (Comprehensive assessment of water management in Agriculture Research Report 18). Colombo: International Water Management Institute;2005.
11. Tabieh M, Al-Karablieh E, Salman A, Al-Qudah H, Al-Rimawi A, Qtaishat T. Farmers' ability to pay for irrigation water in the Jordan Valley. J Water Resource Prot. 2015;7(15):1157. https://doi.org/10.4236/jwarp.2015.715095.
12. Venot J., Molle F, Hassan Y. Irrigated agriculture, water pricing and water savings in the lower Jordan River Basin (in Jordan) (Comprehensive assessment of water management in Agriculture Research Report 18). Colombo: International Water Management Institute;2007.
13. Molle F, Venot J, Hassan Y. Irrigation in the Jordan Valley: Are water pricing policies overly optimistic? Agric Water Manag. 2008;95(4):427–8. https://doi.org/10.1016/j.agwat.2007.11.005.
14. World Bank. The cost of irrigation water in the Jordan Valley. Washington, DC: World Bank; 2016.
15. Ramirez AO, Ward AF, Al-Tabini R, Phillips R. Efficient water conservation in agriculture for growing urban water demands in Jordan. Water Policy. 2011;13(1):102–24. https://doi.org/10.2166/wp.2010.066.
16. Salman AZ, Al-Karablieh E. Measuring the willingness of farmers to pay for groundwater in the highland areas of Jordan. Agric Water Manag. 2004;68(1):61–76. https://doi.org/10.1016/j.agwat.2004.02.009.
17. Doppler W, Salman AZ, Al-Karablieh EK, Wolff H. The impact of water price strategies on the allocation of irrigation water: the case of the Jordan Valley. Agric Water Manag. 2002;55(3):171–82. https://doi.org/10.1016/S0378-3774(01)00193-7.
18. Al-Assaf A, Salman AZ, Fisher FM, Al-Karablieh E. A trade-off analysis for the use of different water sources for irrigation (The case of Southern Shounah in the Jordan Valley). Water Int. 2007;32(2):244–3. https://doi.org/10.1080/02508060708692204.
19. Molle F. Water scarcity, prices and quotas: a review of evidence on irrigation volumetric pricing. Irrig Drain Syst. 2009;23(1):43–58. https://doi.org/10.1007/s10795-009-9065-y.
20. Ward FA, Pulido-Velazquez M. Water conservation in irrigation can increase water use. Proc Natl Acad Sci USA. 2008;105(47):18215–20. https://doi.org/10.1073/pnas.0805554105.
21. Grafton RQ, Williams J, Perry CJ, Molle F, Ringler C, Steduto P, Udall B, Wheeler SA, Wang Y, Garrick D, Allen RG. The paradox of irrigation efficiency. Science. 2018;361(6404):748–50. https://doi.org/10.1126/science.aat9314.
22. Pérez Blanco C, Loch A, Ward F, Perry CJ, Adamson D. Agricultural water saving through technologies: a zombie idea. Environ. Res. Lett. 202;116(11). https://doi.org/10.1088/1748-9326/ac2fe0.
23. Steduto P, Hsiao TC, Fereres E, Raes D. Crop yield response to water (FAO Drainage and Irrigation Paper 66). Rome: Food and Agriculture Organization of the United Nations (FAO); 2012.
24. Department of Statistics (DOS). Crops statistics: irrigated and non-irrigated areas. 2021. http://dosweb.dos.gov.jo/agriculture/crops-statistics/.
25. Steduto P, Hsiao TC, Raes D, Fereres E, AquaCrop—the FAO crop model to simulate yield response to water: I. Concepts and underlying principles. Agron J. 2009;101(3):426–37. https://doi.org/10.2134/agronj2008.0139s.
26. Payen S, Basset-Mens C, Colin F, Roignant P, Vanniere H, Perret S. E.T.: an operational field water and salt flows model for agricultural LCA illustrated on a Moroccan Mandarin. In: Proceedings of the 10th international conference on life cycle assessment of food. Dublin: University College Dublin School of Biosystems and Food Engineering;2016. pp. 55–60.
27. Pereira FFS, Sánchez-Román RM, González AMGO. Simulation model of the growth of sweet orange (Citrus sinensis L. Osbeck) cv. Natal in response to climate change. Clim Change. 2017;143(1):101–13. https://doi.org/10.1007/s10584-017-1986-0.
28. Katerji N, Campi P, Mastrorilli M. Productivity, evapotranspiration, and water use efficiency of corn and tomato crops simulated by AquaCrop under contrasting water stress conditions in the Mediterranean region. Agric Water Manag. 2013;130:14–26. https://doi.org/10.1016/j.agwat.2013.08.005.
29. Bird DN, Benabdallah S, Gouda N, Hummel F, Koeberl J, La Jeunesse I, Meyer S, Prettenthaler F, Soddu A, Woess-Gallasch S. Modelling climate change impacts on and adaptation strategies for agriculture in Sardinia and Tunisia using AquaCrop and value-at-risk. Sci Total Environ. 2016;543(B):1019–27. https://doi.org/10.1016/j.scitotenv.2015.07.035.
30. Soddu A, Deidda R, Marrocu M, Meloni R, Paniconi C, Ludwig R, Sodde M, Mascaro G, Perra E. Climate Variability and durum wheat adaptation using the AquaCrop Model in Southern Sardinia. Procedia Environ Sci. 2013;19:830–5. https://doi.org/10.1016/j.proenv.2013.06.092.
31. Toumi J, Er-Raki S, Ezzahar J, Khabba S, Jarlan L, Chehbouni A. Performance assessment of AquaCrop model for estimating evapotranspiration, soil water content and grain yield of winter wheat in Tensift Al Haouz (Morocco): application to irrigation management. Agric Water Manag. 2016;163(2013):219–35. https://doi.org/10.1016/j.agwat.2015.09.007.
32. Raes D, Steduto P, Hsiao CT, Fereres E. Reference manual for Aquacrop version 6.0/6.1. Rome: Food and Agricultural Organization of the United Nations (FAO);2018.
33. Martinez J, Reca J. Water use efficiency of surface drip irrigation versus an alternative subsurface drip irrigation method. J Irrig Drain Eng—ASCE. 2014;140(10):04014030. https://doi.org/10.1061/(ASCE)IR.1943-4774.0000745.

34. Noguchi K, Saito H, Saefuddin R, Šimůnek J. Evaluation of subsurface drip irrigation designs in a soil profile with a capillary barrier. Water. 2021;13(9):1300. https://doi.org/10.3390/w13091300.
35. D'Odorico P, Chiarelli DD, Rosa L, Bini A, Zilberman D, Rulli MC. The global value of water in agriculture. Proc Natl Acad Sci USA. 2020;117(36):21985–93. https://doi.org/10.1073/pnas.2005835117.
36. Department of Statistics (DOS). Agricultural prices: marketing costs, selling prices and farm-gate prices of sold quantities on farm or adjacent to it by month. Department of Statistics (DOS);2021. http://dosweb.dos.gov.jo/agriculture/agricultural-prices.
37. Amosson S, Almas L, Girase RJ, Kenny N, Guerrero B, Vimlesh K, Marek T. Economics of irrigation systems. AgriLIFE EXTENSION Texas A&M System;2002.
38. Young AR, Loomis BJ. Determining the economic value of water: concepts and methods. New York: Routledge; 2014.
39. Just R, Hueth D, Schmitz A. The welfare economics of public policy. Cheltenham: Edward Elgar Publishing; 2004.
40. World Bank, Consumer price index (2010 = 100). Jordan: World Bank;2021b. https://data.worldbank.org/indicator/FP.CPI.TOTL?locations=JO.
41. The Hashemite Kingdom of Jordan (THKJ). Jordan 2025: a national vision and strategy;2014b.
42. Abu-Awwad AM. Influence of different water quantities and qualities on lemon trees and soil salt distribution at the Jordan Valley. Agric Water Manag. 2001;52(1):53–71. https://doi.org/10.1016/S0378-3774(01)00122-6.
43. Abu-Sharar MT, Al-Karablieh KE, Haddadin JM. Role of virtual water in optimizing water resources management in Jordan. Water Resour Manag. 2012;26:3977–93. https://doi.org/10.1007/s11269-012-0116-z.
44. Al-Qerem RA, Suleiman AA, Shatanawi MR. Assessing tomato yield and water saving under deficit irrigation in the Jordan Valley. Jordan J Agric Sci. 2012;8(2):209–22.
45. Mithen S, Jenkins E, Jamjoum K, Nuimat S, Nortcliff S, Finlayson B. Experimental crop growing in Jordan to develop methodology for the identification of ancient crop irrigation. World Archaeol. 2008;40(1):295–308. https://doi.org/10.1080/00438240701843561.
46. Castle WS. Citrus root systems: their structure, function, growth, and relationship to tree performance. Proc Int Soc Citriculture. 1978;62–9.
47. S. L. Gulmon, N.C. Turner, Differences in root and shoot development of tomato (Lycopersicon esculentum L.) varieties across contrasting soil environments. Plant Soil. 1978;49:127–36. https://doi.org/10.1007/BF02149914.
48. Ritzema HP, Satyanarayana TV, Raman S, Boonstra J. Subsurface drainage to combat waterlogging and salinity in irrigated lands in India: lessons learned in farmers' fields. Agric Water Manag. 2008;95(3):179–89. https://doi.org/10.1016/j.agwat.2007.09.012.
49. AL-Zu'bi YA, Al-Kharabsheh A. Multicriteria analysis for water productivity in the Jordan Valley. Water Int. 2003;28(4):501–11. https://doi.org/10.1080/02508060308691727.
50. Saefuddin R, Saito H, Šimůnek J. Experimental and numerical evaluation of a ring-shaped emitter for subsurface irrigation. Agric Water Manag. 2019;211:111–22. https://doi.org/10.1016/j.agwat.2018.09.039.
51. Al-Sayaydeh R, Al-Bawalize A, Al-Ajlouni Z, Akash MW, Abu-Elenein J, Al-Abdallat AM. Agronomic evaluation and yield performance of selected barley (Hordeum vulgare L.) Landraces from Jordan. Int J Agron. 2019;1–12.
52. Jordan Meteorological Department (JMD). Rain rates: normal monthly mean. 2021. http://jometeo.gov.jo/%D9%85%D8%B9%D9%84%D9%88%D9%85%D8%A7%D8%AA-%D9%85%D9%86%D8%A7%D8%AE%D9%8A%D8%A9-%D9%88%D8%B2%D8%B1%D8%A7%D8%B9%D9%8A%D8%A9/%D8%A7%D9%84%D9%85%D8%B9%D8%AF%D9%84%D8%A7%D8%AA-%D8%A7%D9%84%D8%B9%D8%A7%D9%85%D8%A9.
53. Al Mahamid J. Integration of water resources of the upper aquifer in Amman-Zarqa basin based on mathematical modeling and GIS, Jordan. Freiberg Online Geosci. 2005;12. https://doi.org/10.23689/fidgeo-879.
54. van Vuuren DP, Edmonds J, Kainuma M, Riahi K, Thomson A, Hibbard K, Hurtt GC, Kram T, Krey V, Lamarque J, Masui T, Meinshausen M, Nakicenovic N, Smith SJ, Rose SK. The representative concentration pathways: an overview. Clim Change. 2011;109(1):5–31. https://doi.org/10.1007/s10584-011-0148-z.
55. Doorenbos J, Kassam AH. Yield response to water (FAO Irrigation and Drainage Paper 33). Rome, Italy: Food and Agriculture Organization of the United Nations (FAO);1979.
56. Raes D, Steduto P, Hsiao, TC, Fereres E. AquaCrop—the FAO crop model to simulate yield response to water: II. Main algorithms and software description. Agron J. 2009;101(3):438–47. https://doi.org/10.2134/agronj2008.0140s.
57. McMaster GS, Wilhelm WW. Growing degree-days: one equation, two interpretations. Agric For Meteorol. 1997;87(4):291–300. https://doi.org/10.1016/S0168-1923(97)00027-0.
58. Furr JR, Taylor CA. Growth of lemon fruits in relation to moisture content of the soil (Technical Bulletin No. 640). Washington, DC: United States Department of Agriculture;1939.
59. Davies JN, Hobson GE. The influence of environment, nutrition, and genotype—The constituents of tomato fruit. Crit Rev Food Sci Nutr. 1981;15(3):205–80. https://doi.org/10.1080/10408398109527317.

60. Nedeva D, Nikolova A. Fresh and dry weight changes and germination capacity of natural or premature desiccated developing wheat seeds. Bul J Plant Physiol. 1999;25(1–2):3–15.
61. Food and Agricultural Organization of the United Nations (FAO). Jordan: AQUASTAT Country Profile;2008.

The Relation between Water Policies and Strategies and Modernization of the Irrigated Sector in Sudan

Tarig El Gamri and Adil M. Elkhidir

Abstract

The first national water policy for Sudan was developed in 1913 to pave-the-way for the construction of Sennar Dam and the Gezira Scheme. The advent of the gravity irrigation in the Gezira Scheme was a huge step toward modernization. The water policies, strategies and plans were then formulated during the period 1952–1956. They were also revised following the Nile's Water Agreement (1959) and in 2000 and 2007. Relevant legislation includes the Water Resources Act (1995) and Irrigation and Drainage Act (1990). However, the national Integrated Water Resources Management (IWRM) strategy and water efficiency plan are yet to be developed. Besides that, the National Adaptation Plan (NAP), National Adaptation Program of Action (NAPA) and other Climate Change documents also advocate for the modernization of irrigation. Recent modernization interventions include the introduction of center pivot systems and hydro-flume. The framework program of the Transitional Government (2019) puts renovation of the existing irrigation projects as one of its highest priorities. The recent water strategy (2021–2031) called for the modernization of irrigation. The Strategy is based on three transformative plans. The National Irrigation Transformation Plan (NITP) called for upgrading and modernizing 1.1 million ha of existing irrigation schemes, development of 400,000 ha of new irrigation schemes and improving water productivity with at least 50% more crop per drop. Some regional initiatives were also supportive. For instance, the Nile Decision Support System (Nile DSS) of the Nile Basin Initiative (NBI) will guide planning for basin future projects, e.g., expansion in irrigated agriculture coupled with enhanced water use efficiency. Thorough investigation in Sudan's heritage in strategic planning in water resources management will provide momentum for the future policy making process. The potential for modernization of irrigation is high provided that political and economic stability is guaranteed, capacity is enhanced, and investment is encouraged.

Keywords

Center pivot system · Irrigation · Modernization · Policies · Strategies

T. E. Gamri (✉)
The National Centre for Research, Khartoum, Sudan
e-mail: tarig_ncr@yahoo.com

A. M. Elkhidir
Faculty of Engineering, University of Khartoum, Khartoum, Sudan

M. Bahadir et al. (eds.), *Agricultural Water Management*, Water and Wastewater Management,
https://doi.org/10.1007/978-3-031-96134-2_18

1 Introduction

It is evident that human development has been potentiated by access to fresh water and by the capacity to exploit its' potential as a productive resource [1]. According to [2], irrigated agriculture is the main water consumer in the Nile Basin at present that utilizes about 78% of the flow of the river. Due to population growth and economic development, water demands are expected to grow for all sectors including irrigated agriculture.

Sudan is situated in the northeastern part of the African continent between Latitude 21.49 to 38.34 E and Longitude 23.8 to 8.45 N [3]. With an area of about 1.88 million km^2, the country is considered the third largest in Africa preceded by Algeria and DR Congo. The Nile Basin constitutes the most prominent topographical feature of the country, since it makes about 67.4% of the country's total area [4]. About 94% of the country area is located in the dry and semi-dry climates [5].

The population of the country is estimated at about 44.4 million people in 2020 and is projected to reach 57.3 million people by the year 2030. With a population growth rate of more than 2% per year projected that the population would reach 80 million people in 2050 [6]. The current armed conflict in the country triggered substantial demographic changes, since about 10 million people were internally displaced while about 2 million people fled outside Sudan.

Recent estimates showed that Sudan is under water poverty with an estimated per capita water availability of 920 m^3/annum. It is worth noting that the amount 1000 m^3/capita/annum is assumed as the water poverty boundary. More than 75% of water abstraction is taken from the River Nile and its tributaries, about 16% from seasonal streams and the remainder from ground water resources.

Long-term water utilization strategies showed that the country is anticipated to witness severe water shortages. Table 1 shows forecast of demand for water up to 2027 based on the Quarter Century National Strategic Plan.

Table 1 Water demand projection 2010–2027 (BCM) [7] (BCM = billion cubic meter)

Year	Irrigation	Domestic supply	Animals and others	Total
2010	27.1	1.1	3.9	32.1
2020	32.6	1.9	5.1	39.6
2025	40.3	2.5	5.3	48.0
2027	42.5	2.8	7.3	52.6

In 2020, the African Union formulated the Framework for Irrigation Development and Agricultural Water Management in Africa (IDAWM) with main emphasis in modernization of irrigation [8]. Moreover, modernization of both rain fed and irrigated agrarian systems is a fundamental policy constituent for many riparian countries including Sudan with a view of building more resilient food systems and maintaining food security.

Since 2018, the situation in Sudan is highly vibrant consequently, substantial policy revision has taken place during the transitional period and the same is expected to take place when a peaceful solution is reached for the present armed conflict.

This work aimed at providing a comprehensive analysis of the national efforts in strategic planning in irrigation water management and hence contributing to the policy development efforts that will guide modernization in the irrigated sector.

2 Irrigated Agriculture

Since about 70% of the area of the country is arid or semi-arid and generally with highly variable rainfall, Sudan has to rely on irrigated agriculture to achieve food security and to produce cash crops. Area wise, Sudan is the second largest country in Africa that practice irrigated agriculture [9]. The sub-sector is gaining more and more importance due to the drought episodes that stricken the country during the previous century.

According to [3], land use in the country is classified as follows:

- Arable land: 90.7 million ha
- Cropped lands: 23.5 million ha
- Irrigated agriculture: 3.5 million ha
- Rain fed agriculture: 12.4 million ha
- Forests: 10.3% of the country's total area
- About 25.6% of Sudan's total land area is rangeland
- About 8.1% of Sudan is set aside as protected areas

It is worth noting that discrepancies among estimates of areas assigned to the different land uses exist.

The huge land resources of the country constitute a substantial potential for agricultural expansion utilizing the uncommitted available water resources in Sudan. The following is a brief overview of the water resources of the country.

Rainfall. Presently, the total annual rainfall of the country is estimated at about 400 BCM [3, 10]. According to [10, 11], the amount and distribution of rainfall plays an important role in the environment and the socio-economic elements of the country, principally pastoralism and rainfed agriculture. The rainy season commences in the southern parts of Sudan during May, while, in the Northern part, the onset of the rainy season is either during late June or beginning of July [11].

The River Nile and its tributaries. Figure 1 shows the major watersheds of the Sudan. The Nile River and its tributaries receive their waters from three main basins [10]:

1. The Ethiopian Plateau EP: the EP provides approximately 86% of the flow of the Nile through the Blue Nile, Atbara and the Sobat Rivers and their tributaries.
2. The Equatorial Lakes Plateau: this Plateau provides the remaining 14% of the Nile discharge. The huge water quantities that are provided by Bahr El-Ghazal and Bahr El-Jebel Rivers are lost in the extensive swamps.
3. The Atbara River is comprised of two main tributaries which are Upper Atbara and Setit Rivers. The two branches originate at the

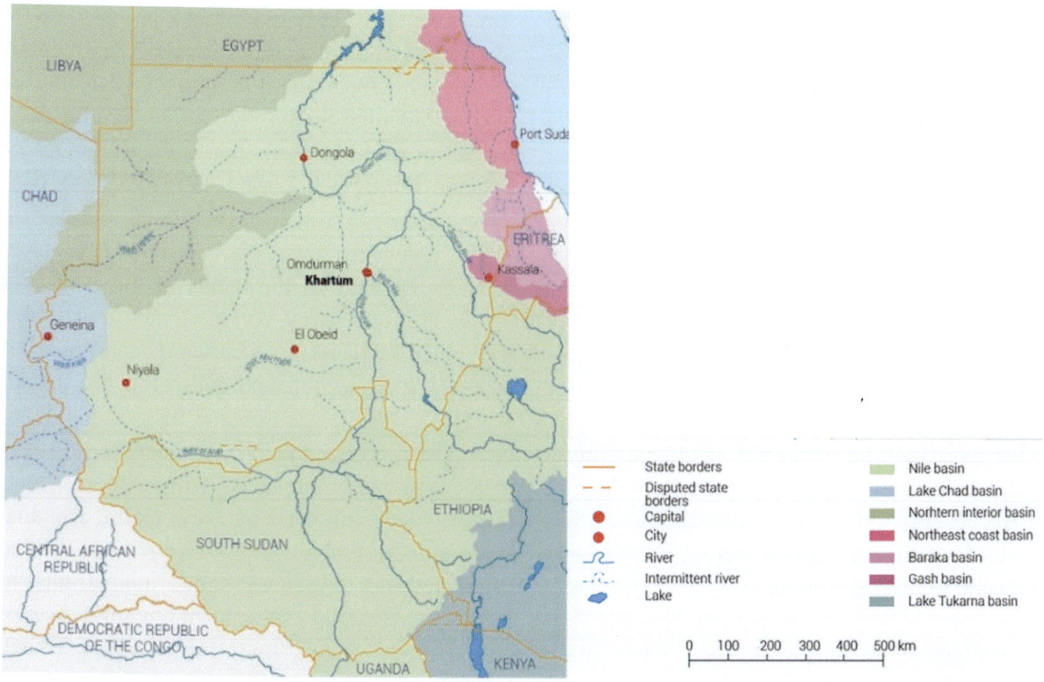

Fig. 1 Major watersheds of Sudan [12]

northern part of the Ethiopian Plateau. The average annual flow of the River Atbara is estimated at about 12 BCM.

Seasonal streams (Wadis and Khors). There are several seasonal streams within or outside the Nile watershed known as Wadis and Khors, which are characterized by high irregular short duration flows. Approximately 300 Wadis are spread in the Red Sea Region and the Savannah Belt with intermittent flow regimes. The gross discharge of wadis in the country is estimated to fluctuate between 5 to 7 BCM/annum.

Groundwater. Main groundwater aquifers in Sudan comprise Nubian Sandstone, Umm Ruwaba Formation and alluvial deposits. The Basement Complex Formations (impermeable) underlay approximately 50% of the area of the country. Renewable annual groundwater resources are about 4.2 BCM.

The area under modern irrigated agriculture is estimated at about 3.5 million ha out of approximately 90.7 million ha of arable lands. The Nile system constitutes the source of about 93% of the irrigation water, of which the Blue Nile accounted for about 67%. Gravity irrigation constitutes the principal irrigation method even though pumps serve part of the irrigated area. Figure 2 exhibits irrigation projects in Lower Blue Nile and White Nile Sub-Basin.

About 70% of the labour force in the rural areas is employed in agriculture and agro industries, being the main source of income in rural arcas [13].

The irrigation sector is of great importance to the Sudanese economy. Continued growth of this sector will result in increasing demand for the Nile Waters [14]. According to [15], water is the most limiting factor that constrains future expansion.

According to [14], the agricultural sector has declined during the period between 1990 and 2020 due to several factors:

- Negligence by the government since more attention went to the oil and services sectors;
- Enactment of inappropriate policies and strategies, lack of funding and inadequate research and capacity;
- Internal political conflicts and instabilities, particularly in the Western and Eastern agricultural hubs. It is worth mentioning that the current war spreads over 70% of the country's area.

Water is the most determining natural resource for expanding and maintaining irrigated agriculture, as 85% of Sudan's 90.7 million ha arable land remains uncultivated because of either physical or human water scarcities [12]. For example, only about 50% of the area of the large-scale irrigation schemes (1.2 million ha) is cultivated each year (mainly human). Similarly, about one-third of the area of small to medium scale pump irrigated schemes along the Nile system (900,000 ha) is under cultivation every year (mainly human). Similarly, the seasonal flood irrigation schemes of Gash, Tokar and Abu Habil are estimated to cover a total area of about 500,000 ha. However, the actual cultivated area infrequently go above 120,000 ha per annum (physical/human).

Cultivation of cotton on the deltas of the seasonal rivers of Gash and Baraka in early 1860s marks the commencement of the employment of irrigation for modern commercial farming in Sudan. Then the Sudan Plantations Syndicate (a British private company) constructed the Gezira Scheme during the period 1911–1928 to produce cotton. On a later development and during 1920s, privately own pump-irrigated projects were established on the banks of the Bluc and White Niles to grow cotton, fruits, vegetables and other food crops. Afterward, a number of schemes were established following the design criteria of the Gezira Scheme including Rahad, New Halfa and Suki. The production of sugar in the country commenced with Guneid Sugar Factory 1962 followed by New Halfa, Kenana, West Sinnar, Asalaya and White Nile [17]. The sugar industry triggered some innovations to improve irrigation efficiencies viz. the use of long furrow, siphon, hydo-flume and cut-back techniques.

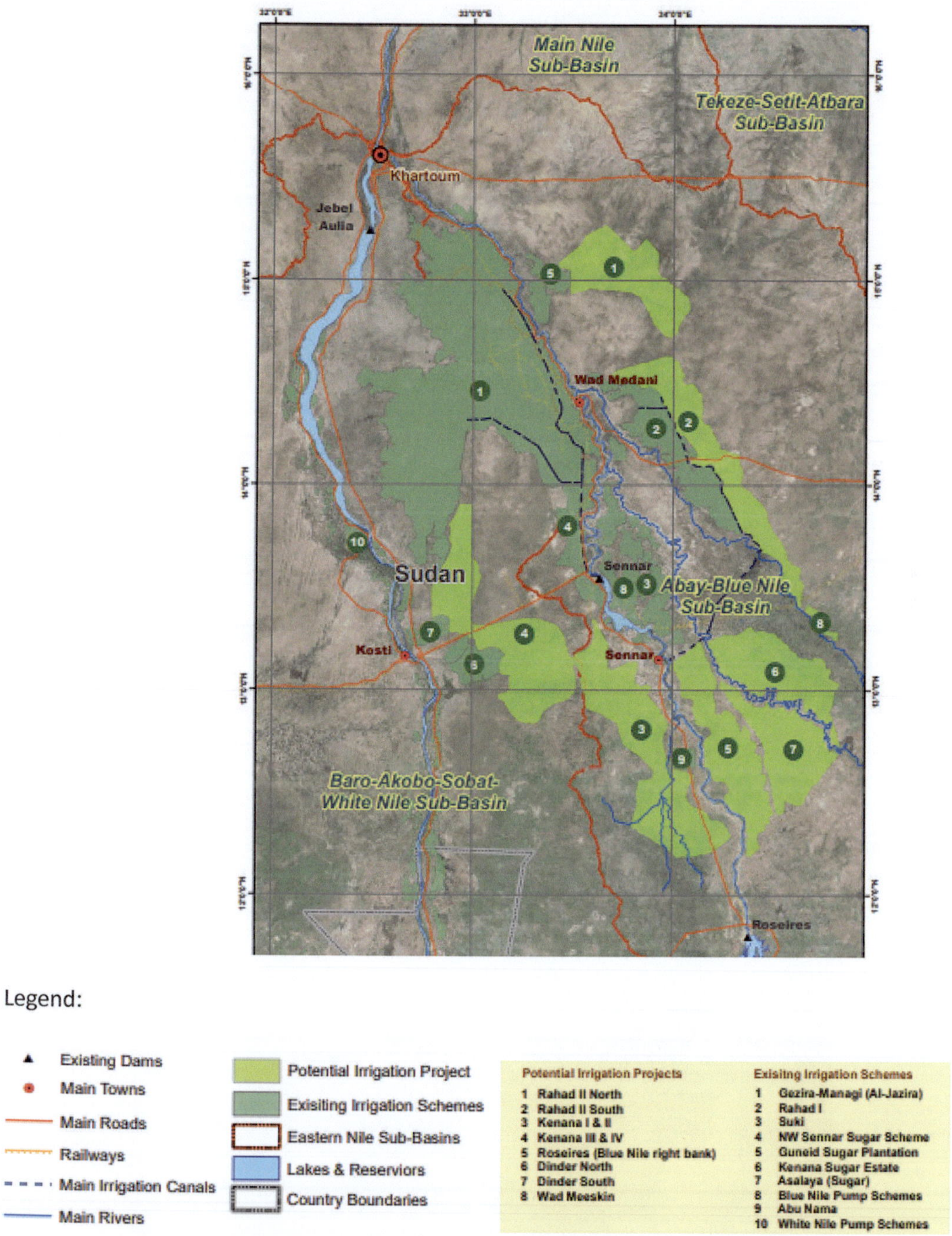

Fig. 2 Irrigation projects in Lower Blue Nile and White Nile Sub-Basin [16]

A series of multi-purpose dams (irrigation and hydro-power) were constructed on the River Nile and its tributaries: Sinnar 1925, Jebel Aulia 1937, Khasm El Girba 1962, Roseries 1966 (heightened in 2015), Merowe 2006 and Atbara and Setiet 2016. Figures 3, 4, 5, 6, 7 and 8 show these dams, respectively. The resettlement projects associated with Merowe Dam are

Fig. 3 Sennar Dam [19]

Fig. 4 Jebel Aulia Dam [19]

Fig. 5 Khasm El Girba Dam [19]

Fig. 6 Roseires Dam [15]

Fig. 7 Downstream of Merowe Dam [20]

characterized by relatively extensive use of the Center Pivot irrigation system [18].

As shown in Fig. 9, productivity under irrigated agriculture is among the lowest in Africa. However, [12] concluded that significant improvement in water use efficiency is attainable. For instance, during the 2019/2020 season, some pioneer farmers in the Gezira scheme achieved a wheat yield of 6 ton/ha by improving water distribution. Similarly in the Gash Delta, the productivity of the major sorghum crop increased from 0.9 to 2.0 ton/ha, while water consumption was reduced by 30% [14].

Very low irrigation efficiencies were reported in some of the major schemes in the country [22]. This was attributed to the unsatisfactory maintenance due to lack of adequate finance [23]. The authors concluded that 20% of total irrigation water may be saved by rehabilitating and developing the irrigation systems coupled with improvement of water management. In addition to that, estimation of the actual crop water requirements and selection of the optimum crop patterns to reduce water losses were recommended.

According to [17], the following options may be adopted to maximize returns from irrigated agriculture in the Sudan:

- Construction of new projects or extensions to existing schemes.

Fig. 8 Upper Atbara and Setit Dam Complex [21]

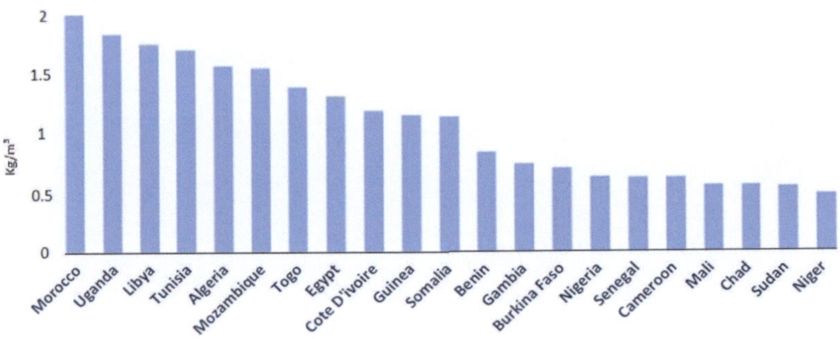

Fig. 9 Productivity of irrigated agriculture in some African countries [12]

- Crop intensification to increase the area cropped each year.
- Selection of the most suitable cropping patterns to maximize productivity.

The authors also stressed the importance of developing encouraging national policies and plans.

3 Evolution of Water Policy in Sudan

The first national water policy for Sudan was crafted during the colonial era in 1913 to allow for the construction of Sennar Dam and the Gezira Irrigation Scheme [24]. It is worth noting that non-conducive policies prevail throughout the whole production sectors particularly agriculture, hence leading to low productivity and loss of competitiveness [25]. Moreover, [26] attributed the low level of implementation of national strategies and programs to institutional instability and inadequate capacities and investment among others.

In accordance with the Water Resources Act (WRA, 1995), the National Council for Water Resources (NCWR) was created with powers to tackle most of the issues pertinent to water allocation, pollution control, monitoring, information management, economic and financial management, stakeholder participation, flood and drought management, and management of the infrastructural facilities [27]. Activation of the NCWR will pave the way toward mainstreaming and implementation of IWRM, the author concluded. It is worth noting that state water policies are deficient, nonetheless affected by the federal water policies.

3.1 Policy Environment Before Independence

Introduction of irrigated agriculture using the floodwaters of the Gash River and the Baraka River for commercial farming (in the late 1860s) was one of the first efforts to modernize irrigation in Sudan. According to [10], both streams originate from Eritrea, and their deltas were used to produce cotton. During the period 1924–1926, canals were constructed in the Gash Delta with a view of controlling the floods, however, sandstorms made canals impractical in Baraka Delta. The availability of groundwater encouraged pump irrigation to supplement deficiencies of flood and cultivation of a second crop. The two schemes were governed by the following Acts:

- Gash Farmers' Act, 1928
- Delta Tokar Farmers' Protection Act, 1943

Extensive surface (gravity) irrigated agriculture commenced during the British colonial era (1898–1956). The agrarian policy focused on the expansion of cotton production. Introduction of motorized pumping systems took place at the start of the twentieth century, to replace the traditional practices of flood irrigation and water wheel (Sagia).

The first national water policy of 1913 was crafted to pave the way for the construction of Sennar Dam and the Gezira Scheme. The Gezira Scheme, which is situated between the Blue Nile and the White Nile is the country's first-born and principal gravity irrigation scheme. The scheme started in 1925 and gradually expanded to cover an area of around 880,000 ha distributed among 114,000 tenancies. The scheme used to be the backbone of the national economy by serving as a major source of foreign currency and state revenues [28]. Moreover, the scheme maintained the national food security and generated livelihood for the 2.7 million people who reside within its boundaries.

3.2 Policy Environment After Independence

During the period between 1956 and 1977, irrigated agriculture expanded from 1.17 million ha to more than 1.68 million ha. The 1980s was an era of renovation, to boost the performance of the irrigation sub-sector. During 1990s, some smaller schemes were approved to the private sector, while the four strategic schemes (Gezira and Managil, New Halfa, and Rahad) remained under government control.

3.2.1 Nile Waters Master Plan 1980–2000

The objective of the Nile Water Study commissioned by the Ministry of Irrigation was to prepare a Master Plan for water utilisation for the period between 1980 and 2000. The main planning criterion was the optimum economic use of the Nile water. Due consideration had been given to the existing developments, comprehensive assessment of the available resources (land, water and human), adoption of the pre-selected projects and development of alternative programs. The elements of the Master Plan include the following [15]:

- Irrigation development of the Nile and its tributaries.
- Possible increase in the flow of the Nile by conservation measures.
- Additional seasonal and long-term storage.
- The use of Nile Waters for hydropower generation.

During the early stages of the project, it was recognized that heightening Roseires Dam is key for achieving the project objectives over the planning period.

The planned irrigation developments include vertical expansion and intensification of existing schemes, e.g., Gezira Scheme, which will increase the total annual cultivated area from 4 million feddans in 1980 to 6.4 million feddans in the year 2000 (1 feddan = 0.42 ha).

3.2.2 Sudan's National Comprehensive Strategy (1992–2002)

Instigated by the severe impacts of the drought cycles of the seventies and eighties, the strategy assigned the National Centre for Research (NCR) to study desertification and desert control measures [28]. Consequently, NCR established the Desert Research Department in 1992, which was promoted into a full research institute in 2003 and specialized research station (Rawakeeb Research Station) (RRS). The research output contributed to the modernization of irrigation as follows:

- Development of low-cost building materials for canal lining using locally available aggregates, i.e., gravel and water (wadi) transported sand that eliminated seepage losses [29]. It is worth mentioning that seepage losses from irrigation canals at RRS were estimated at about 30% of the abstracted water [30]. Different seepage control techniques were used including cast-in situ and precast linings and production of concrete pipes [31]. According to [32], adoption of canal lining at Rawakeeb Research Station tremendously improved irrigation efficiencies and crop productivity. Figure 10 shows a lined section and field outlet pipe made of local aggregates at RRS.
- Introduction of water harvesting and supplementary irrigation greatly improved forage production in the semi desert climates of west Omdurman [33].

3.2.3 National Water Policy and Strategy 1992

According to [23], this policy was crafted to compliment the National Comprehensive Strategy (1992–2002). The following policy objectives were found to be most relevant to modernization of irrigation:

- Efficient utilization and development of surface and groundwater resources.
- Developing the water sector manufacturing industry to supply water pumps, drilling equipment, pipes, tanks, and spare parts.
- Developing economic criteria for the utilization of water in such a manner as to maintain a balance between the costs on one hand and the economic and social return on the other.

Fig. 10 A lined section and field outlet pipe made of local aggregates

3.2.4 The Water Resources Act (WRA) of 1995

The Water Resources Act (WRA) of 1995 is considered the main Act dealing with the water resources in Sudan. Other Acts are supplemental to it, like the Groundwater and Wadis Directorate Act, 1998, the Irrigation and Drainage Act, 1990, the Gezira Scheme Act, 2005 (amended 2014). [26] concluded that the Act is designed with emphasis on water development rather than sustainable water management, and the Dublin principles were not considered. The Water Law of 1995 delegated states to manage aquifers and watersheds within their respective jurisdiction, therefore, state water corporations were established. Some states developed state water laws (Gedarif, North Kordofan and South Darfur states). The Interim National Constitution mandated state authorities to manage surface and groundwater waters within the state's jurisdiction. A framework law was prepared to guide states to develop their water laws.

As mentioned in Sect. 3 of this chapter, the NCWR was created in accordance with WRA with powers to tackle water allocation, stakeholder participation, management of the infrastructural facilities among others.

3.2.5 National Water Policy 2000

The policy was formulated by a multi-disciplinary team guided by an international consultant and based on a very wide consultation and engagement of the relevant institutions [23]. The policy embraced IWRM as a mean to alleviate the water management crisis for the first time in the country.

The most relevant policy principles and objectives include the following:

1. The improvement of water use efficiency in agriculture shall be a priority.
2. The licensing of water use must include a rigorous assessment of water resources to be utilized.
3. Water pricing must promote equitable and efficient water use. In this regard, the document stressed that political support is needed to implement legislation aiming at substantial changes in agricultural water charges.
4. The use of water in irrigation projects should be monitored and evaluated using appropriate Monitoring & Evaluation Systems.

3.2.6 The National IWRM Plan

The Ministry of Water Resources and Electricity (MWRE) endeavored to develop the Plan. The process was financially assisted by UNEP with technical support from the Water Resources Commission of South Africa. The initial phase of the process, i.e., the vision/policy development phase was successfully concluded [24]. Unfortunately, the process was halted due to lack of interest.

It is worth mentioning that IWRM proved to be an appropriate approach for achieving efficient utilization of water resources and to resolve conflicts between the competing water users. Despite the above-mentioned draw backs, the National Water Policy of 2000 and the National Plan for Environmental Management of 2007 fabulously embraced the concepts of IWRM. In addition to that the following documents support IWRM: The Interim National Constitution 2005, The Quarter Century Strategic Plan, The National Population Policy and The National Water Supply and Sanitation Policy 2010.

3.2.7 The Quarter Century Comprehensive National Strategy QCS (2002–2027)

In 2002, the government started a process to develop strategic plans for the whole sectors. The idea was unrealistic, since by that time the country was facing serious security, economic and political challenges. Beside that, the then regime was facing global and regional isolation, economic sanctions and was classified as state sponsor of terror. The long-life span of the strategy is another drawback, since the capacity for strategic planning was too weak.

3.2.8 Integrated Water Policy and Strategy (IWPS) 2007

Inspired by the QCS (2002–2027), a process to develop a strategy for water resources from 2002 to 2027 was initiated [23]. The vision of the strategy was stated as follows: "Efficient utilization of water resources to realize the socio-economic development on integrated, balanced and sustainable basis". As the vision implies, modernization of irrigation systems is well entrenched in the strategy. One of the main justifications to formulate the new strategy is to achieve efficient utilization of water in agriculture. In addition to that, the strategy strives at capacity development and stakeholder participation including establishment of Water User's Associations.

Formulation of the Water Policy of 2007 was grounded on the Transitional Constitution of Sudan of 2005, the water policies of 1977, 1992, and 2000 in addition to the macroeconomic and social development policies, strategies and plans [33]. Other directives take account of advancement of the role of women in water resources management and devising incentives for the sustainable use of the resource. However, state water policies are still underway.

3.2.9 Five-Year Strategic Plan (2007–2011)

The strategy constitutes the second phase of the QCS (2002–2027), which was formulated with accompanying policies to promote adaptation to climate change in the water sector. The strategic objectives that were directly linked to modernization of the irrigation systems may be summarized as follows:

- Completion of major irrigation projects and improvement of infrastructure.
- Improvement of agricultural production and productivity (more crops per drop).
- Promotion of scientific research, introduction of new technology and capacity building.
- Achievement of water and food security.

3.2.10 Agricultural Revival Programme 2008

The Programme gives priority to those projects, which benefit people in rural areas such as water harvesting projects and can give quick results within a short period of time.

The Programme is based on nine key indicators of success; however, the following were found to be most pertinent to modernization of the irrigation systems:

- Creating an appropriate atmosphere for sustainable agricultural development.
- Development and modernization of agricultural systems.
- Development of international partnership specially the component on the transfer of new technologies and management systems.

On the other hand, the strategic objectives, which are related to modernization of the irrigation systems include following:

- Increasing crop productivity and water use efficiency.
- Achievement of a national balanced growth with a view to encourage settlement in the rural areas.

The relevant program policies include:

- Allocate at least 20% of public expenditure for building and modernization of agricultural and livestock infrastructure and advancement of technological innovation.
- Reform existing irrigation schemes and those under establishment.
- Ensure per capita water requirements and provision of irrigation for poverty reduction and food security in rural areas.
- Modernize irrigation systems including the application of agricultural precision technologies.

Despite the great efforts exerted to prepare the Program document it doomed to a tragic failure. The Secretary General of the Program attributed that to the unstable economic situation, devaluation of the national currency and inflation that has led to increased cost of agricultural production [34].

3.2.11 The National Adaptation Programme of Action (NAPA)

The NAPA consultation process identified priority adaptation activities and needs for the five ecological zones of Old Sudan (before cessation of South Sudan) including the following [24]:

- Introduction of new water harvesting/spreading techniques making use of intermediate technologies.
- Rehabilitation of existing dams and improvement of water infrastructure.
- Construction of dams and water storage facilities in some wadis, particularly in western Sudan.
- Introduction of water-conserving agricultural land management practices.

Based on the NAPA, some projects were designed and implemented during the period (2010–2014), and (2014–2017) these projects focused on strengthening the adaptive capacities of small-scale farmers and pastoralists with a view of enhancing their resilience to climatic shocks. Central to the interventions is the introduction of irrigated agriculture especially in Bara locality of North Kordofan State and Fashagha Locatity of the Gedarif State (on the Banks of Atbara River), where rainfed agriculture and raising of small ruminants were the dominant livelihoods.

The regulation of Atbara River flow due to the construction of the twin dams of Upper Atbara and Setit facilitated installation of small irrigation pumps. By doing so, the project created sustainable livelihoods for about 200 farmers, improved the nutritional status of the residents in the area through availing vegetables and other food crops. In addition, the projects assisted in bridging the feed gaps during the frequent drought years by the use of crop residues [35]. It is worth noting that irrigated agriculture was also introduced in the upper terraces in Fashagha using solar pumping to harness groundwater, and similar impacts were also evident [36].

3.2.12 The National Adaptation Plan (NAP)

The prime objective of Sudan-NAP is to minimize vulnerability to the impacts of climate change by augmenting adaptive capacity and resilience building for state and federal institutions. According to [37], the NAP prioritized potential adaptation interventions at the state level. Although it is visible throughout the states' plans, some states explicitly identified the need to modernize the irrigation system as one of their top priorities including North Darfur, North Kordofan, Kassala, River Nile, Khartoum and Gezira. Most of the states identified introduction of modern water harvesting and water conservation techniques as their top priorities.

Based on the NAP, some projects were implemented in the various states in Sudan including the following:

- Enhancing the resilience of communities living in climate change vulnerable areas using Ecosystem Based Approaches to Adaptation (2018–2022).
- The Rural Livelihoods' Adaptation to Climate Change in the Horn of Africa—GEF/ AFDB (2019–2022).
- Strengthening adaptation planning processes and capacity for implementation of adaptation actions in agricultural and water sectors GCF/FAO (2020–2022).

4 Policy Environment Post December Revolution 2018

After the revolution, Sudan found great global sympathy with the change that was achieved. Hence, countries and international organizations showed great interest to resume cooperation with the new Sudan [38]. Inspired by the Charter of the Coalition of Forces of Freedom and Change, the Constitutional Document and the other initiatives, the Ministry of Cabinet

Affairs developed a Framework of Work for the Transitional Government. The second priority was on averting economic crisis and laying the foundation for sustainable development including the following sub priorities:

- Development and strengthening agriculture, livestock, and industrial sectors.
- Rehabilitation of existing irrigated and rain-fed projects.

4.1 Sudan Water Sector Livelihoods Transforming Strategy (SWS-LTS) 2021–2031

SWS-LTS is the first strategy that holistically deals with the existing challenges and harnesses the prospects that occur at present to allow for a secure innovative demand to achieve sustainable renovation of the Sudan water sector [16]. The development of the Strategy is based on wide range of consultations including national stakeholders and international partners. In addition, the strategy provides a framework for communication and engagement with other sectors, development and funding institutions, and research and academia.

The ambition of the SWS-LTS will endeavor to contribute to the four key priorities of the Government of Sudan as follows: (1) hunger eradication and maintenance of food security for no less than seven million people in rural farming and herding communities, (2) to generate attractive and pleasing employment openings for about two million people, especially rural youth and women, (3) coverage of the whole country with safe, sufficient and accessible water supply services, and (4) to assist in the maintenance of peaceful co-existence midst all competing water users especially in rural areas. These priorities are greatly analogous to the emphasis of the Sustainable Development Goals (SGDs) 1, 6 and 8.

The Strategy is based on three harmonizing Transformative Plans:

- The National Irrigation Transformation Plan (NITP)
- The National Water Supply Transformation Plan (NWS-TP):
- The National Water Resources Management Transformation Plan (NWRM-TP)

4.1.1 The National Irrigation Transformation Plan (NITP)

According to [16], the overarching goal of the TP can be stated as follows: "Resilient and sustainable, improved and new irrigation facilities and services for all". The Plan is composed of five Transformative packages as illustrated below:

- Transformative package I: Upgrading and modernizing 1.1 million ha of existing irrigation schemes.
- Transformative package II: Development of 400,000 ha of new irrigation schemes.
- Transformative package III: Improving water productivity with at least 50% more crop per drop.
- Transformative package IV: Inclusive, gender-aware capacity building, institutional strengthening and solution transfer.
- Transformative package V: Facilitative policy, transparent, and participatory governance.

It was concluded that attaining 50% increment in water productivity within the specified time frame is an enormous challenge since a balance should be reached to achieve other priorities as well [14].

5 Role of Regional Cooperation

Realizing the fact that boundaries for a river basin create a natural unit for water resources management, [38] advocated for the adoption of integrated watershed management and establishment of river basin organizations (RBOs). Nevertheless, the legal framework for Nile Cooperation is yet to be approved.

Recognizing the significance of basin (regional) cooperation, [40] suggested the following basin-wide adaptation measures:

- Coordinated reservoir operation.
- Promotion of basin-wide agricultural trade with other regions.
- Power and transport interconnection.
- Establishment of a joint mechanism for resource solicitation.
- Joint research.

The National Plan for the Development and Utilization of Water Resources stressed on basin-wide cooperation and improvement of water use efficiency in the irrigation sector to boost water supplies of the country [41].

Moreover, the Nile Decision Support System (Nile DSS) will empower the riparian countries to weigh trade-offs and consequences for basin future projects [40].

Prominent potential 'win–win' benefits include development and trade of clean energy (hydropower), improvement and expansion in irrigated and rainfed agriculture coupled with enhanced water use efficiency, in addition to water resources conservation and protection of the environment. Other far-reaching benefits of basin wide cooperation include realizing durable regional economic integration and maintenance of peace and security through the basin [42].

6 International and National Investments

The country could attract only few mega international and national investments in the agricultural sector due to the fragile political and economic situation, and lack infrastructure and stable investment policies.

6.1 Amtaar Investment

Amtaar is a joint Sudan-UAE investment that introduced the innovative floppy sprinkler irrigation system harnessing the groundwater to irrigate an area 6,000 ha to produce forage crops and corn. The total annual produce of the project, which is located in the Northern Sudan, is estimated at about 230,000 ton of dry forage, which is shipped to the UAE on weekly basis.

6.2 DAL Group

DAL Agricultural Services was founded in 1984. In 2006, a large modern dairy farm was established in the Southern part of Khartoum. To ensure continuous fodder supply, Al Waha Farm was established in 2010 with a capacity of 80 million tons using 127 center pivots [43]. Figure 11 exhibits Alfa Alfa Harvesting.

Fig. 11 Alfa Alfa harvesting [43]

In 2020, an agreement was signed with the African Development Bank to lend The DAL Group $75 million to contribute to improving the national food security and to create job opportunities. In the same year, the Group reached an agreement with IHC Food Holding of UAE to establish an agro business that includes developing about 100,000 acres in the Northern Sudan with a total cost of $225 million [44].

7 Conclusions

Sudan has got a huge agricultural potential, since its arable lands dominates an area of about 91 million ha, beside that the country water resources are still underutilized. Despite the fact that the potential for irrigated agriculture is very high, the area under modern irrigated agriculture is estimated at about 3.5 million ha. The early 1860s mark the inauguration of the employment of irrigation for commercial farming in Sudan through cultivation of cotton on the deltas of the seasonal rivers of Gash and Baraka. During the period 1925–2016, multi-purpose dams (irrigation and hydro-power) were constructed on the River Nile and its tributaries: Sinnar 1925, Jebel Awlia 1937, Khasm El Girba 1962, Roseries 1966 (heightened in 2015), Merowe 2006 and Atbara and Setiet 2016. The deterioration of the agricultural sector during the period 1990–2020 has been attributed to the lack of governmental support, inappropriate policies and strategies and internal armed conflicts and instabilities. The first national water policy was crafted during the colonial era in 1913 to allow for the construction of Sennar Dam and the Gezira irrigation scheme. The scheme used to be the back bone of the national economy, maintained the national food security and generated livelihood for the 2.7 million people, who reside within its boundaries. After independence, in 1956, several national water policies and strategies were enacted including Nile Waters Master Plan 1980–2000, National Water Policy and Strategy 1992, National Water Policy 2000, Integrated Water Policy and Strategy (IWPS) 2007. Other national development and sectoral policies and strategies were also found to tackle the issue of modernization of irrigation. Such policies include Sudan's National Comprehensive Strategy (1992–2002), the Quarter Century Comprehensive National Strategy QCS (2002–2027), Agricultural Revival Programme 2008, the National Adaptation Plan (NAP) 2016 among others. Inspired by the spirit of the December Revolution 2018, a new comprehensive water strategy has been crafted, namely Sudan Water Sector Livelihoods Transforming Strategy (SWS-LTS) 2021–2031. The new strategy embraces a National Irrigation Transformation Plan.

Strategic planning in the country was highly elaborated. However, the low level of implementation of national strategies and programs was attributed to conflicts, institutional instability and inadequate capacities and investment among others.

References

1. Global Water Partnership (GWP). Integrated water resources management. Sweden: Global Water Partnership, TAC Background paper; 2000.
2. NBI. Policy brief 2022–01: irrigation development in the Nile Basin. Entebbe, Uganda: NBI; 2020.
3. Government of Sudan (GoS) and UNEP. Sudan first state of environment and outlook report 2020: environment for peace and sustainable development. In collaboration with UK Aid and HCENR. Nairobi, Kenya: UNEP; 2020. p. 323.
4. MoI. Sudan land of opportunities: facts and figures. Sudan: Ministry of Information (MoI); 2011.
5. Republic of Sudan REDD+ Programme (REDD+). Communication strategy and work plan for the REDD+ readiness programme. Reducing emissions from deforestation and forest degradation REDD+-Sudan. (REDD+ Sudan); 2017.
6. Nile Basin Initiative (NBI). State of the Nile River basin report 2012. NBI; 2012
7. Ministry of Irrigation and Water Resources (MOIWR). Summary of the country strategy on integrated water resources management. Sudan: MOIWR; 2007.
8. African Union. Framework for irrigation development and agricultural water management in Africa (IDAWM). Addis Ababa, Ethiopia: African Union; 2020.

9. FAO. AQUASTAT country profile Sudan. Food and Agriculture Organization of the United Nations (FAO). Rome, Italy; 2015.
10. Adam AM, Abdo GM. Chapter 2: water and hydrology (surface and groundwater) 41–81. In: Salih AM, editor. Water resources in Sudan. Water Research Centre, University of Khartoum; 2017
11. Abdalla AK. Chapter 1: water and climate 1–40. In: Salih AM, editor. Water Resources in Sudan. Water Research Centre, University of Khartoum; 2017.
12. https://water.fanack.com/sudan/water-resources-in-sudan/
13. Republic of Sudan REDD+ Programme (REDD+). In-depth analysis of drivers of deforestation and forest/ range degradation. Consultancy Report by Hassan and Tag Consultants; 2018
14. Mulder E, van Steenbergen F. Sudan water policy review and analysis. water-PIP technical report series. Delft, the Netherlands: IHE Delft Institute for Water Education; 2021.
15. GDRS-MI (Government of the Democratic Republic of the Sudan, Ministry of Irrigation). Nile waters master plan summary (consultancy Report prepared by: Goyne Et Bellier, Sir Alexander Gibb and Partners, Hunting Technical Services Limited and Sir M. McDonalds and Partners). Sudan: Ministry of Irrigation; 1977.
16. Eastern Nile Technical Regional Office (ENTRO). Atlas of maps. Addis Ababa, Ethiopia: ENTRO; 2015
17. Elkhidir AM, Khalid AOM. Chapter 5: water and agriculture nexus 114–178. In: Salih AM, editor. Water resources in Sudan. Water Research Centre, University of Khartoum; 2017.
18. Mustafa ZAE. Hydraulic evaluation of center pivot and floppy irrigation systems using different energy sources on water productivity for Maize (Zea mays L.) in River Nile State. PhD Thesis submitted to the Department of Agricultural Engineering, Faculty of Agriculture, University of Khartoum; 2021.
19. Abdelrahman AM. Grand Ethiopian renaissance dam (GERD) and the hydroelectric power in Sudan. Unpublished Report; 2020.
20. Sharfi EMA, Murdis M. Chapter 6: water and energy nexus. In: Salih AM, editor. Water resources in Sudan. Water Research Centre, University of Khartoum; 2017. pp. 179–265.
21. https://tractebel-engie.com/files/attachments/.1700/wl1200hp1200q85_Dam_Complex_Upper_Atbara_lowres.jpg. Browsed Sept. 2024.
22. Saeed AB, El Gamri T. Efficient use and appropriate management of irrigation water in Sudan. Assessed Proc. the Nile Basin development Forum. 17–19 Nov. Khartoum, Sudan: Friendship Hall; 2008. 236–252.
23. Ahmed SI, Eldaw AK. Issues and challenges in water resources development and utilization in the Sudan. In: Ibrahim G, Zaroug M, editors, Proc. forum on the future of water res. in the Sudan: strengthening public and political awareness on water issues. Future studies centre and the min. of irrigation and water res. 31 Aug.–1 Sept 2006. Arab Organization for Agric. Khartoum, Sudan: Development Conf. Hall; 2006. pp 37–54.
24. Adam AM. Chapter 16: water governance 409–427. In: Salih AM, editor. Water resources in Sudan. Water Research Centre, University of Khartoum; 2017a.
25. El Gamri T. Towards climate resilience in the Nile Basin: the case of the Sudan. J Faculty Arts (Arts J). 2014;32:87–114.
26. Higher Council for Environment and Natural Resources (HCENR). National adaptation programme of action (NAPA). Sudan: HCENR, Ministry of Environment and Physical Development; 2007.
27. Adam AM. Chapter 23: water laws and regulations in Sudan 568–595. In: Salih AM, editor. Water resources in Sudan. Water Research Centre, University of Khartoum; 2017b.
28. El Gamri T, El Nasikh MH. Desertification studies in the Sudan. Annal Arid Zones. 2003; 42(3–4):309–331.
29. El Gamri T. Prospects and constraints of desert agriculture: experience from west omdurman. Environmental monitoring and assessment J Vol 99–2004 (57–73). (Special Issue: Alternative Ways to Combat Desertification). The Netherlands: Kluwer Academic Publishers; 2004.
30. El Gamri T. M.Sc. Agric., Dept. of Agric. Eng., Fac. of Agric., U. of K. Title of the thesis: Development and evaluation of low cost building materials for canal lining; 1998.
31. El Gamri T, A/hameed N. Use of precast concrete canal-lets for irrigation canal lining. Sudan: Sudan Academy of Sciences (SASJ,) Sudan Academy of Sciences (2014)
32. Ahmed TM, El Gamri T, Abdelmagid IM. Impacts of canal lining on irrigation efficiencies and crop productivity. Al Buhuth (Sudanese J Sci Res). 2009; 12(1):70–82.
33. Mohamed ZY, Saeed AB, El Gamri T. Use of rain water harvesting and supplementary irrigation for forage production under dry land conditions. SASJ. 2011; 4(1):47–62. Sudan: Sudan Academy of Sciences.
34. Abdalla SH, Mohamed KA. Water policy of Sudan: National and Co-basin Approach; 2007.
35. www.alrakoba.net. الأمين العام للنهضة الزراعية المهندس عبد الجبار حسين: مساحات القطن تدنت من (1,2) مليون إلى مئة ألف فدان هذا العام - صحيفة الراكوبة. In Arabic. Accessed 2022.
36. Africa Regional Synthesis Report (ARSR). Case study: the climate change adaptation project, Sudan. Africa Regional Synthesis Report for the Eighth World Water Forum. Brasilia, Brazil; 2018.
37. El Gamri T. Adaptation at a Glance. Higher council for environment and natural resources, UNDP and DFATD–Canada; 2016.

38. Higher Council for Environment and Natural Resources (HCENR). National Adaptation Plan (NAP). Sudan: HCENR, Ministry of Environment, Natural Resources and Physical Development; 2016.
39. Ministry of Cabinet Affairs (MCA). Framework of work for the transitional government. MCA: Republic of the Sudan; 2019.
40. Cap-Net. Integrated water resources managements plans: training manual and operational guide. Cap-Net; 2008.
41. Ministry of Water Resources and Electricity (MWE). 2014 National plan for the development and utilization of water resources (In Arabic). MWE; 2014.
42. Nile Basin Initiative NBI. A long river journey 20 years of cooperation under the NBI. Entebbe, Uganda: NBI; 2019.
43. https://dalgroup.com/agriculture/. Agriculture: building for the future. Accessed 2022. https://en.wikipedia.org/wiki/DAL_Group. DAL Group. Accessed 2022.

Sustainable Irrigation and Irrigation Indices: A Theoretical Review

Yusuf Ucar and Otilija Miseckaite

Abstract

Irrigation, which is the most important practice that increases agricultural production in arid and semi-arid regions, has stimulating effects on both the economy in the regions, where production is made, and the global world economy by ensuring the sustainability of food production. From this point of view, irrigation, which is a compulsory and indispensable practice for meeting food needs, has started to be questioned in recent years due to climate change and its negative effects on ecosystems, especially on freshwater resources. Due to the nature of agricultural activities, this questioning is often carried out along with the impact of agricultural production on ecosystems, which makes it difficult to understand the impact of irrigation on ecosystems. On the other hand, most of the research to determine the effectiveness of irrigation has focused on performance analysis of irrigation systems or determination of water use efficiency (WUE) and water productivity (WP). However, in order to achieve sustainable irrigation, the economy, environment, rural communities, and natural resources affected by irrigation need to be assessed in a holistic approach. To make this assessment, technical, economic, social, and institutional indicators capable of representing the irrigated area need to be defined.

Keywords

Environment · Irrigation indices · Sustainable agriculture · Sustainable irrigation

1 Introduction

The term "sustainability" was first used by the German scientist, Hans Carl von Carlowitz [1], to mean "no more trees should be cut than grown" in an article discussing the sustainability of forests that were heavily cut down at that time for various reasons [2]. Although of sustainability vary across disciplines, the clearest definition of sustainable development was provided by the Brundtland Report from the World Commission on Environment and Development (WCED) [3]. According to WCED [3], "Sustainable development is development that meets the needs of the present without compromising the ability of future generations to meet their own needs". The World Development Report [4] explains sustainable development as: "*To*

Y. Ucar (✉)
Faculty of Agriculture, Department of Agricultural Structures and Irrigation, Isparta University of Applied Sciences, Isparta, Türkiye
e-mail: yusufucar@isparta.edu.tr

O. Miseckaite
Agriculture Academy, Department of Water Engineering, Faculty of Engineering, Vytautas Magnus University, Kaunas, Lithuania
e-mail: otilija.miseckaite@vdu.lt

M. Bahadir et al. (eds.), *Agricultural Water Management*, Water and Wastewater Management,
https://doi.org/10.1007/978-3-031-96134-2_19

ensure a stability among economic development, social development and environmental protection, positioning them as interdependent and reinforcing pillars of sustainable development". However, Durning [5] defines it as: "*Sustainable development goals are about enhancing the quality of life for all people around the world, without increasing the use of natural resources beyond Earth's carrying capacity.*" Sustainability is generally relies on four parameters such as environmental, social, economic, and institutional [6, 7].

Economists and politicians [8] now frequently use the term of sustainability, which was known and used by ecologists and environmentalists until about 20 years ago. Although economists and politicians have recently used the concept of sustainability often, the agricultural sector, which produces economic value by utilizing basic natural resources such as soil and water, remains at the very heart of sustainability. Reganold et al. [9] stated that sustainable agriculture is not just a traditional method like organic or ecological farming, since organic or ecological farming alone does not guarantee sustainable production. For agriculture to be truly sustainable, it must produce quality and sufficient food. They added that it should aim to protect while being both environmentally safe and economically viable. Agricultural sustainability focuses on environmental sustainability, reducing negative impacts on the environment and health, developing and utilizing ecosystem resources and protecting biodiversity and soil quality [10]. Achieving sustainability in agriculture depends on a balancing environmental, economic and social factors, Determining whether this balance has been achieved established relies on well-defined sustainability indicators and the representativeness. Latruffe et al. [11] explained that agricultural sustainability can be measured using environmental, economic and social indicators. Environmental indicators include many parameters due to their popularity, economic indicators have relatively fewer; and social indicators fall somewhere in between. Using these indicators, agricultural sustainability can be evaluated across several levels: (i) natural resources, (ii) crop production, (iii) economic benefit, and (iv) the community in a rural areas [12]. While each of these dimensions is valuable the most important in agricultural sustainability is the sustainability of ecosystems, which encompasses natural resources such as soil and water. Wilderer [13] stated that ecosystems should serve as a models; therefore, there should be a reduction in raw material and energy consumption leading to less waste and higher profits. Additionally, water resources should be evaluated within this framework. According to McDermid et al. [14], approximately 70% global freshwater use and 90% of consumption is attributed to irrigation. Achieving sustainability in irrigation, which is quite complicated, depends primarily on determining the effects of irrigation on the ecological system, and establishing indicators (irrigation indices) that can objectively measure these effects.

In this study, the concept of sustainability is defined in agriculture and irrigation within the framework of the relevant literature, and the potential negative impacts of irrigation on the ecosystems of agricultural areas are emphasized. In addition, the measures to be taken for sustainable irrigation in terms of irrigation indices, which have recently been frequently used and indicate whether sustainable irrigation is successful or not, are discussed.

2 Environmental Impact of Irrigation

2.1 Agro-Hydrology

Ecological balance is interfered with by anthropogenic activities such as irrigation. The main intervention in the hydrological balance through irrigation, which is one of the most important technological applications to increase agricultural production, is in the form of taking water from sources such as rivers, lakes or underground water reserves, and transferring it to the agricultural areas. The second important effect of irrigation is the special hydrological zone that irrigation water creates, where it is applied. This special hydrological zone, also called irrigation hydrology, consists of precipitation, irrigation, soil water, groundwater, groundwater discharge, surface runoff, pumping, evaporation and transpiration [15], drainage water and deep percolation. Depending on the basin, from which irrigation water is drawn, it can cause water scarcity and environmental damage, especially during dry seasons [16]. One of the most typical examples of the adverse impact of irrigation on the environment is the Aral Sea Basin. Irrigation activities carried out in this region without considering the sustainable irrigation approach have negatively affected the hydrological balance and caused irreparable environmental damage. Jin et al. [17], in their study in the Aral Sea basin, reported a decrease of 20 km^3/year in the water storage, and the Aral Sea surface area declined from 32,000 km^2 to 10,000 km^2 between 2000 and 2015. The researchers stated that in places, where the ecological balance is disturbed, such as the Aral Sea, vegetation decreases and severe dust storms occur due to the increase in surface temperature and decrease in precipitation [18, 19]. Undoubtedly, the effect of irrigation is important in this change of ecological balance in the Aral Sea Basin, but climate change, whose impact has been felt more in recent years, also plays a crucial role. For this reason, it is especially important to develop and to implement sustainable irrigations practices in arid and semi-arid areas like the Aral Sea Basin.

Evapotranspiration, an important parameter of the hydrological cycle, consists of evaporation from the soil and transpiration from leaves, and it is stated that the share of evaporation in evapotranspiration is 10–20% [15]. This rate varies depending on the irrigation method, irrigation frequency, and the irrigation water amount. Irrigation-triggered evapotranspiration decreases in the long term because of declines in short-wave net radiation, and the change in irrigation-induced precipitation outside the irrigation area is quite small (0.01–0.1 mm/day), and this mostly occurs in mid-latitude regions [20]. The change in irrigated and non-irrigated areas significantly affects the usability potential of surface and groundwater [21]. Undoubtedly, this situation varies depending on the evaporative demands of the plants grown. Failure to meet the evaporative demand of the plant due to water scarcity, which means less application of irrigation water due to water scarcity, reduces plant production, and this may endanger food safety. One of the parameters of irrigation hydrology is surface and subsurface drainage water, which is the source of especially excess irrigation water in agricultural areas. In agricultural fields, a suitable drainage system is required to control the salt accumulation in the root zone and water table [22]. It is possible to classify agricultural drainage into two main categories: surface and subsurface drainage. It should be decided which method will be applied in these areas according to the soil, topography, rainfall, irrigation method and plant type of the agricultural area. Surface drainage is a method mostly used on sloped lands, where the upper soil layer is impermeable. It is cheaper and easier to construct. Another drainage method that will allow both to remove excess water and to control salinity by controlling the water table in the plant root area is the subsurface drainage method, which is remarkably effective in controlling salinity in arid and semi-arid regions. However, compared to surface drainage, more careful design criteria and more investment costs are needed.

The other two components of agro-hydrology, runoff, and deep percolation, vary depending on the irrigation methods and the irrigation water amount. Runoff arises from poor control of irrigation water or the lack of infrastructure services such as land leveling. Surface runoff can also be seen in sprinkler and drip irrigation methods that are designed without considering the infiltration rate of the soil. Since surface runoff is visible, it is easier to monitor and to take precautions. Deep percolation occurs when more irrigation water is applied in the effective root zone than the water storage capacity of the soil, regardless of irrigation methods. Deep percolation occurs more in light-textured soil than in heavy-textured soils. Both agro-hydrological parameters are difficult to monitor and to evaluate the effect of water efficiency, and their reduction is especially important for sustainable irrigation. To reduce/prevent surface runoff and deep percolation, (i) not more irrigation water should be applied than the crop needs, which can be calculated from data from soil moisture monitoring or directly from water consumption equations; (ii) the infiltration rate of the soil should be used in the selection of sprinkler and drippers; and (iii) land leveling should be done in agricultural lands, where surface irrigation is common.

2.2 Waterlogging and Salinity

One of the most important problems in agricultural lands that have been irrigated for a long time is the increase in salinity in the soil and groundwater. Especially in inadequate drainage conditions, the capillary rise of groundwater toward the plant root zone causes unsuitable conditions for cultivation in the plant root zone and a decrease in total yield. Waterlogging and soil salinization are the most significant factors that threaten the sustainability of irrigated agriculture [23]. Soil salinization and waterlogging affect over 20% of the total global irrigated areas presently, and this problem could expand to over 50% of the total global irrigated areas by 2050 [24].

Waterlogging and sanitation in irrigated areas are called "twin threats", because they occur in sequential processes [25]. Although these twin hazards in irrigated areas occur simultaneously, their formation mechanisms are different from each other. While the cause of waterlogging in inadequate drainage conditions in arid and semi-arid regions is excess irrigation water, this can also cause salinization. In humid and semi-humid regions, especially in areas without drainage facilities, the cause of waterlogging may be excessive rainfall and impermeable soil layers. In such regions, if there is a drainage facility, the salt that tends to accumulate in the root area is washed away from the root area by rainfall. As can be understood from these explanations, waterlogging is a problem arising from the interaction of factors, such as (i) lack of effective surface drainage, (ii) ineffective use of groundwater, (iii) leakage from irrigation channels, (iv) application of excessive irrigation water, (v) cultivation of plants that are not suitable for the region, (vi) differences in the region, where agricultural production is carried out, (vii) inappropriate planting planning, (viii) lack of irrigation scheduling, and (ix) lack of land improvements required for irrigated agriculture [26]. In areas with waterlogging, due to the shallower development of the root systems of plants, the soil volume that can absorb water and nutrients is reduced, and this has a negative impact on productivity. Waterlogging also reduces soil aeration and root respiration due to the decrease in oxygen in the root zone, which may reduce productivity and result in the complete death of the roots and, therefore, the plants, especially in cereals [27, 28]. Another negative effect on plant development is that it reduces soil fertility by preventing nitrogen fixation by legumes [29]. Apart from this, waterlogging also causes disruption of the stability of roads and buildings, and may cause the spread of human diseases [30]. The quality of irrigation water is expressed by electrical conductivity (EC), which determines the concentration and composition of salt dissolved in the water. In the quality assessment of irrigation water, the total amount of dissolved solids in the water, sodium hazards like sodium adsorption ratio (SAR) and residual sodium carbonate (RSC) as well as ion toxicity, such as boron and Cl, considered [31]. About one billion ha of the world's 13.2 billion ha of land is covered by saline and/or sodic soils, with 25 to 30% of irrigated areas affected by salt, and these soils have low commercial value as their production capacity decreases due to land degradation [31]. Salinity in soils is examined in two groups, primary and secondary salinity, according to their formation mechanisms. In contrast to the primary salinity that occurs in arid areas, secondary salinity refers to the salinization of soil due to irrigated agriculture. Especially in regions where water resources are limited, the use of low-quality water in irrigation is one of the most principal factors in salinization. Using low-quality water for irrigation accelerates the salinization of agricultural lands because it contains elevated levels of dissolved matter. Even if poor quality water is not used in irrigation, applying too much water causes waterlogging and carries the salt upwards through capillarity, causing the soil to become salinized.

In arid and semi-arid regions, the type of plant grown, soil structure and topography, availability and quality of irrigation water, and farmer preferences play a key role in choosing the irrigation method. The selected irrigation method should both increase water use efficiency by maximizing the yield from the unit area and create minimum salinity in the agricultural lands. Each irrigation method has different advantages or disadvantages in ensuring the sustainability of irrigation in terms of salinity. In flood, border and sprinkler irrigation methods, salt accumulation on the surface is not possible due to downward water movement in conditions where the groundwater is not high. In these methods, if there is salt accumulation in the upper layers of the soil, it accumulates in the deeper layers through washing. In these methods, salt accumulation increases as one descends from the surface. If there is an effective subsurface drainage system, it is possible to remove salt from the plant root area without accumulation. Otherwise, there is an increase in underground salinity [32]. The furrow irrigation method, which is among the surface irrigation methods, differs from the others in that the entire soil surface is not wet. The advantages of this method over others are that the entire soil surface is not wet, and that it is the only surface irrigation method that can be used on sloping lands, where furrows are formed horizontally. In the furrow irrigation method, the highest salt accumulation occur in the bed between the two furrows, when irrigated in two furrows. If alternative furrows are irrigated, the plants will not be harmed, because the salt accumulation will accumulate on the edge of the unirrigated furrow away from the plant. If sloping beds are used, the highest salt accumulation occurs in the center of the bed, near the top of the bed, and the plants remain at a safe distance from the salt [32, 33]. Salt accumulated on furrow ridges can be removed from time to time by rainfall. In this case, reliable cultivation can be done in terms of salinity. In regions with insufficient rainfall, additional precautions should be taken to remove the salt accumulated on the furrow ridges.

Different dynamics are effective in soil salinity in surface and subsurface drip irrigation, where irrigation water is used more effectively. With this method, poor quality water can be used safely in irrigation, because a wet and more moisture-stable zone is created immediately around the dripper. In this method, salt accumulation occurs right at the border of the wet soil zone wetted by the dripper. The highest salt concentrations are at the soil surface and in the middle of any two emitters that are at the edge of the moist soil profile [32]. Plant cultivation can be done safely within the created wet zone. In regions, where the quality of irrigation water is not good or annual precipitation is below 300 mm, application of additional washing water to remove the salt accumulated around the wet zone is mandatory for sustainable irrigation [34]. Subsurface drip irrigation, a special application form of the drip irrigation method, contributes to the sustainability of irrigation by decreasing evaporation losses and deep infiltration, and enlarging water productivity. In the subsurface drip irrigation method, water carries the salt in the soil to the upper layers of the soil by capillarity, and the water evaporates to the atmosphere leaves the salt behind. Compared with the surface drip irrigation method, the salt concentration in the upper layer of the soil is twice as high in the subsurface drip irrigation [35]. In both the surface and the subsurface drip irrigation method, salinity dynamics in the soil vary depending on the amount of irrigation water applied, irrigation interval, soil structure and rainfall amount, as well as differences depending on the dripper flow rate. Especially in the use of saline water, the use of the surface drip irrigation method instead of subsurface drip irrigation [35] is thought to be more important in the sustainability of irrigation due to salinity management.

As discussed above, irrigation is the most crucial factor in secondary salinization of soil. As seen in many historical records, if necessary, precautions are not taken, especially in irrigated lands in arid and semi-arid regions, soil salinity negatively affected crop growth and productivity in the short term, and salinity has caused agricultural lands to deteriorate to a level that could not be cultivated in the long term. For sustainable agriculture, it is necessary to prevent the formation of secondary salinity caused by irrigation. For this purpose, it is necessary to install drainage systems in irrigated areas to control groundwater and salinity, to avoid applying excessive irrigation water, to choose the irrigation method by considering plant-soil-land conditions, and to comply with irrigation schedules.

2.3 Public Health

Wastewater, which is considered an alternative to freshwater applied in agriculture, is used for irrigation because it is cheap, easily accessible, and rich in certain nutrients [36]. Meanwhile, as Scott et al. [37] point out, in various developing countries, wastewater is not well collected and treated before being discharged to the environment, so environmental impacts are less considered. Particularly in arid and semi-arid agricultural areas, the use of untreated or minimally treated wastewater can cause critical health problems for both farmers and food consumers [38]. Water exposed to geochemical pollution such as arsenic [39], to agricultural activities such as pesticide and fertilizer applications, and to industrial pollution such as cadmium [40] can cause neurological disorders, respiratory diseases, fertility disorders, endocrine and dermal problems.

2.4 Nitrate Leaching

Various nutrients are needed for plant development, and the most needed nutrient among these nutrients is nitrogen. Until the mid-nineteenth century, the nitrogen needs of plants were met with organic fertilizers or the natural transformation of nitrogen. However, the use of mineral fertilizers began to increase due to the necessity of increasing the productivity obtained from the unit area to meet the food needs [41, 42]. The increase in the use of mineral fertilizers, excessive rainfall and unconscious irrigation have begun to pollute both groundwater and surface water resources [43]. Cui et al. [44] stated that many countries have established their own standards in terms of nitrate. For example, according to the standards of the World Health Organization (WHO), nitrate content in drinking water should not exceed 50 mg/L. The European Union states that the maximum nitrate content is 50 mg/L, and NO_3-N content is 5.6 mg/L. The US EPA, Japan, and Canada state that the maximum NO_3-N content in groundwater should be 10 mg/L [45–47]. Nitrate pollution in surface water resources and oceans creates anoxic zones in these areas, triggering eutrophication, and it harms plants and animals [48]. In addition, excessive nitrate accumulation in drinking water can cause various diseases in terms of public health.

Although nitrate is also transported and accumulated by wind erosion, one of the most important transport routes is excessive rainfall, and the other one is uncontrolled irrigation. Nitrate that is not utilized by plants and/or adsorbed by clay particles or soil organic matter and accumulates in the root zone moves downward with excessive irrigation water and excessive rainfall, polluting groundwater and surface waters with runoff. This is higher in light textured soils [49]. The degree of nitrate pollution from irrigation can vary between irrigated areas. Delgado [50] and Povilaitis et al. [51] reported that N losses through leaching are site-specific and influenced by local variations in rainfall, soil water holding capacity, drainage characteristics and mineralization-nitrification rates of organic N in the soil.

To reduce nitrate leaching, it is necessary to avoid over-application of both irrigation water and nitrate. In irrigated areas, it is possible to calculate the irrigation water needs of plants instantaneously by empirical equations such as Class A Pan and Penman Monteith or by monitoring soil moisture, while nitrogen application is applied depending on previous applications and is not calculated instantaneously. Quemada et al. [52] reported that proper irrigation management (58%) was more successful in preventing nitrate leaching than improved fertilizer management (39%) and improved fertilizer technologies (24%) in orchard. As can be seen, preventing nitrate accumulation should be one of the most important goals in sustainable irrigation. By preventing excessive irrigation water use, it is possible both to preserve the ecological balance of clean water resources and to prevent water quality by reducing nitrate leakage. While taking these measures, considering parameters such as the type of plant grown, the water retention capacity of the soil, the amount of soil organic nitrogen and rainfall will help to increase the effectiveness of the measures taken.

2.5 Dams and Ponds

Dams are built to provide flood protection, drinking and potable water, energy production and irrigation water [53]. The first dam construction dates to 5000 years ago [54]. Dams, which are mostly built for multi-purpose use, provide irrigation water for agriculture among these purposes. As with some engineering structures, dams have certain negative impacts on the environment. In addition to their contribution to economic development and social welfare, dams have impacts on water quality, aquatic ecology, land use, vegetation, terrestrial wildlife, and air quality [55].

In the irrigation regime, especially when surface water resources are utilized, dams have a particularly important role in terms of providing water regulation in the summer months, when rainfall is low, but irrigation water demand is high. From the perspective of irrigation alone, the dilemma is that if dams are not built, there is no access to a reliable water supply for irrigation. However in contrast, the construction of dams can have negative impacts on the ecosystem. These include changes to the riverbank due to excavation, decreasing water quality due to accumulation of drainage water, presence of pollutants from agricultural activities that harm fish, plankton as well as microorganisms (53) and changes to the natural water flow in downstream reaches (56). It is imperative to evaluate this dilemma in terms of sustainable irrigation and, therefore, sustainable food production and to take measures accordingly.

3 Sustainable Irrigation

Archaeological records of irrigation go back to 6000 BC, which is defined as the balanced and controlled storage of water needed for agricultural plants in the plant root zone that cannot be met by rainfall and is one of the key parameters of increasing productivity in agriculture. In the nineteenth century, discoveries in basic science fields such as physics, chemistry and biology led to the emergence of sub-fields of study such as soil chemistry, soil physics and plant physiology, whose main purpose is to ensure the planning and operation of sustainable irrigation systems [57]. Irrigated agriculture provides more than 40% of the world's food production from 20% of cultivated land, making irrigation an important contributor to global food security [58, 59]. The world's irrigated area covers over 275 million ha, with an estimated annual growth rate of 1.3% [60], and a projected global irrigated land of 393 Mha by 2080 [61].

Today, the main objective of water management policies is to ensure sustainability. The basic principle in this regard is to ensure the supply–demand balance for water in a certain region in parallel with the basic principle of the concept of sustainability. Especially in regions where water resources are scarce and water demand is difficult to meet, reducing water demand is the most fundamental necessity to ensure supply–demand balance. Increasing productivity by reducing the demand for water in the agricultural sector, which is the largest user of freshwater resources, depends on the implementation of effective water and irrigation management by increasing water use efficiency [62]. The sustainability of irrigation, which is affected by parameters such as general water management policies, irrigation management policies, agricultural practices, and farmer behaviors, depends on a good understanding of these parameters and the development of sustainable irrigation policies by taking them into account (Fig. 1). According to van

Fig. 1 Parameters affecting sustainable irrigation

Maanen et al. [63], while sustainable irrigation is defined, as "*the water withdrawal does not exceed the capacity of environmental flow requirements nor deplete groundwater resources in any region*", unsustainable irrigation is defined, as "*water consumption for irrigation is higher than local water availability causing loss of environmental flows and depletion the freshwater stocks*". Cai et al. [64] emphasize that sustainable irrigation water management should simultaneously achieve two goals: (i) sustaining irrigated agriculture for food security, and (ii) protecting the associated natural environment. In addition to these definitions, "sustainability", which is a complex concept, includes many parameters such as technical, economic, and social factors when evaluated in terms of "agricultural water management". In this respect, agricultural water management is not only a part of traditional water management, but also a concept that incorporates several interrelated parameters such as food production, rural development, and natural resource management [65]. The main objective of sustainable agricultural water management is to increase the irrigation water use efficiency (WUE) or water productivity (WP). Irrigation efficiency is one of the main concepts used in determining overall irrigation performance, evaluating irrigation water use, and in particular, determining/increasing the benefits obtained from the water resources used in irrigation [66]. The concept of water productivity has recently been used instead of irrigation efficiency and is traditionally defined as the ratio of the benefits derived from activities such as crop production, fisheries and forestry to the water used to produce these products [67]. By increasing water efficiency, (i) it will be possible to meet the food needs of the growing population, (ii) to allocate the water saved from agricultural irrigation for environmental uses and drinking water in urban areas, and (iii) to contribute to economic development and to improve income distribution [67]. Increasing water productivity is also possible by improving the performance of irrigation systems. Irrigation system performance can be evaluated based on the parcel, irrigation canal and irrigation system [68] and includes the physical efficiency of the irrigation scheme and irrigation scheme management efficiency, until the water is delivered from the source to the plant. Indicators such as water transmission efficiency, water application efficiency, water storage efficiency and irrigation efficiency [69] as well as indicators such as net water supply ratio and total water supply ratio are used to determine the performance of irrigation systems. The efficiency of water and/or irrigation management in irrigation systems depends on the level of cooperation between water users and the water management authority. Ensuring sustainability in irrigation depends on eliminating the negative effects of irrigation on the ecological system. Sustainably irrigated agriculture also depends on ecologically compatible farming practices on farmland, and therefore, irrigation and drainage must be linked to advanced irrigation techniques and management for the best use of water. Practices at the farm and watershed levels are similar depending on technical, social, and financial situations [70]. Montazar [71] stated that measures such as deficit irrigation strategies, water-saving irrigation practices, accurate evapotranspiration estimates for irrigation scheduling, water-efficient crop management, irrigation with wastewater, surface irrigation improvements, and improvements in small-scale irrigation systems can be considered to increase water use efficiency. Many scientific studies have taken the view that irrigation sustainability can be achieved by increasing WUE or WP. It is not just a matter of providing irrigation water at the desired time and place, and of the desired quality. While these parameters are essential for irrigation sustainability, they are not sufficient on their own. Therefore, the impact of the measures to be developed will be limited as the assessments made by considering only water resources or only irrigation action is far from an integrated approach. In this respect, considering irrigation only as an economic activity that increases agricultural production and considering sustainability as the conservation or effective management of natural resources is far from contributing to the solution of the problem. Socioeconomic and sociotechnical conditions should also be taken into consideration within an integrated approach. In studies assessing biophysical constraints to irrigation expansion

[72, 73], areas where rainfed agriculture is already practiced are identified as potential areas for irrigation expansion, and sustainable irrigation can contribute to this. It was estimated that if the total biophysical potential for sustainable irrigation were fully utilized, sustainably irrigated crop production could almost double [72].

In the light of all this information, success in sustainable irrigation can be achieved by considering the following items:

i. To increase the irrigation rate by reducing the difference between the area opened for irrigation and the actual irrigated area,
ii. To consider plant-soil-environmental conditions while determining the irrigation scheduling,
iii. To increase water application and water conveyance efficiency,
iv. To increase water, use efficiency (WUE) or water productivity (WP),
v. To establish a monitoring and evaluation system from the parcel scale to the irrigation scheme scale, considering the specific characteristics of each irrigation system,
vi. To ensure that socio-technical measures are also considered within the concept of sustainable irrigation,
vii. To promote the use of marginal waters in agriculture,
viii. To grow drought resistant plants, and
ix. To apply precision irrigation techniques.

4 Irrigation Indices

In the management of water resources, one of the main objectives is to seek policies that reduce the negative effects of current and future events and meet the water needs of people and the environment as much as possible. To achieve these goals, it is necessary to determine to what extent the goals are achieved by making performance measurements. Thus, it is possible to make improvements in policies by comparing the target and actual performance. The sustainability index shows the performance of alternative policies from the perspective of water users and the environment, and it is a measure of vulnerability reduction [74]. In the study conducted by Loucks [75], one of the basic studies on ensuring the sustainability of water resources, reliability, resilience, and vulnerability indicators were taken into consideration and within this framework, it was aimed to measure alternative plans and policies with the help of the above indicators and the sustainability of the systems. Irrigation indices are often evaluated together with sustainability in irrigated agricultural areas. In this way, the intertwining of the source of water and the irrigated agricultural areas contributes to the evaluation. For this reason, irrigation indices and indices used in the evaluation of irrigated agricultural areas can often be evaluated together. Despite this association, since the dominant parameter here is irrigation, it is thought that using the terminology of irrigation indices is more accurate in terms of reflecting reality. However, while determining whether sustainability in irrigation has been achieved, performance criteria that can be measured from the field scale to the irrigation scheme scale, from the irrigation scheme scale to the basin scale need to be developed. In measuring sustainability in irrigation, objectives must first be determined, converted into performance criteria and finally into irrigation indices that can be calculated mathematically.

The most important parameter used in determining sustainability in irrigated agricultural areas is water productivity [76]. Although research in sustainable irrigation has largely focused on indicators that are remarkably effective in irrigation management, such as water productivity (WP), water use efficiency (WUE) and water saving efficiency (WSE) [77], these indicators are not sufficient for determining the sustainability of irrigation. Undoubtedly, one of the key indicators of sustainable irrigation is the efficient use of water, which aims to obtain maximum yield from a unit of water. The most important measure of this efficiency is increasing WP, which

Fig. 2 **a** Surface and **b** drip irrigation methods for apple in Boğazova-Eğirdir-Isparta, Türkiye (Photos credited by Yusuf Ucar)

is influenced by many factors such as the type of plant growth, region, water quality, irrigation time planning, farmer habits, the way the irrigation network is operated, and irrigation method (Fig. 2). Increasing WP contributes to the increase in social and economic indicators by reducing the negative impact of irrigation on the environment.

In the holistic assessment of irrigation, it is imperative to use agricultural, environmental, social, economic, and socio-technical indicators together with water use indicators such as WP or WUE [75, 77–85]. The indices used in the evaluation of irrigated agriculture have recently been transformed into irrigation indices (Table 1). Recent research supports the assessment of the sustainability of irrigation water use in evaluating the concept of sustainable irrigation, thus enabling the Water-Energy-Food Nexus (WEFN) perspective to be operationalized at the local scale. With the WEFN, it is possible to assess water use or water withdrawal for different crops and to analyze their multidimensional impact on water resources. Thus, reliable results can be obtained with a holistic assessment in terms of sustainable irrigation [82].

5 Conclusions

It is vital that irrigated agricultural areas, which account for about 40% of total food production, contribute reliably to food production today and in the future. Despite the significant contribution of irrigated agriculture to global food production, it can cause some irreversible degradation in ecological systems. In order to maintain sustainable food production in irrigated areas, it is imperative for the concept of sustainable irrigation to identify the negative effects of irrigation on ecosystems and to restore them negative effects before they become irreversible. Although it is possible to give many examples from historical periods regarding the negative impact of irrigation, and there are some opinions that sustainable irrigation cannot be achieved, it seems possible to achieve sustainable irrigation by using today's technological potential and knowledge.

To achieve sustainable irrigation, it is first necessary to popularize the concept of "irrigation is not just irrigation". Based on this conceptual integrity, it is necessary to evaluate irrigation in a comprehensive approach and to position the "irrigation indices" to be developed within this conceptual integrity. It is viable to consider the

Table 1 Some indices used in sustainable irrigation

References	Title	Indicators
Loucks [75]	Quantifying trends in system sustainability	Reliability, Resilience Vulnerability
Eigeland and Hooper [78]	The use of multiple indicators to maximize sustainability in irrigation	Resources, Social, Economics, Environment, Externalities
Manoliadis [79]	Analysis of irrigation systems using sustainability-related criteria	Environmental Indices Efficiency, Sustainability, Relative cost
Cai [80]	A framework for sustainability analysis in water resources management and application to the Syr Darya Basin	Risk criteria, Environmental criteria, Equity criteria, Economic acceptability criteria
Kellett et al. [81]	Indicator frameworks for assessing irrigation sustainability	Healthy soil, Physical water Resources, Farming and irrigation methods, Site design, Water service provision, Landscape integrity, Climate
Tiewtoy et al. [82]	Irrigation sustainability assessment of selected Projects in Tha Chin Basin, Thailand	Physical, Socio-economic, Environmental, Institutional
de Vito et al. [83]	An index-based approach for the sustainability assessment of practice based on the water-energy-food nexus framework, irrigation	Irrigation water footprint index Energy footprint for irrigation index Irrigation water-cost footprint index
Borsato et al. [84]	Weak and Strong Sustainability of Irrigation: A Framework for Irrigation Practices Under Limited Water Availability	Deficit, Overuse or over expenditure, Reliability, Resilience
Cánovas-Molina et al. [85]	A proposal for an Irrigation Sustainability Index for agricultural basins: application in a semi-arid river basin	Environmental index Socioeconomic and institutional sustainability Irrigation sustainability

following aspects in the development of irrigation indices to ensure that the results obtained from these indices are realistic.

1. Sustainability of irrigation depends on available water resources and irrigation involves using available water resources in agricultural areas outside their natural locations. Also, the contribution of irrigation to agricultural production is indisputably large. In this case, irrigation is the main factor in irrigated agriculture; therefore, the term "irrigation indices" should be used instead of "irrigated agriculture indices."
2. With irrigation indices, it should be possible to make evaluations on a wide range from a

parcel to basin basis. There should be transitivity between the parcel-irrigation scheme and watershed indices.
3. Real-time evaluation should be possible with irrigation indices. In this way, problems can be addressed immediately, and obstacles to water use can be removed.
4. It is not possible to understand irrigation and to ensure its sustainability with approaches that are developed to evaluate only one part of irrigation (soil, water resource, salinization, social and economic situation, etc.). Irrigation indices should be able to evaluate not only the agricultural area, but also all parameters affected by irrigation, such as environmental, socio-economic, socio-cultural, socio-technical, agricultural production and energy.
5. Irrigation agriculture has a multidimensional structure, and the characteristics of these systems may vary from region to region. Although universal equations for water productivity (WP) can be developed, especially the socio-economic and socio-technical dimensions differ according to region. Therefore, it is necessary to develop indices for sustainability in each irrigation system in accordance with the local conditions.

References

1. von Carlowitz HC. Sylvicultura oeconomica. Reprint published 2000. Bergakademie, Freiberg; 1713. ISBN 3–86012–115–4.
2. Wiersum KF. 200 years of sustainability in forestry: lessons from history. Environ Manage. 1995;19:321–9.
3. WCED. World commission on environment and development (WCED). New York, NY, USA: Our Common Future; Oxford University Press; 1987. p. 1987.
4. World Bank. World development report 1992: development and the environment; 1992. https://documents1.worldbank.org/curated/en/995041468323374213/pdf/105170REPLACEMENT0WDR01992.pdf. Accessed 30 March 2024.
5. Durning AT. How much is enough? WW Norton & Company: The consumer society and the future of the earth; 1992.
6. Burns TR. The sustainability revolution: a societal paradigm shift. Sustainability. 2012;4(6):1118–34.
7. Ruggerio CA. Sustainability and sustainable development: a review of principles and definitions. Sci Total Environ. 2021;786: 147481.
8. Károly K. Rise and fall of the concept sustainability. J Environ Sustain. 2011;1(1):1.
9. Reganold JP, Papendick RI, Parr JF. Sustainable agriculture. Sci Am. 1990;262(6):112–21.
10. Raj A, Jhariya MK, Khan N, Banerjee A, Meena RS. Ecological intensification for sustainable development. Ecol Intensif Nat Resour Sustain Agric; 2021: 137–170.
11. Latruffe L, Diazabakana A, Bockstaller C, Desjeux Y, Finn J, Kelly E, Uthes S. Measurement of sustainability in agriculture: a review of indicators. Stud Agric Econ. 2016;118(3):123–30.
12. Bathaei A, Štreimikienė D. A systematic review of agricultural sustainability indicators. Agriculture. 2023;13(2):241.
13. Wilderer PA. Sustainable water resource management: the science behind the scenes. Sustain Sci. 2007;2(1):1–4.
14. McDermid S, Nocco M, Lawston-Parker P, Keune J, Pokhrel Y, Jain M, Yokohata T. Irrigation in the Earth system. Nat Rev Earth Environ. 2023;4(7):435–53.
15. Allen RG, Clemmens AJ, Willardson LS. Agro-hydrology and irrigation efficiency. ICID Working Group Water and Crops; 2005.
16. Taye MT, Haile AT, Fekadu AG, Nakawuka P. Effect of irrigation water withdrawal on the hydrology of the Lake Tana sub-basin. J Hydrol Reg Stud. 2021;38: 100961.
17. Jin Q, Wei J, Yang ZL, Lin P. Irrigation-induced environmental changes around the Aral sea: an integrated view from multiple satellite observations. Remote Sens. 2017;9(9):900.
18. Su Y, Li X, Feng M, Nian Y, Huang L, Xie T, Chen F. High agricultural water consumption led to the continued shrinkage of the Aral Sea during 1992–2015. Sci Total Environ. 2021;777: 145993.
19. Wang X, Cui B, Chen Y, Feng T, Li Z, Fang G. Dynamic changes in water resources and comprehensive assessment of water resource utilization efficiency in the Aral Sea basin, Central Asia. J Environ Manage. 2024;353: 120198.
20. Sorooshian S, AghaKouchak A, Li J. Influence of irrigation on land hydrological processes over California. J Geophys Res Atmos. 2014;119(23):13–137.
21. Nakayama T. Simulation of the effect of irrigation on the hydrologic cycle in the highly cultivated Yellow River Basin. Agric For Meteorol. 2011;151(3):314–27.

22. Konukcu F, Gowing JW, Rose DA. Dry drainage: a sustainable solution to waterlogging and salinity problems in irrigation areas? Agric Water Manag. 2006;83(1–2):1–12.
23. Houk E, Frasier M, Schuck E. The agricultural impacts of irrigation induced waterlogging and soil salinity in the Arkansas Basin. Agric Water Manag. 2006;85(1–2):175–83.
24. Singh A. Soil salinization management for sustainable development: a review. J Environ Manage. 2021;277: 111383.
25. Singh A. Groundwater modelling for the assessment of water management alternatives. J Hydrol. 2013;481:220–9.
26. Bowonder B, Ramana KV, Rajagopal R. Waterlogging in irrigation projects. Sadhana. 1986;9:177–90.
27. Araki H. Water uptake of soybean (Glycine max L. Merr.) during exposure to O2 deficiency and field level CO2 concentration in the root zone. Field Crops Res. 2006;96(1):98–105.
28. Teakle NL, Flowers TJ, Real D, Colmer TD. Lotus tenuis tolerates the interactive effects of salinity and waterlogging by 'excluding' Na^+ and Cl^- from the xylem. J Exp Bot. 2007;58(8):2169–80.
29. Mirschel W, Wenkel KO, Schultz A, Pommerening J, Verch G. Dynamic phenological model for winter rye and winter barley. Eur J Agron. 2005;23(2):123–35.
30. Askri B, Bouhlila R, Job JO. Development and application of a conceptual hydrologic model to predict soil salinity within modern Tunisian oases. J Hydrol. 2010;380(1–2):45–61.
31. Shahid SA, Zaman M, Heng L. Introduction to soil salinity, sodicity and diagnostics techniques. In: Zaman M, Shahid SA, Heng L, editors. Guideline for salinity assessment, mitigation and adaptation using nuclear and related techniques. Cham, Switzerland: Springer; 2018
32. Zaman M, Shahid SA, Heng L (2018) Irrigation systems and zones of salinity development. In: Zaman M, Shahid SA, Heng L, editors. Guideline for salinity assessment, mitigation and adaptation using nuclear and related techniques; 2018, pp 91–111
33. Devkota M, Gupta RK, Martius C, Lamers JPA, Devkota KP, Sayre KD, Vlek PLG. Soil salinity management on raised beds with different furrow irrigation modes in salt-affected lands. Agric Water Manag. 2015;152:243–50.
34. Güngör Y, Erözel Z, Yıldırım O. Sulama. Ankara Üniversitesi Basımevi. Yayın No: 1525. Ankara; 2002.
35. Rafie RM, El-Boraie FM. Effect of drip irrigation system on moisture and salt distribution Patterns under North Sinai Conditions. Egypt J Soil Sci. 2017;57(3):247–60.
36. Keraita B, Drechsel P, Amoah P. Influence of urban wastewater on stream water quality and agriculture in and around Kumasi, Ghana. Environ Urban. 2003;15(2):171–8.
37. Scott CA, Faruqui NI, Raschid-Sally L (eds.). Wastewater use in irrigated agriculture: confronting the livelihood and environmental realities. CABI; 2004.
38. Drechsel P, Scott CA, Raschid-Sally L, Redwood M, Bahri A. Wastewater irrigation and health. Assessing and mitigating risk in low-income Countries. London: Earthscan; 2009.
39. Senanayake N, Mukherji A. Irrigating with arsenic contaminated groundwater in West Bengal and Bangladesh: a review of interventions for mitigating adverse health and crop outcomes. Agric Water Manag. 2014;135:90–9.
40. Simmons RW, Pongsakul P, Saiyasitpanich D, Klinphoklap S. Elevated levels of cadmium and zinc in paddy soils and elevated levels of cadmium in rice grain downstream of a zinc mineralized area in Thailand: implications for public health. Environ Geochem Health. 2005;27(5–6):501–11.
41. Fuentes Yagüe JL. El suelo y los fertilizantes. Ministerio de Agricultura, Pesca y Alimentación. Mundi-Prensa. Madrid; 1999.
42. Betrán J. Revisión de la fertilización nitrogenada. Fertilización nitrogenada. Guía de actualización. Informaciones Técnicas. Centro de Investigación y Tecnología agroalimentaria. Zaragoza; 2006.
43. Quílez D, Yagüe MR, Isla R. Lavado de nitrato y riego. Fertilización nitrogenada. Guía de actualización. Informaciones Técnicas. Centro de Investigación y Tecnología agroalimentaria. Ed. Gobierno de Aragón. Zaragoza; 2006.
44. Cui M, Zeng L, Qin W, Feng J. Measures for reducing nitrate leaching in orchards: a review. Environ Pollut. 2020;263: 114553.
45. Dunn SM, Darling WG, Birkel C, Bacon JR. The role of groundwater characteristics in catchment recovery from nitrate pollution. Hydrol Res. 2012;43(5):560–75.
46. Huno SK, Rene ER, van Hullebusch ED, Annachhatre AP. Nitrate removal from groundwater: a review of natural and engineered processes. J Water Supply Res Technol—AQUA. 2018;67(8):885–902.
47. Rahmati O, Samani AN, Mahmoodi N, Mahdavi M. Assessment of the contribution of N-fertilizers to nitrate pollution of groundwater in western Iran (Case Study: Ghorveh-Dehgelan Aquifer). Water Qual Expo Health. 2015;7:143–51.
48. Diaz RJ. Overview of hypoxia around the world. J Environ Qual. 2001;30(2):275–81.
49. Alva AK, Paramasivam S, Fares A, Delgado JA, Mattos D Jr, Sajwan K. Nitrogen and irrigation management practices to improve nitrogen uptake efficiency and minimize leaching losses. J Crop Improv. 2006;15(2):369–420.
50. Delgado JA. NLEAP simulation of soil type effects on residual soil nitrate-nitrogen in the San Luis Valley and potential use for precision agriculture. In: Robert PC, Rust RH, Larson WE, editors. Proceedings of the fourth international conference on precision agriculture. Madison, WI: American Society of Agronomy; 1999. pp. 1367–1378.
51. Povilaitis A, Rudzianskaite A, Miseviciene S, Gasiunas V, Miseckaite O, Živatkauskiene I. Efficiency

of drainage practices for improving water quality in Lithuania. Trans ASABE. 2018;61(1):179–96.
52. Quemada M, Baranski M, Nobel-de Lange MNJ, Vallejo A, Cooper JM. Meta-analysis of strategies to control nitrate leaching in irrigated agricultural systems and their effects on crop yield. Agr Ecosyst Environ. 2013;174:1–10.
53. Tahmiscioğlu MS, Anul N, Ekmekçi F, Durmuş N. Positive and negative impacts of dams on the environment. In: International congress on river basin management; 2007. pp. 759–769.
54. Schmutz S, Moog O. Dams: ecological impacts and management. In: Schmutz S, Sendzimir J, editors. Riverine ecosystem management: science for governing towards a sustainable future; 2018. pp. 111–127.
55. Alla YMK, Liu L. Impacts of dams on the environment: a review. Int J Environ Agric Biotechnol. 2021;6(1).
56. Wu H, Chen J, Xu J, Zeng G, Sang L, Liu Q, Yin Z, Dai J, Yin D, Liang J, Ye S. Effects of dam construction on biodiversity: a review. J Clean Prod. 2019;221:480–9.
57. Sojka RE, Bjorneberg DL, Entry JA. Irrigation: an historical perspective. Encycl Soil Sci. 2002;1(1070):745–9.
58. Turral H, Svendsen M, Faures JM. Investing in irrigation: reviewing the past and looking to the future. Agric Water Manag. 2010;97(4):551–60.
59. World Bank. Water in agriculture; 2024. https://www.worldbank.org/en/topic/water-in-agriculture#1. Accessed 13 April 2024.
60. Hedley CB, Knox JW, Raine SR, Smith R. Water: advanced irrigation technologies. In: van Alfen NK, editors. Encyclopedia of agriculture and food systems. United States: Elsevier; 2014. pp. 378–406.
61. Fischer G, Tubiello FN, Van Velthuizen H, Wiberg DA. Climate change impacts on irrigation water requirements: effects of mitigation, 1990–2080. Technol Forecast Soc Chang. 2007;74(7):1083–107.
62. Alcon F, García-Bastida PA, Soto-García M, Martínez-Alvarez V, Martin-Gorriz B, Baille A. Explaining the performance of irrigation communities in a water-scarce region. Irrig Sci. 2017;35:193–203.
63. van Maanen N, Andrijevic M, Lejeune Q, Rosa L, Lissner T, Schleussner CF. Accounting for socioeconomic constraints in sustainable irrigation expansion assessments. Environ Res Lett. 2022;17(7): 075004.
64. Cai X, McKinney DC, Rosegrant MW. Sustainability analysis for irrigation water management in the Aral Sea region. Agric Syst. 2003;76(3):1043–66.
65. Iglesias A, Garrote L. Adaptation strategies for agricultural water management under climate change in Europe. Agric Water Manag. 2015;155:113–24.
66. Howell TA. Irrigation efficiency. Encyclopedia of water science. Marcel Dekker, New York. 2003;1(1):467–472.
67. Molden D, Oweis T, Steduto P, Bindraban P, Hanjra MA, Kijne J. Improving agricultural water productivity: between optimism and caution. Agric Water Manag. 2010;97(4):528–35.
68. Ucar Y. Performance assessment irrigation schemes according to comparative indicators: a case study of Isparta, Turkey. Eur J Sci Res. 2011;52(1):82–90.
69. Irmak S, Odhiambo LO, Kranz WL, Eisenhauer DE. Irrigation efficiency and uniformity, and crop water use efficiency. Biological Systems Engineering: Papers and Publications; 2011. p. 451.
70. Loiskandl W, Nolz R. Requirements for sustainable irrigated agriculture. Agronomy. 2021;11(2):306.
71. Montazar A. Irrigation tools and strategies to conserve water and ensure a balance of sustainability and profitability. Agronomy. 2021;11(10):2037.
72. Rosa L, Rulli MC, Davis KF, Chiarelli DD, Passera C, D'Odorico P. Closing the yield gap while ensuring water sustainability. Environ Res Lett. 2018;13(10): 104002.
73. Beltran-Pea A, Rosa L, D'Odorico P. Global food self-sufficiency in the 21st century under sustainable intensification of agriculture. Environ Res Lett. 2020;15: 095004. https://doi.org/10.1088/1748-9326/AB9388.
74. Sandoval-Solis S, McKinney DC, Loucks DP. Sustainability index for water resources planning and management. J Water Resour Plan Manag. 2011;137(5):381–90.
75. Loucks DP. Quantifying trends in system sustainability. Hydrol Sci J. 1997;42(4):513–30.
76. Antunes P, Santos R, Cosme I, Osann A, Calera A, De Ketelaere D, Nagarajan S. A holistic framework to assess the sustainability of irrigated agricultural systems. Cogent Food Agric. 2017;3(1):1323542.
77. Arouna A, Dzomeku IK, Shaibu AG, Nurudeen AR. Water management for sustainable irrigation in rice (Oryza sativa L.) production: a review. Agronomy. 2023;13(6):1522.
78. Eigeland N, Hooper B. The use of multiple indicators to maximise sustainability in irrigation. Aust J Environ Manag. 2000;7(4):204–12.
79. Manoliadis OG. Analysis of irrigation systems using sustainability-related criteria. J Environ Qual. 2001;30(4):1150–3.
80. Cai X, McKinney DC, Lasdon LS. A framework for sustainability analysis in water resources management and application to the Syr Darya Basin. Water Resour Res. 2002;38(6):21–1–21–14.
81. Kellett BM, Bristow KL, Charlesworth PB. Indicator frameworks for assessing irrigation sustainability. CSIRO Land and Water; 2005. CSIRO Land and Water Technical Report No. 01/05. January 2005.
82. Tiewtoy S, Clemente RS, Perret SR, Singh Babel M, Weesakul S. Irrigation sustainability assessment of selected projects in tha Chin Basin, Thailand. Irrig Drain. 2011;60(3):296–307.

83. de Vito R, Portoghese I, Pagano A, Fratino U, Vurro M. An index-based approach for the sustainability assessment of irrigation practice based on the water-energy-food nexus framework. Adv Water Resour. 2017;110:423–36.
84. Borsato E, Rosa L, Marinello F, Tarolli P, D'Odorico P. Weak and strong sustainability of irrigation: a framework for irrigation practices under limited water availability. Front Sustain Food Syst. 2020;4:17.
85. Cánovas-Molina A, García-Frapolli E, Ruggerio CA. A proposal of an Irrigation Sustainability Index for agricultural basins: application in a semi-arid river basin. Irrig Sci. 2023;41(1):173–82.

Agricultural Water Management—Global Perspective

Havva Eylem Polat

Abstract

While the world's population is increasing rapidly, we are also faced with the negative effects of climate change. Meeting the increasing food needs in parallel with population growth is directly related to agricultural production and management of natural resources that provide input to production. Water is one of the most important natural resources for life, and it is indispensable for agriculture. However, negative impacts such as environmental pollution, drought, excessive precipitation and thus floods limit the use of water. In order to achieve the desired level of production and efficiency in agriculture, the amount of water required for the growth of crops, if cannot be met by precipitation, should be given to the root zone in a controlled manner, which is called irrigation. Research in the past revealed that water is used more than necessary for agricultural irrigation worldwide. In order not to waste water, an appropriate agricultural water management system should be applied according to climate, soil, topographic and crop conditions. In this study, a situation assessment has been made for some prominent countries' agricultural production, and suggestions are presented regarding proper agricultural water management there.

Keywords

Agriculture · Climate change · Drought · Irrigation · Water management · Water restriction · Global water concerns

1 Introduction

About three quarters of the world's population lives in developing countries. It is estimated that this population will reach 8 billion in 2025 and approximately 10 billion in 2050, and 90% of the population growth will take place in these developing countries. In order to feed this increasing population, the production per unit area should increase approximately by two times. It is estimated that 670 million people will struggle with hunger in 2030. This number is approximately 78 million more than the 2030 estimates, which started in 2015 with a scenario, where the COVID 19 pandemic did not occur yet. At the same time, this number is 8% of the world's population in 2015 [1]. The number of people suffering from hunger in the world is increasing every year. The United Nations Sustainable Development Goals aim to end all forms of hunger and malnutrition by 2030, and to ensure that all people, especially

H. E. Polat (✉)
Faculty of Agriculture, Agricultural Structures and Irrigation Department, Ankara University, Ankara, Türkiye
e-mail: epolat@ankara.edu.tr

M. Bahadir et al. (eds.), *Agricultural Water Management*, Water and Wastewater Management,
https://doi.org/10.1007/978-3-031-96134-2_20

children, have adequate nutrition throughout the year. For this purpose, it is necessary to invest in infrastructure and technology, especially in order to increase the productivity needed in agricultural production [2].

Water Resources provide the economic growth and environmental sustainability. Development and economic growth put pressure on resources and make water security an issue for people and nature. Great uncertainties remain regarding the amount of water needed to meet the demand for food, energy and other human consumption, and to protect Earth's ecosystems. These uncertainties increase with the effect of climate change. By 2050, agricultural activities must produce 60% more food globally and 100% more in developing countries [3, 4].

Due to population growth, urbanization, and climate change, competition for water resources is expected to increase, with a particular impact on agriculture. Currently, agriculture accounts for 70% of all freshwater withdrawals globally [3]. Along with cross-sectoral reallocation of water and reduction of water use in agriculture, improvements in water use efficiency and in water distribution systems should also be undertaken. Increasing the efficiency of water use in agriculture requires the use water distribution systems of adequate capacity and advanced technologies—digital agriculture. Water scarcity and climate change will modify the water use in agriculture in the forthcoming years [5]. Irrigation scheduling is the first step to fill the water management in production to determine the volume of water and the time to irrigate the crop. With the digitalization of agriculture, multiple and different values are measured at the same time [6]. It is possible to build models to assess crop development, soil resources and water availability to support decision-making and even agriculture automation. Such models can evaluate evapotranspiration [7] or soil properties and their dynamics, and even assess water demands in paddy fields. Crop-derived data acquired by IoT sensors (internet of things) and combined with data collected by additional IoT devices such as drones and remote sensing devices, making it possible to correlate multiple parameters and better approximating crop growth dynamics, which aids a broader crop understanding. IoT sensing technologies make room for innovative ways to better use natural resources such as water for irrigation [8, 9].

As the current growth rates of global agricultural water demand are unsustainable, the agricultural sector needs to increase water use efficiency by reducing water losses and most importantly increasing water-related crop yields. Agricultural water pollution, which may be exacerbated by increased intensive farming activities, can be reduced through a combination of tools, including stricter regulations, practices and well-targeted appropriations. The development of water resources and effective use of irrigated agricultural lands, which are among the most important elements of productivity increase in agricultural production, are also important in terms of food supply. The bottlenecks to be experienced due to population growth and the pressures caused by climate change necessitate more efficient use of water resources per unit area. In particular, it is necessary to establish the necessary infrastructure and technology, to provide the desired amount and quality of water at the desired time and place, to allocate it on a sectoral basis, to improve it after use, to reuse, to protect the environment, to provide social justice, to support economic development, and to operate in accordance with international laws [10].

In this study, agricultural water managements of developed and developing world countries, which are prominent in terms of agricultural production, are evaluated. In the evaluation, the criteria listed below are taken into consideration that directly affect water management and water consumption:

- Soil and water resources
- Irrigated area – irrigable area
- Infrastructure potential in agricultural irrigation
- Water allocation status
- Crop pattern and amount of water used in irrigation
- Waste water use in irrigation
- Renewable water resources
- Irrigation water quality

In the study, the data of FAO, AQUASTAT, OECD, and World Bank 2000–2022 are used [5, 11, 12]. Considering the prominent countries in agriculture, their irrigation management and water consumption are reviewed, evaluated and summarized.

2 State of Agricultural Land and Irrigation

According to the data of AQUASAT 2022 [11], the cultivated areas of the European Union, United Kingdom, United States of America, Türkiye and some other countries are given in Fig. 1. According to Fig. 1, India, China, and the United States are the countries with the largest areas in terms of both permanent crops and arable land areas.

Permanent crops areas in different countries are given in Fig. 2. There is an increase in the areas of permanent crops between 2000 and 2020. There has been an increase in cultivated areas in order to meet the increasing food need with the increase in population. The increase in cultivated areas shows that the water used in agriculture also increases. While this area increased significantly between 2015–2020 in Spain and Italy, it tended to decrease in Brazil as of 2005. In Central Asian countries, there has been a significant increase over the years. It can be said that Central Asian countries have better conditions in agricultural production.

While looking at the arable land areas in the world, it is seen that France, India, Spain, Sudan and the United States of America take the lead (Fig. 3). In the last two decades, the share of arable land areas tends to decrease in Italy, Argentina and some African countries.

Water and soil resources, drought, desertification, urban sprawl, and improper agricultural practices, including irrigation and grazing methods, are adversely impacted by climate change. EU countries are experiencing an increasing impact of drought. The migration of urban and rural populations, excessive construction, and land sealing have detrimental effects on agricultural lands. Pollution and degradation of water and soil resources are exacerbated by extreme weather events such as severe droughts and floods. In Italy, especially in the last 10 years, unexpected extreme weather conditions (flood, storm,

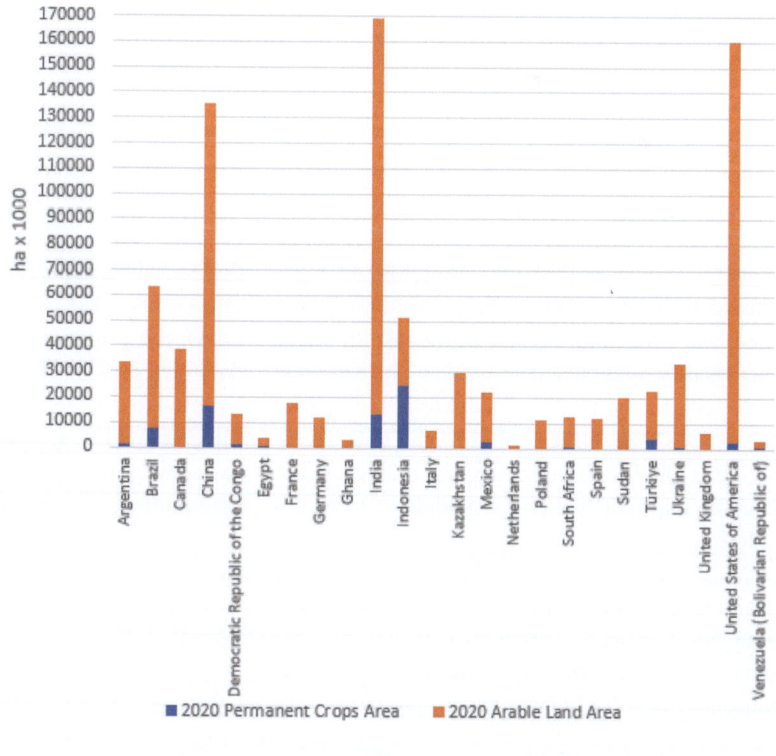

Fig. 1 Cultivated area, permanent crops and arable areas in some countries

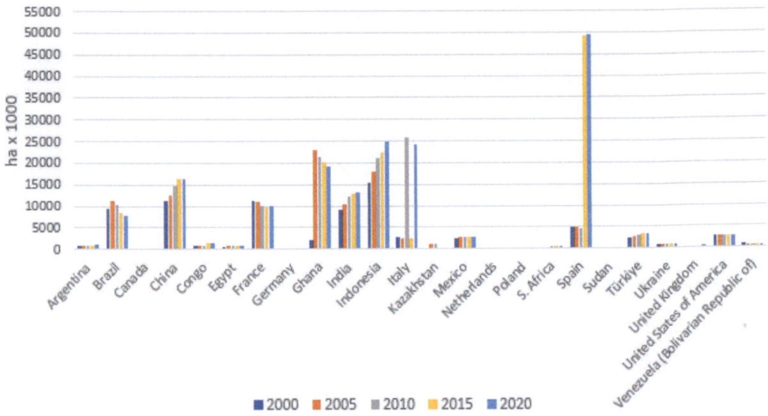

Fig. 2 Change of Permanent Crops Area in some countries between 2000–2020 years

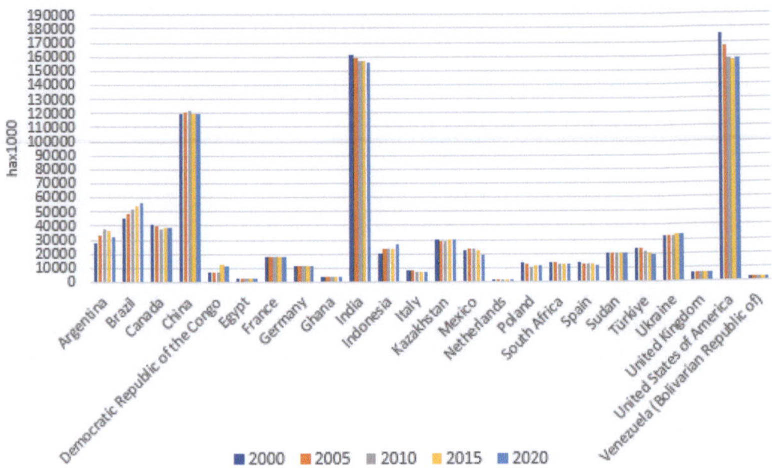

Fig. 3 Change of Arable Land Area in some countries between 2000–2020 years

etc.) and drought effects have adversely affected agricultural lands [13]. There are difficulties in meeting the water needs of the crops in the arable lands. Water has now become a resource that should be used limitedly. Therefore, changes should be made in agricultural water management and aquaculture types. In the last 20 years, increasing irrigation and pesticide use practices in Italy with intensive crop production, the change of seasons, the effects of drought and excessive precipitation have caused the arable land to become unproductive [14]. In addition to reconsidering agricultural techniques and irrigation practices, studies should be carried out to ensure digitalization in agriculture and to realize climate adaptation projects. Although it is said that the most important factor in the decrease of arable land in Africa is climate change, problems such as wrong irrigation methods and over-irrigation, deterioration and erosion of soils have been much more effective in agricultural areas. When deforestation and overgrazing are added to all these, there has been a decrease in arable land that turns into deserts [15]. Although Argentina has fertile lands, the pollution of water and soil resources over time and the severe effects of climate change have led to a decrease in arable land in the country [16]. In addition, increased pesticide use and excessive irrigation cause soil erosion and soil degradation. In particular, it can be said that while it is among the leading countries in the world in corn and soybean exports, the irregularity of land and water use policies also contributed to this decrease [1].

2.1 Water for Agriculture

Although three quarters of the world is covered with water, the amount of fresh water suitable for human use is less than 1%. Together with the climate, food, environment and energy, water ensures the sustainability of life. Food security will become an increasingly pressing global issue over the coming decades [17]. According to the World Health Organization 2020 Report, about 750 million people had difficulty in accessing food and water and in 2019 [5, 18]. As the most important input in agriculture, water has an impact on access to healthy and safe food. Agriculture uses the major share of global freshwater resources. It is known that water for agriculture and food production is used inefficiently in many parts of the world [1, 5].

Irrigated agricultural lands constitute 20% of the total cultivated land. In irrigated agriculture production, at least two times more yield per unit land is obtained than dry agriculture. Irrigated agriculture allows for product diversification. In this way, it contributes to 40% of the total food production in the world. It is estimated that the population will exceed 10 billion by 2050 and, therefore, agricultural production will need to increase by 70% by 2050.

Presently, agriculture accounts for 70% of global freshwater withdrawal. The water consumption of crops and evaporation events vary based on climate and soil conditions, further contributing to increased water usage in agriculture.

The use and movement of water can be defined in two ways as physical and virtual [19]. The physical movement of water is the allocation of surface and underground water resources for use in the agricultural, environmental, industrial and urban sectors. Concentration of food production in water-abundant regions and trade in water-scarce regions are also defined as virtual movement of water.

Efficient use of water in agriculture can be achieved by making improvements in water use and water distribution systems [20]. The use of advanced technology for water transmission and distribution systems is important in this respect. Digitization of agriculture is an important key in agricultural water management. Soil moisture sensors, monitoring via satellite and internet, irrigation timing planning, irrigation programming and the use of automation in irrigation are rapidly replacing traditional methods. In this way, the effective use and management of water will become easier.

Agricultural water management is not just about plant production. Since animal husbandry is a branch of agriculture, water consumption in animal husbandry should also be taken into consideration. Water for livestock is used to grow animal feed, for direct consumption by livestock, and for processing livestock. Direct water consumption by livestock is less than 1–2% of total water use. However, water quality is extremely important for livestock production [21]. In addition, clean natural and purified water resources are needed in aquaculture production [22].

2.2 Irrigation Water Management

The most important element in the management of water resources is to prevent losses during the transmission and distribution of water in agricultural irrigation, to reduce excess water demand in irrigated areas, to ensure effective use of water and to reduce risk. For this reason, ensuring the sustainable use of soil and water resources by protecting them from agricultural point of view requires the best management of current and future irrigation projects. In Türkiye, different institutions and organizations have authority in the management of water resources and the implementation and support of irrigation projects. The approaches developed by these institutions within the framework of their own areas of responsibility, the relevant legislation and the diversity of their experiences also make their approaches to irrigation projects different.

In order to establish a balance in water allocations and to reduce the pressure on the amount allocated for agricultural irrigation in particular, it is necessary to develop all water resources throughout the country for the short-term future, to plan irrigation infrastructures and to ensure the use of appropriate irrigation techniques. Activating the use of water in agriculture is an important measure that comes to the forefront with the necessity of correct irrigation practices and will prevent unnecessary consumption. The fact that the long-term average of irrigation efficiency in Türkiye did not reach 50% further increases the importance of this measure and modern irrigation techniques in the future. Irrigation efficiency is the most important indicator for water use efficiency, and its high level largely depends on the suitability of the irrigation infrastructure and the selection of appropriate irrigation methods. Completion of important infrastructure works in terms of irrigation, such as connecting parcels with irrigation canals and service roads, increasing the size of parcels to a scale that will provide economic production, fulfilment of on-farm development services through land consolidation works, will provide a very high level of savings in irrigation water with the transition to closed irrigation systems. The irrigation system encompasses all the required facilities and structures from the water source to the targeted irrigation area, ensuring water delivery to the crop root zone. Irrigation methods, such as sprinkler and drip irrigation, heavily rely on pressurized irrigation systems where water is conveyed at a specific pressure through enclosed lines up to the field's starting point. Otherwise, the required operating pressure will negatively affect the cost and additional power will be needed by pumping. However, the planning and operation of pressurized irrigation methods require a certain level of knowledge, and the ways of providing the pressure required by the system may adversely affect the irrigation cost.

Surface irrigation practices in EU countries, except of The Netherlands, tend to decrease between 2000 and 2020 (Fig. 4). Again, surface irrigation methods are more common in United Kingdom than in other countries. Since technology and innovations are not used much in Indonesia, Congo, Sudan and Ghana, surface irrigation practices are preferred. Some problems arise in agricultural water management, as large amounts of water are consumed in surface irrigation practices. Considering the climate change and the water scarcity, it is important to gradually abandon surface irrigation practices and switch to sprinkler and drip irrigation systems.

Sprinkler irrigation systems are increasingly used when evaluated between 2000 and 2020 (Fig. 5). This type of irrigation is widespread in EU countries, Africa and Asia. Adequate irrigation with less water provides an accurate water management in areas with water constraints. For the sustainability of water resources, it is important to prefer pressurized irrigation methods.

In the sprinkler irrigation method, the irrigation water is conveyed to the sprinkler heads placed on the field through closed pipes, sprayed from the

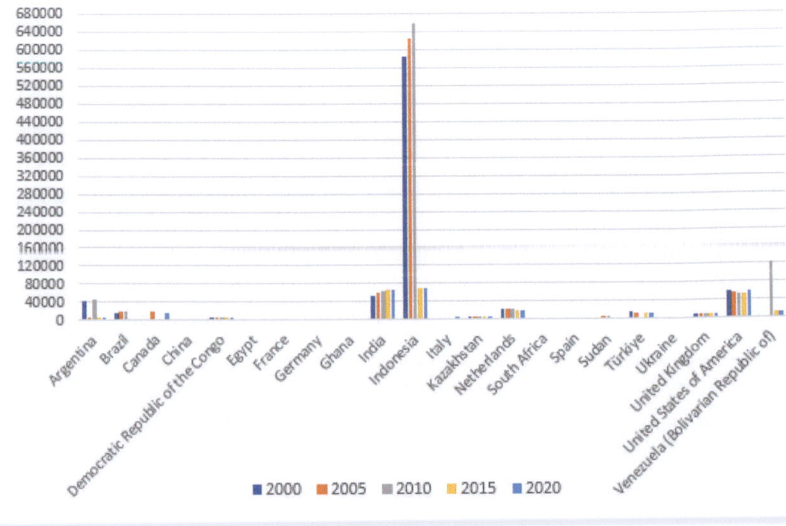

Fig. 4 Area equipped for full control irrigation—surface irrigation

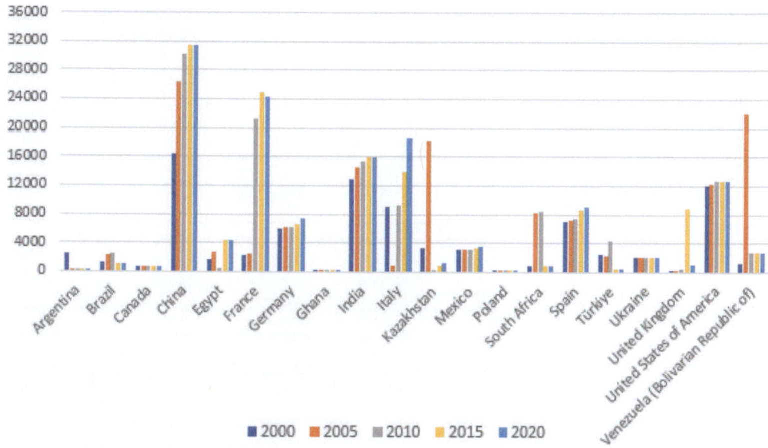

Fig. 5 Area equipped for full control irrigation—sprinkler irrigation

Fig. 6 Sprinkler irrigation systems (Cihanbeyli, Konya, Türkiye—original photo by Ahmet Cengiz Yıldırım)

nozzles to the atmosphere with a certain pressure, and provided to reach the soil surface, the water reaching the soil surface infiltrates into the soil and is stored in the crop root zone [20].

In classical sprinkler systems, irrigation water is transmitted to the field by pipes and sprayed into the atmosphere with high pressure from the sprinkler heads on the pipes (Fig. 6). Generally, the main pipeline is fixed, the lateral pipelines with sprinkler heads on the surface are mobile, and the transportation of these lateral pipelines from one stop to another requires significant labor. In conventional sprinkler irrigation systems, water loss is less than in surface irrigation systems.

Center pivot sprinkler irrigation systems move around a center (Fig. 7). The diameter of the moving circle can reach up to 1000 m. It is a more economic irrigation system than all irrigation systems, including drip irrigation, in areas with 48, 6 ha or more, where field crops are grown. In regions where groundwater is used for irrigation, placing the center of the irrigation system where the well is located can reduce the total cost of the system per unit area. The inability to irrigate some areas in the corners of the parcels is the most important disadvantage of this system, but this has been minimized thanks to the developments in recent years. The primary factors driving the preference for this system are its absence of continuous labor requirements during operation and its capacity to effectively irrigate areas with slopes of up to 12%.

The definition of total renewable water resources refers to the long-term average sum of internal renewable water resources and external natural renewable water resources. It corresponds to the maximum theoretical annual amount of water actually available for a country at a given time. According to Fig. 8, the total amount of renewable water resources in the USA is higher

Fig. 7 Center pivot irrigation systems, (**a**: Ceylanpınar Agriculture Center-Şanlıurfa, Türkiye [23]. **b**: Haymana Research Center of Ankara University, Türkiye, photo by Ahmet Cengiz Yıldırım)

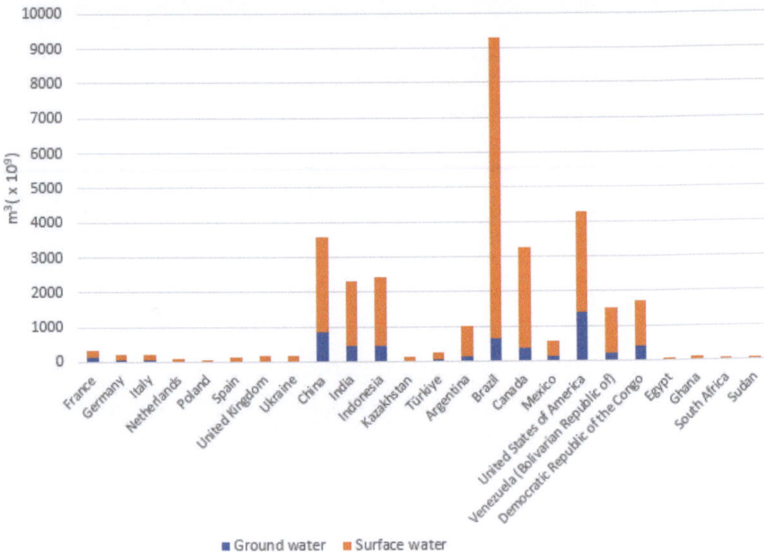

Fig. 8 Total renewable water resources according some countries in 2020

than in other countries. Brazil is abundant in surface water resources and relies heavily on them for irrigation purposes. The existence of renewable water resources will also facilitate measures to be taken against water shortages.

2.3 Monitoring and Evaluation

Monitoring and evaluation studies in irrigation facilities are of great importance, as they produce basic data that will help to develop policies and strategies in many strategic issues such as the planning of agricultural production, ensuring food security, and the sustainable use of soil and water resources. Likewise, monitoring and evaluation of irrigation projects from technical, social, economic and environmental aspects, especially during the operation phase, is one of the most important factors in the success of these projects.

The activities carried out in each season in the irrigation facilities opened for operation should be monitored by the relevant institution/organization, if they are independent on the type of operation organization and the scale of the irrigated area. Monitoring results, including the technical inspection, including the administrative inspection, as well as the irrigation results, should be evaluated and reported for each irrigation season [19]. These studies, which give clues about the performance of irrigation systems, are one of the most important tools of irrigation management. Evaluations allow to determine the efficiency and development opportunities of irrigation projects, to provide data that will help in the design of new systems, and to compare different methods, systems and business models that will form the basis for making economic decisions. According to the Annual Evaluation Reports, it is imperative to include measures in the annual work programs of the relevant institutions/organizations that aim to enhance performance indicators negatively impacting the sustainable use of the irrigation facility. Furthermore, necessary legislative arrangements should also be established to support these efforts.

The dynamic conditions of both domestic and foreign markets, influenced by price fluctuations and supply–demand factors, necessitate precise production planning in terms of the right product, location, timing, and quantity. Furthermore, the impacts of climate change and water demands from other sectors require controlling water consumption to maximize its benefits. Unfortunately, the lack of infrastructure to generate seasonal data hinders effective planning studies. Without the essential data, analysis studies cannot be conducted, leading to the inability to develop effective policies and strategies based on these analyses.

The ability to improve water management in agriculture is typically constrained by inadequate policies, major institutional under-performance, and financing limitations. Critical public and private institutions (encompassing agricultural and water ministries, basin authorities, irrigation agencies, water users' and farmer organizations) generally lack the enabling environment and necessary capacities to effectively carry out their functions.

For example, basin authorities often hold limited ability to enforce water allocations and to convene stakeholders. Institutions charged with developing irrigation often limit themselves to capital-intensive larger scale schemes and tend to rely on public sector-based approaches rather than developing opportunities for small-scale private financing and irrigation management. Farmers and their organizations are also often responding to highly distorted incentive frameworks in terms of water pricing and agricultural support policies, which further hinder positive developments in the sector.

Moreover, most governments and water users fail to invest adequately in the maintenance of irrigation and drainage systems. While inadequate management and operation may play a part in the poor performance of irrigation and drainage systems, it is especially the failure to sufficiently maintain systems that results in their declining performance and the subsequent need for rehabilitation. This failure to provide adequate funds for maintenance of irrigation and drainage systems has resulted in the "build-neglect-rehabilitate-neglect" cycle commonly observed in the sector.

The agricultural water management sector is undergoing a transformative phase to embrace modern and sustainable service provision, despite the existing constraints. It advocates for a unified water approach, aiming to establish resilient water services and sustain water resources while effectively addressing broader social and economic water-related impacts. This transformation involves redefining governance and service provision, supporting watershed management, and implementing greening initiatives within the sector. Achieving these objectives necessitates providing better incentives for innovation, fostering reforms, and promoting accountability in the agricultural water management sector [24].

The following precautions should be taken into consideration in the use and management of agricultural waters [25, 26]:

- The holistic management of water resources at the basin scale has been taken as a basis.
- Enforcing pollution prevention activities and establishing distinct quality standards for each water resource based on its intended use.
- Responsibility for preventing pollution in water resources has been given to relevant institutions.
- The fragmented structure in the allocation of water resources has been eliminated.
- A water allocation registry has been introduced to record water allocations.
- Responsibilities have been brought to the competent authorities regarding the prevention of illegal use and network losses in irrigation systems.

- Regulations regarding drainage waters have been introduced.
- The issue of charging a price over the amount in the use of water has been introduced.
- Ensuring the recovery and reuse of used water has been legally regulated.
- The formation and duties of the Water Management High Council have been regulated at the level of law.
- With the Water Information System, it was regulated that all kinds of information regarding water resources should be gathered under a single system.

2.4 Water Quality

Surface water resources are polluted due to domestic, industrial, agricultural activities, mining and geothermal activities and atmospheric pollutant transport [27]. The first indicator of pollution is moss and it causes congestion especially in sprinkler and drip irrigation. In this case, farmers use groundwater for irrigation, causing decreases in groundwater levels and threatening groundwater resources in basins. On the other hand, concentrations of metal and semi-metal compounds originating from other activities, especially mining activities, reduce the quality of drinking and utility water and irrigation water in dams.

Generally, the concentrations of salt and sodium in waters are considered in the planning of irrigation projects. However, rapid urbanization, industrialization, intensive use of fertilizers and pesticides in agriculture cause significant pollution of water resources. For these reasons, it is important to monitor water resources in irrigation planning and operation by considering more polluting parameters due to their impact on soils and crops, and therefore, on all living beings.

3 Conclusions

Today, the problems experienced in the capacity and quality of water resources emphasize that we need to make optimum use of water resources and this is inevitable. Switching to water-saving irrigation methods such as mechanized sprinkler and drip irrigation instead of irrigation methods that lead to excessive water use in irrigated agriculture and irrigating more agricultural areas with the saved water or allocating the saved water to other sectors (drinking water or industry) should not be seen as a stand-alone solution. In addition, other measures should be taken Furthermore, to maximize income from each unit of water, additional measures must be implemented. Apart from adopting water-saving irrigation technologies in agricultural operations, conducting irrigation scheduling and optimization studies can also be crucial. This involves providing training to technical staff on these matters and raising awareness among farmers about the significance of efficient water management for enhanced productivity and profitability.

In successful irrigation management, key factors include ensuring water availability, efficient delivery, correct irrigation practices, and optimizing crop patterns according to water potential. Additionally, strategic production planning and policy development are essential for sustainable income and water use.

The use of automated control technologies in the irrigation of crops will greatly reduce waste the water resources, to increase the utilization rate, and at the same time, to provide the best growth environment for crops increasing the production and improving the quality of food. IoT plays a key role in agriculture, since the sensors can measure several parameters continuously by using the cloud processing power [6]. It is feasible to develop models that assess crop development, soil resources, and water availability, serving as valuable tools to facilitate the decision-making process in agriculture or even contribute to agriculture automation.

These models assess evapotranspiration, soil properties and dynamics, and water demands in paddy areas [7]. Data from crop obtained by IoT sensors and meshed up with data gathered by additional IoT devices, such as drones and remote sensing devices, make it possible to correlate a multitude of parameters that helps in a broader crop understanding and to better approach the crop growth dynamics. The IoT sensing technologies make room for innovative ways to better use natural resources like water for irrigation [8, 9].

The use of new technologies contributes positively to agricultural water management. The use of automated control technology in the irrigation of crops will greatly reduce to waste the water resources, to increase the utilization rate, and at the same time to provide the best growth environment for crops to increase production and improve the quality of food [28].

Studies show that Central Asia will have better conditions in agricultural water management in the future. Good agricultural practices together with new technologies such as the internet of things (IoT) and artificial intelligence enable more efficient use of water.

As a result, the most suitable climate, growing conditions and, accordingly, ET values should be determined for each crop. Irrigation time scheduling should be done according to crop ET values and water should be used as much as necessary.

Acknowledgements The author would like to thank Ankara University Internet Library, which provided data and resources for this review and research study. Thanks for helping graphic works and photos to Dr. Sertan AVCI and Ahmet Cengiz YILDIRIM. Also thanks to the Exceed Swindon project, which encouraged the author to study on this book chapter.

References

1. FAO, IFAD, UNICEF, WFP and WHO. The State of Food Security and Nutrition in the World 2022. Repurposing food and agricultural policies to make healthy diets more affordable. Rome: FAO. 2022.https://doi.org/10.4060/cc0639en. https://www.fao.org/3/cc0639en/cc0639en.pdf
2. UN Agricultural Technology for Sustainable Development Report. 2019. https://digitallibrary.un.org/search?f1=author&as=1&sf=title&so=a&rm=&m1=p&p1=UN.+Secretary-General&ln=en
3. D'Odorico P, Chiarelli DD, Rosa L, Bini A, Zilberman D, Rulli MC. The global value of water in agriculture. Proc Natl Acad Sci USA. 2020;117(36):21985–93. https://doi.org/10.1073/pnas.2005835117.
4. UN World Water Development Report. 2017. https://www.unwater.org/publications/un-world-water-development-report-2017
5. FAO. The State of Food and Agriculture—Overcoming the water challenges in Agriculture. 2020. https://www.fao.org/documents/card/en/c/cb1447en/
6. Kamienski C, Soininen JP, Taumberger M, Dantas R, Toscano A, Cinotti TS, Maia RF, Neto AT. Smart water management platform: IoT-based precision irrigation for agriculture. Sensors. 2019;2019(19):276.
7. Allen RG, Pereira LS, Raes D, Smith M. Crop Evapotranspiration—Guidelines for computing crop water requirements. In FAO Irrigation and Drainage Paper 56; Rome, Italy: FAO 1998.
8. Jabro JD, Stevens WB, Iversen WM, Allen BL, Sainju UM. Irrigation scheduling based on wireless sensors output and soil-water characteristic curve in two soils. Sensors. 2020;2020(20):1336.
9. Kim Y, Robert G, Evans WMI. Control and remote sensing of an irrigation system using ZigBee wireless network. IEEE Trans Instrum Meas. 2008;2008(57):1379–87.
10. Kodal S, Ahi Y. Tarımda Su Verimliliği. Anahtar Journal. 2018;354:30–7.
11. AQUASTAT. 2022. https://www.fao.org/aquastat/en/
12. Worldbank. 2022. https://www.worldbank.org/en/topic/water-in-agriculture#1
13. Zhang T, Tan Q, Wang S, Zhang T, Hu K, Zhang S. Assessment and management of composite risk in irrigated agriculture under water-food-energy nexus and uncertainty. Agric Water Manag. 2022;262: 107322. https://doi.org/10.1016/j.agwat.2021.107322.
14. Worldbank. Arable land (% of land area)—Italy, The World Bank IBRD-IDA Data. 2020. https://data.worldbank.org/indicator/AG.LND.ARBL.ZS?locations=IT
15. UNEP. UN Environment Programme. 2018. https://www.unep.org/news-and-stories/story/fridayfact-every-minute-we-lose-23-hectares-arable-land-worldwide-drought#:~:text
16. Munoz LA, Gómez MC, Jiménez D, Hasiner E, Schroeder K, Treguer D, Juergenliemk A, Horst A, Andy Jarvis A, Kropff W. Digital Agriculture Profile—Argentina. FAO 2021/CB3955EN/1/04.21. 2021. https://www.fao.org/3/cb3955en/cb3955en.pdf
17. Fischer G. Transforming the global food system. Nature. 2018;562:501–2. https://doi.org/10.1038/d41586-018-07094-6.
18. FAO. SMART Irrigation – SMART WASH: Solutions in Response to the Pandemic Crisis in Africa. Land and Water Discussion Paper No. 16. Rome: FAO. 2020. www.fao.org/3/cb1306en/CB1306EN.pdf
19. Mekonnen M, Hoekstra A. The green, blue and grey water footprint of crops and derived crop products. Hydrol Earth Syst Sci. 2011;15:1577–600. https://doi.org/10.5194/hess-15-1577-2011.
20. Molden D, Oweis T, Steduto P, Bindraban PS, Hanjra MA, Kijne J. Improving agricultural water productivity: between optimism and caution. Agric Water Manag. 2010;97(4):528–35. https://doi.org/10.1016/j.agwat.2009.03.023.
21. FAO. Water use in livestock production systems and supply chains—guidelines for assessment (Version

1). Livestock Environmental Assessment and Performance (LEAP) Partnership. Rome: FAO. 2019. www.fao.org/3/ca5685en/ca5685en.pdf
22. FAO. The state of world fisheries and aquaculture: opportunities and challenges. Rome: FAO. 2014. www.fao.org/3/a-i3720e.pdf
23. TİGEM. General Directorate of Agriculture Production Centres, tigem.gov.tr/Galeri. 2025.
24. Franks T, Garces-Restrepo C, Putuhena F. Developing capacity for agricultural water management: current practice and future directions. Irrig Drain. 2008;57:255–67. https://doi.org/10.1002/ird.433.
25. Iglesias A, Garrot L. Adaptation strategies for agricultural water management under climate change in Europe. Agric Water Manag. 2015;155:113–24.
26. Wang S, Tan Q, Zhang T, Zhang, T. Water management policy analysis: Insight from a calibration-based inexact programming method, Agric Water Manag. 2022;269;107682. ISSN 0378-3774. https://doi.org/10.1016/j.agwat.2022.107682.
27. WHO. Hygiene: UN-Water GLAAS findings on national policies, plans, targets and finance. Geneva, WHO. 2020. www.unwater.org/publications/hygiene-un-water-glaas-findings-on-national-policies-plans-targets-and-finance/. Licence: CC BY-NC-SA 3.0 IGO.
28. Valipour M. Future of agricultural water management in Africa. Arch Agron Soil Sci. 2015;61(7):907–27. https://doi.org/10.1080/03650340.2014.961433.

MIX
Papier aus verantwortungsvollen Quellen
Paper from responsible sources
FSC® C105338

If you have any concerns about our products,
you can contact us on
ProductSafety@springernature.com

In case Publisher is established outside the EU,
the EU authorized representative is:
Springer Nature Customer Service Center GmbH
Europaplatz 3, 69115 Heidelberg, Germany

Printed by Libri Plureos GmbH
in Hamburg, Germany